# Handbook of Mixture Analysis

# Chapman & Hall/CRC
# Handbooks of Modern Statistical Methods

Series Editor

**Garrett Fitzmaurice**, *Department of Biostatistics, Harvard School of Public Health, Boston, MA, U.S.A.*

The objective of the series is to provide high-quality volumes covering the state-of-the-art in the theory and applications of statistical methodology. The books in the series are thoroughly edited and present comprehensive, coherent, and unified summaries of specific methodological topics from statistics. The chapters are written by the leading researchers in the field, and present a good balance of theory and application through a synthesis of the key methodological developments and examples and case studies using real data.

**Handbook of Survival Analysis**
*Edited by John P. Klein, Hans C. van Houwelingen, Joseph G. Ibrahim, and Thomas H. Scheike*
**Handbook of Mixed Membership Models and Their Applications**
*Edited by Edoardo M. Airoldi, David M. Blei, Elena A. Erosheva, and Stephen E. Fienberg*
**Handbook of Missing Data Methodology**
*Edited by Geert Molenberghs, Garrett Fitzmaurice, Michael G. Kenward, Anastasios Tsiatis, and Geert Verbeke*
**Handbook of Design and Analysis of Experiments**
*Edited by Angela Dean, Max Morris, John Stufken, and Derek Bingham*
**Handbook of Cluster Analysis**
*Edited by Christian Hennig, Marina Meila, Fionn Murtagh, and Roberto Rocci*
**Handbook of Discrete-Valued Time Series**
*Edited by Richard A. Davis, Scott H. Holan, Robert Lund, and Nalini Ravishanker*
**Handbook of Big Data**
*Edited by Peter Bühlmann, Petros Drineas, Michael Kane, and Mark van der Laan*
**Handbook of Spatial Epidemiology**
*Edited by Andrew B. Lawson, Sudipto Banerjee, Robert P. Haining, and María Dolores Ugarte*
**Handbook of Neuroimaging Data Analysis**
*Edited by Hernando Ombao, Martin Lindquist, Wesley Thompson, and John Aston*
**Handbook of Statistical Methods and Analyses in Sports**
*Edited by Jim Albert, Mark E. Glickman, Tim B. Swartz, Ruud H. Koning*
**Handbook of Methods for Designing, Monitoring, and Analyzing Dose-Finding Trials**
*Edited by John O'Quigley, Alexia Iasonos, Björn Bornkamp*
**Handbook of Quantile Regression**
*Edited by Roger Koenker, Victor Chernozhukov, Xuming He, and Limin Peng*
**Handbook of Environmental and Ecological Statistics**
*Edited by Alan E. Gelfand, Montserrat Fuentes, Jennifer A. Hoeting, Richard L. Smith*
**Handbook of Mixture Analysis**
*Edited by Sylvia Frühwirth-Schnatter, Gilles Celeux, and Christian P. Robert*

For more information about this series, please visit:
https://www.crcpress.com/go/handbooks

# Handbook of Mixture Analysis

*Edited by*

Sylvia Frühwirth-Schnatter
Gilles Celeux
Christian P. Robert

CRC Press
Taylor & Francis Group
Boca Raton London New York

CRC Press is an imprint of the
Taylor & Francis Group, an **informa** business

CRC Press
Taylor & Francis Group
6000 Broken Sound Parkway NW, Suite 300
Boca Raton, FL 33487-2742

© 2019 by Taylor & Francis Group, LLC
CRC Press is an imprint of Taylor & Francis Group, an Informa business

No claim to original U.S. Government works

Printed on acid-free paper
Version Date: 20181129

International Standard Book Number-13: 978-1-4987-6381-3 (Hardback)

**Library of Congress Cataloging-in-Publication Data**

Names: Frühwirth-Schnatter, Sylvia, 1959- editor. | Celeux, Gilles, editor. | Robert, Christian P., 1961- editor.
Title: Handbook of mixture analysis / edited by Sylvia Frühwirth-Schnatter, Gilles Celeux, and Christian P. Robert.
Description: Boca Raton, Florida : CRC Press, [2019] | Includes bibliographical references and index.
Identifiers: LCCN 2018040389| ISBN 9781498763813 (hardback : alk. paper) | ISBN 9780429055911 (e-book).
Subjects: LCSH: Mixture distributions (Probability theory) | Distribution (Probability theory).
Classification: LCC QA273.6 .H3465 2019 | DDC 519.2/4--dc23
LC record available at https://lccn.loc.gov/2018040389

**Visit the Taylor & Francis Web site at**
**http://www.taylorandfrancis.com**

**and the CRC Press Web site at**
**http://www.crcpress.com**

# Contents

# Preface

Mixture models have been around for over 150 years now and they are found in many branches of statistical modelling as a versatile and multifaceted tool. Historically, they are found in works by two major continental statisticians of the nineteenth century, Adolphe Quételet and Louis-Adolphe Bertillon, and were later studied by two great British intellectuals, Karl Pearson and Walter Weldon, who later founded the journal *Biometrika* with Francis Galton. Pearson provided a method for estimating the parameters of a normal mixture. Despite or because of this highly respectable ancestry, mixtures remain an exciting and lively area of research, as clearly supported by the present volume. The significant and diverse contributions of the authors here give the lie to the dire warning that "mixtures, like tequila, are inherently evil and should be avoided at all costs" expressed by Larry Wasserman (on his now defunct *Normal Deviate* blog).

To borrow the oft-repeated aphorism from George Box, "All models are wrong, but some are useful". Mixture models are particularly useful, if mostly wrong, as handy substitutes for more elaborate techniques of density estimation, unsupervised clustering, capturing unobserved heterogeneity, etc. Mixture modelling can be applied to a wide range of data: univariate or multivariate, continuous or categorical, cross-sectional, time series, networks, as illustrated in this handbook. While not claiming to provide an exhaustive coverage of this rich field, we think the exposition provided by the following chapters gives a motivating picture and can serve to grasp the advantages of using mixtures for modelling as well as engage in methodological or theoretical research.

Over the last 30 years, only very few authoritative monographs on mixture models have been published. Hence, we were very delighted when we were invited by Rob Calver of CRC Press in the spring of 2014 to edit a handbook on mixture analysis. Our first editorial meeting took place in Paris in November 2014 and we drafted an ambitious proposal with 24 chapters. For each of these chapters we had ideal candidates in mind whom we invited to join the project and to develop the chapter. We are extremely grateful that nearly all of them accepted our invitation and, as a result, that many outstanding scholars in the field have contributed to the handbook and its breadth.

Regrettably, a few had to leave the project early on for diverse and understandable reasons. Most sadly, in the early days of the project, in May 2015, Bruce Lindsay passed away. Bruce's contributions to mixture inference are numerous and deep, with major advances in nonparametric maximum likelihood and geometric representations that gave new perspectives on identifiability. We are all missing this great statistician and wonderful man.

An important milestone in putting this handbook together was a two-day workshop in Vienna in December 2015 which many contributors attended, either in person or via Skype. Mike Kuhn even travelled all the way from Valparaíso, Chile. We had many inspiring discussions and also enjoyed visiting together the charming Viennese Christmas markets in the evening. Besides discussing formats and notations, one important outcome of this meeting was the collective decision to publish the handbook in three parts: "Foundations and Methods" (Part I), "Mixture Modelling and Extensions" (Part II) and "Selected Applications" (Part III).

In July 2016 the three of us met again in Paris for nearly two weeks for the second editorial meeting. Despite the wonderful hot weather, we worked hard on our own chapter on model selection for mixtures and gave feedback to the other authors on the preliminary versions of all the other chapters. Finally, September 2017 saw our final editorial meeting in Vienna, where we merged all the chapters into one single volume and started editing the chapters and unifying the notation. It was truly an amazing feeling when we printed a first version of the handbook at the end of the meeting and celebrated the feat in the famous Café Landtmann on the Ringstraße.

In its final form, the Handbook comprises 19 chapters written by 35 contributors. Part I introduces and motivates mixture models, explains the EM algorithm and discusses Bayesian methods from both theoretical and computational viewpoints. A chapter on nonparametric Bayesian mixture models extends the common finite mixture framework to mixtures with countably infinite number of components, as in the nonparametric case. A chapter on model selection which discusses this challenging topic from both a frequentist and a Bayesian perspective and makes the important distinction between density estimation and clustering applications concludes Part I. Part II extends the perspective on using mixture modelling by presenting the special features of model-based clustering, including several applications; the modelling of count data; the specifics of skewed and heavy-tailed components; the challenges met when dealing with high-dimensional data; the extension to mixtures of experts, hidden Markov models and time series analysis, with applications in economics; and mixtures with nonparametric components. Part III highlights the strengths of mixture modelling for selected applications, namely industry, image segmentation and finance. The final two chapters illustrate how the objects of this modelling range from the infinitely small, as in DNA and cRNA data in genomics, to the infinitely large, when analysing objects in astronomy such as stars and galaxy clusters.

As already expressed, we are most sincerely grateful to all contributors for their input to this book and for patiently waiting for the final outcome, as well as for their comments on other parts of the book. We want to add a very special thank you to David Hunter, who carried on with the redaction of their common chapter after Bruce Lindsay passed away.

We further thank Rob Calver and Lara Spieker of CRC Press for their boundless patience through the many missed deadlines and their support in the completion of the handbook. We are also grateful to WU Wien (Vienna University of Economics and Business) for supporting the Workshop in Vienna in December 2015 and hosting the last editorial meeting in September 2017, and to CRC Press for its financial support for the second editorial meeting in Paris in July 2016.

Finally, we are greatly indebted to our spouses Rudi Frühwirth, Maya, and Brigitte Plessis for their love, understanding and constant support for our research activities throughout the years.

Sylvia Frühwirth-Schnatter                                    Gilles Celeux & Christian Robert

Vienna, July 2018                                                            Paris, July 2018

# *Editors*

**Sylvia Frühwirth-Schnatter** is Professor of Applied Statistics and Econometrics at the Department of Finance, Accounting, and Statistics, Vienna University of Economics and Business, Austria. She has contributed to research in Bayesian modelling and MCMC inference for a broad range of models, including finite mixture and Markov switching models as well as state space models. She is particularly interested in applications of Bayesian inference in economics, finance, and business. She began working on finite mixture and Markov switching models 20 years ago and has published more than 20 articles in this area in leading journals such as the *Journal of the American Statistical Association, Journal of Computational and Graphical Statistics*, and *Journal of Applied Econometrics*. Her monograph *Finite Mixture and Markov Switching Models* (2006) was awarded the Morris-DeGroot Prize 2007 by the International Society of Bayesian Analysis. In 2014, she was elected Member of the Austrian Academy of Sciences.

**Gilles Celeux** is Director of Research Emeritus with INRIA Saclay-Île-de-France, France. He has conducted research in statistical learning, model-based clustering and model selection for more than 35 years and has led INRIA teams. His first paper on mixture modelling was written in 1981 and he has been one of the co-organizers of the summer working group on model-based clustering since 1994. He has published more than 40 papers in international journals of statistics and written two textbooks in French on classification. He was Editor-in-Chief of *Statistics and Computing* between 2006 and 2012 and has been Editor-in-Chief of the *Journal of the French Statistical Society* since 2012.

**Christian P. Robert** is Professor of Mathematics at CEREMADE, Université Paris-Dauphine, PSL Research University, France, and Professor of Statistics at the Department of Statistics, University of Warwick, UK. He has conducted research in Bayesian inference and computational methods covering Monte Carlo, MCMC, and ABC techniques for more than 30 years, writing *The Bayesian Choice* (2001) and *Monte Carlo Statistical Methods* (2004) with George Casella. His first paper on mixture modelling was written in 1989 on radiograph image modelling. His fruitful collaboration with Mike Titterington on this topic spans two enjoyable decades of visits to Glasgow, Scotland. He has organized three conferences on the subject of mixture inference, with the last one at the International Centre for Mathematical Sciences leading to the edited book *Mixtures: Estimation and Applications* (2011), co-authored with K. L. Mengersen and D. M. Titterington.

# Contributors

**Clair Alston-Knox**
Griffith University, Australia

**Christophe Ambroise**
UMR MIA-Paris, AgroParisTech, INRA,
Université Paris-Saclay, Laboratoire de
Mathématiques et Modélisation d'Évry
(LaMME), Université d'Évry-Val-d'Essonne,
UMR CNRS 8071, ENSIIE, France

**Julyan Arbel**
Laboratoire Jean Kuntzmann, Université
Grenoble Alpes, INRIA Grenoble-Rhône-Alpes,
France

**Gilles Celeux**
INRIA Saclay, France

**Earl Duncan**
ACEMS, Queensland University of Technology,
Australia

**Eric D. Feigelson**
Pennsylvania State University, USA

**Florence Forbes**
Université Grenoble Alpes, Inria, CNRS,
Grenoble INP, LJK, Grenoble, France

**Sylvia Frühwirth-Schnatter**
Department of Finance, Accounting, and
Statistics, Vienna University of Economics and
Business, Austria

**Elisabeth Gassiat**
Laboratoire de Mathématiques d'Orsay,
Université Paris-Sud, CNRS, Université
Paris-Saclay France

**Isobel Claire Gormley**
School of Mathematics and Statistics,
University College Dublin, Ireland

**Clara Grazian**
University of Oxford, UK and Università degli
Studi "Gabriele d'Annunzio", Italy

**Peter J. Green**
University of Bristol, UK and UTS, Sydney,
Australia

**Bettina Grün**
Johannes Kepler University Linz, Austria

**David R. Hunter**
Department of Statistics, Penn State
University, USA

**Kaniav Kamary**
INRIA Saclay, France

**Dimitris Karlis**
Department of Statistics, Athens University of
Economics and Business, Greece

**Sylvia Kaufmann**
Study Center Gerzensee, Foundation of the
Swiss National Bank, Switzerland

**Michael A. Kuhn**
Millennium Institute of Astrophysics,
Universidad de Valparaíso, Chile

**Prabhani Kuruppumullage Don**
Department of Statistics, Penn State
University, USA

**Jeong Eun Lee**
Auckland University of Technology, New
Zealand

**Bruce G. Lindsay**
Department of Statistics, Penn State
University, USA

**John M. Maheu**
McMaster University, Canada

**Gertraud Malsiner-Walli**
Department of Finance, Accounting, and
Statistics, Vienna University of Economics and
Business, Austria

**Jean-Michel Marin**
Université de Montpellier, France

**Kerrie L. Mengersen**
School of Mathematical Sciences, Queensland
University of Technology, Australia

**Peter Müller**
UT Austin, USA

**Thomas Brendan Murphy**
School of Mathematical Sciences, University
College Dublin, Ireland

**Damien McParland**
School of Mathematical Sciences, University
College Dublin, Ireland

**Christian P. Robert**
Université Paris-Dauphine, CEREMADE,
Paris, France and University of Warwick, UK

**Stéphane Robin**
UMR MIA-Paris, AgroParisTech, INRA,
Université Paris-Saclay, France

**David Rossell**
Universitat Pompeu Fabra, Spain

**Judith Rousseau**
University of Oxford, UK and Université
Paris-Dauphine PSL, France

**Azam Shamsi Zamenjani**
University of New Brunswick, Canada

**Mark F. J. Steel**
University of Warwick, UK

**Nicole White**
Institute for Health and Biomedical Innovation,
Queensland University of Technology, Australia

# List of Symbols

## Mathematical Notation

| | |
|---|---|
| $\mathbb{N}$ | natural numbers |
| $\mathbb{N}_0$ | natural numbers including zero |
| $\mathbb{R}$ | real numbers |
| $\mathfrak{S}(G)$ | set of all permutations of $\{1, \ldots, G\}$ |
| $\mathfrak{s}$ | permutation of $\{1, \ldots, G\}$ |
| $\mathcal{V}_G$ | unit simplex of dimension $G$ |
| $I$ | identity matrix |
| $I_d$ | identity matrix of dimension $d \times d$ |
| $\mathbf{1}$ | vector of ones |
| $\text{Diag}\,(x_1, \ldots, x_d)$ | diagonal matrix with diagonal elements $x_1, \ldots, x_d$ |
| $A^\top, a^\top$ | transpose of a matrix $A$ or a vector $a$ |
| $A^{-1}$ | inverse of a matrix $A$ |
| $|A|$ | determinant of a matrix $A$ |
| $\text{tr}\,(A)$ | trace of a matrix $A$ |
| $B(\alpha, \beta)$ | beta function |
| $\Gamma(\alpha)$ | gamma function |
| $\mathbb{I}_A(x)$ | indicator function of the set $A$ |
| $\mathbb{I}(x = a)$ | indicator function for an event |

## Mixture Notation

| | |
|---|---|
| $G$ | number of groups/states in a finite mixture/hidden Markov model |
| $g$ | index for groups/components in a finite (or infinite) mixture |
| $g, h$ | index for states of a finite (or infinite) Markov mixture |
| $f_g(\cdot|\cdot)$ | density of the $g$th component |
| | |
| $\eta$ | weight distribution in a finite (or infinite) mixture |
| $\eta_g$ | weight of the $g$th mixture component |
| $\xi$ | transition matrix of a hidden Markov chain |
| $\xi_{hg}$ | probability of moving from state $h$ to state $g$ |
| $\theta$ | vector of all parameters of a mixture model |
| $\theta_g$ | parameter vector of $g$th component |
| $\mu_g$ | mean vector of $g$th component |
| $\sigma_g^2, \Sigma_g$ | variance–covariance matrix of $g$th component |
| $\beta_g$ | regression parameter of $g$th component |
| | |
| $d$ | dimension of data space |
| $n$ | number of cases/observations |
| $y$ | data (vector)/outcome/dependent observation |
| $y_i$ | data for $i$th case |
| $\mathbf{y}$ | collection of values $\{y_1, \ldots, y_n\}$ |
| $x$ | independent observations/regressors |
| $x_i$ | independent observations/regressors for $i$th case |
| $\mathbf{x}$ | collection of values $\{x_1, \ldots, x_n\}$ |
| | |
| $z$ | latent indicator/latent state (takes values in $1, \ldots, G$) |
| $z_i$ | group to which $i$th case belongs |
| $z_{ig}$ | $z_{ig} = 1$, if case $i \in$ group $g$, $z_{ig} = 0$ otherwise |
| $\mathbf{z}$ | collection of values $\{z_1, \ldots, z_n\}$ |
| $\mathcal{S}_G^n$ | lattice $\{1, \ldots, G\}^n$ of all $G^n$ classifications $\mathbf{z}$ |
| $n_g$ | number of cases/observations in the $g$th group |
| $n_{hg}$ | number of transitions from state $h$ to $g$ |
| $G_+$ | number of non-empty groups/clusters |
| $\mathcal{C}$ | partition of $n$ objects into $G_+$ clusters |
| $C_g$ | cluster $g$ |
| | |
| $e_1, \ldots, e_G$ | hyperparameters in Dirichlet prior |
| $e_0$ | hyperparameter in a symmetric Dirichlet prior |
| $\alpha$ | precision parameter in Dirichlet process prior |
| $H_0$ | base measure in Dirichlet process prior |
| | |
| L | likelihood for mixture model |
| $\mathrm{L}_c$ | complete-data likelihood |
| $\mathrm{L}_o$ | observed-data likelihood |
| $\ell$ | log likelihood for mixture model |
| $\ell_c$ | complete-data log likelihood |
| $\ell_o$ | observed-data log likelihood |

# Probability & Statistics

| | |
|---|---|
| $P(A)$ | probability of event $A$ |
| $X, Y$ | random variables |
| $E(Y)$ | expectation of $Y$ |
| $V(Y)$ | variance of $Y$ |
| $\text{Cov}(Y)$ | variance–covariance matrix of $Y$ |
| | |
| $\mathcal{B}er(p)$ | Bernoulli distribution |
| $\mathcal{B}e(a, b)$ | beta distribution |
| $\mathcal{B}(N, p)$ | binomial distribution |
| $\chi^2_\nu$ | $\chi^2$ distribution |
| $\mathcal{D}(e_1, \ldots, e_G)$ | Dirichlet distribution |
| $\mathcal{D}_G(e_0)$ | symmetric Dirichlet distribution |
| $\mathcal{DP}(\alpha, H_0)$ | Dirichlet process |
| $\mathcal{E}(\lambda)$ | exponential distribution (with expectation $1/\lambda$) |
| $\mathcal{G}(a, b)$ | gamma distribution (with expectation $a/b$) |
| $\mathcal{IG}(a, b)$ | inverse gamma distribution |
| $\mathcal{M}(N, p_1, \ldots, p_G)$ | multinomial distribution |
| | |
| $\mathcal{N}(\mu, \sigma^2)$ | normal distribution |
| $\mathcal{N}_d(\mu, \Sigma)$ | $d$-variate normal distribution |
| $\phi(\cdot)$ | pdf of standard normal distribution |
| $\Phi(\cdot)$ | cdf of standard normal distribution |
| $\phi(\cdot|\mu, \sigma^2)$ | pdf of $\mathcal{N}(\mu, \sigma^2)$ |
| $\Phi(\cdot|\mu, \sigma^2)$ | cdf of $\mathcal{N}(\mu, \sigma^2)$ |
| | |
| $\mathcal{P}(\mu)$ | Poisson distribution |
| $t_\nu(\mu, \sigma^2)$ | Student $t$ distribution |
| $\mathcal{U}[a, b]$ | uniform distribution on $[a, b]$ |
| $\mathcal{W}(\nu, S)$ | Wishart distribution |
| $\mathcal{W}^{-1}(\nu, S)$ | inverse Wishart distribution |
| | |
| $p(\theta)$ | prior density of $\theta$ |
| $p(\theta|\mathbf{y})$ | posterior density of $\theta$ given $\mathbf{y}$ |
| | |
| $\overline{y}$ | sample mean |
| $s^2_y$ | sample variance (divided by $n$) |
| $S_y$ | sample covariance matrix |

# Part I

# Foundations and Methods

# 1

# Introduction to Finite Mixtures

**Peter J. Green**

*University of Bristol, UK and UTS Sydney, Australia*

## CONTENTS

## 1.1  Introduction and Motivation

Mixture models have been around for over 150 years, as an intuitively simple and practical tool for enriching the collection of probability distributions available for modelling data. In this chapter we describe the basic ideas of the subject, present several alternative representations and perspectives on these models, and discuss some of the elements of inference about the unknowns in the models. Our focus is on the simplest set-up, of finite mixture models, but we discuss also how various simplifying assumptions can be relaxed to generate the rich landscape of modelling and inference ideas traversed in the rest of this book.

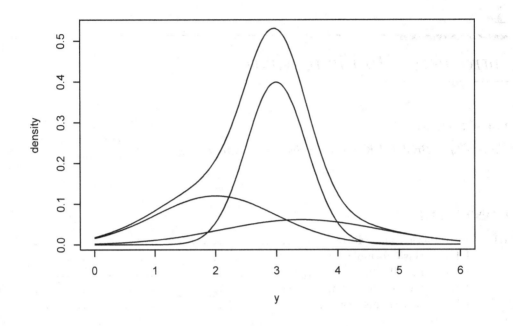

**FIGURE 1.1**
Example of a mixture of three normal densities, giving a unimodal leptokurtic density with slight negative skew. The model is $0.3 \times \mathcal{N}(2, 1) + 0.5 \times \mathcal{N}(3, 0.5^2) + 0.2 \times \mathcal{N}(3.4, 1.3^2)$.

### 1.1.1   Basic formulation

Sometimes a simple picture tells nearly the whole story: the basic finite mixture model is illustrated by Figure 1.1. It assumes that data $y$ are drawn from a density modelled as a *convex combination* of *components* each of specified parametric form $f$:

$$y \sim \sum_{g=1}^{G} \eta_g f(\cdot | \theta_g); \tag{1.1}$$

we use the conditioning bar "|" without necessarily implying a Bayesian viewpoint on inference. The usual set-up is that data are modelled as a simple random sample $\mathbf{y} = (y_1, y_2, \ldots, y_n)$ of variables drawn independently from (1.1), but of course this form of density could find a role in many other data-modelling contexts.

In inference about a mixture model, the component density $f$ is fixed and known; the component-specific parameters $\theta_g$ (often vector-valued) and the weights $\eta_g$, non-negative and summing to 1, are usually (but not always) considered to be unknown, and the number of components $G$ is also sometimes unknown. There may be partial equality among $\theta_g$; for example, in the case of a normal mixture where $f(\cdot) = \phi(\cdot | \mu, \sigma^2)$, the normal density, we might sometimes be interested in location mixtures with a constant variance, so $\theta_g = (\mu_g, \sigma^2)$ or sometimes in a scale mixture $\theta_g = (\mu, \sigma_g^2)$.

The model (1.1) exhibits an inherent *exchangeability* in that it is invariant to permutation of the component labels. This exchangeability is considered by many authors to imply a kind of *unidentifiability*; we discuss this issue below in Section 1.3.

Another, mathematically equivalent, way to write the model (1.1) is in integral form,

$$y \sim \int f(\cdot | \theta) H(d\theta). \tag{1.2}$$

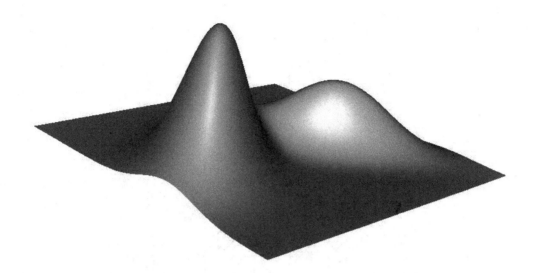

**FIGURE 1.2**

Example of a mixture of two bivariate normal densities. The model is $0.7 \times$ $\mathcal{N}\left(\begin{pmatrix} -1 \\ 1 \end{pmatrix}, \begin{pmatrix} 1 & 0.7 \\ 0.7 & 1 \end{pmatrix}\right) + 0.3 \times \mathcal{N}\left(\begin{pmatrix} 2.5 \\ 0.5 \end{pmatrix}, \begin{pmatrix} 1 & -0.7 \\ -0.7 & 1 \end{pmatrix}\right).$

The finite mixture model (1.1) corresponds to the discrete case where the mixing measure $H$ places probability mass $\eta_g$ on the atom $\theta_g$, $g = 1, 2, \ldots, G$.

It can be useful to allow slightly more generality:

$$y \sim \sum_{g=1}^{G} \eta_g f_g(\cdot | \theta_g), \tag{1.3}$$

where different components are allowed different parametric forms; the full generality here is seldom used outside very specific contexts, but cases where the $f_g$ differ only through the values of measured covariates are common; see Section 1.2. Of course, such an assumption modifies the degree of exchangeability in the indexing of components.

Many of the ideas in mixture modelling, including all those raised in this chapter, apply equally to univariate and multivariate data; in the latter case, observed variates $y_i$ are random vectors. Figure 1.2 illustrates a bivariate normal mixture with two components. Mixture models are most commonly adopted for continuously distributed variables, but discrete mixtures are also important; an example is provided in Figure 1.3. We continue to use $f$ for what will now be a probability mass function (or density function with respect to counting measure).

Finally, mixture models can equally well be defined as convex combinations of distribution functions or measures; this is little more than a notational variation, and we will use the customary density function representation (1.1) throughout this chapter.

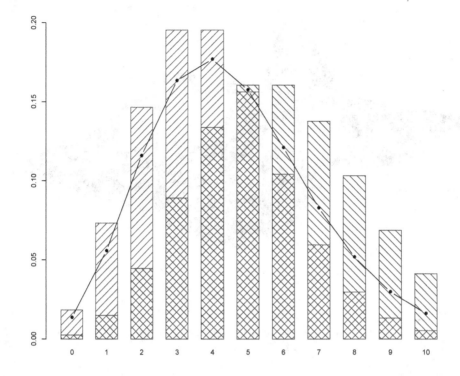

**FIGURE 1.3**
A mixture of two Poisson distributions: $0.7 \times \mathcal{P}(4) + 0.3 \times \mathcal{P}(6)$.

## 1.1.2  Likelihood

Modern inference for mixture models – whether Bayesian or not – almost always uses the likelihood function, which in the case of $n$ independently and identically distributed (i.i.d.) observations from (1.1) has the form

$$p(\mathbf{y}|\theta, G) = \prod_{i=1}^{n} \sum_{g=1}^{G} \eta_g f(y_i|\theta_g). \tag{1.4}$$

Most of the unusual, and sometimes challenging, features of mixture analysis stem from having to handle this product-of-sums form.

## 1.1.3  Latent allocation variables

An alternative starting point for setting up a mixture model is more probabilistic. Suppose the population from which we are sampling is heterogeneous: there are multiple groups, indexed by $g = 1, 2, \ldots, G$, present in the population in proportions $\eta_g$, $g = 1, 2, \ldots, G$. When sampling from group $g$, observations are assumed drawn from density $f(\cdot|\theta_g)$. Then we can imagine that an observation $y$ drawn from the population is realized in two steps: first, the group $z$ is drawn from the index set $g = 1, 2, \ldots, G$, with $\mathrm{P}(z = g) = \eta_g$; and secondly, given $z$, $y$ is drawn from $f(\cdot|\theta_z)$.

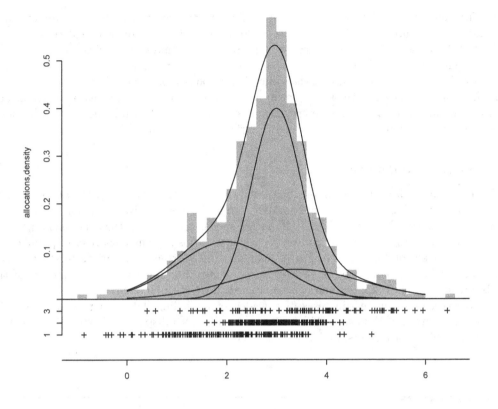

**FIGURE 1.4**

The same mixture of three normal densities as in Figure 1.1, together with a sample of size 500, sampled according to (1.5), histogrammed, and plotted against the allocation variables.

This two-stage sampling gives exactly the same model (1.1) for the distribution of $y$. In a random sample, we have

$$y_i | z_i \sim f(\cdot | \theta_{z_i}) \quad \text{with} \quad \mathrm{P}(z_i = g) = \eta_g, \tag{1.5}$$

independently for $i = 1, 2, \ldots, n$. (In the case corresponding to (1.3), $f(\cdot | \theta_{z_i})$ would be replaced by $f_{z_i}(\cdot | \theta_{z_i})$.) The $z_i$ are *latent* (unobserved) random variables; they are usually called *allocation variables* in the mixture model context. This construction is illustrated in Figure 1.4, for the same three-component univariate normal mixture as in Figure 1.1.

This alternative perspective on mixture modelling provides a valuable "synthetic" view of the models, complementary to the "analytic" view in (1.1). It is useful to be able to keep both of these perspectives, irrespective of whether the groups being modelled by the different components $g = 1, 2, \ldots, G$ have any reality (as substantively interpretable subgroups) in the underlying population or not. In particular, very commonly the latent variable representation (1.5) is key in computing inference for mixture models, as we will see below, and later in the volume.

This duality of view does bring a peril, that it can be rather tempting to attempt to deliver inference about the allocation variables $z_i$ even in contexts where the mixture model (1.1) is adopted purely as a convenient analytic representation.

More abstractly, the synthetic view of mixture modelling provides a prototype example of a much broader class of statistical models featuring latent variables; mixtures provide a useful exemplar for testing out methodological innovation aimed at this broader class.

Titterington et al. (1985) draw a clear distinction between what they call "direct applications" of mixture modelling, where the allocation variables represent real subgroups in the population and (1.5) is a natural starting point, and "indirect applications" where the components are more like basis functions in a linear representation and we begin with (1.1).

A different way to think about the same distinction is more focused on outcomes. Direct applications are based intrinsically on an assumption of a heterogeneous population, and delivering inference about that heterogeneity is a probable goal, while indirect applications essentially amount to a form of semi-parametric density estimation.

### 1.1.4   A little history

The creation-myth of mixture modelling involves two of the great intellectuals of the late nineteenth century: the biometrician, statistician and eugenicist Karl Pearson and the evolutionary biologist Walter Weldon. The latter speculated in 1893 that the asymmetry he observed in a histogram of forehead to body length ratios in female shore crab populations could indicate evolutionary divergence. Pearson (1894) fitted a univariate mixture of two normals to Weldon's data by a method of moments, choosing the five parameters of the mixture so that the empirical moments matched those of the model.

However, numerous earlier uses of the idea of modelling data using mixtures can be found in the work of Holmes (1892) on measures of wealth disparity, Newcomb (1886) who wanted a model for outliers, and Quetelet (1852).

## 1.2   Generalizations

The basic mixture model (1.1) sets the scene and provides a template for a rich variety of models, many covered elsewhere in this book. So this section is in part an annotated contents list for the later chapters in the volume.

### 1.2.1   Infinite mixtures

It is by no means logically necessary or inevitable that the number of components $G$ in (1.1) be finite. Countably infinite mixtures

$$y \sim \sum_{g=1}^{\infty} \eta_g f(\cdot|\theta_g)$$

are perfectly well defined and easily specified, although in practice seldom used. Adopting such a model avoids various difficulties to do with choosing $G$, whether fixed in advance, or inferred from data (see Chapters 4, 6, 7, 8 and 10). And of course, using a *population* model in which the number of components is unlimited in no way implies that a *sample* uses more than a finite number. Whether $G$ is finite or infinite, there will in general be "empty components", components $g$ for which $z_i \neq g$ for all $i = 1, 2, \ldots, n$.

The kind of countably infinite mixture that has been most important in applications is the Dirichlet process mixture (Lo, 1984), and its relatives, that form a central methodology in Bayesian nonparametric modelling (see Chapter 6). In Dirichlet process mixtures, the allocation variables $z_i$ are no longer drawn (conditionally) independently from a distribution $(\eta_g)$, but rather jointly from a Pólya urn process: ties among the $z_i, i = 1, 2, \ldots, n$, determine a clustering of the indices $1, 2, \ldots, n$, in which the probability of any given set of clusters is given by

$$\frac{e_0^{G_+} \prod_j (n_j - 1)!}{e_0(e_0 + 1) \cdots (e_0 + n - 1)},$$

where $G_+$ is the number of non-empty clusters, whose sizes are $n_1, n_2, \ldots, n_{G_+} \geq 1$, and $e_0 > 0$ is a parameter controlling the degree of clustering. This is not the usual approach to defining the Dirichlet process mixture model, further exposed in Chapters 6 and 17, but Green & Richardson (2001) show that it is equivalent to the standard definitions. They also explain how this model arises as a limit of finite mixture models (1.1) with appropriate priors on $\{\eta_g, \theta_g\}$, namely where the $\theta_g$ are *a priori* i.i.d., and $(\eta_1, \eta_2, \ldots, \eta_G) \sim \mathcal{D}(e_0/G, e_0/G, \ldots, e_0/G)$, considered in the limit as $G \to \infty$.

One of the reasons why this is sometimes called a "nonparametric" mixture model is that the complexity of the model automatically adapts to the sample size. In a Dirichlet process mixture model applied to $n$ i.i.d. data, the prior mean number of components is $\sim e_0 \log(n/e_0)$ for large $n$. However, this cannot be directly compared to a finite mixture model where $G$ is fixed, since that model allows "empty components", that is, components in the model to which, in a particular sample, no observation is allocated; in the Dirichlet process mixture, only non-empty components are counted.

### 1.2.2   Continuous mixtures

Another approach to infinity in the number of components starts from the integral form (1.2). Whereas the finite mixture model corresponds to the case where $H$ is discrete, in general, $H(\cdot)$ can be *any* probability distribution on the parameter space $\Theta$, and the resulting model remains well defined.

Continuous mixtures are sometimes called *compound probability distributions*; they are, if you like, distributions with random parameters. One context where these models arise, perhaps disguised, is in simple Bayesian analysis, where the mixing distribution corresponds to the prior, and the base/component distribution to the likelihood.

Several standard models fall into this class; these include exponential family models mixed by their conjugate "priors", in Bayesian language, so giving an explicit unconditional density.

A beta mixture of binomial distributions (each with the same index parameter) is a beta-binomial distribution, often used for data overdispersed relative to the binomial:

$$p(y) = \int_0^1 \binom{n}{y} \theta^y (1-\theta)^{n-y} \times \frac{\theta^{\alpha-1}(1-\theta)^{\beta-1}}{B(\alpha,\beta)} d\theta = \binom{n}{y} \frac{B(\alpha+y, \beta+n-y)}{B(\alpha,\beta)},$$

for $y = 0, 1, 2, \ldots, n$; see Figure 1.5. When there are more than two categories of response, this becomes a Dirichlet-multinomial distribution.

A gamma mixture of Poisson distributions is a negative-binomial distribution,

$$p(y) = \int e^{-\theta} \frac{\theta^y}{y!} \times \frac{\theta^{\alpha-1}\beta^\alpha e^{-\beta\theta}}{\Gamma(\alpha)} d\theta = \frac{\Gamma(\alpha+y)}{y!\Gamma(\alpha)} \pi^y (1-\pi)^\alpha,$$

for $y = 0, 1, 2, \ldots$, where $\pi = (1+\beta)^{-1}$, and again a common use is for overdispersion.

For continuous responses, a commonly-used continuous mixture is the scale mixture of normals model,

$$p(y) = \int \phi(y|\mu, \sigma) H(d\sigma),$$

which preserves the symmetry about $\mu$ of the density, but varies the shape and tail behaviour according to $H$. Familiar examples include the $t$ and double-exponential (Laplace)

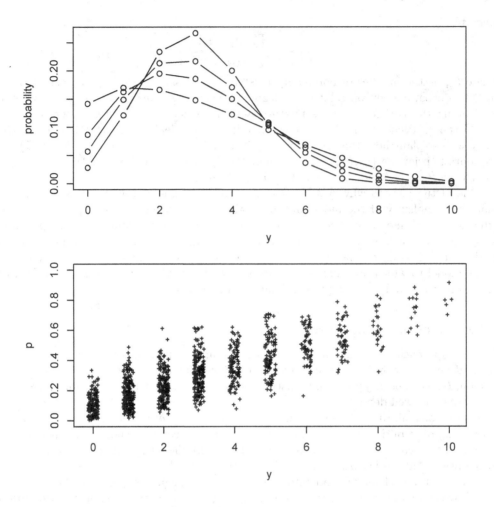

**FIGURE 1.5**
Upper panel: The binomial distribution $\mathcal{B}(10, 0.3)$, and beta-binomial distributions with $(\alpha, \beta) = (6, 14)$, $(3, 7)$ and $(1.5, 3.5)$ (in decreasing order of the height of the mode). Lower panel: 1000 draws from the distribution of $(y, \pi)$, where $\pi \sim \mathcal{B}e(1.5, 3.5)$ and $y|\pi \sim \mathcal{B}(10, \pi)$; the points are jittered randomly in the horizontal direction to avoid overplotting.

distributions, which arise when the mixing distribution $H$ is inverse gamma or exponential, respectively.

Another use of continuous mixtures for continuous distributions is provided by the nice observation that any distribution on $[0, \infty)$ with non-increasing density can be represented as a mixture over $\theta$ of uniform distributions on $[0, \theta]$; see, for example, Feller (1970, p. 158).

Aspects of continuous mixtures are covered in detail in Chapter 10.

### 1.2.3   Finite mixtures with nonparametric components

A different kind of mixture modelling with nonparametric aspects other than the Bayesian nonparametric modelling mentioned above takes the mixture model in the form of (1.3), or its alternative representation in terms of distribution functions, and replaces one or more

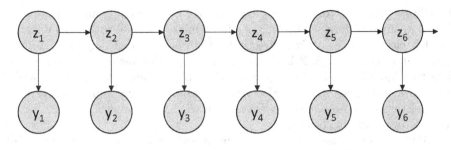

**FIGURE 1.6**
Graphical representation of a hidden Markov model (as a directed acyclic graph).

of the components by a nonparametric density, perhaps subject to qualitative constraints. Some examples include the mixture of a known density with an unknown density assumed only to be symmetric, or non-decreasing. Another would be an arbitrary finite mixture of copies of the same unknown parametric density with different unknown location shifts.

Most of the research work on such models has focused on theoretical questions such as identifiability and rates of convergence in estimation, rather than methodology or application. These models are considered in detail in Chapter 14.

### 1.2.4 Covariates and mixtures of experts

In regression modelling with mixtures, we want to use analogues of (1.1) to model the conditional distribution of a response $y$ given covariates $x$ (scalar or vector, discrete or continuous, in any combination). There are many ways to do this: the weights $\eta_g$ could become $\eta_g(x)$, and/or the component densities could become $f(\cdot|\theta_g, x)$ or $f(\cdot|\theta_g(x))$, and these $\eta_g(x)$ and $\theta_g(x)$ could be parametric or nonparametric regression functions, as the modeller requires.

The case where both $\eta_g(x)$ and $\theta_g(x)$ appear as functions of the covariate,

$$y|x \sim \sum_{g=1}^{G} \eta_g(x) f(\cdot|\theta_g(x)),$$

is known in the machine learning community as a *mixture of experts* model (Jacobs et al., 1991), the component densities $f(\cdot|\theta_g(x))$ being the "experts", and the weights $\eta_g(x)$ the "gating networks". The special cases where only $\eta_g$ or only $\theta_g$ depend on $x$ have been called the "gating-network" and the "expert-network" mixture of experts models. Mixture of experts models are fully discussed in Chapter 12.

### 1.2.5 Hidden Markov models

In contexts where responses $y_t, t = 1, 2, \ldots, T$, are delivered sequentially in time, it is natural to consider mixture models analogous to (1.1) but capturing sequential dependence. This is most conveniently modelled by supposing that the allocation variables, now denoted by $z_t$, form a (finite) Markov chain, say on the states $\{1, 2, \ldots, G\}$ and with transition matrix $\xi$. This defines a *hidden Markov model*,

$$y_t|z_1, z_2, \ldots, z_t \sim f(\cdot|\theta_{z_t}),$$
$$\mathrm{P}(z_t = g|z_1, z_2, \ldots, z_{t-1} = h) = \xi_{hg}. \tag{1.6}$$

Figure 1.6 displays the structure of this model as a directed acyclic graph. This set-up can naturally and dually be interpreted either as a finite mixture model with serial dependence, or as a Markov chain observed only indirectly (as the word "hidden" suggests) through the "emission distributions" $f(\cdot|\theta_{z_t})$. It is also a template for a rich variety of models extending the idea to (countably) infinite state spaces, and to continuous state spaces, where these are usually called *state-space* models. Also very important in many applications are hidden Markov models where the order of dependence in the state sequence $z_1, z_2, \ldots, z_t, \ldots$ is greater than 1 (or of variable order).

Hidden Markov models, and other time-series models involving mixtures, are fully discussed in Chapter 13 with an emphasis on economic applications, whereas Chapter 17 discusses applications in finance and Chapter 18 applications in genomics.

## 1.2.6   Spatial mixtures

Where data are indexed spatially rather than temporally, mixture models can play an analogous role, and as usual the models proposed can often be viewed as generalizations or translations from time to space. For problems where the observational units are discrete locations or non-overlapping regions of space, the Hidden Markov random field model of Green & Richardson (2002) was introduced as an analogue of (1.6) where the serial Markov dependence among $z_t$ is replaced by a Potts model, a discrete Markov random field, with $t$ now indexing locations or regions in space. Letting $\mathbf{z} = \{z_t\}$ be the collection of all unknown allocations, we take

$$p(\mathbf{z}) \propto \exp\left\{\psi \sum_{t' \sim t} \mathbb{I}(z_t = z_{t'})\right\},$$

where $t' \sim t$ denotes that spatial locations $t$ and $t'$ are contiguous in some sense, and $\psi$ is an interaction parameter, positive for positive association. In the context of Green & Richardson (2002), the observations $y_t$ were counts of cases of a rare disease in regions indexed by $t$, modelled as conditionally independent given $\mathbf{z}$, with

$$y_t|z_t \sim \mathcal{P}(\theta_{z_t} E_t),$$

where $(\theta_g)_{g=1}^G$ are parameters, *a priori* i.i.d., and $(E_t)$ are expected counts based on population size, adjusted for, say, age and sex. Again, there are dual interpretations, as a mixture model with hidden spatial dependence, or as a spatial process observed with noise. In the latter view, this model can then be associated with the rich variety of image models built on hidden spatial Markov processes for the "true scene" (Geman & Geman, 1984; Besag, 1986).

An alternative to modelling dependence among the allocation variables directly through a discrete stochastic process is to introduce spatial dependence between the *distributions* of the allocation variables, that is, between the weights at different spatial locations. Models of this kind were introduced by Fernández & Green (2002), who assume that conditionally independent, spatially indexed observations $y_t$ are modelled as

$$y_t \sim \sum_{g=1}^G \eta_{tg} f_t(\cdot|\theta_g),$$

where the (location-specific) weights $\eta_{tg}$ are modelled as a spatial stochastic process; they have something of the flavour of (gating-network) mixtures of expert models where the "covariate" is spatial location. Fernández & Green (2002) consider two specific forms for the spatial weights $\eta_{tg}$: one a multinomial logit transformation of (latent) independent

Gaussian processes, the other using a grouped-continuous model, also based on a Gaussian process. For the same rare-disease mapping context, the component densities were given by

$$f_t(y|\theta_g) = e^{-\theta_g E_t} \frac{(\theta_g E_t)^y}{y!}.$$

Spatial mixtures, and in particular their application to image analysis, are discussed in Chapter 16.

## 1.3   Some Technical Concerns

### 1.3.1   Identifiability

Frühwirth-Schnatter (2006) observes that the density (1.1) remains invariant on either (a) including an additional component with arbitrary parameter $\theta$ but weight $\eta = 0$, or (b) replicating any of the components, giving the new one(s) the same parameter $\theta$ and sharing with it the existing weight $\eta$, and that this violates identifiability. In commenting on this observation, Rousseau & Mengersen (2011) write: "This non-identifiability is much stronger than the non-identifiability corresponding to permutations of the labels in the mixture representation." I would go further and argue that the label-permutation invariance is best not thought of as non-identifiability at all.

Leaving aside special situations where the mixture representation (1.1) is adopted in a situation where the component labels $g$ represent real pre-identified groups in a population – and where, for example, a Bayesian analysis would use a prior specification in which different priors could be used for different groups – the model is indeed invariant to permutation of the labels $g = 1, 2, \ldots, G$. But it seems to me to be wrong to refer to this as lack of identifiability; it is not a property of the mixture model (1.1), but a side-effect of the representation used therein.

When the mixing distribution $H$ has finite support, the mixture model (1.1) is equivalent to the integral representation (1.2). The only difference is that in (1.1) the atoms of the distribution $H$ have been numbered, $g = 1, 2, \ldots$. But there is no natural or canonical way to do this numbering: the labels $g$ are artificial creatures of the choice of representation, and serve only to pair the parameter values $\theta_g$ with their corresponding weights $\eta_g$.

If we focus, more properly, on inference for the mixing distribution $H$, then identifiability is possible: for example, Teicher (1963) showed that finite mixtures of normal or gamma distributions are identifiable (but interestingly, this is in general not true for binomial distributions; see also Chapter 12, Section 12.5.1). Focusing on $H$ does more than solve the problem of invariance to label permutation; it even fixes the "stronger" non-identifiability that is obtained by allowing zero-weight or coincident components, since such variants do not affect the mixing distribution $H$ either.

The objective of Rousseau & Mengersen (2011) is to study the impact of this "non-identifiability" on asymptotic inference; it would be interesting to see whether the difficulties they uncover also apply in the case of asymptotic inference about $H$.

### 1.3.2   Label switching

A related but different problem, but one with the same root cause, is the phenomenon of "label switching". This has attracted a lot of attention from researchers and generated many papers. It shows up particularly in iterative methods of inference for a finite mixture model, especially Markov chain Monte Carlo (MCMC) methods for Bayesian mixture analysis. However, the problem can be stated and solved without even mentioning iteration.

Consider the model (1.1), with estimates $\widehat{\theta}_g$ and $\widehat{\eta}_g$ of the unknown quantities. Then not only is the model (1.1) invariant to permutation of the labels of components, so is the *estimated model*

$$y \sim \sum_{g=1}^{G} \widehat{\eta}_g f(\cdot | \widehat{\theta}_g). \tag{1.7}$$

Given that both numberings of components are arbitrary, there is no reason whatsoever for the numbering of the components in (1.1) and (1.7) to be coherent. For example, we cannot say that $\widehat{\theta}_1$ estimates $\theta_1$!

In an MCMC run, the lack of coherence between the numberings can apply separately at each iteration, so commonly "switchings" are observed.

Numerous solutions to this perceived problem have been proposed, but the only universal valid and appropriate approach to both apparent non-identifiability and label switching is to recognize that you can only make inference about the mixing distribution $H$. You can estimate (or find the posterior distributions of) the number of (distinct) atoms of $H$ that lie in a given set, or the total weight of these atoms, or the weight of the component with the largest variance, or indeed any other property of the mixture distribution $\sum_{g=1}^{G} \eta_g f(\cdot | \theta_g)$, but you cannot unambiguously estimate, say, $\theta_1$ or $\eta_3$.

If the group labels do have a substantive meaning, for example, group $g = 1$ is "the group with the largest variance", then $\theta_1$ and $\eta_1$ do become individually meaningful. But you still cannot expect this substantive meaning to automatically attach to the labelling of the groups in the *estimated* mixture $\sum_{g=1}^{G} \widehat{\eta}_g f(\cdot | \widehat{\theta}_g)$.

## 1.4    Inference

This opening chapter is not the place to go into inference for mixture models in any detail, but this section attempts to give the flavour of some of the algorithmic ideas for basic inference in finite mixture models.

### 1.4.1    Frequentist inference, and the role of EM

An enormous variety of methods have been considered over the years for frequentist inference about unknowns in a mixture model. For the standard problem of estimation based on a simple random sample from (1.1), Pearson's oft-cited work (Pearson, 1894) used the method of moments to estimate the five free parameters in a two-component univariate normal mixture, and for at least a century thereafter novel contributions to such methods for mixtures continued to appear. A second big group of methods is based on a suitable distance measure between the empirical distribution of the data sample and the mixture model, to be minimized over choice of the free parameters. Except for some very particular special cases, both the method of moments and minimum-distance methods require numerical optimization, there being no explicit formulae available.

In current practice the dominant approach to inference is by maximum likelihood, based on maximization of (1.4), again using numerical methods. Subject to solving the numerical issues, such an approach is immediately appealing on many grounds, including its continuing applicability when the basic model is elaborated, through the introduction of covariates for example.

The usual numerical approach to maximum likelihood estimation of mixture models, advocated for example throughout the book by McLachlan & Peel (2000), uses the EM algorithm (Dempster et al., 1977). Mixture modelling and the EM algorithm are identified together so strongly that mixtures are often used as examples in tutorial accounts of EM. This is highly natural, not only because of statisticians' typical conservatism in their toolkits of numerical methods, but also because the natural area of applicability of EM, to "hidden data" problems, maps so cleanly onto the mixture model set-up.

Referring back to the latent allocation variable formulation (1.5), it is clear that if the $z_i$ were known, we would have separate independent samples for each component $g$, so that estimation of $\theta_g$ would be a standard task. Conversely, if the parameters $\theta_g$ were known, allocating observations $y_i$ to the different components could be done naturally by choosing $z_i = g$ to maximize $f(y_i|\theta_g)$. One could imagine a crude algorithm that alternates between these two situations; this could be seen as a "decision-directed" or "hard classification" version of the EM algorithm, as applied to mixtures.

The actual EM algorithm is the "soft-classification" analogue of this. The $z_i$ are treated as the "missing data", and represented by the indicator vectors with components $z_{ig}$, where $(z_{ig} = 1) \equiv (z_i = g)$. In the E step of the algorithm $z_{ig}$ is replaced by its conditional expectation $\mathrm{E}(z_{ig}|\theta, \eta, y_i) = \tau_{ig}$ given the current values of the parameters and weights, that is,

$$\tau_{ig} = \frac{\eta_g f(y_i|\theta_g)}{\sum_{g'} \eta_{g'} f(y_i|\theta_{g'})}, \qquad (1.8)$$

and in the M step, $\eta_g$ and $\theta_g$ are updated to maximize the corresponding expected complete-data log likelihood,

$$\sum_i \sum_g \tau_{ig} \log(\eta_g f(y_i|\theta_g)),$$

which can often be done in closed form. When the $\eta_g$ and $\theta_g$ vary independently for each component $g$, this maximization may be done separately for each. The EM algorithm, though sometimes slow, enjoys the usual advantages that each iteration increases, and under certain conditions maximizes, the likelihood. We see that in this algorithm, the latent allocation variables $z_i$ play a crucial role, irrespective of whether the groups have a substantive interpretation.

Maximum likelihood inference for mixture models is covered in full detail in Chapters 2 and 3.

### 1.4.2 Bayesian inference, and the role of MCMC

This elevation of the allocation variables to essential ingredients of a numerical method is also seen in the most natural Markov chain Monte Carlo method for Bayesian inference of the parameters and weights. Diebolt & Robert (1990, 1994) proposed data augmentation and Gibbs sampling approaches to posterior sampling for parameters and weights of finite mixture distributions. Given the current values of $\{\eta_g\}$ and $\{\theta_g\}$, the Gibbs sampler update for the allocation variables $z_i$ is to draw them independently from $\mathrm{P}(z_i = g|\theta, \eta, y_i) = \tau_{ig}$, where $\tau_{ig}$ is as in (1.8); this is simply Bayes' rule. Note the close parallel with EM here.

MCMC updating of the $\eta_g$ and $\theta_g$ requires Gibbs sampling, or more generally Metropolis–Hastings updates, using the full conditionals of these variables given the data **y** and the current values of the allocation variables **z**. The details here depend on the form of the base density $f$ and the priors on the parameters; for normal mixtures under normal-inverse gamma priors as in Diebolt & Robert (1994), the Gibbs sampling update is straightforward and involves only sampling from Dirichlet, inverse gamma and normal densities.

There is an obvious parallel between the EM algorithm for the maximum likelihood solution and the MCMC algorithm for posterior sampling in the conjugate-prior Bayesian model, with the same kind of central role for the latent allocation variables.

Bayesian inference for mixture models is discussed in detail in Chapter 4, while Chapter 5 is devoted to posterior sampling.

### 1.4.3 Variable number of components

One of the many appeals of a Bayesian approach to mixture analysis is that it supports a particularly clean way of dealing with uncertainty in the number $G$ of components. In a Bayesian set-up, inference about $G$ can be performed simultaneously with that about the weights $\eta_g$ and component parameters $\theta_g$, whereas in frequentist inference, $G$ has to be treated as a model indicator, an unknown of a different character than the other unknowns, the traditional parameters. For the rest of this subsection, we consider only the Bayesian approach.

When $G$ is not treated as known we are in one of those situations where "the number of unknowns is one of the things you don't know", a type of problem that Roeder & Wasserman (1997) termed "transdimensional". The objective of the inference will be to deliver the joint posterior distribution of $\{G, (\eta_g, \theta_g)_{g=1}^G\}$. There are numerous possible approaches to the computational task of computing this posterior, of which sampling-based approaches are much the most popular. As often happens with MCMC methods for transdimensional problems, those for mixture analysis fall into two categories: within-model and across-model samplers.

Within-model sampling involves running separate MCMC posterior simulations for each of a range of values of $G$ from 1 up to some pre-assigned maximum of interest. Each individual simulation is a standard fixed-$G$ analysis, along the lines of Section 1.4.2. In addition to delivering a fixed-$G$ posterior sample, the simulation must calculate, or rather estimate, the marginal likelihood $p(\mathbf{y}|G)$ of the data, given $G$. After all runs are complete, the separate inferences can be combined across $G$ using the fact that

$$p(G|\mathbf{y}) = \frac{p(G)p(\mathbf{y}|G)}{\sum_G p(G)p(\mathbf{y}|G)},$$

where $p(G)$ is the prior distribution of $G$. In across-model sampling, a single MCMC sampler is constructed, targeting the joint posterior of $\{G, (\eta_g, \theta_g)_{g=1}^G\}$, using for example the method of Carlin & Chib (1995), reversible jump MCMC (Richardson & Green, 1997) or a birth-and-death process (Stephens, 2000); see Chapter 7 of this volume for a full account.

Both within- and across-model simulations can be used to deliver exactly the same conclusions, since both ultimately estimate the same joint posterior in a simulation-consistent way. One or another approach may be more computationally efficient for the same precision of a particular quantity of interest, depending on circumstances.

### 1.4.4 Modes versus components

Motivation for mixture modelling often stems from visual exploratory data analysis, where a histogram of data suggests multimodality. But of course there need be no simple relationship between the number of modes in a mixture such as (1.1) and the number $G$ of components. Indeed, Figure 1.1 displays a three-component mixture with only one mode, while there is nothing in (1.1) to prevent even a single component having any number of modes. If component densities in (1.1) are unimodal, then the mixture will have $G$ modes, provided the $\theta_g$ are sufficiently different.

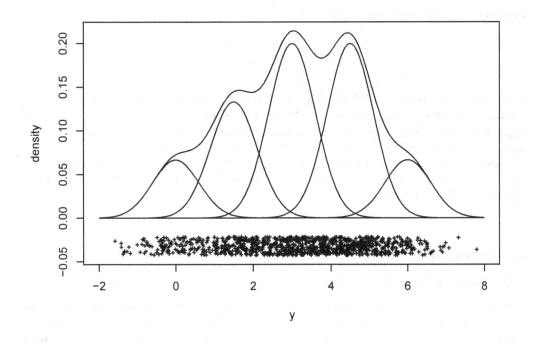

**FIGURE 1.7**
A normal mixture model with five components, three modes, and for which a sample of 1000 draws seem to form one cluster. The model is $0.1 \times \mathcal{N}(0, 0.6^2) + 0.2 \times \mathcal{N}(1.5, 0.6^2) + 0.3 \times \mathcal{N}(3, 0.6^2) + 0.3 \times \mathcal{N}(4.5, 0.6^2) + 0.1 \times \mathcal{N}(6, 0.6^2)$.

Rigorous statistical theory and methodology taking explicit note of modality is rare, though recent work by Chacón (2015) provides a population background for this. Améndola et al. (2017) discuss the history of the problem of enumerating the possible numbers of modes for a mixture of $G$ Gaussian distributions in $d$ dimensions, and give improved lower bounds, and the first upper bound, on the maximum number of non-degenerate modes.

### 1.4.5 Clustering and classification

In a rather similar way, mixture models may be adopted under an informal belief that the data are clustered, with observations within each cluster drawn from some simple parametric distribution. It is a small step apparently to regard "subpopulations" as "clusters", but again there may be no simple relationship between inferred components in a fitted mixture model and clusters in the original data. The term "cluster" commonly suggests both a degree of *homogeneity* within a cluster and a degree of *separation* between clusters, while typically *not* implying any specific within-cluster distribution. A fitted mixture model may need more components than there are apparent clusters in the data simply because the assumed parametric form of the component densities is incorrect, so that each homogeneous cluster requires multiple components to fit it well. See the artificial example in Figure 1.7.

Nevertheless, finite mixture modelling does provide one of the few simple and rigorous model-based approaches to clustering, a theme that is explored in detail in Chapter 8.

For convincing model-based inference about clusters, using a within-cluster data distribution that is itself likely to be a mixture, it seems necessary to adopt a second level of indexing of weights and parameters, as in

$$y \sim \sum_{g=1}^{G} \sum_{h=1}^{H_g} \eta_{gh} f(\cdot | \theta_{gh}),$$

where $g$ indexes clusters and $h$ components within clusters. The demands of homogeneity within clusters and separation between clusters can be met by appropriately modelling the $\theta_{gh}$; for example, in a Bayesian formulation, we might require $\theta_{gh} - \theta_{g'h'}$ to have small variance if $g = g'$ and large variance otherwise, perhaps most conveniently achieved with a hierarchical formulation such as $\theta_{gh} = \alpha_g + \beta_{gh}$, with a suitably "repelling" prior on the $\alpha_g$. A more intricate version of this basic idea has been thoroughly investigated recently in Malsiner-Walli et al. (2017).

## 1.5 Concluding Remarks

In this opening chapter, I hope to have set the scene on some key ideas of mixture modelling, stressing the underlying unity of the main concepts, as they apply to different data structures and in different contexts. The full versatility of mixtures as a modelling tool, the wide range of specific methodologies, the rich and subtle theoretical underpinnings of inference in the area, and ideas for further research will only become fully clear as the rest of the book unfolds.

# Bibliography

AMÉNDOLA, C., ENGSTRÖM, A. & HAASE, C. (2017). Maximum number of modes of Gaussian mixtures. Preprint, arXiv:1702.05066.

BESAG, J. (1986). On the statistical analysis of dirty pictures. *Journal of the Royal Statistical Society, Series B* **48**, 259–279.

CARLIN, B. P. & CHIB, S. (1995). Bayesian model choice through Markov chain Monte Carlo. *Journal of the Royal Statistical Society, Series B* **57**, 473–484.

CHACÓN, J. E. (2015). A population background for nonparametric density-based clustering. *Statistical Science* **30**, 518–532.

DEMPSTER, A. P., LAIRD, N. M. & RUBIN, D. B. (1977). Maximum likelihood from incomplete data via the EM algorithm (with discussion). *Journal of the Royal Statistical Society, Series B* **39**, 1–38.

DIEBOLT, J. & ROBERT, C. P. (1990). Estimation des paramètres d'un mélange par échantillonnage bayésien. *Notes aux Comptes-Rendus de l'Académie des Sciences I* **311**, 653–658.

DIEBOLT, J. & ROBERT, C. P. (1994). Estimation of finite mixture distributions by Bayesian sampling. *Journal of the Royal Statistical Society, Series B* **56**, 363–375.

FELLER, W. (1970). *An Introduction to Probability Theory and its Applications*, vol. 1. New York: John Wiley.

FERNÁNDEZ, C. & GREEN, P. J. (2002). Modelling spatially correlated data via mixtures: a Bayesian approach. *Journal of the Royal Statistical Society, Series B* **64**, 805–826.

FRÜHWIRTH-SCHNATTER, S. (2006). *Finite Mixture and Markov Switching Models.* New York: Springer-Verlag.

GEMAN, S. & GEMAN, D. (1984). Stochastic relaxation, Gibbs distributions and the Bayesian restoration of images. *IEEE Transactions on Pattern Analysis and Machine Intelligence* **6**, 721–741.

GREEN, P. J. & RICHARDSON, S. (2001). Modelling heterogeneity with and without the Dirichlet process. *Scandinavian Journal of Statistics* **28**, 355–375.

GREEN, P. J. & RICHARDSON, S. (2002). Hidden Markov models and disease mapping. *Journal of the American Statistical Association* **92**, 1055–1070.

HOLMES, G. K. (1892). Measures of distribution. *Journal of the American Statistical Association* **3**, 141–157.

JACOBS, R., JORDAN, M., NOWLAN, S. J. & HINTON, G. E. (1991). Adaptive mixture of local experts. *Neural Computation* **3**, 19–87.

LO, A. Y. (1984). On a class of Bayesian nonparametric estimates: I. Density estimates. *Annals of Statistics* **12**, 351–357.

MALSINER-WALLI, G., FRÜHWIRTH-SCHNATTER, S. & GRÜN, B. (2017). Identifying mixtures of mixtures using Bayesian estimation. *Journal of Computational and Graphical Statistics* **26**, 285–295.

MCLACHLAN, G. J. & PEEL, D. (2000). *Finite Mixture Models*. New York: John Wiley.

NEWCOMB, S. (1886). A generalized theory of the combination of observations so as to obtain the best result. *American Journal of Mathematics* **8**, 343–366.

PEARSON, K. (1894). Contribution to the mathematical theory of evolution. *Philosophical Transactions of the Royal Society of London A* **185**, 71–110.

QUETELET, A. (1852). Sur quelques proprietés curieuses que présentent les résultats d'une série d'observations, faites dans la vue de déterminer une constante, lorsque les chances de rencontrer des écarts en plus et en moins sont égales et indépendantes les unes des autres. *Bulletin de l'Académie Royale des Sciences, des Lettres et des Beaux-Arts de Belgique* **19**, 303–317.

RICHARDSON, S. & GREEN, P. J. (1997). On Bayesian analysis of mixtures with an unknown number of components. *Journal of the Royal Statistical Society, Series B* **59**, 731–792.

ROEDER, K. & WASSERMAN, L. (1997). Contribution to the discussion of the paper by Richardson and Green. *Journal of the Royal Statistical Society, Series B* **59**, 782.

ROUSSEAU, J. & MENGERSEN, K. (2011). Asymptotic behaviour of the posterior distribution in overfitted mixture models. *Journal of the Royal Statistical Society, Series B* **73**, 689–710.

STEPHENS, M. (2000). Bayesian analysis of mixture models with an unknown number of components – an alternative to reversible jump methods. *Annals of Statistics* **28**, 40–74.

TEICHER, H. (1963). Identifiability of finite mixtures. *Annals of Mathematical Statistics* **34**, 1265–1269.

TITTERINGTON, D. M., SMITH, A. F. M. & MAKOV, U. E. (1985). *Statistical Analysis of Finite Mixture Distributions*. New York: John Wiley.

# 2

## EM Methods for Finite Mixtures

**Gilles Celeux**

*INRIA Saclay, France*

## CONTENTS

## 2.1   Introduction

Even in the simplest situation of a two-component mixture where only the mixing proportion $\eta_1$ is missing, the likelihood equation of a mixture model is not available in closed form. Obviously, in such a simple situation the maximum likelihood estimate (MLE) of the mixture parameters can be derived easily with standard optimization algorithms such as Newton–Raphson. But the number of parameters $\theta$ in a mixture model grows rapidly with the dimension $d$ of variables and with the number $G$ of components. This means the Newton–Raphson algorithm becomes expensive both mathematically and computationally for evaluating the observed information matrix of the vector parameter $\theta$. Moreover, this algorithm does not increase the likelihood of being maximized at each of its iterations. It is thus no surprise that the Newton–Raphson algorithm is far from being the most exploited algorithm to derive the MLE of a finite mixture model.

By contrast, the EM algorithm (Dempster et al., 1977) stands as the most popular algorithm for the derivation of the MLE or the posterior mode for hidden structure models. Since mixture models are typical cases of hidden structure models, where the allocation variables $z_i$, $i = 1, \ldots, n$, are missing (as outlined in Chapter 1), EM applies to this setting. As a matter of fact, and as noted in Chapter 1, mixtures are often used as a prime illustration of the implementation of an EM algorithm. Indeed, mixture structures allows one to clearly highlight the rationale, the advantages, and the possible drawbacks of the EM algorithm (see McLachlan & Peel, 2000). In this chapter, we restrict our attention to maximum likelihood estimation for mixtures, which still leads to a clear description of the EM algorithm.

## 2.2    The EM Algorithm

### 2.2.1    Description of EM for finite mixtures

Often the presence of missing data in the model of interest makes maximum likelihood (ML) inference difficult. This is clear when considering the mixture model, since the observed-data or observed log likelihood $\mathbf{y} = (y_1, \ldots, y_n)$,

$$\ell_o(\theta) = \sum_{i=1}^{n} \log \left[ \sum_{g=1}^{G} \eta_g f_g(y_i \mid \theta_g) \right],$$

involves the logarithm of a sum. The idea of the EM algorithm in this context is to make use of the missing labels $z_i$ by maximizing at each iteration the conditional expectation of the complete-data or completed log likelihood

$$\ell_c(\theta, \mathbf{z}) = \sum_{i=1}^{n} \sum_{g=1}^{G} z_{ig} \log(\eta_g f_g(y_i \mid \theta_g)),$$

given the observed data and a current value of the parameter, towards the derivation of the MLE of its vector parameter in a simple way.

The EM algorithm takes advantage of the simple relation connecting the observed and the completed likelihoods,

$$\ell_o(\theta) = \ell_c(\theta, \mathbf{z}) - \sum_{g=1}^{G} \sum_{i=1}^{n} z_{ig} \log \tau_{ig},$$

where $\tau_{ig}$ is the conditional probability that $y_i$ arises from component $g$, for $i = 1, \ldots, n$ and $g = 1, \ldots, G$, given the parameter value $\theta$. This leads to

$$\ell_o(\theta) = Q(\theta \mid \theta^{(s)}) - H(\theta \mid \theta^{(s)}),$$

where $H(\theta \mid \theta^{(s)}) = \sum_{g=1}^{G} \sum_{i=1}^{n} \tau_{ig}^{(s)} \log \tau_{ig}$ and $\theta^{(s)}$ is the current parameter value. From Jensen's inequality, we have that $H(\theta \mid \theta^{(s)}) \leq H(\theta^{(s)} \mid \theta^{(s)})$. Thus if $Q(\theta \mid \theta^{(s)}) \geq Q(\theta^{(s)} \mid \theta^{(s)})$ then $\ell_o(\theta) \geq \ell_o(\theta^{(s)})$ and the likelihood value increases.

Therefore, starting from an arbitrary initial value $\theta^{(0)}$, the EM algorithm can be summarized as follows.

**E step** Compute $Q(\theta, \theta^{(s)}) = \mathrm{E}(\ell_c(\theta, \mathbf{z}) \mid \mathbf{y}, \theta^{(s)})$, where the expectation is taken with respect to $p(\mathbf{z}|\mathbf{y}, \theta^{(s)})$.

**M step** Determine $\theta^{(s+1)} = \arg\max_\theta Q(\theta, \theta^{(s)})$.

In the mixture context, we have

$$Q(\theta|\theta^{(s)}) = \sum_{i=1}^{n} \sum_{g=1}^{G} \tau_{ig}^{(s)} \left\{ \log \eta_g + \log f_g(y_i|\theta_g) \right\},$$

$\tau_{ig}^{(s)}$ being the conditional probability that $y_i$ arises from component $g$, for $i = 1, \ldots, n$ and $g = 1, \ldots, G$, for the current parameter value $\theta^{(s)}$. Thus, the E step reduces to the computation of these conditional probabilities:

**E step**

$$\tau_{ig}^{(s)} = \frac{\eta_g^{(s)} f_g(y_i \mid \theta_g^{(s)})}{\sum_{g'=1}^{G} \eta_{g'}^{(s)} f_{g'}(y_i \mid \theta_{g'}^{(s)})}, \tag{2.1}$$

for $i = 1, \ldots, n$ and $g = 1, \ldots, G$.

Note that the above M step depends on the mixture model at hand. For instance, in the case where the density $f_g(\cdot \mid \theta_g)$ is a multivariate Gaussian density with $\theta_g = (\mu_g, \Sigma_g)$, where $\mu_g$ is the mean and $\Sigma_g$ the covariance matrix of a Gaussian distribution, this M step amounts to the updates

**M step**

$$\eta_g^{(s+1)} = \frac{\sum_{i=1}^{n} \tau_{ig}^{(s)}}{n},$$

$$\mu_g^{(s+1)} = \frac{\sum_{i=1}^{n} \tau_{ig}^{(s)} y_i}{\sum_{i=1}^{n} \tau_{ig}^{(s)}},$$

$$\Sigma_g^{(s+1)} = \frac{\sum_{i=1}^{n} \tau_{ig}^{(s)} (y_i - \mu_g^{(s+1)})(y_i - \mu_g^{(s+1)})^{\top}}{\sum_{i=1}^{n} \tau_{ig}^{(s)}},$$

for $g = 1, \ldots, G$.

The EM algorithm enjoys nice practical features, which explains its widespread use. The M step is often of closed form for the very reason that the EM algorithm takes the missing labels into account, in a way easy to code that does not involve numerical difficulties since it bypasses inverting large-dimensional matrices. As apparent from the formulas in the M step described above, the updated estimates are standard ML estimates where the observations are weighted with the conditional probabilities ($\tau_{ig}$). For some specific mixture models, it may still happen that the M step is not available in closed form. Examples of such situations are found in Celeux & Govaert (1995) for Gaussian mixture models where the component covariance matrices are assumed to share the same eigenvectors but have different eigenvalues. Nonetheless, the increase in the computational burden most often remains limited.

Most importantly, Dempster et al. (1977) provide a proof that the observed log likelihood is increasing at each iteration, $\ell_o(\theta^{(s+1)}) \geq \ell_o(\theta^{(s)})$, with equality if and only if $Q(\theta^{(s+1)}|\theta^{(s)}) = Q(\theta^{(s)}|\theta^{(s)})$. As remarked above, this is a direct consequence of Jensen's inequality applied to the convex function $H(\theta \mid \theta^{(s)})$ and it stands as a fundamental convergence property of the EM algorithm.

## 2.2.2   EM as an alternating-maximization algorithm

In the mixture context, Hathaway (1986) has shown that the EM algorithm can be viewed as an alternate optimization algorithm aiming to maximize the fuzzy clustering criterion

$$\mathcal{F}_c(\theta, \mathbf{s}) = \ell_c(\theta, \mathbf{s}) + H(\mathbf{s}) \tag{2.2}$$

where $\mathbf{s} = (s_{ig})_{i=1,\dots,n}^{g=1,\dots,G}$ denotes a fuzzy clustering matrix of the $n$ observations in $G$ clusters,

$$\ell_c(\theta, \mathbf{s}) = \sum_{i=1}^{n} \sum_{g=1}^{G} s_{ig} \log(\eta_g f_g(y_i \mid \theta_g))$$

is the completed log likelihood associated to $\mathbf{s}$, and

$$H(\mathbf{s}) = -\sum_{i=1}^{n} \sum_{g=1}^{G} s_{ig} \log s_{ig}$$

is the entropy of the fuzzy clustering matrix $\mathbf{s}$. More specifically:

(a) Since $\sum_{g=1}^{G} s_{ig} = 1$ for $i = 1, \dots, n$, maximizing $\mathcal{F}_c(\theta, \mathbf{s})$ with respect to $\mathbf{s}$ for fixed $\theta$ leads, after standard Lagrangian manipulation, to

$$s_{ig} = \tau_{ig} = \frac{\eta_g f_g(y_i \mid \theta_g)}{\sum_{g'=1}^{G} \eta_{g'} f_{g'}(y_i \mid \theta_{g'})},$$

for $i = 1, \dots, n$ and $g = 1, \dots, G$. This is exactly the E step of EM.

(b) Maximizing $\mathcal{F}_c(\theta, \mathbf{s})$ in $\theta$ for a fixed value of $\mathbf{s}$ leads to maximizing $\ell_c(\theta, \mathbf{s})$ since $H(\mathbf{s})$ does not depend on $\theta$. Thus, this is exactly the M step of EM.

This interpretation of the EM algorithm as an alternating-maximization algorithm has been extended to a general context in Neal & Hinton (1998): EM can be regarded as an alternating-optimization algorithm to maximize the criterion

$$\mathcal{F}_c(P, \theta) = E_P(\log p(\mathbf{y}, \mathbf{z} \mid \theta)) + H(P), \tag{2.3}$$

where $P$ is a probability distribution over the space of the missing data $\mathbf{z}$ and

$$H(P) = -E_P(\log P)$$

is the entropy of $P$. Moreover, the criterion $\mathcal{F}_c$ can be related to the Kullback–Leibler divergence $KL$ between $P(\mathbf{z})$ and the conditional distribution $p(\mathbf{z} \mid \mathbf{y}, \theta)$ of the missing data $\mathbf{z}$ knowing the observed data $\mathbf{y}$ and the parameter $\theta$:

$$\mathcal{F}_c(\mathrm{P}, \theta) = \ell_o(\theta) + KL(P(\mathbf{z}), p(\mathbf{z} \mid \mathbf{y}, \theta)), \tag{2.4}$$

where

$$KL(P(\mathbf{z}), p(\mathbf{z} \mid \mathbf{y}, \theta)) = \int P(\mathbf{z}) \log \frac{P(\mathbf{z})}{p(\mathbf{z} \mid \mathbf{y}, \theta)} d\mathbf{z}.$$

Relation (2.4) can be exploited to define variational approximations of the EM algorithm when the E step is intractable (see, for instance, Govaert & Nadif, 2008; or Titterington, 2011). The idea is to restrict the distribution $p(\mathbf{z})$ to factorize with respect to well-chosen groups $z_t$, $t = 1, \dots, K$, so that

$$P(\mathbf{z}) = \prod_{t=1}^{K} P(z_t) \tag{2.5}$$

and to seek the distribution of the form (2.5) for which $KL(P(\mathbf{z}), p(\mathbf{z} \mid \mathbf{y}, \theta))$ is minimum. Using (2.3), it is easy to see that this distribution yields a lower bound on $\mathcal{F}_c(P, \theta)$ among the distributions $P$ of the form (2.5). Obviously, the factorization in (2.5) is chosen to ensure the tractability of this minimization.

When needed, the variational Bayesian EM algorithm is a popular algorithm to approximate the posterior distribution of parameterized models (see, for instance, Bishop, 2006; Chapter 10, this volume; or Titterington, 2011).

## 2.3  Convergence and Behavior of EM

General results on the theoretical behavior of the EM algorithm can be found in Dempster et al. (1977) and Wu (1983). Such results cannot be detailed here, but the most important aspects are recalled in the following. From the increase of the observed log likelihood at each iteration of the EM algorithm, it follows that the MLE of $\theta$ is a fixed point of the EM algorithm. More precisely, denoting by $EM$ the stepwise operator of the EM algorithm (that is, $\theta^{(s+1)} = EM(\theta^{(s)})$), we have the following theorem.

**Theorem 2.1** *For $\theta^* \in \operatorname{argmax} \ell_o(\theta)$, we have almost surely that*

$$
\begin{aligned}
\ell_o(EM(\theta^*)) &= \ell_o(\theta^*), \\
Q(EM(\theta^*)|\theta^*) &= Q(\theta^*|\theta^*),
\end{aligned}
$$

*and, if $\theta^*$ is unique, $EM(\theta^*) = \theta^*$.*

Under standard regularity conditions it can further be shown that the fixed points of the EM algorithm are stationary points of the observed log likelihood. Unfortunately, these stationary points can either be local maxima or saddle-points of the observed log likelihood, meaning that the algorithm does not necessarily produce a global maximum. Obviously if $\ell_o(\theta)$ is unimodal with a single stationary point, then $(\theta^{(s)})$ converges towards the unique maximizer $\theta^*$ of $\ell_o(\theta)$. Moreover, we have the following result.

**Theorem 2.2** *Under standard regularity conditions, any fixed point $\theta^*$ of the EM algorithm satisfies the relation*

$$
D(EM)(\theta^*) = (D^{20}Q(\theta^*|\theta^*))^{-1} D^{20}H(\theta^*|\theta^*), \tag{2.6}
$$

*$D(EM)(\theta^*)$ being the Jacobian matrix of the operator $EM$ at $\theta^*$, and $D^{20}Q(\theta^*|\theta^*)$ $(D^{20}H(\theta^*|\theta^*))$ being the Hessian matrix of $Q(\theta^*|\theta^*)$ $(H(\theta^*|\theta^*))$ with respect to its first argument at $\theta^*$.*

Thus the convergence rate of the EM algorithm towards a fixed point $\theta^*$ is determined by the eigenvalues of $D(EM)$. Moreover, from (2.6), we have

$$
I(\theta|\mathbf{y}) = -D^{20}Q(\theta^*|\theta^*) + D^{20}H(\theta^*|\theta^*), \tag{2.7}
$$

where $I(\theta|\mathbf{y})$ is the empirical observed information on $\theta$, $-D^{20}Q(\theta^*|\theta^*)$ is complete information, and $-D^{20}H(\theta^*|\theta^*)$ is missing information, and $D(EM)(\theta^*)$ can be regarded as the "ratio" of the missing information over the complete information. The following result can also be deduced from (2.7).

**Theorem 2.3** *For any fixed point* $\theta^*$ *of the EM algorithm,*

$$D^2 \ell_o(\theta^*) = D^{20} Q(\theta^* | \theta^*) \left[ I_d - D(EM)(\theta^*) \right], \tag{2.8}$$

$I_d$ *being the identity matrix.*

This result means that, provided the matrix $D^{20}Q(\theta^*|\theta^*)$ is negative definite, a fixed point $\theta^*$ of the EM algorithm is attractive (with its eigenvalues belonging to $[0, 1]$) if and only if it is a local maximum of $\ell_o$. From (2.7), the greater the largest eigenvalue of $D(EM)(\theta^*)$, the slower the convergence of the EM algorithm towards an attractive fixed point $\theta^*$.

In the specific framework of mixtures with components from an exponential family, Redner & Walker (1984) refine these results to show that, under regularity conditions:

(a) the unique solution of the log likelihood equations almost surely exists;

(b) for $n$ large enough $(\theta^{(s)})$ linearly converges towards this solution, provided the initial position $\theta^{(0)}$ of the EM algorithm is not too far this optimal solution.

The important regularity conditions of Redner & Walker (1984) are that the mixing proportions are positive and the Fisher information matrix when evaluated at the true $\theta$ is positive definite.

The results of Redner & Walker (1984) are helpful in highlighting some important features of the EM algorithm, in particular concerning its practical ability to derive the ML estimates of the mixture parameters. First, as for most iterative algorithms used to optimize a non-convex function, the EM algorithm solution depends on its initial position. This dependence can be severe when the log likelihood function includes many local maxima and saddle-points, since the EM algorithm stops at the first fixed point it reaches.

Second, the rate of linear convergence to an attractive fixed point is determined by the largest eigenvalue, always smaller than 1, of $D(EM)$. Thus the larger the missing information, the slower the EM algorithm. In practice, the EM algorithm is well known for often having dramatically slow convergence speeds, even when the log likelihood function is convex. An example of this slow convergence is discussed in detail in Campillo & Le Gland (1989). We report in Figure 2.1 two different situations these authors dealt with. On the left-hand side, the missing information is low and the EM algorithm converges in five iterations. On the right-hand side, the missing information is high and the EM algorithm has not yet converged after 200 iterations, with only the first 12 iterations shown in Figure 2.1.

Moreover, it happens that some mixture distributions, like Gaussian distributions with free component-specific covariance matrices, have unbounded likelihood functions. In such cases, the EM algorithm is jeopardized by degenerate solutions (spurious local maximizers).

Despite these possible drawbacks, the EM algorithm generally does a good job of deriving the MLE or the posterior mode of a mixture model. The algorithm remains, without a doubt, *the* reference algorithm for ML estimation in a mixture model. However, improved versions of EM may have to be used in order to avoid the know pitfalls of the original algorithm (slow convergence, possibly high dependence on initialization, spurious maximizers). In the following section, several ways to address these pitfalls are presented.

## 2.4   Cousin Algorithms of EM

All the algorithms presented in this section share with the EM algorithm the characteristic of making use of completed data by approximating the missing labels knowing the

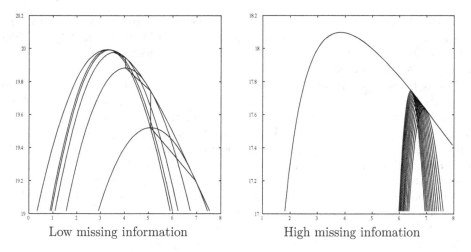

Low missing information          High missing infomation

**FIGURE 2.1**

EM convergence behavior for a low missing information case (left) and EM convergence behavior for a high missing information case (right). The observed log likelihood function is shown as a solid line against the parameter to be estimated; the iterations of EM are shown as dashed lines, and the successive graphs of the function $Q(\theta^{(s+1)} \mid \theta^{(s)})$ are shown as dotted lines.

observed data and a current value of the model parameters. This is achieved in different ways, depending on the focus of interest.

### 2.4.1 Stochastic versions of the EM algorithm

Starting from an initial value, the Stochastic EM (SEM) algorithm (Celeux & Diebolt, 1985) replaces the missing labels by simulating them from their conditional distribution, knowing the observed data and a current value of the mixture parameters. Iteration $s$ of the SEM algorithm is as follows.

**E step** The E step is the same as for EM and consists of computing the conditional probabilities $\tau_{ig}^{(s)}$ that $y_i$ originates from component $g$ for $i = 1, \ldots, n$ and $g = 1, \ldots, G$, for the current parameter value $\theta^{(s)}$ as in (2.1).

Next is the stochastic step:

**S step** Simulate the missing labels $z_{ig}^{(s)}$ according to the multinomial distribution with parameters $(\tau_{i1}^{(s)}, \ldots, \tau_{iG}^{(s)})$. This step results in a completed sample $(y_i, z_i^{(s)})$, $i = 1, \ldots, n$.

**M step** This step consists of maximizing the completed log likelihood $\ell_c(\theta, \mathbf{z})$ to obtain the next maximizer $\theta^{(s+1)}$. Typically, this step is analogous to the M step of EM. It simply leads to replacing the $\tau_{ig}^{(s)}$ conditional probabilities with the $z_{ig}^{(s)}$ in the likelihood equations.

Thus, SEM generates a Markov chain that is aperiodic, irreducible, and ergodic under mild conditions (Diebolt & Celeux, 1993). Its stationary distribution $\Psi_n$ is approximatively centered at the ML estimator of $\theta$. Under regularity conditions, Nielsen (2000) showed that if $X$ denotes a random vector drawn from the stationary distribution $\Psi_n$ and $\theta_0$ is the true value of the parameter $\theta$, then $\sqrt{n}(X - \theta_0)$ converges to a Gaussian distribution with mean zero and regular variance matrix $I(\theta_0)^{-1}[I_d - \{I_d - F(\theta_0)\}^{-1}]$, where $I(\theta_0)$ denotes the observed Fisher information matrix of $\theta_0$, $I_d$ the identity matrix, and $F(\theta_0)$ the expected fraction of missing information.

Thus, a natural parameter estimate derived from an SEM sequence is the mean of the iterated values, typically obtained after a *burn-in* period (SEMmean). An alternative estimate is to consider the parameter value leading to the largest log likelihood in an SEM sequence (SEMmax). In practice, both pointwise estimators perform similarly.

In a mixture context, where the E step is most often simple, the main appeal of the SEM algorithm is avoiding being stuck at the first stationary point of the log likelihood function it reaches. At each iteration there is a non-zero probability of accepting an updated estimate $\theta^{(s+1)}$ with log likelihood value lower than at $\theta^{(s)}$. For this very reason, the SEM algorithm avoids being stuck at a saddle-point or a spurious maximizer of the likelihood function. Thus the SEM algorithm can further be exploited to get a good starting value for the EM algorithm, if need be. Starting from SEMmean or SEMmax positions, an EM algorithm will likely converge to the MLE in a few iterations.

Different algorithms based on simulations as in the SEM algorithm are available. Some are mentioned below.

### The MCEM algorithm

This algorithm proposes a Monte Carlo implementation of the E step of the EM algorithm (Wei & Tanner, 1990). It replaces the computation of $Q(\theta|\theta^{(s)})$ by the derivation of an empirical version $Q_{(s+1)}(\theta|\theta^{(s)})$, based on $m$ draws of the missing vector $\mathbf{z}$ from its current conditional distribution $p(\mathbf{z}|\mathbf{y}, \theta^{(s)})$. If $m = 1$, MCEM reduces to SEM, while for large values of $m$, MCEM behaves like EM. Usually, in order to achieve a compromise between speed and precision, MCEM is run in a "simulated annealing" way with small values of $m$ in the first iterations and increasingly larger values of $m$ as $s$ increases. With a suitable rate of convergence of $m$ to infinity, MCEM can be proven to converge to a sensible local maximum of the likelihood function (Fort & Moulines, 2003). In the mixture context, MCEM, which remains a rather slow algorithm, appears to be of little use, since, in practice, the E step corresponds to the computation of $Q(\theta|\theta^{(s)})$ and this reduces to the derivation of the conditional probabilities $\tau_{ig}^{(s)}$ that $y_i$ is generated from component $g$, $i = 1, \ldots, n$ and $g = 1, \ldots, G$, for the current parameter value $\theta^{(s)}$.

### The Simulated Annealing EM algorithm

This algorithm implements the intuition given above, namely, operating like an SEM algorithm at the start and finishing like an EM algorithm. More precisely, the $s$th iteration update of $\theta$ with Simulated Annealing EM is

$$\theta^{(s+1)} = (1 - \gamma_{s+1})\theta_{EM}^{(s+1)} + \gamma_{s+1}\theta_{SEM}^{(s+1)}.$$

where $\theta_{EM}^{(s+1)}$ $(\theta_{SEM}^{(s+1)})$ is the updated value of $\theta$ under EM (SEM) and $(\gamma_s)$ is a sequence of non-negative real numbers decreasing to zero with initial value $\gamma_0 = 1$. A slow convergence rate of $\gamma_s$ is necessary for good performance. The conditions $\lim_{s \to \infty}(\gamma_s/\gamma_{s+1}) = 1$ and $\sum_s \gamma_s = \infty$ are necessary to ensure the almost sure convergence of the Simulated Annealing EM algorithm to a local maximizer of the observed log likelihood whatever the starting point. This was established in the context of finite mixtures from some exponential family by Celeux & Diebolt (1992). From a practical point of view, it is important that $\gamma_s$ remains close to $\gamma_0 = 1$ during the early iterations in order to allow the algorithm to avoid suboptimal stationary values of $\ell_o(\theta)$.

*The Stochastic Approximation EM algorithm*

This algorithm is analogous to the Simulated Annealing EM algorithm and works as follows, starting from an arbitrary initial value $\theta^{(0)}$.

**Simulation step** As in SEM, this step makes use of $m(s)$ realizations $\mathbf{z}_j, j = 1, \ldots, m(s)$, of the missing label vector that are simulated from the multinomial distribution with parameters $(\tau_{i1}^{(s)}, \ldots, \tau_{iG}^{(s)})$ for $i = 1, \ldots, n$, where $m(s)$ is an increasing sequence of integers starting from $m(1) = 1$.

**Stochastic approximation step** This step computes the current approximation of $Q(\theta, \theta^{(s)})$ according to

$$\hat{Q}_{s+1}(\theta) = \hat{Q}_s(\theta) + \gamma_s \left( \frac{1}{m(s)} \sum_{j=1}^{m(s)} \ell_c(\theta, \mathbf{z}_j) - \hat{Q}_s(\theta) \right), \tag{2.9}$$

where $(\gamma_s)$ is a sequence of non-negative real numbers decreasing to zero.

**M step** This step derives $\theta^{(s+1)} = \arg\max_\theta \hat{Q}_{s+1}(\theta)$.

If the sequence $(\gamma_s)$ is such that $\sum_{s=1}^{\infty} \gamma_s = \infty$ and $\sum_{s=1}^{\infty} \gamma_s^2 < \infty$, then under regularity conditions, Delyon et al. (1999) proved that the Stochastic Approximation EM algorithm converges to a local maximum of the observed log likelihood. From a practical point of view, this algorithm is expected to behave as the Simulated Annealing EM algorithm and it is important that the sequence $\gamma_s$ decreases slowly to zero while the number of simulations could be set to one at each iteration, that is, $m(s) = 1$ for any $s$.

The goal of both the Simulated Annealing and the Stochastic Approximation EM algorithms is to provide pointwise estimates that are expected to converge to the MLE, while the SEM algorithm generates a Markov chain whose stationary distribution is expected to be centered on the MLE. In practice, there are no significant differences between the estimates provided by the Simulated Annealing or the Stochastic Approximation EM algorithms and the estimates provided by an SEM algorithm followed by a few iterations of EM.

## 2.4.2 The Classification EM algorithm

The algorithm now presented is particularly relevant in the model-based clustering framework where the mixture model is considered in a clustering task where each mixture component is associated to a cluster of the data; see Chapter 8 for a comprehensive review. The Classification EM (CEM) algorithm estimates both the mixture parameters and the missing labels by maximizing the complete-data log likelihood $\ell_c(\theta, \mathbf{z})$. The $s$th iteration of the CEM algorithm proceeds as follows.

**E step** The E step is similar to those in the EM and SEM algorithms. It consists of computing the conditional probabilities $\tau_{ig}^{(s)}$ that $y_i$ arises from component $g$ for $i = 1, \ldots, n$ and $g = 1, \ldots, G$, given the current parameter vector $\theta^{(s)}$.

The classification step consists of the following derivation of labels $\mathbf{z}^{(s)}$ at iteration $s$:

**C step** Assign each observation $y_i$ to the mixture component maximizing $\tau_{ig}^{(s)}$ over $g =$

$1, \ldots, G$, that is,

$$z_{ig}^{(s)} = \begin{cases} 1, & \text{if } g = \arg \max_{g'} \tau_{ig'}^{(s)}, \\ 0, & \text{otherwise.} \end{cases}$$

**M step** This step consists of maximizing the completed log likelihood $\ell_c(\theta, \mathbf{z}^{(s)})$ to obtain the next maximizer value $\theta^{(s+1)}$. This M step is formally identical to the M step of SEM.

It is important to note that the CEM algorithm does not maximize the observed-data log likelihood but the complete-data log likelihood. Thus, the CEM algorithm is expected to provide biased estimates of the mixture parameters. The more the mixture components are overlapping, the larger the bias of CEM estimates becomes. On the other hand, the CEM algorithm is a $k$-means type algorithm and as such converges rapidly to a fixed point. But, in most cases, it can be expected to be quite sensitive to the initial value. Moreover, the SEM algorithm can also be considered as a stochastic version of both EM and CEM algorithms, despite these algorithms not maximizing the same criterion.

Note further that Celeux & Govaert (1992) proposed a specific Simulated Annealing version of the CEM algorithm. This algorithm substitutes the E step with the AE ("annealing") step where the conditional probabilities $\tau_{ig}$, $i = 1, \ldots, n$; $g = 1, \ldots, G$, are replaced by the scores

**AE step**

$$\rho_{ig}^{(s)} = \frac{\{\eta_g^{(s)} f_g(y_i \mid \theta_g^{(s)})\}^{1/t_s}}{\{\sum_{g'=1}^{G} \eta_{g'}^{(s)} f_{g'}(y_i \mid \theta_{g'}^{(s)})\}^{1/t_s}},$$

for $i = 1, \ldots, n$ and $g = 1, \ldots, G$, the sequence $(t_s)$ being a decreasing sequence of non-negative numbers starting from $t_0 = 1$.

For $t_s = 1$ this algorithm is exactly the SEM algorithm and when $t_s$ decreases from 1 to 0 as the iteration index increases, the algorithm morphs from the SEM into the CEM algorithm. In practical situations, it is recommended to have the sequence $(t_s)$ slowly decreasing to 0. As mentioned above, the CEM algorithm and its stochastic versions are mostly useful in the model-based clustering context. However, the empirically verified fact that the CEM algorithm converges quickly to solutions that are often reasonable makes this algorithm potentially useful for maximizing the observed log likelihood. In particular, this applies to settings where the mixture components are expected to be well separated (see Celeux & Govaert, 1993) or to use the CEM algorithm to initialize the EM algorithm, as shown in Section 2.6.

---

## 2.5   Accelerating the EM Algorithm

Since the EM algorithm can suffer from convergence problems, numerous methods that speed up its convergence have been proposed; see McLachlan & Krishnan (2008, Chapter 4) for an excellent review. In Chapter 3 of this volume, an acceleration procedure based on the "Aitken acceleration" method is presented. In this section, we only present an alternative acceleration technique specific to the mixture framework.

As expressed in (2.6), the rate of convergence of EM towards a stationary point of the likelihood function is governed by the fraction of missing information in the data. A large amount of missing information can induce a slow convergence of EM. Several authors have proposed variants of the EM algorithm for counteracting such slow convergence. For instance, Fessler & Hero (1995) proposed the Space-Alternating Generalized EM (SAGE) algorithm, which updates the parameters sequentially by alternating between several small hidden-data spaces defined by the algorithm designer. In the same spirit, Meng & van Dyk (1997) conceived a general Alternating Expectation-Conditional Maximization (AECM) algorithm which couples acceleration of the convergence by allowing the data augmentation scheme to vary within and between iterations with the simplification of the computation by incorporating model reduction into the M step.

We now describe the Componentwise EM algorithm for Mixtures (CEMM) algorithm, specific to the mixture context, which basically updates only one component at a time, leaving the other parameters unchanged (Celeux et al., 2001). The CEMM algorithm was inspired by the SAGE algorithm. Improved convergence rates are reached by updating the parameters sequentially on small groups of observations associated with small missing data spaces rather than one large complete data space. The idea is that less informative missing data spaces lead to smaller root-convergence factors and hence faster converging algorithms. For simplicity, the CEMM algorithm is described for a multivariate Gaussian mixture model.

CEMM considers the decomposition of the parameter vector $\theta = (\theta_g, \eta_g)_{g=1,\ldots,G}$ with $\theta_g = (\mu_g, \Sigma_g)$. Therefore the component updated at iteration $s$ is as follows. At iteration $(s)$, set

$$g = s + \left\lfloor \frac{s}{G} \right\rfloor G + 1,$$

where $\lfloor x \rfloor$ denotes the integer part of $x$.

**E step** Compute, for $i = 1, \ldots, n$,

$$\tau_{ig}^{(s)} = \frac{\eta_g^{(s)} f_g(y_i \mid \theta_g^{(s)})}{\sum_{g'=1}^{G} \eta_{g'}^{(s)} f_{g'}(y_i \mid \theta_{g'}^{(s)})}.$$

**M step** Set

$$\eta_g^{(s+1)} = \frac{\sum_{i=1}^{n} \tau_{ig}^{(s)}}{n},$$

$$\mu_g^{(s+1)} = \frac{\sum_{i=1}^{n} \tau_{ig}^{(s)} y_i}{\sum_{i=1}^{n} \tau_{ig}^{(s)}},$$

$$\Sigma_g^{(s+1)} = \frac{\sum_{i=1}^{n} \tau_{ig}^{(s)} (y_i - \mu_g^{(s+1)})(y_i - \mu_g^{(s+1)})^{\top}}{\sum_{i=1}^{n} \tau_{ig}^{(s)}},$$

and, for $\ell \neq g$, define $\theta_\ell^{(s+1)} = \theta_\ell^{(s)}$.

The advantage of the CEMM algorithm over the SAGE algorithm is that it uses the new information as soon as it is available rather than waiting until all parameters have been updated. Actually, the SAGE algorithm is not exactly a componentwise algorithm because the mixing proportions are then updated at the same iteration, which involves the whole complete data structure. For this reason, it may fail to be significantly faster than the standard EM algorithm. This shows the main intuition of the componentwise EM algorithm that Celeux et al. (2001) proposed for mixtures. No iteration requires the whole complete data space as missing data space. It can therefore be expected to converge faster in various situations. However, it is important to note that with the CEMM algorithm, the mixing proportions $\eta_g$ do not necessarily sum to 1, hence the algorithm may leave the parameter space. But it has been proven in Celeux et al. (2001) that this algorithm is guaranteed to

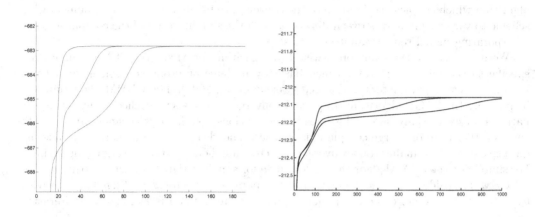

**FIGURE 2.2**
Comparison of the log likelihood at each cycle for EM (solid line), SAGE (dashed line) and
CEMM (dotted line) for a well-separated component Gaussian mixture case (left) and in
an overlapping component Gaussian mixture case (right).

return to the parameter space at convergence by using the proximal interpretation of EM
presented in Chrétien & Hero (2008).

In Figure 2.2, we reproduce from Celeux et al. (2001) two numerical situations highlighting the typical behavior of the CEMM algorithm, when compared to the EM and SAGE
algorithms. The graph on the left shows that the EM algorithm is faster than the CEMM
algorithm when the mixture components are well separated, while the graph on the right
shows that the CEMM algorithm exhibits significant improvement over the EM algorithm
with overlapping mixture components. In the latter situation the EM algorithm can be expected to encounter slow convergence situations. Thus, the CEMM algorithm can be useful
in some settings. An intuitive explanation is that the componentwise strategy prevents the
CEMM algorithm from staying too long at critical points (typically saddle-points) where
the standard EM algorithm is likely to get trapped.

## 2.6   Initializing the EM Algorithm

The choice of $\theta^{(0)}$ is clearly decisive for the EM algorithm, especially when the choice of a
sensible number of components $G$ is required. In fact, information criteria used to select a
mixture models are all based on the maximum likelihood values; see Chapter 7, Section 7.3.2
for a comprehensive review. Several strategies have been proposed to initialize EM for estimating the mixture parameters and they are available in most mixture software packages.
Initialization strategies can be distinguished by the importance they give to randomness.

## 2.6.1 Random initialization

Some procedures available in the MIXMOD software (http://www.mixmod.org) make intensive use of random initializations and have been proposed in Biernacki et al. (2003) (see also Berchtold (2004)):

(a) The *SEM procedure* involves starting EM with the parameter value providing the largest likelihood from a long run of the SEM algorithm.

(b) The *CEM procedure* involves repeating the CEM algorithm a large number of times from random initial positions and starting EM with the parameter value providing the largest likelihood from these runs of the CEM algorithm.

(c) The *small EM procedure* involves using a large number of short runs of EM. This means stopping and restarting EM at a random location before it exhibits convergence. After a number of repetitions, EM is run from the parameter value providing the largest likelihood from these short runs of EM.

Moreover, it is recommended to try several starts from the CEM and the small EM procedure. Among these procedures, small EM is often preferred (see Biernacki et al., 2003), and it is the default initialization procedure in the MIXMOD software. But none of these procedures has been shown to outperform the other ones. Moreover, they can exhibit disappointing performance in high-dimensional settings because the domain parameter to be explored becomes very large. The same difficulty may appear when the number $G$ of mixture components is large.

## 2.6.2 Hierarchical initialization

At the opposite end of the spectrum of initialization methods, deterministic initialization procedures such as the hierarchical procedure proposed in the Mclust software (http://www.stat.washington.edu/mclust/) do not use random starting solutions. Such hierarchical initialization can be outperformed by the above-mentioned random procedures since they do not extensively explore many initial positions. But they nonetheless provide more stable solutions and they may be preferred in high-dimensional settings or for a large number of mixture components. A recursive initialization procedure aiming at getting the best of both worlds has been proposed in Celeux & Baudry (2015) and is described in the next subsection.

## 2.6.3 Recursive initialization

A problem that often occurs in a mixture analysis with different numbers of components is that some solutions may be suboptimal or spurious. This problem is shared with the random initialization procedures of Section 2.6.1. The recursive procedures presented in Celeux & Baudry (2015) aim to avoid these irrelevant parameter estimates. Assume that the user wants to choose a mixture model with a number $G$ of components within the range $\{G_{\min}, \ldots, G_{\max}\}$. A recursive initialization involves splitting *at random* one of the $G$ components into two components to get a $(G+1)$ solution:

1. First, the $G_{\min}$ solution is thoroughly designed using, for instance, the *small EM* procedure repeated a large number of times.

2. From $G = G_{\min}$, the initial position of the $(G+1)$-component mixture results from splitting one of the $G$-component mixture into two components.

The resulting procedures differ in the way the component to be split is chosen. In Celeux & Baudry (2015), three such strategies are considered.

(a) *Random choice* (RC): pick the split mixture component at random (Papastamoulis et al., 2016).

(b) *Optimal sequential choice* (OSC): pick the split mixture component by optimizing a splitting criterion.

(c) *Complete choice* (CC): split all $G$ components into two components and pick the split mixture component leading to the largest likelihood.

Other splitting strategies, such as the one in Fraley et al. (2005) which is designed to deal with large data sets, are possible. Concerning strategy OSC, many sensible splitting criteria are possible, but numerical experiments reported in Celeux & Baudry (2015) show that there is little incentive to choose among these criteria. Moreover, the same numerical experiments show that strategy CC outperforms the other ones, while strategy OSC shows no advantage compared to strategy RC. Finally, CC is not so expensive and can thus be recommended.

## 2.7    Avoiding Spurious Local Maximizers

Deriving the maximum likelihood parameter estimate of a Gaussian mixture faces an important difficulty since the likelihood function of Gaussian mixtures with unrestricted component covariance matrices is unbounded. Thus, spurious local maximizers of the likelihood are found in the parameter space. The EM algorithm can thus fail because of these singularities depending on the starting values, models, and numbers of components (see, for instance, McLachlan & Peel, 2000, Section 3.10). Such spurious maximizers can be avoided by imposing some constraints on the mixture parameters. The simplest and most popular constraints assume equal mixing proportions or component covariance matrices. But such constraints may sound unrealistic and cannot be regarded as a general answer to the spurious maximizer problem.

Bayesian regularization can be seen as a more general solution of this issue; see Ciuperca et al. (2003) or Fraley & Raftery (2007). It involves replacing the MLE with the maximum *a posteriori* (MAP) estimate, obtained by maximizing the penalized log likelihood $\ell_o(\theta) + \log p(\theta)$, with $p(\theta)$ being a prior distribution on the vector parameter $\theta$. Since the Bayesian framework is only introduced to avoid degeneracies in the estimation of the component covariance matrices, there is no need to design prior distributions on the mixing proportions and the component mean vectors.

Following Fraley & Raftery (2007), it is convenient to use an inverse Wishart exchangeable conjugate prior distribution for the component-specific covariance matrices, that is, $\Sigma_g \sim \mathcal{W}^{-1}(\nu, \Lambda)$ for $g = 1, \ldots, G$. Obviously, the choice of the hyperparameters $\nu$ and $\Lambda$ is important. The choice $\nu = d + 2$ was recommended in Fraley & Raftery (2007) and applied in Celeux & Baudry (2015). In Fraley & Raftery (2007), the scale matrix $\Lambda$ was set to

$$\Lambda = \frac{1}{G^{1/d}} S_y, \tag{2.10}$$

with $S_y$ being the empirical covariance matrix of the data $\mathbf{y}$. Another choice, advocated in Celeux & Baudry (2015), is

$$\Lambda = \frac{\sigma_0^{1/d}}{|S_y|^{1/d}} S_y,$$

where $\sigma_0$ is a small positive number. This choice implies that $|\Lambda| = \sigma_0$. The larger $\sigma_0$, the larger the regularization becomes. Thus, this tuning parameter allows for control of the regularization, and weaker regularization than the fixed hyperparameter $\Lambda$ defined in (2.10).

With such choices, the MAP estimate of $\theta$ is derived via the EM algorithm. In this framework, the formulas of the EM algorithm are indeed unchanged, except for the updating of the covariance matrices in the M step. It then becomes (at the $(s+1)$th iteration):

$$\Sigma_g^{(s+1)} = \frac{\Lambda + \sum_{i=1}^{n} \tau_{ig}^{(s)} (y_i - \mu_g^{(s+1)})(y_i - \mu_g^{(s+1)})^\top}{\nu + \sum_{i=1}^{n} \tau_{ig}^{(s)} + d + 2}$$

(see Fraley & Raftery, 2007, for details). Assuming diagonal or spherical component covariance matrices does not guard against spurious maximizers, and regularization of the covariance matrices is also desirable in these settings. For diagonal component covariance matrices $\Sigma_g = \text{Diag}(B_{g1}, \dots, B_{gd})$, for instance, the conjugate inverse gamma prior $B_{gj} \sim \mathcal{IG}(\nu/2, \zeta_j/2)$ can be used for the diagonal elements $B_{gj}$, $j = 1, \dots, d$, $g = 1, \dots, G$, with hyperparameters $\nu = d + 2$ and

$$\zeta_j = (\sigma_0)^{1/d} \frac{s_j}{s_1 \cdots s_d},$$

where $s_j$ is the empirical variance of the variable $j$ and $\sigma_0$ is a small positive number. The updating of the $B_{gj}$s at the $(s+1)$th iteration of the EM algorithm is as follows:

$$B_{gj}^{(s+1)} = \frac{\zeta_j + \sum_{i=1}^{n} \tau_{ig}^{(s)} (y_{ij} - \mu_{gj}^{(s+1)})^2}{\nu + \sum_{i=1}^{n} \tau_{ig}^{(s)} + 2}.$$

Finally, with covariance matrices $\Sigma_g = \lambda_g I$ being proportional to the identity matrix in each component $g = 1, \dots, G$, the conjugate inverse gamma prior $\lambda_g \sim \mathcal{IG}(\nu/2, \zeta/2)$ can be applied, with the hyperparameters $\nu = d + 2$ and $\zeta = 2(\sigma_0)^{1/d}$, with $\sigma_0$ again being a small positive number. The updating of the $\lambda_g$ at the $(s+1)$th iteration of the EM algorithm is as follows:

$$\lambda_g^{(s+1)} = \frac{\zeta + \sum_{i=1}^{n} \tau_{ig}^{(s)} (y_i - \mu_g^{(s+1)})^\top (y_i - \mu_g^{(s+1)})}{\nu + d \sum_{i=1}^{n} \tau_{ig}^{(s)} + d + 2}.$$

Obviously, in each case, the choice of the hyperparameter $\sigma_0$ ends up being quite influential. For instance, it is important that the $\sigma_0$ thus chosen does not hide the data structure. This hyperparameter is indeed driving the strength of the regularization. A careful sensitivity analysis must then be performed to choose values $\sigma_0$ ensuring stable and meaningful estimates.

---

## 2.8   Concluding Remarks

The EM algorithm is a well-established algorithm and indeed *the* algorithm of reference for deriving the MLE or performing posterior mode estimation of mixture models. We expect it to retain its dominating position for many years to come. Indeed, Chapter 3 presents an expansion of the EM algorithm to a more general mathematical structure (the *product-of-sum* structure). Despite many of its characteristics being well documented, research on the EM and related algorithms remains very active. An important question which is still open concerns deciding when a fixed point of the EM is near the global optimum or a consistent solution of the likelihood equations, or a poor local optimum of the likelihood. With this question in mind, Balakrishnam et al. (2017) developed a theoretical framework for quantifying when and how quickly the EM algorithm converges to a small neighborhood of a given global optimum of the population likelihood (obtained in the limit with infinite data). In particular, they give precise characterizations of the region of convergence for symmetric mixtures of two Gaussians with isotropic covariance matrices. This important theoretical work confirms that to guarantee a well-behaved EM algorithm, we require a good initialization.

# Bibliography

BALAKRISHNAM, S., WAINWRIGHT, M. J. & YU, B. (2017). Statistical guarantees for the EM algorithm: From population to sample-based analysis. *Annals of Statistics* **45**, 77–120.

BERCHTOLD, A. (2004). Optimisation of mixture models: Comparison of different strategies. *Computational Statistics* **19**, 385–406.

BIERNACKI, C., CELEUX, G. & GOVAERT, G. (2003). Choosing starting values for the EM algorithm for getting the highest likelihood in multivariate Gaussian mixture models. *Computational Statistics and Data Analysis* **41**, 561–575.

BISHOP, C. M. (2006). *Pattern Recognition and Machine Learning*. New York: Springer-Verlag.

CAMPILLO, F. & LE GLAND, F. (1989). MLE for partially observed diffusions: Direct maximization vs. the EM algorithm. *Stochastic Processes and Applications* **33**, 245–274.

CELEUX, G. & BAUDRY, J.-P. (2015). EM for mixtures – random initialization could be hazardous. *Statistics and Computing* **25**, 713–726.

CELEUX, G., CHRÉTIEN, S., FORBES, F. & MKHADRI, A. (2001). A component-wise EM algorithm for mixtures. *Journal of Computational and Graphical Statistics* **10**, 699–712.

CELEUX, G. & DIEBOLT, J. (1985). The SEM algorithm: a probabilistic teacher algorithm derived from the EM algorithm for the mixture problem. *Computational Statistics Quarterly* **2**, 73–82.

CELEUX, G. & DIEBOLT, J. (1992). A Stochastic Approximation type EM algorithm for the mixture problem. *Stochastics and Stochastics Reports* **41**, 119–134.

CELEUX, G. & GOVAERT, G. (1992). A classification EM algorithm for clustering and two stochastic versions. *Computational Statistics and Data Analysis* **14**, 315–332.

CELEUX, G. & GOVAERT, G. (1993). Comparison of the mixture and the classification maximum likelihood in cluster analysis. *Journal of Statistical Computation and Simulation* **47**, 127–146.

CELEUX, G. & GOVAERT, G. (1995). Gaussian parsimonious clustering models. *Pattern Recognition* **28**, 781–793.

CHRÉTIEN, S. & HERO, A. O. (2008). On EM algorithms and their proximal generalizations. *ESAIM: Probability and Statistics* **12**, 308–326.

CIUPERCA, G., RIDOLFI, A. & IDIER, J. (2003). Penalized maximum likelihood estimator for normal mixtures. *Scandinavian Journal of Statistics* **30**, 45–59.

DELYON, B., LAVIELLE, M. & MOULINES, E. (1999). On a stochastic approximation version of the EM algorithm. *Annals of Statistics* **27**, 94–128.

DEMPSTER, A., LAIRD, N. & RUBIN, D. (1977). Maximum likelihood from incomplete data via the EM algorithm (with discussion). *Journal of the Royal Statistical Society Series B* **39**, 1–38.

DIEBOLT, J. & CELEUX, G. (1993). Asymptotic properties of a stochastic EM algorithm for estimating mixing proportions. *Communications in Statistics: Stochastic Models* **9**, 599–613.

FESSLER, J. A. & HERO, A. O. (1995). Penalized maximum-likelihood image reconstruction using space-alternating generalized EM algorithms. *IEEE Transactions in Image Processing* **4**, 1417–29.

FORT, G. & MOULINES, E. (2003). Convergence of the Monte Carlo EM for curved exponential families. *Annals of Statistics* **31**, 1220–1259.

FRALEY, C., RAFTERY, A. & WEHRENS, R. J. (2005). Incremental model-based clustering for large datasets with small clusters. *Journal of Computational and Graphical Statistics* **14**, 529–546.

FRALEY, C. & RAFTERY, A. E. (2007). Bayesian regularization for normal mixture estimation and model-based clustering. *Journal of Classification* **24**, 155–181.

GOVAERT, G. & NADIF, M. (2008). Block clustering with Bernoulli mixture models: Comparison of different approaches. *Computational Statistics and Data Analysis* **52**, 3233–3245.

HATHAWAY, R. J. (1986). Another interpretation of the EM algorithm for mixture distributions. *Statistics and Probability Letters* **4**, 53–56.

McLACHLAN, G. & PEEL, D. (2000). *Finite Mixture Models*. New York: John Wiley.

McLACHLAN, G. J. & KRISHNAN, T. (2008). *The EM Algorithm and Extensions*. New York: John Wiley. Second Edition.

MENG, X.-L. & VAN DYK, D. A. (1997). The EM algorithm – an old folk-song sung to a fast new tune (with discussion). *Journal of the Royal Statistical Society Series B* **59**, 511–567.

NEAL, R. M. & HINTON, G. E. (1998). A view of the EM algorithm that justifies incremental, sparse and other variants. In *Learning in Graphical Models*, M. I. Jordan, ed. Dordrecht: Kluwer Academic Publishers, pp. 355–358.

NIELSEN, S. F. (2000). The stochastic EM algorithm: Estimation and asymptotic results. *Bernoulli* **6**, 457–489.

PAPASTAMOULIS, P., MARTIN-MAGNIETTE, M.-L. & MAUGIS-RABUSSEAU, C. (2016). On the estimation of mixtures of Poisson regression models with large numbers of components. *Computational Statistics and Data Analysis* **93**, 97–106.

REDNER, R. & WALKER, H. (1984). Mixture densities, maximum likelihood and the EM algorithm. *SIAM Reviews* **26**, 195–239.

TITTERINGTON, D. M. (2011). The EM algorithm, variational approximations and expectation propagation for mixtures. In *Mixtures: Estimation and Applications*, K. L. Mengersen, C. P. Robert & D. M. Titterington, eds. Chichester: Wiley, pp. 1–29.

WEI, G. & TANNER, M. (1990). A Monte Carlo implementation of the EM algorithm and the poor man's data augmentation algorithm. *Journal of the American Statistical Association* **85**, 699–704.

WU, C. (1983). On the convergence properties of the EM algorithm. *Annals of Statistics* **11**, 95–103.

# 3

# An Expansive View of EM Algorithms

**David R. Hunter, Prabhani Kuruppumullage Don, and Bruce G. Lindsay**
*Penn State University, USA*

## CONTENTS

## 3.1 Introduction

In their most basic form, EM algorithms provide a general method for finding local maxima of a likelihood function when we may conceive of some subset of the full data as having been unobserved in the experiment. These algorithms are the computational workhorses of maximum likelihood estimation in mixture models, which provide perhaps the best-known class of examples in which EM algorithms are effective; see also Chapter 2 for a comprehensive review. In adopting the "expansive" view in this chapter, we explain the EM mechanism without reference specifically to mixture models or even maximum likelihood. Indeed, we argue for a definition of EM that includes any algorithm based on the same basic principle that guarantees the algorithms' success in their original framework. Then, returning to the topic of finite mixtures, we describe several algorithms we consider instances of EM that extend this framework, each one of them related to mixture models. In this way, we hope to both elucidate the workings of EM and demonstrate the broad applicability of these ideas, which we feel is a testament to the genius and simplicity of the EM scheme.

The name "EM algorithm" can trace its origin to the seminal paper of Dempster et al. (1977). Yet as with so many episodes in science, the true story of the algorithm's origins are somewhat murkier. For one thing, the "EM algorithm" is not an algorithm at all: as Dempster et al. (1977) acknowledge in a footnote, "our use of the term 'algorithm' can be criticized because we do not specify the sequence of computing steps actually required to carry out a single E- or M-step." Indeed, the iteratively applied E and M steps of Dempster et al. (1977) are best viewed as a recipe for creating algorithms, and for this reason we refer throughout this article not to "the EM algorithm" but rather to "EM algorithms" more broadly. For another thing, there are numerous examples of EM algorithms that pre-date the Dempster et al. (1977) article. Any attempt to catalog them here would be incomplete, so we instead refer interested readers to the book-length treatment of EM by McLachlan & Krishnan (2007). Yet it was Dempster et al. (1977) who first articulated the general framework for creating EM algorithms, an ingenious innovation that we feel deservedly places their paper among the most highly cited statistical articles in history.

While many treatments of EM algorithms begin with the missing-data framework and concepts such as "completed-data log likelihood" and "observed-data log likelihood", we take a different approach here. We survey the EM landscape from high altitude, beginning with a development of the essential *product-of-sums* form taken by many difficult-to-maximize likelihoods for which EM algorithms provide a helpful tool. We eventually return to Earth in Section 3.3 and explain how the more general development relates to the specific maximum-likelihood-with-missing-data paradigm. We conclude with several examples as well as some thoughts on convergence criteria.

## 3.2    The Product-of-Sums Formulation

We start with a maximization problem involving a real-valued objective function $L(\theta)$ of the possibly vector-valued parameter $\theta$. Typically, a global maximum is desired; but here, we will generally mean a local maximum, both for theoretical reasons (e.g. likelihood theory typically guarantees only that some local maximum is a consistent estimator) and for practical ones (most optimization methods, including those described here, attain at best a local maximum).

An EM algorithm is a specialized algorithm, in that it can only be used on objective functions with a particular mathematical structure one might call *product-of-sums*. To wit, let us suppose that $L(\theta)$ has the following product representation:

$$L(\theta) = \prod_{i=1}^{n} L_i(\theta). \tag{3.1}$$

Note that while we have used a single index $i$ in the product, it can be thought of as shorthand for a set of indices, say $(i, j, k)$, over which products are taken. In addition, we assume that each $L_i(\theta)$ has the summation representation

$$L_i(\theta) = \sum_{g=1}^{G_i} L_{ig}(\theta) \quad \text{or} \quad L_i(\theta) = \int L_{ig}(\theta) \, dg, \tag{3.2}$$

where $L_{ig}(\theta)$ is a nonnegative function of $\theta$ for all $i$ and $g$. Since we may take the logarithm of the $L(\theta)$ function without changing the $\theta$ values that optimize it, it will be convenient to define

$$\ell(\theta) = \log L(\theta) = \sum_{i=1}^{n} \log L_i(\theta) = \sum_{i=1}^{n} \log \left( \sum_{g=1}^{G_i} L_{ig}(\theta) \right). \tag{3.3}$$

We might call the form of $\ell(\theta)$ in equation (3.3) *sum-of-log-of-sums*. We allow the possibility that $\ell(\theta)$ takes on the value of $-\infty$ when $L(\theta) = 0$, although that issue generally has no bearing on an EM algorithm as long as one initiates the algorithm using some $\theta^{(0)}$ satisfying $L(\theta^{(0)}) > 0$.

### 3.2.1 Iterative algorithms and the ascent property

To start an EM algorithm requires an initial parameter value, which we denote $\theta^{(0)}$. Starting with $s = 0$, the algorithm applies a rule that maps $\theta^{(s)}$ to $\theta^{(s+1)}$, then increases $s$ by 1, and then repeats the process until some stopping criterion is achieved (see Section 3.5 for some thoughts about stopping). The key to any EM algorithm is the fact that, in the product-of-sums case, it is possible to create the rule mapping $\theta^{(s)}$ to $\theta^{(s+1)}$ in such a way that

$$L(\theta^{(s+1)}) \geq L(\theta^{(s)}) \tag{3.4}$$

is guaranteed. Inequality (3.4) is called the ascent property of an EM algorithm.

The EM approach is specific to the product-of-sums situation and involves the creation and maximization, for each iteration $s$, of a surrogate function of $\theta$ that depends on $\theta^{(s)}$. We denote this surrogate function by $Q(\theta \mid \theta^{(s)})$. One can think of $Q(\theta \mid \theta^{(s)})$ as a local approximation to $L(\theta)$ in a neighborhood of $\theta^{(s)}$, where this approximation has an important property called *minorization* that we describe in the next subsection.

### 3.2.2 Creating a minorizing surrogate function

Referring once again to equation (3.3) defining $\ell(\theta)$, let us ignore the summation over $i$ for the present and focus on the expression

$$\log L_i(\theta) = \log \left( \sum_{g=1}^{G_i} L_{ig}(\theta) \right)$$

for a particular $i$. As a special case of Jensen's inequality, if we define

$$w_{ig}^{(s)} = \frac{L_{ig}(\theta^{(s)})}{\sum_{h=1}^{G_i} L_{ih}(\theta^{(s)})} \tag{3.5}$$

for $g = 1, \ldots, G_i$, then

$$\log L_i(\theta) - \log L_i(\theta^{(s)}) \geq \sum_{g=1}^{G_i} w_{ig}^{(s)} \log \left[ \frac{L_{ig}(\theta)}{L_{ig}(\theta^{(s)})} \right]. \tag{3.6}$$

Inequality (3.6), which is the key to constructing the surrogate function for an EM algorithm, may also be derived directly from the concavity of the logarithm function since, by definition, any concave function $\varphi(\cdot)$ must satisfy

$$\varphi \left( \sum_{g=1}^{G_i} w_{ig}^{(s)} \left[ \frac{L_{ig}(\theta)}{L_{ig}(\theta^{(s)})} \right] \right) \geq \sum_{g=1}^{G_i} w_{ig}^{(s)} \varphi \left[ \frac{L_{ig}(\theta)}{L_{ig}(\theta^{(s)})} \right],$$

because $w_{i1}^{(s)}, \ldots, w_{iG_i}^{(s)}$ are nonnegative constants that sum to unity. If we now sum over $i$, the left-hand side of inequality (3.6) becomes $\ell(\theta) - \ell(\theta^{(s)})$.

Therefore, if we define

$$Q(\theta \mid \theta^{(s)}) = \sum_{i=1}^{n} \sum_{g=1}^{G_i} w_{ig}^{(s)} \log L_{ig}(\theta)$$

$$= \sum_{i=1}^{n} \sum_{g=1}^{G_i} \left[ \frac{L_{ig}(\theta^{(s)})}{\sum_{h=1}^{G_i} L_{ih}(\theta^{(s)})} \right] \log L_{ig}(\theta), \tag{3.7}$$

then we may conclude that

$$\ell(\theta) - \ell(\theta^{(s)}) \geq Q(\theta \mid \theta^{(s)}) - Q(\theta^{(s)} \mid \theta^{(s)}). \tag{3.8}$$

In other words, at iteration $s$, the increase in $\ell(\theta)$ must always be at least as large as the increase in $Q(\theta \mid \theta^{(s)})$. In particular, if we find $\theta$ that makes the latter increase as large as possible by maximizing with respect to $\theta$, we will guarantee some increase in the value of $\ell(\theta)$ above $\ell(\theta^{(s)})$. Thus, definition (3.7) together with maximization of $Q(\cdot \mid \theta^{(s)})$ guarantees the characteristic ascent property (3.4) of all EM algorithms. We will see in Section 3.3 that creating the $Q(\theta \mid \theta^{(s)})$ function of equation (3.7) is called the "E step" of an EM algorithm. The key to the success of the EM paradigm is the fact that $Q(\theta \mid \theta^{(s)})$ is generally much easier to maximize directly than the original objective function $\ell(\theta)$ of equation (3.3).

*Minorization* is a key characteristic guaranteed by inequality (3.8). A function is said to minorize another function at the point $\theta_0$ if it provides a uniform lower bound and if equality is attained at $\theta_0$. In other words, inequality (3.8) implies that the function

$$Q(\theta \mid \theta^{(s)}) - Q(\theta^{(s)} \mid \theta^{(s)}) + \ell(\theta^{(s)}) \tag{3.9}$$

is a minorizer of $\ell(\theta)$ at $\theta^{(s)}$. The general theory of minorization-maximization (MM) algorithms (Hunter & Lange, 2004), of which any EM algorithm is a special case, shows that iteratively minorizing at the current parameter value and then maximizing the resulting minimizer will guarantee the ascent property (3.4).

The same techniques explained above may be applied to the case in which $\ell(\theta)$ is a sum-of-log-of-integrals instead of a sum-of-log-of-sums. In this case, equation (3.7) is replaced by

$$Q(\theta \mid \theta^{(s)}) = \sum_{i=1}^{n} \int \left[ \frac{L_{ig}(\theta^{(s)})}{\int L_{ih}(\theta^{(s)}) \, dh} \right] \log L_{ig}(\theta) \, dg.$$

*Technical notes*

Earlier, we insisted only that $L_{ig}(\theta)$ be nonnegative, not strictly positive; yet it appears at first glance that zero values may create problems in (3.6). Fortunately, such problems are easily avoided. First, there is no loss of generality in restricting attention to the subset of $\Theta$ consisting of those $\theta$ for which $\ell(\theta)$ is finite, and the left-hand side of (3.8) is always well defined in this set $\Theta$. Second, we are justified in ignoring any summands on the right-hand side of (3.6) for which $L_{ig}(\theta^{(s)}) = 0$ because the limit of $t \log(1/t)$ as $t \to 0$ is zero. This convention works even when $L_{ig}(\theta)$ is also zero, for omitting such $g$ from both sides of (3.6) changes nothing. Finally, in case $L_{ig}(\theta)$ is ever zero while $L_{ig}(\theta^{(s)})$ is not, inequality (3.6) remains true if we allow the right-hand side to take the value $-\infty$.

## 3.3    Likelihood as a Product of Sums

The product representation over variable $i$ in (3.1) typically arises in likelihood methods because independent observations $Y_1, Y_2, \ldots, Y_n$ lead to such a product representation. In the statistics framework, the inner summation representation (3.2) arises in a more specialized way. When one has a known parametric joint density $f(y, z; \theta)$ for two variables $Y$ and $Z$, then the marginal density for $Y$ typically has the representation

$$f(y; \theta) = \sum_z f(y, z; \theta) \quad \text{or} \quad \int f(y, z; \theta) \, dz,$$

where the choice of sum or integral depends on whether the density function $f(y, z; \theta)$ is a probability mass function or a Lebesgue density on the $Z$ space. Thus, in this context, the variable $z$ plays the role of summation index $g$.

Therefore, if we are given a hypothetical data set $(Y_1, Z_1), (Y_2, Z_2), \ldots, (Y_n, Z_n)$, each pair of which is independent of the other pairs, then the objective (log likelihood) function could be written as

$$\ell(\theta) = \log \left( \prod_{i=1}^{n} f_i(y_i; \theta) \right) \tag{3.10}$$

$$= \sum_{i=1}^{n} \log \left( \sum_{z} f_i(y_i, z; \theta) \right). \tag{3.11}$$

Moreover, the density for the $i$th pair $(Y_i, Z_i)$ is $f_i(y, z)$, but the statistician only observes the data $(Y_1, \ldots, Y_n)$, and so must use the log likelihood given in (3.10). For this reason, we sometimes refer to $\ell(\theta)$ as the observed-data log likelihood function; other chapters in this book use the notation $\ell_o(\theta)$ instead of $\ell(\theta)$. Just as we call $(Y_1, \ldots, Y_n)$ the observed data, we call $(Z_1, \ldots, Z_n)$ the missing data.

Mixture models provide a wide variety of observed- and missing-data examples well suited to EM algorithms. In the finite mixture case, we posit a population composed of a certain number, say $G$, of distinct subgroups. Let $\eta_g$ denote the proportion of the population consisting of subgroup $g$, $1 \leq g \leq G$. We assume that variable(s) $Y$ is measured for each individual in a sample from the population, and that $Y$ is distributed according to a density $f_g(y)$ for individuals in the $g$th subgroup. If the data $y_1, \ldots, y_n$ are observed without their corresponding subgroup labels $z_1, \ldots, z_n$, the "missing data" paradigm fits perfectly: there is a clear sense in which each observed $y_i$ is merely a function of the complete $(y_i, z_i)$ and each $z_i$ is missing. In this example, we may rewrite equation (3.11) as

$$\ell(\theta) = \sum_{i=1}^{n} \log \left[ \sum_{g=1}^{G} \eta_g f_g(y_i) \right]. \tag{3.12}$$

Although a growing body of literature considers the model of equation (3.12) without specifying the parametric form of $f_g(y)$ (see Chauveau et al., 2015, for a survey of some of this work) we often assume a parametric form for the density function, so we may replace $f_g(\cdot)$ by $f(\cdot; \theta_g)$. In this context, the $L_{ig}(\theta)$ function of equation (3.3) becomes $\eta_g f(y_i; \theta_g)$. The weights defined in equation (3.5) are given by

$$w_{ig}^{(s)} = \frac{\eta_g^{(s)} f(y_i; \theta_g^{(s)})}{\sum_{h=1}^{G} \eta_h^{(s)} f(y_i; \theta_h^{(s)})},$$

and the surrogate function for the mixture model is

$$Q^{(s)}(\theta) = \sum_{i=1}^{n} \sum_{g=1}^{G} w_{ig}^{(s)} \log \eta_g + \sum_{i=1}^{n} \sum_{g=1}^{G} w_{ig}^{(s)} \log f(y_i; \theta_g). \tag{3.13}$$

Notice that in (3.13), each element of $\theta = (\eta_1, \ldots, \eta_G, \theta_1, \ldots, \theta_G)$ appears alone in its own separate sum. This means that $Q^{(s)}(\theta)$ is much easier to maximize than $\ell(\theta)$. This is typical of EM algorithms, which often replace a single difficult maximization with a series of much easier ones. The surrogate function $Q^{(s)}(\theta)$ is sometimes called the completed-data log likelihood and denoted by $\ell_c^{(s)}(\theta)$ or simply $\ell_c(\theta)$, though it should always be remembered

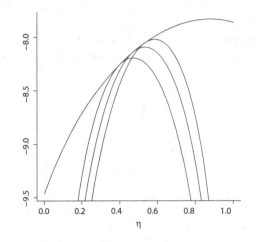

**FIGURE 3.1**
The solid curve is the observed-data log likelihood function $\ell(\eta)$ for the two-component mixture model $\eta\mathcal{N}(0,1)+(1-\eta)\mathcal{N}(1,1)$ with $n = 6$. The data are $-1.0, -0.5, 0.0, 0.5, 0.8, 1.6$. The dotted curves are the functions that minorize $\ell(\eta)$ at the points $\eta^{(0)} = 0.4$, $\eta^{(1)} = 0.472$, and $\eta^{(2)} = 0.534$.

that this completed-data log likelihood function depends on the current iterate $\theta^{(s)}$ even when this fact is not reflected explicitly by the notation.

There is no generic maximizer of (3.13) with respect to $\theta_g$, though often when we know the functional form of $f(\cdot;\theta_g)$ an explicit maximizer can be derived. On the other hand, the maximizer with respect to the $\eta_g$ parameters, which must sum to unity, is always the same for any finite mixture model:

$$\eta_g^{(s+1)} = \frac{1}{n}\sum_{i=1}^{n} w_{ig}^{(s)}. \tag{3.14}$$

Figure 3.1 illustrates the $\ell(\theta)$ function and several of its minorizing functions defined by equation (3.9) for a simple finite mixture model example. Consider a mixture problem with $G = 2$ in which both subgroup densities are completely known: $f_1(\cdot)$ and $f_2(\cdot)$ are both normal with unit variance and means 0 and 1, respectively. Here, the only unknown parameter is $\eta \equiv \eta_1$ (since $\eta_2$ is just $1 - \eta_1$). Starting from $\eta^{(0)} = 0.4$, the figure depicts the first three iterations of an EM algorithm.

## 3.4  Non-standard Examples of EM Algorithms

This section presents several examples, each related in some way to mixture models, that illustrate the main theme of this chapter. That is, although an EM algorithm often follows a precise recipe once the observed and missing data are defined, this standard paradigm is not necessary to define an EM algorithm. All that one needs is the product-of-sums, or sum-of-log-of-sums, representation.

### 3.4.1 Modes of a density

Li et al. (2007) introduced a Modal EM (MEM) algorithm that finds the local maxima, or modes, of a given density. Consider a two-component normal mixture density $f(y) = \eta_1 f_1(y) + \eta_2 f_2(y)$, where $y$ is real-valued, $\eta_1$ and $\eta_2$ are the prior probabilities of the two mixture components, and $f_g(y) = \phi(y|\mu_g, \sigma^2)$ is the density of component $g$. Given an initial value $y^{(0)}$, MEM solves a local maximum of the mixture density, $f(y)$. We see that

$$\log f(y) = \log\{\eta_1 f_1(y) + \eta_2 f_2(y)\}$$

is in the log-of-sums format. Taking $n = 1$ and following the steps explained in Section 3.2.2, we can obtain the surrogate $Q\left(y|y^{(0)}\right)$ as

$$Q\left(y|y^{(0)}\right) = w_1(y^{(0)}) \log \phi(y|\mu_1, \sigma^2) + w_2(y^{(0)}) \log \phi(y|\mu_2, \sigma^2),$$

where

$$w_1(y^{(0)}) = 1 - w_2(y^{(0)}) = \frac{\eta_1 \phi(y^{(0)}|\mu_1, \sigma^2)}{\eta_1 \phi(y^{(0)}|\mu_1, \sigma^2) + \eta_2 \phi(y^{(0)}|\mu_2, \sigma^2)}.$$

Solving for $y$, we have

$$y = w_1(y^{(0)})\mu_1 + w_2(y^{(0)})\mu_2.$$

Thus, the MEM algorithm in this case iterates between calculating $w_1(y^{(s-1)})$ and $w_2(y^{(s-1)})$ at the E step, and calculating $y^{(s)}$ at the M step, until convergence.

### 3.4.2 Gradient maxima

The gradient function in its original form (Lindsay, 1995) can be used to test whether a latent distribution, say $\mathcal{H}_0$, is the nonparametric maximum likelihood estimator in an infinite-dimensional space of mixing distributions. In the space of distribution functions, define a path from $\mathcal{H}_0$ to any other distribution $\mathcal{H}_1$ as $\mathcal{H}_\alpha = (1 - \alpha)\mathcal{H}_0 + \alpha\mathcal{H}_1$. Notice that for every $\alpha$, this construction generates an intermediate distribution. Let $L^*(\alpha) = L(\mathcal{H}_\alpha)$ be the likelihood along the above path. Then the derivative of $\log L^*(\alpha)$ at $\alpha = 0$ is the *directional derivative* corresponding to the path from $\mathcal{H}_0$ to $\mathcal{H}_1$ and it has the form

$$D_{\mathcal{H}_0}(\mathcal{H}_1) = \sum_{i=1}^{d} n(i) \left( \frac{L_i(\mathcal{H}_1)}{L_i(\mathcal{H}_0)} - 1 \right),$$

where $d$ is the number of distinct observations and $n(i)$ is the multiplicity of the $i$th observation. The *gradient function* is defined as a special case of the directional derivative with degenerate $\mathcal{H}_1$ at $\phi$ and it has the form

$$D_{\mathcal{H}_0}(\phi) := D_{\mathcal{H}_0}(\Delta_\phi) = \sum_{i=1}^{d} n(i) \left( \frac{L_i(\phi)}{L_i(\mathcal{H}_0)} - 1 \right).$$

Lindsay (1995) also shows that any $\mathcal{H}$ is a maximum likelihood estimator if and only if $D_{\mathcal{H}}(\phi) \geq 0$ for all $\phi$.

Let us return again to the finite mixture model example of equation (3.12). Equation (3.13) gives the surrogate function for the EM algorithm and equation (3.14) gives the formula for the M step of the mixing parameter update $\eta_g^{(s+1)}$. In this setting, the gradient function $D(\phi^\star)$ for a single parameter value $\phi^\star$ is

$$D(\phi^\star) = \sum_{i=1}^{n} \left( \frac{f(y_i; \phi^\star)}{\sum_{h=1}^{G} \eta_h f(y_i; \phi_h)} - 1 \right) = -n + \sum_{i=1}^{n} \frac{f(y_i; \phi^\star)}{\sum_{h=1}^{G} \eta_h f(y_i; \phi_h)}.$$

As Lindsay (1995) explained, if, for some $\phi^\star$, $D(\phi^\star) > 0$, then a higher likelihood can be obtained, improving the model fit. The $\phi^\star$ that maximize $D(\phi^\star)$ will be the best candidates for improvement, and adding the constant $n$ does not change the maximizer. Therefore, we use an EM algorithm on $D(\phi^\star) + n$. For notational simplicity, let

$$\alpha_i = \frac{1}{\sum_{h=1}^{G} \eta_h f(y_i; \phi_h)}.$$

Then

$$\log[D(\phi^\star) + n] = \log \sum_{i=1}^{n} \alpha_i f(y_i; \phi^\star),$$

which is in the log-of-sums format. The surrogate function for maximizing $D(\phi^\star)$ is

$$Q^{(s)}(\phi^\star) = \sum_{i=1}^{n} w_i^{(s)} \log f(y_i; \phi^\star),$$

where

$$w_i^{(s)} = \frac{\alpha_i f\left(y_i; \phi^{\star(s)}\right)}{\sum_{i=1}^{n} \alpha_i f\left(y_i; \phi^{\star(s)}\right)}.$$

Once we know the form of $f(y_i; \phi)$, we can maximize $Q^{(s)}(\phi^\star)$ for $\phi^\star$ to find candidates, if they exist, that would improve our current solution.

### 3.4.3 Two-step EM

Govaert & Nadif (2003) first proposed the block mixture model for clustering rows and columns simultaneously. Let $Y = \{y_{ij} : i \in I \text{ and } j \in J\}$ be the data matrix, where $I$ is the set of $R$ row indices and $J$ is the set of $C$ column indices. Suppose that the row labels $Z_i = z_i$ are drawn independently from the density $p(z)$ on the set $\{1, 2, \ldots, G_1\}$, and the column labels $W_j = w_j$ are drawn independently from the density $q(w)$ on $\{1, 2, \ldots, G_2\}$. Let $f(y_{ij}; \theta_{z_i, w_j})$ be the density of the $ij$th observation conditioned on the true labels $\mathbf{z}$ and $\mathbf{w}$. The unconditional density of $Y$ is

$$f(Y) = \sum_{z_1}^{G_1} \cdots \sum_{z_R}^{G_1} \sum_{w_1}^{G_2} \cdots \sum_{w_C}^{G_2} \left( \prod_{i=1}^{R} \prod_{j=1}^{C} f(y_{ij}; \theta_{z_i, w_j}) \right) \left( \prod_{i=1}^{R} p(z_i) \right) \left( \prod_{j=1}^{C} q(w_j) \right). \quad (3.15)$$

It can be shown (Govaert & Nadif, 2003, 2005; Wyse & Friel, 2012) that a standard EM algorithm cannot be directly applied in this setting due to the large number of operations involved. As an alternative to the full likelihood of equation (3.15), Kuruppumullage Don (2014) proposed estimating the parameters using the composite likelihood

$$\log L_{RC}(\theta) = \underbrace{\sum_{i=1}^{R} \log \left\{ \sum_{z=1}^{G_1} p_z \left( \prod_{j=1}^{C} \sum_{w=1}^{G_2} q_w f(y_{ij}; \theta_{zw}) \right) \right\}}_{\text{cl}_1(\theta)}$$

$$+ \underbrace{\sum_{j=1}^{C} \log \left\{ \sum_{w=1}^{G_2} q_w \left( \prod_{i=1}^{R} \sum_{z=1}^{G_1} p_z f(y_{ij}; \theta_{zw}) \right) \right\}}_{\text{cl}_2(\theta)}, \quad (3.16)$$

where $q_w = \mathrm{P}(w_j = w)$ and $p_z = \mathrm{P}(z_i = z)$. Since the $\mathrm{cl}_1(\theta)$ term in equation (3.16) is in the log-of-sums form, we use the surrogate construction explained in Section 3.2.2. Letting

$$L_i(\theta) = \sum_{z=1}^{G_1} p_z \left( \prod_{j=1}^{C} \sum_{w=1}^{G_2} q_w f(y_{ij}; \theta_{zw}) \right),$$

the minorizing surrogate of $\mathrm{cl}_1(\theta) = \sum_{i=1}^{R} \log L_i(\theta)$ is

$$Q_1(\theta|\theta^{(s)}) = \sum_{i=1}^{R} \sum_{z=1}^{G_1} w_{iz}^{(s)} \log \left\{ p_z \left( \prod_{j=1}^{C} \sum_{w=1}^{G_2} q_w f(y_{ij}; \theta_{zw}) \right) \right\}, \qquad (3.17)$$

where

$$w_{iz}^{(s)} = \frac{p_z^{(s)} \left( \prod_{j=1}^{C} \sum_{w=1}^{G_2} q_w^{(s)} f(y_{ij}; \theta_{zw}^{(s)}) \right)}{\sum_{z'=1}^{G_1} p_{z'}^{(s)} \left( \prod_{j=1}^{C} \sum_{w=1}^{G_2} q_w^{(s)} f(y_{ij}; \theta_{z'w}^{(s)}) \right)}.$$

Equation (3.17) can be rewritten as

$$Q_1(\theta|\theta^{(s)}) = \sum_{i=1}^{R} \sum_{z=1}^{G_1} w_{iz}^{(s)} \log p_z + \sum_{i=1}^{R} \sum_{z=1}^{G_1} w_{iz}^{(s)} \sum_{j=1}^{C} \underbrace{\log \left( \sum_{w=1}^{G_2} q_w f(y_{ij}; \theta_{zw}) \right)}_{\mathrm{cl}_1^*}.$$

We see that although $Q_1(\theta|\theta^{(s)})$ can be directly solved for $p_z$, solving it for $q_w$ and $\theta_{zw}$ is still problematic due to the log-of-sums form of $\mathrm{cl}_1^*$. We therefore minorize a second time, following the same construction, and define

$$Q_2(\theta|\theta^{(s)}) = \sum_{i=1}^{R} \sum_{z=1}^{G_1} w_{iz}^{(s)} \log p_z + \sum_{i=1}^{R} \sum_{z=1}^{G_1} w_{iz}^{(s)} \sum_{j=1}^{C} \sum_{w=1}^{G_2} w_{izw}^{(s)} \log q_w$$

$$+ \sum_{i=1}^{R} \sum_{z=1}^{G_1} w_{iz}^{(s)} \sum_{j=1}^{C} \sum_{w=1}^{G_2} w_{izw}^{(s)} \log f(y_{ij}; \theta_{zw}),$$

where

$$w_{izw}^{(s)} = \frac{q_w^{(s)} f(y_{ij}; \theta_{zw}^{(s)})}{\sum_{w'=1}^{G_2} q_{w'}^{(s)} f(y_{ij}; \theta_{zw'}^{(s)})}.$$

Notice that minorization is a transitive operation: since $Q_1(\cdot|\theta^{(s)})$ minorizes $\mathrm{cl}_1(\cdot)$ at $\theta^{(s)}$ and $Q_2(\cdot|\theta^{(s)})$ minorizes $Q_1(\cdot|\theta^{(s)})$ at $\theta^{(s)}$, we conclude that $Q_2(\cdot|\theta^{(s)})$ minorizes $\mathrm{cl}_1(\cdot)$ at $\theta^{(s)}$. Thus, we could define an EM algorithm by repeatedly creating $Q_2(\theta|\theta^{(s)})$ in the E step, then maximizing it in the M step, and at each iteration we would guarantee the ascent property in the $\mathrm{cl}_1(\theta)$ function.

However, recall that the composite log likelihood function $\log L_{RC}(\theta)$ actually consists of $\mathrm{cl}_1(\theta) + \mathrm{cl}_2(\theta)$. Therefore, to search for a composite maximum likelihood estimator, we should repeat the steps above to obtain a computationally tractable minorizing function for $\mathrm{cl}_2(\theta)$, then add this minimizer to $Q_2(\theta|\theta^{(s)})$. The resulting sum gives a minorizer of $\log L_{RC}(\theta)$ whose maximizer provides the next parameter estimate $\theta^{(s+1)}$.

## 3.5    Stopping Rules for EM Algorithms

As popular as EM algorithms are for their computational simplicity and guaranteed monotone ascent property, they have the weakness of a linear rate of convergence, which in practice can require many iterations to estimate parameters with reasonable accuracy (Böhning et al., 1994). Commonly used stopping rules to decide whether an algorithm has reached its maximum are based on assessing the relative change of log likelihood or parameter values. That is, the algorithm is stopped if $\left|\ell^{(s+1)} - \ell^{(s)}\right| \leq \epsilon$ or $\left\|\theta^{(s+1)} - \theta^{(s)}\right\| \leq \epsilon$, where $\ell^{(s)} = \ell(\theta^{(s)})$ is the log likelihood calculated at the $s$th iteration and $\epsilon$ is a small positive constant.

It is worth noting that such stopping rules capture the idea of lack of progress rather than numerical accuracy. Böhning et al. (1994) developed a technique to predict the value of the log likelihood at the maximum likelihood solution, say $\hat{\ell}$, by using Aitken acceleration on log likelihood estimates. This acceleration device is applicable to any log likelihood sequence with linear convergence. The Aitken accelerated estimate of $\hat{\ell}$ at the $s$th iteration is

$$\hat{\ell}^{(s)} = \ell^{(s-1)} + \frac{1}{1 - c^{(s)}} \left( \ell^{(s)} - \ell^{(s-1)} \right),$$

where

$$c^{(s)} = \frac{\left( \ell^{(s+1)} - \ell^{(s)} \right)}{\left( \ell^{(s)} - \ell^{(s-1)} \right)}$$

is the estimated linear rate of convergence at the $s$th iteration. That is, $c^{(s)}$ is an estimate of

$$\lim_{s \to \infty} \frac{\hat{\ell} - \ell^{(s)}}{\hat{\ell} - \ell^{(s-1)}}, \tag{3.18}$$

which can be shown to exist as a positive constant for most EM algorithms. Indeed, the existence of the positive limit (3.18) is the defining characteristic of a linear rate of convergence.

Using the convergence properties of sequences $\ell^{(s)}$ and $\hat{\ell}^{(s)}$, Böhning et al. (1994) developed a useful stopping rule that actually reflects the numerical accuracy of the estimates. This rule is

$$\text{Stop EM if } 0 < \left( \hat{\ell}^{(s)} - \ell^{(s)} \right) < \epsilon.$$

Lindsay (1995) discussed this stopping rule, noting that if the tolerance value is small, say, $\epsilon < 0.005$, then we are pursuing a numerical accuracy in the parameter estimates that is minor relative to the magnitude of their likelihood confidence intervals.

## 3.6    Concluding Remarks

EM algorithms are the main computational workhorses for maximum likelihood calculations in mixture model contexts, and this fact does not change if we expand the notion of what defines an EM algorithm. In this chapter, we have described such an expansion, presenting several examples of algorithms that we consider EM since they are built from functions having the sum-of-log-of-sums form. In this expansive view, the E step still relies on Jensen's inequality (an inequality based on expectation) to construct a minorizing function to maximize in the M step; thus, these algorithms are also expectation-maximization algorithms even though they do not fit the standard pattern first introduced in the seminal paper by Dempster et al. (1977). Each of the examples we consider here is rooted in mixture models, even though none of them uses a standard EM algorithm. Furthermore, each enjoys all of the theoretical properties (both good and bad) of more traditionally defined EM algorithms. In particular, the linear rate of convergence that characterizes EM algorithms may be exploited to engineer a stopping criterion that strives to ensure numerical accuracy in parameter estimates rather than reacting to slow progress of the algorithm.

# Bibliography

BÖHNING, D., DIETZ, E., SCHAUB, R., SCHLATTMANN, P. & LINDSAY, B. G. (1994). The distribution of the likelihood ratio for mixtures of densities from the one-parameter exponential family. *Annals of the Institute of Statistical Mathematics* **46**, 373–388.

CHAUVEAU, D., HUNTER, D. R. & LEVINE, M. (2015). Semi-parametric estimation for conditional independence multivariate finite mixture models. *Statistics Surveys* **9**, 1–31.

DEMPSTER, A. P., LAIRD, N. M. & RUBIN, D. B. (1977). Maximum likelihood from incomplete data via the EM algorithm (with discussion). *Journal of the Royal Statistical Society, Series B* **39**, 1–38.

GOVAERT, G. & NADIF, M. (2003). Clustering with block mixture models. *Pattern Recognition* **36**, 463–473.

GOVAERT, G. & NADIF, M. (2005). An EM algorithm for the block mixture model. *IEEE Transactions on Pattern Analysis and Machine Intelligence* **27**, 643–647.

HUNTER, D. R. & LANGE, K. (2004). A tutorial on EM algorithms. *American Statistician* **58**, 30–37.

KURUPPUMULLAGE DON, P. (2014). *Estimation and Model Selection for Block Clustering with Mixtures: A Composite Likelihood Approach*. Ph.D. thesis, Pennsylvania State University, USA.

LI, J., RAY, S. & LINDSAY, B. G. (2007). A nonparametric statistical approach to clustering via mode identification. *Journal of Machine Learning Research* **8**, 1687–1723.

LINDSAY, B. G. (1995). *Mixture Models: Theory, Geometry, and Applications*. Hayward, CA: Institute of Mathematical Statistics.

MCLACHLAN, G. J. & KRISHNAN, T. (2007). *The EM Algorithm and Extensions*. New York: John Wiley.

WYSE, J. & FRIEL, N. (2012). Block clustering with collapsed latent block models. *Statistics and Computing* **22**, 415–428.

# 4

# Bayesian Mixture Models: Theory and Methods

**Judith Rousseau, Clara Grazian, and Jeong Eun Lee**

*University of Oxford, UK and Université Paris-Dauphine PSL, France; University of Oxford, UK and Università degli Studi "Gabriele d'Annunzio", Italy; Auckland University of Technology, New Zealand*

## CONTENTS

## 4.1 Introduction

Mixture models have the form

$$p(y|H,\psi) = \int_\Theta f(y|\theta,\psi)H(d\theta),$$

so that $\psi$ is a parameter common to all components and $H$ is a probability measure on $\Theta$ called the mixing distribution. Of particular interest is the case where $H$ is a discrete distribution, which is the case treated in this chapter. In other words,

$$H(d\theta) = \sum_{g=1}^{G} \eta_g \delta_{\theta_g}(d\theta),$$

where $G$ might be finite or infinite.

In this chapter we present some aspects of Bayesian inference in the context of mixture models. Mixture models are particularly useful for density estimation, parameter estimation in the context of heterogeneous populations, as well as clustering. The advantage of the

Bayesian approach is that in all these problems the quantities of interest are easily accessible via the posterior distribution, as each of these quantities is a function of the parameters (and possibly the data). In Section 4.3 we will describe the asymptotic behaviour of the corresponding posterior distributions.

As with all Bayesian approaches, one of the critical aspects of the method is the determination of the prior distribution on the parameters. Section 4.2.2 summarizes what we view as important aspects of prior choices in mixture models.

## 4.2  Bayesian Mixtures: From Priors to Posteriors

### 4.2.1  Models and representations

In the Bayesian approach $Y|\theta \sim P_\theta$ and a prior distribution is defined on $\Theta$ to model the uncertainty on $\theta : \theta \sim P$, so that $(Y, \theta)$ has joint distribution given by $P_\theta \times P$. Given $n$ realizations $\mathbf{y} = (y_1, \ldots, y_n)$ of $Y$, we can define the posterior distribution as the conditional probability of $\theta$ given $\mathbf{y}$. If the model is dominated by a common measure $\mu$ (say, Lebesgue measure), then the posterior distribution is computed using Bayes' formula,

$$p(\theta|\mathbf{y}) = \frac{p(\mathbf{y}|\theta)p(\theta)}{\int_\Theta p(\mathbf{y}|\theta)p(\theta)d\nu(\theta)},$$

where $p(\cdot)$ is the prior density with respect to some measure $\nu$ and $p(\mathbf{y}|\theta) = \prod_{i=1}^n f_\theta(y_i)$, with $f_\theta$ being the density of $P_\theta$ with respect to the measure $\mu$, is the likelihood. Basics (and less basics) on Bayesian methods can be found, for instance, in the textbooks of Robert (2007), Berger (1985) and Bernardo & Smith (1994).

In the context of mixture models, the unknown parameter is $(H, \psi)$. There are essentially three families of models. First, finite mixtures with a fixed number $G$ of components,

$$p(y|\eta, \psi, \theta) = \sum_{g=1}^G \eta_g f(y|\theta_g, \psi), \tag{4.1}$$

where $G$ is known. The prior in this case consists of a prior distribution on $\psi$ and on $(\eta, \theta)$, where $\eta = (\eta_1, \ldots, \eta_G) \in \mathcal{V}_G := \{x \in [0,1]^G; \sum_{g=1}^G x_g = 1\}$, the $G$-dimensional unit simplex, and $\theta = (\theta_1, \ldots, \theta_G) \in \Theta^G$. The most common way to construct such priors is by considering $p = p_\eta \otimes h^{\otimes G}$, where $h$ is a probability density on $\Theta$ with respect to some given measure and $p_\eta$ is a probability density on $\mathcal{V}_G$ with respect to Lebesgue measure. Hence, under this prior distribution the $\theta_g$ are independently and identically distributed and independent of $\eta$. Typically, $p_\eta$ is chosen to be the density of a Dirichlet distribution.

A second class are finite mixture models with an unknown number $G$ of components. The model is the same as in (4.1); however, this time the unknown parameter is $(G, \eta_1, \ldots, \eta_G, \theta_1, \ldots, \theta_G, \psi)$. We then construct a prior in a hierarchical way as

$$G \sim P_G,$$
$$(\eta_1, \ldots, \eta_G, \theta_1, \ldots, \theta_G, \psi)|G \sim P(\cdot|G).$$

Inference for both $G$ and $(\eta_1, \ldots, \eta_G, \theta_1, \ldots, \theta_G, \psi)$ is described in Chapter 1.

The third class are infinite mixtures, which corresponds to mixture (4.1) with $G = +\infty$. There exist many instances of prior distributions in this context, the most common being the Dirichlet process mixture model, where the mixing distribution $\eta$ follows a Dirichlet process; see Chapter 6 for more details.

If interest lies in density estimation or prediction, then the posterior distribution of $(f(y|\theta,\eta,\psi), y \in \mathcal{Y})$, considered as a function of $\theta, \psi$ and $\eta$, provides point estimates such as the posterior predictive density $\hat{f}(y) = \mathrm{E}(f(y|\theta,\eta,\psi)|\mathbf{y})$, together with measures of uncertainty. In particular, it is common practice to construct pointwise credible intervals of the form $(f_1(y), f_2(y))$ such that, for all $y \in \mathcal{Y}$, $\mathrm{P}(f_1(y) \le f(y|\theta,\eta,\psi) \le f_2(y)|\mathbf{y}) = 1 - \alpha$, for a given credibility $1 - \alpha$, although a more satisfying measure of uncertainty would be a credible band in the form $\mathrm{P}(a_1(y) \le f(y|\theta,\eta,\psi) \le a_2(y), \forall y \in \mathcal{Y}|\mathbf{y}) = 1 - \alpha$, which is more difficult to construct.

Mixture models are also used in model-based clustering approaches (see Chapter 8 for details) using the latent allocation variables $z_i$ for each $i = 1,\ldots,n$:

$$y_i|z_i = g, \theta, \psi, \eta \overset{ind}{\sim} f(\cdot|\theta_g, \psi), \quad z_i \overset{i.i.d}{\sim} \mathcal{M}(1, \eta_1, \ldots, \eta_G), \qquad (4.2)$$

where $\mathcal{M}(\cdot)$ stands for the multinomial distribution with parameter $(\eta_1,\ldots,\eta_G)$. Given observations $\mathbf{y} = (y_1,\ldots,y_n)$, various quantities can be used to infer on the cluster structure of the data, among them the posterior distribution on the clusters themselves formed from $\mathbf{z} = (z_1,\ldots,z_n)$, the posterior classifiers $\mathrm{P}(z_i = \cdot|\mathbf{y})$ for all $i = 1,\ldots,n$, and the maximum *a posteriori* (MAP) estimator of $\mathbf{z}$, that is, the mode of the posterior distribution $p(\mathbf{z}|\mathbf{y})$ of $\mathbf{z}$.

In some other cases the goal is to perform inference on the parameters of the components. If an exchangeable prior is chosen on the componentwise parameters $(\eta, \theta) = (\eta_1,\ldots,\eta_G,\theta_1,\ldots,\theta_G)$, the posterior distribution is invariant under permutations of the component indices and component labels are neither identifiable nor of interest. However, this causes some issues when conducting inference on the parameters $(\eta, \theta)$. As introduced in Chapter 1, this phenomenon is called label switching and has been well studied; see Celeux et al. (2000), Stephens (2000), Frühwirth-Schnatter (2001) and Jasra et al. (2005). There are several issues related to label switching. A major issue is that when sampling from the posterior distribution, it is important to ensure that the sampler visits the whole support of the posterior distribution, in particular the $G!$ equivalent modes. This means that post-processing is necessary to obtain relevant point estimators of the various parameters of the components $(\theta_g, \eta_g)$, $g \le G$. Indeed, consider for instance the posterior mean of each $\theta_g$. When each equivalent mode has been visited, one obtains $\mathrm{E}(\theta_g|\mathbf{y}) = \sum_{\mathfrak{s}} \mathrm{E}(\theta_{\mathfrak{s}(g)}|\mathbf{y})$ where $\mathfrak{s}$ ranges over the set $\mathfrak{S}(G)$ of permutations of $(1,\ldots,G)$. This is meaningless, in particular since each component has the same posterior mean.

An illustration of label switching is given in Figures 4.1 and 4.2, which show posterior approximations (obtained via a Metropolis–Hastings algorithm) of the location parameters of a Gaussian mixture model with observations simulated from $0.7 \times \mathcal{N}(-1, 0.5) + 0.3 \times \mathcal{N}(1, 0.5)$. These figures exhibit the multimodality of the marginal posterior distributions of the location parameters (with similar features for the other parameters of the Gaussian mixture) and make it clear that the Metropolis–Hastings Markov chain visits both modes several times along the iterations.

To achieve component-specific inference and, in particular, to give a meaning to each component, numerous relabelling methods have been proposed; see, for instance, Richardson & Green (1997), Celeux et al. (2000), Stephens (2000), Frühwirth-Schnatter (2001), Jasra et al. (2005), Geweke (2007), Marin & Robert (2008), Rodriguez & Walker (2014) and van Havre et al. (2015), among many others. The R package label.switching (Papastamoulis, 2013) incorporates some of these label switching removal methods.

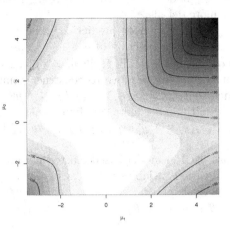

**FIGURE 4.1**
Approximation on a grid of values of the posterior distribution for the location parameters $(\mu_1, \mu_2)$ in the mixture model $\eta_1 \mathcal{N}(\mu_1, \sigma_1) + (1 - \eta_1)\mathcal{N}(\mu_2, \sigma_2)$ with simulations from $0.7 \times \mathcal{N}(-1, 0.5) + 0.3 \times \mathcal{N}(1, 0.5)$.

### 4.2.2  Impact of the prior distribution

As with all Bayesian approaches, one of the critical aspects of the method is the determination of the prior distribution on the parameters. There is a large literature on possible choices of prior distributions in the context of mixture models; see, for instance, Frühwirth-Schnatter (2006) for references. In this section we summarize what we view as important aspects of prior choices in mixture models.

#### 4.2.2.1  Conjugate priors

The most popular families of prior distributions in the context of mixture models are the conjugate priors under the complete-data likelihood $p(\mathbf{y}, \mathbf{z}|\theta, \eta, \psi)$,

$$p(\mathbf{y}, \mathbf{z}|\theta, \eta, \psi) = \prod_{g=1}^{G} \prod_{i:z_i=g} f(y_i|\theta_g, \psi)\eta_g^{n_g}, \quad n_g = \sum_{i=1}^{n} \mathbb{I}(z_i = g).$$

It then appears that a conjugate family can be constructed by considering a Dirichlet prior on the weights $(\eta_1, \ldots, \eta_G)$, a conjugate prior $p(\theta_g)$ for each emission model $\prod_{i:z_i=g} f(y_i|\theta_g, \psi)$ given the parameter $\psi$ and the allocation variables $z_i$, when they exist, and a conjugate prior on $\psi$ associated to either the marginal likelihood

$$\prod_{g=1}^{G} \int_{\Theta} \prod_{i:z_i=g} f(y_i|\theta_g, \psi)dp(\theta_g),$$

or the conditional likelihood $\prod_{g=1}^{G} \prod_{i:z_i=g} f(y_i|\theta_g, \psi)$. The main advantage for using such prior distributions is computational, since a Gibbs sampling algorithm can then be used to simulate from the posterior distribution (Diebolt & Robert, 1994); see Chapter 5 for more details. Although analytical convenience is not to be neglected, in particular in contexts of large data sets, using such prior distributions as a default prior raises a few issues, such as calibration or choice of the associated hyperparameters in the absence of substantive

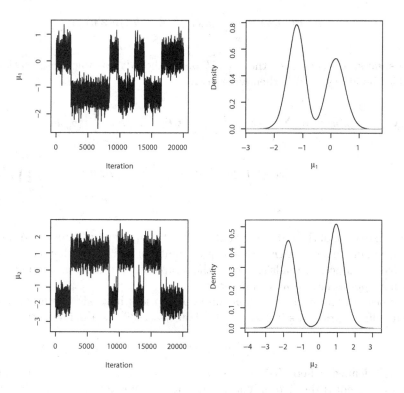

**FIGURE 4.2**
Markov chains obtained via a Metropolis–Hastings algorithm (left) and approximations of the posterior densities (right) of $\mu_1$ (top) and $\mu_2$ (bottom), for data generated from the mixture $0.7 \times \mathcal{N}(-1, 0.5) + 0.3 \times \mathcal{N}(1, 0.5)$.

prior information. This problem becomes more acute when accounting for the fact that in such cases improper priors cannot be used because they typically imply improper posteriors in mixture models, as will be shown in Section 4.2.2.2. For the same reason, they cannot be replaced by vague priors, since the scale parameters associated with such priors will be highly influential on the posterior distribution. It is therefore not a good strategy to use a conjugate prior with a large scale as a weakly informative prior in the context of mixture models.

### 4.2.2.2 Improper and non-informative priors

While impropriety of the prior distribution does not automatically imply impropriety of the posterior distribution (Berger, 1985), mixture models are known to prohibit the use of improper priors based on (*a priori*) independent $\theta_g$s. To see this, we decompose the observed likelihood $p(\mathbf{y}|\eta, \theta, \psi)$ as a sum over all possible partitions of the data into $G$ groups. More precisely, in the case of independent improper priors for each component,

$$p(\theta_1, \dots, \theta_G) \propto \prod_{g=1}^{G} p(\theta_g), \quad \int p(\theta_1) d\theta_1 = \infty, \tag{4.3}$$

using the latent representation (4.2), the observed likelihood can be written as

$$p(\mathbf{y}|\eta, \psi, \theta) = \sum_{\mathbf{z} \in \mathcal{S}_G^n} p(\mathbf{y}|\mathbf{z}, \theta_1, \dots, \theta_G) p(\mathbf{z}|\eta), \tag{4.4}$$

where the summation runs over the set $\mathcal{S}_G^n$ of all the $G^n$ possible classifications $\mathbf{z}$. The marginal likelihood $p(\mathbf{y}|\eta, \psi)$ is then written as

$$\int_{\Theta^G} p(\mathbf{y}|\eta, \theta, \psi) p(\theta) d\theta = \sum_{\mathbf{z} \in \mathcal{S}_G^n} p(\mathbf{z}|\eta) \int_{\Theta^G} p(\mathbf{y}|\mathbf{z}, \theta_1, \dots, \theta_G) \prod_{g=1}^{G} p(\theta_g) d\theta_g$$

$$\geq p(\mathbf{z}_0|\eta) \int_{\Theta} p(\theta_1) d\theta_1 \times \int_{\Theta^{G-1}} p(\mathbf{y}|\mathbf{z}_0, \theta_2, \dots, \theta_G) \prod_{g=2}^{G} p(\theta_g) d\theta_g$$

$$= +\infty,$$

where $\mathbf{z}_0$ is a partition of the data such that the first[1] component is empty, and the posterior distribution is not defined. Although the probability of the event that no observation is allocated to one component tends to zero as the sample size increases, independent improper priors are not appropriate to make inference.

Other types of non-informative (and improper) priors have also been considered. The well-known Jeffreys prior was introduced by Jeffreys (1939) as a default prior and defined as

$$p^{\mathrm{J}}(\theta) \propto |I(\theta)|^{1/2},$$

whenever well defined, where $I(\cdot)$ stands for the expected Fisher information matrix and the symbol $|\cdot|$ denotes the determinant. While it does not represent the ultimate answer in the search for a non-informative prior, it could often be considered as a benchmark prior distribution to be compared with other proposals.

Bernardo & Girón (1988) derived Jeffreys priors for mixtures weights where components have disjoint supports, showing that the Dirichlet distribution is a reasonable approximation and is used, for instance, in Figueiredo & Jain (2002).

The closed-form derivation of the Fisher information matrix is almost inevitably impossible for mixture models, since the $(i, j)$th entry of the Fisher information matrix in the setting of mixture models involves an integral of the form

$$\int_y \frac{\partial^2 \log\left[\sum_{g=1}^{G} \eta_g f(y|\theta_g, \psi)\right]}{\partial \theta_i \partial \theta_j} \left[\sum_{g=1}^{G} \eta_g f(y|\theta_g, \psi)\right]^{-1} \mathrm{d}y,$$

and this integral can rarely be solved analytically. Although this prior does not model the components' parameters as independent, it still leads to an improper posterior distribution whatever the number of observations, as proven by Grazian & Robert (2018) in the context of regular location–scale mixture models. Finally, Rubio & Steel (2014) derived a closed-form expression for the Jeffreys prior of a location–scale mixture with two disjoint components, which unfortunately again produces an improper posterior distribution.

This set of impropriety results does not imply that any improper prior should be avoided, but rather that particular attention must be paid in this setting. Other attempts to define prior distributions on the parameters of mixture models that only add a small amount of information are available. Some proposals are based on a reparameterization of the model. Mengersen & Robert (1996), for example, propose a reparameterization that divides the

---

[1] In connection with Chapter 1, note that "first" is rarely meaningful in such a context.

parameters into global (or reference) parameters and departures from these references, and allows for improper priors on the global parameters. This reparameterization has been used for exponential components by Gruet et al. (1999) and for Poisson components by Robert & Titterington (1998). Moreover, Roeder & Wasserman (1997) propose, for a Gaussian mixture model, a Markov prior which assumes that the parameter $\theta_g$ is a perturbation of the parameter of the previous component $\theta_{g-1}$, for all $g \geq 2$. Recently, Kamary et al. (2018) proposed an alternative parameterization in terms of moments of the marginal distribution of the observations that allows for improper priors as well.

Another possibility is to consider a hierarchical prior, as proposed by Grazian & Robert (2018) in the context of Gaussian mixture models and other types of location–scale mixtures. Let $\theta_g = (\mu_g, \sigma_g)$ be, respectively, the mean and the standard deviation of the $g$th Gaussian component. Grazian & Robert (2018) propose a hierarchical prior defined by first constructing an independent prior for each component,

$$\mu_g \overset{i.i.d.}{\sim} \mathcal{N}(\mu_0, \zeta_0), \quad \sigma_g \overset{i.i.d.}{\sim} \frac{1}{2}\mathcal{U}[0, \zeta_0] + \frac{1}{2}\frac{1}{\mathcal{U}[0, \zeta_0]},$$

then using the Jeffreys prior associated with the weights,

$$\eta \sim p^J(\eta_1, \ldots, \eta_G | \mu, \sigma),$$

and, finally, putting a non-informative prior on the hyperparameters $\mu_0$ and $\zeta_0$:

$$p(\mu_0, \zeta_0) \propto \frac{1}{\zeta_0}.$$

As in Mengersen & Robert (1996), the parameters in the mixture model are considered tied together; on the other hand, the link among the parameters is created not by a reparameterization of the model, but in a hierarchical way, which coherently allows for the definition of non-informative priors. Grazian & Robert (2018) prove that the posterior distribution derived from this hierarchical representation of the mixture model is proper and may be used for inference.

### 4.2.2.3 Data-dependent priors

A different approach to tackling the issue of improper or, more generally, non-informative priors is to consider data-dependent priors, as proposed for instance by Diebolt & Robert (1994), Raftery (1996), Richardson & Green (1997) and Wasserman (2000).

Diebolt & Robert (1994) and Wasserman (2000) propose to turn an improper prior $p(\theta_g)$ into a proper posterior by changing the distribution of the data set. The idea is as follows. Instead of working with the observed likelihood $p(\mathbf{y}|\theta, \eta, \psi) = \sum_{\mathbf{z} \in \mathcal{S}_G^n} p(\mathbf{y}, \mathbf{z}|\theta, \eta, \psi)$, where $\mathcal{S}_G^n = \{1, \ldots, G\}^n$, the authors propose to derive the posterior distribution from the pseudo-likelihood $\tilde{p}(\mathbf{y}|\theta, \eta, \psi) = \sum_{\mathbf{z} \in \mathcal{S}_G^{n,0}} p(\mathbf{y}, \mathbf{z}|\theta, \eta, \psi)$, where $\mathcal{S}_G^{n,0} = \{\mathbf{z} \in \mathcal{S}_G^n; \min_{j \leq G} \sum_{i=1}^n \mathbb{I}(z_i = j) \geq n_0\}$, in which $n_0$ is the minimal training sample size in each emission model $\prod_{i:z_i=g} f(y_i|\theta_g, \psi)p(\theta_g)$ to ensure that the posterior distribution within each component is proper. Wasserman (2000) shows that this approach corresponds to using a "data-dependent prior" and that the posterior distribution has good asymptotic properties, if the initial improper prior $p(\theta_g)$ is well chosen.

Raftery (1996) and Richardson & Green (1997) propose a different type of data-dependent prior. They consider conjugate priors as described in Section 4.2.2.1 with hyperparameters that are data-dependent. In particular, the empirical mean is used for the location hyperparameters of $p(\theta_g)$, and the scaling hyperparameters are related to the range of the data.

As a side remark, the R package bayesm (Rossi & McCulloch, 2010) is designed to perform Bayesian inference for Gaussian mixtures using a data-dependent prior, by automatically defining hyperparameter values from the model.

### 4.2.2.4  Priors for overfitted mixtures

In Sections 4.2.2.2 and 4.2.2.3 we described proposals to construct non-informative or vaguely informative priors in mixture models with a known number of components. When the number of components is not determined *a priori*, it can be inferred using a hierarchical prior model which includes a probability distribution on the number of components; see Section 4.3.4 and Chapter 7 for more details. An alternative approach is to saturate the model and consider a large number of components $G$ with an informative prior distribution on $(\eta_1, \ldots, \eta_G, \theta_1, \ldots, \theta_G)$. This approach has been suggested by Frühwirth-Schnatter (2006) and formalized by Rousseau & Mengersen (2011). The main difficulty in such context is the non-identifiability of the parameters. Indeed, a density in the form

$$f(y|\eta^*, \psi^*) = \sum_{g=1}^{G^*} \eta_g^* f(y|\theta_g^*, \psi^*)$$

can be represented in the saturated model $f(y|\eta, \psi) = \sum_{g=1}^{G} \eta_g f(y|\theta_g, \psi)$, for $G^* < G$, by emptying the extra components, that is,

$$\eta_g = \eta_g^*, \quad \theta_g = \theta_g^*, \quad g \leq G^*, \quad \eta_{G^*+1} = \ldots = \eta_G = 0,$$

or by merging the extra components to existing ones, for instance,

$$\eta_g = \eta_g^*, \quad \theta_g = \theta_g^*, \quad g \leq G^*, \quad \theta_{G^*+1} = \ldots = \theta_G = \theta_{G^*}^*.$$

The idea behind the informative prior proposed by Rousseau & Mengersen (2011) is to select a prior which will favour one of the two types of configurations: either emptying or merging the extra components. In particular, a Dirichlet prior on $(\eta_1, \ldots, \eta_G) \sim \mathcal{D}(e_1, \ldots, e_G)$ with $\max_g e_g < r/2$, with $r$ being the dimension of $\theta_g$, favours emptying the extra components, while a Dirichlet prior of the form $\mathcal{D}(e_1, \ldots, e_G)$ with $\min_g e_g > r/2$ favours merging of the extra components. This is described in more detail in Section 4.3.3. In the former case, the posterior distribution on the weights $(\eta_1, \ldots, \eta_G)$ concentrates on small values for $\eta_{G^*+1}, \ldots, \eta_G$, with the convention that $\eta_1 \geq \ldots \geq \eta_G$. This allows for an interpretable posterior distribution on the parameters. This idea has since been used in various contexts, including model-based clustering (Malsiner-Walli et al., 2016, 2017) and in more complex models. For instance, Durante et al. (2017) have extended this type of idea when the data consist of networks to obtain a flexible but parsimonious representation of their distribution, and Yang & Dunson (2014) use this idea to define a Bayesian aggregation procedure. Grazian & Robert (2018) show that Jeffreys' prior on the weights of a mixture model automatically deals with the problem of overfitting, concentrating the posterior distribution on the meaningful components, and prove that it may be interpreted under the representation proposed by Rousseau & Mengersen (2011).

An alternative way to favour the configuration where extra components are emptied out is to select a repulsive prior on $(\theta_1, \ldots, \theta_G)$. This approach has been pursued, for instance, by Petralia et al. (2012); see also Chapter 6, Section 6.5.

In practice it is often the case that the components have symmetric roles so that it is natural to choose $e_1 = \ldots = e_G = e_0$. Within the constraint $e_0 < r/2$, the choice of $e_0$ is quite influential for finite sample sizes. Small values of $e_0$ lead to efficient ways of emptying the extra components in moderate sample sizes, but the Gibbs sampler to simulate the posterior distribution of the parameters suffers from slow mixing. To overcome this difficulty and to better investigate the posterior distribution, van Havre et al. (2015) propose a prior parallel tempering algorithm to allow for sampling efficiently from the posterior distributions under different values of $e_0$ simultaneously and thus to explore the behaviour of the posterior distribution under these different values of $e_0$. This can be used, for instance, as a pre-processing of the model, allowing one to select only a few reasonable candidates for the number of components $G$ and then rely on a more formal model selection procedure on $G$.

## 4.3 Asymptotic Properties of the Posterior Distribution in the Finite Case

As explained in Section 4.1, mixture models can be used for density estimation, for classification or clustering, or for parameter estimation. Depending on the specific use, different aspects of the posterior distribution need to be studied. In the following section we will describe the asymptotic behaviour of the posterior distribution in terms of the marginal density of the observations, which is a starting point of most asymptotic analysis.

### 4.3.1 Posterior concentration around the marginal density

In this section we describe some general results on the concentration of the posterior distribution around the true marginal density of the observations. Let $p_0$ be a mixture defined by (4.1) and consider a prior distribution on the parameters $\eta, \theta, \psi$ and possibly $G$. Let $d(p, p_0)$ be either the $L_1$-distance ($d(p, p_0) = \int_{\mathcal{Y}} |p(y) - p_0(y)| dy$) or the Hellinger distance ($d(p, p_0)^2 = \int_{\mathcal{Y}} (\sqrt{p(y)} - \sqrt{p_0(y)})^2 dy$) between $p$ and $p_0$. The posterior distribution concentrates at rate $\epsilon_n = o(1)$ around $p_0$ if

$$P\left(d(p, p_0) \leq \epsilon_n | \mathbf{y}\right) = 1 + o_p(1), \quad \mathbf{y} = (y_1, \ldots, y_n) \overset{i.i.d.}{\sim} p_0. \tag{4.5}$$

It is common knowledge that estimating the marginal density $p$ in a mixture model is much easier than recovering the parameters. As we could not find a general result on posterior concentration for the marginal density $p$ on finite mixture models, we present here such a general result, including its proof.

To do so, we denote by $KL(p_1, p_2)$ the Kullback–Leibler divergence between the two densities $p_1$ and $p_2$, and $V(p_1, p_2) = \int p_1(y)[\log(p_1(y)/p_2(y))]^2 dy - KL(p_1, p_2)^2$. Consider a prior distribution $P$ on $(\eta, \theta, \psi) = (\eta_g, \theta_g, \psi, g \leq G)$ with positive and continuous density $p$ on the interior of $\mathcal{V}_G \times \Theta^G \times \Xi$, with $\mathcal{V}_G$ denoting the unit simplex of dimension $G$ and $\psi \in \Xi$. Assume also that there exist $e_1, \ldots, e_G \geq 0$ such that

$$p(\eta_1, \ldots, \eta_G) \geq c_0 \eta_1^{e_1 - 1} \times \ldots \times \eta_G^{e_G - 1}, \quad \forall (\eta_1, \ldots, \eta_G) \in \mathcal{V}_G. \tag{4.6}$$

**Theorem 4.1** *Let $p_0$ be a mixture density of the form (4.1), with $\Theta \subset \mathbb{R}^{r_1}$ and $\Xi \subset \mathbb{R}^{r_2}$,*

$$p_0(y) = \sum_{g=1}^{G} \eta_g^* f(y|\theta_g^*, \psi^*), \quad (\eta_1^*, \ldots, \eta_G^*) \in \mathcal{V}_G, \quad \theta_g^* \in \text{int}(\Theta), \quad \psi^* \in \text{int}(\Xi),$$

*where* $\text{int}(\cdot)$ *denotes the interior set, and let $P$ be a prior distribution satisfying the above conditions. Make the following assumptions:*

(a) *For all $c > 0$, there exist $H_1 > 0$ and $R_n \leq n^{H_1}$ such that*

    (i) $P(\Theta_n^c) \leq e^{-c \log n}$ *and* $P(\Xi_n^c) \leq e^{-c \log n}$, *with* $\Theta_n = \{\|\theta\| \leq R_n\} \cap \Theta$ *and* $\Xi_n = \{\|\theta\| \leq R_n\} \cap \Xi$;

    (ii) *there exist constants* $L, H_2 \geq 0$ *such that, for all* $\theta, \theta' \in \Theta_n$ *and* $\psi, \psi' \in \Xi_n$,

$$\|f(\cdot|\theta, \psi) - f(\cdot|\theta', \psi')\|_1 \leq L n^{H_2}(\|\theta - \theta'\| + \|\psi - \psi'\|).$$

(b) *There exists a constant* $H_3 \geq 0$ *such that*

$$\left\{ \sum_g |\eta_g - \eta_g^*| + \|\theta_g - \theta_g^*\| + \|\psi - \psi^*\| \leq n^{-H_3} \right\} \subset S_n,$$

*with*

$$S_n = \{(\theta, \eta, \psi); KL(p_0, p(\cdot|\theta, \eta, \psi)) + V(p_0, p(\cdot|\theta, \eta, \psi)) \leq 1/n\}.$$

*Then there exists $M > 0$ such that*

$$P\left( \|p_0 - p(\cdot|\theta, \eta, \psi)\|_1 \leq M \frac{\sqrt{\log n}}{\sqrt{n}} \,\middle|\, \mathbf{y} \right) = o_{p_0}(1).$$

*Proof* The proof is a simple adaptation of Theorem 5.1 of Ghosal et al. (2000), which itself is an application of their Theorem 2.1. Assumption (b) implies that condition (2.9) in Theorem 2.4 of Ghosal et al. (2000) is satisfied with every sequence $\epsilon_n^2 \geq \epsilon_0^2 n^{-1} \log n$, choosing $\epsilon_0$ large enough as soon as $(\theta^*, \psi^*, \eta^*)$ is in the interior of $\Theta^G \times \Xi \times \mathcal{V}_G$. If $p_0$ corresponds to parameters that are on the boundary (i.e. there exists $g \leq G$ such that $\eta_g^* = 0$), then the at most polynomial decrease of the prior on $\eta$ (see (4.6) above) also leads to condition (2.9).

Condition (a)(i) implies condition (2.3) of Theorem 2.1 of Ghosal et al. (2000) by choosing $c$ accordingly. To verify condition (2.2) of Theorem 2.1 of Ghosal et al. (2000), we use condition (a)(ii) which implies that

$$\begin{aligned}
D\left(\epsilon_n, \Theta_n \times \Xi_n \times \mathcal{V}_G, d(\cdot, \cdot)\right) &\leq D\left(\epsilon_n n^{-H_3}, \Theta_n \times \Xi_n \times \mathcal{V}_G, \|\cdot\|\right) \\
&\lesssim (n^{H_3} R_n \epsilon_n)^{kr_1 + k - 1 + r_2} \lesssim e^{(kr_1 + k - 1 + r_2)(H_2 + H_3)\log n + \log(1/\epsilon_n)} \\
&\lesssim e^{(kr_1 + k - 1 + r_2)(H_2 + H_3 + 1/2)\log n} \lesssim e^{n\epsilon_n^2},
\end{aligned}$$

when $\epsilon_n \asymp \sqrt{\log n / n}$.                                                 □

Although Theorem 4.1 is of interest in its own right, due to the fact that $p_0$ might be the object of interest, for instance in the context of prediction, it also opens up a number of important consequences on the asymptotic behaviour of the posterior distribution of the parameters $(\eta, \theta, \psi)$.

Theorem 4.1 is typically a starting point for obtaining more refined results, such as posterior concentration rates on other quantities or Bernstein–von Mises theorems. To recover the parameter $(\eta, \theta, \psi)$ from the marginal densities $p(y|\eta, \theta, \psi)$, one needs an identifiability result on the parameter. In the case of mixture models two types of non-identifiability can occur. The first is a (weak) non-identifiability on the labels of the components; see Chapter 1 and Section 4.2.1. A more severe form of non-identifiability can also occur in presence of overfitting. When the number of components is well specified, that is, when the true number of components satisfies $G^* = G$, and under regularity conditions on the emission densities $f(y|\theta_g, \psi)$, the mixture model is a regular model and, up to the label-switching issue, is identifiable. Then all the usual asymptotic results are valid: namely, the Bernstein–von Mises theorem on the parameters, the Laplace approximation of the marginal density of the observations and $1/\sqrt{n}$ convergence rates of Bayesian estimators such as posterior means (after post-processing for label switching). On the other hand, Theorem 4.1 does not require such strong regularity conditions and remains valid even for an overspecified number of components.

### 4.3.2 Recovering the parameters in the well-behaved case

When the number of components is well specified, as noted in Titterington et al. (1985), most mixtures of continuous distributions are identifiable, with the exception of mixtures of $\mathcal{U}[0, \theta]$, $\theta > 0$, and so are many mixtures of discrete distributions. These finite mixture models are typically regular under weak conditions and both the maximum likelihood estimator of the parameter and the posterior distribution are asymptotically normal with the usual means and variances. In this section we briefly recall the conditions that are required to obtain asymptotic normality of the posterior distribution in finite mixture models and we state the result. Consider model (4.1).

**C1** The function $(\theta_g, \psi) \to \log f(y|\theta_g, \psi)$ is twice continuously differentiable, with its second derivative satisfying

$$\mathbb{E}_{\eta^*, \theta^*, \psi^*} \left( \sup_{|\theta_g - \theta_g^*| \leq \delta} \sup_{|\psi - \psi^*| < \delta} \|D^2 \log f(y|\theta_g, \psi)\| \right) < +\infty, \quad \forall g \leq G,$$

$$\mathbb{E}_{\eta^*, \theta^*, \psi^*} \left( \|D \log f(y|\theta_g^*, \psi^*)\| \right) < +\infty, \quad \forall g \leq G.$$

**C2** The prior distribution has positive and continuous density at $(\eta^*, \theta^*, \psi^*)$.

**C3** The functions

$$f(y|\theta_g^*, \psi) - f(y|\theta_1^*, \psi), \quad 1 < g \leq G, \quad \nabla_{\theta_g} f(y|\theta_g^*, \psi), \nabla_\psi f(y|\theta_g^*, \psi), \quad 1 \leq g \leq G,$$

are linearly independent as functions of $y$.

Conditions **C1** and **C2** are the usual regularity conditions leading to a local quadratic approximation of the log likelihood around the true parameter. Condition **C3** is more specific to mixture models and ensures that the Fisher information matrix at the true parameter is positive definite. The functions $f(y|\theta_g^*, \psi) - f(y|\theta_1^*, \psi)$ appear in the derivatives of the marginal density $p(\cdot|\theta, \eta, \psi)$ with respect to the parameter $\eta$ and $\nabla_{\theta_g} f(y|\theta_g^*, \psi), \nabla_\psi f(y|\theta_g^*, \psi)$ appear when differentiating with respect to $\theta_g$ and $\psi$, $g \leq G$. They are satisfied for most exponential families, be they discrete or continuous.

Because of the lack of identifiability due to label switching, the posterior distribution on $(\eta, \theta, \psi)$ is multimodal with $G!$ identical groups (or almost identical if the prior is not symmetric), hence, to describe the shape of the asymptotic distribution of the parameters in a meaningful way, it is helpful to consider some specific permutations of the labels of the groups. There are many ways to fix the permutation. Given that within each mode the posterior distribution concentrates around the true parameter when $n$ is large enough, the modes are well separated and, given a mode, the permutation which minimizes the distance between $(\theta, \eta)$ and the mode projects the parameter $(\theta, \eta)$ onto the given mode. In Theorem 4.2 below we therefore consider the posterior distribution of the parameter obtained after such a permutation, which we denote by $(\mathfrak{s}^*(\theta), \mathfrak{s}^*(\eta))$.

Let $(\eta^*, \theta^*, \psi^*)$ be the true parameter associated with a given permutation. Denote the score vector by $\nabla \ell_n(\eta^*, \theta^*, \psi^*)$ and define $\hat{\eta}_n = \nabla_\eta \ell_n(\eta^*, \theta^*, \psi^*)/\sqrt{n}$, $\hat{\theta}_n =$

$\nabla_\theta \ell_n(\eta^*, \theta^*, \psi^*)/\sqrt{n}$ and $\hat{\psi}_n = \nabla_\psi \ell_n(\eta^*, \theta^*, \psi^*)/\sqrt{n}$. Note that $(\hat{\theta}_n, \hat{\eta}_n, \hat{\psi}_n)$ is asymptotically equivalent to the maximum likelihood estimator (for the same permutation of the labels defining $(\eta^*, \theta^*)$). We then have the following "Bernstein–von Mises" theorem.

**Theorem 4.2** *Assume that the true density function has the form (4.1) with distinct parameter values $\theta_g^*$, $g \le G$, and positive weights $\eta_g^*$, and that $(\eta^*, \theta^*, \psi^*)$ is an interior point of $\Theta \times \Xi \times \mathcal{V}_G$. Assume that the conclusion of Theorem 4.1 holds, together with conditions $C1$–$C3$. Then the Fisher information matrix $I(\eta^*, \theta^*, \psi^*)$ is positive definite and the posterior distribution of $\left( \sqrt{n}(\mathfrak{s}^*(\eta) - \hat{\eta}_n), \sqrt{n}(\mathfrak{s}^*(\theta) - \hat{\theta}_n), \sqrt{n}(\psi - \hat{\psi}_n) \right)$ converges in total variation to the normal distribution $\mathcal{N}(0, I(\eta^*, \theta^*, \psi^*)^{-1})$, where $(\mathfrak{s}^*(\eta), \mathfrak{s}^*(\theta))$ is the vector obtained after the permutation of the labels such that $\|\mathfrak{s}^*(\eta) - \hat{\eta}_n\| + \|\mathfrak{s}^*(\theta) - \hat{\theta}_n\|$ is the smallest among all possible permutations $\mathfrak{s}$ of the labels.*

Theorem 4.2 is a direct application of Theorem 10.1 of van der Vaart (1998) where condition **C1** implies differentiability in quadratic mean and **C3** positivity of the Fisher information matrix. Note that other asymptotic approximations of the posterior distribution have been proposed in the literature, under stronger conditions; in particular, Crawford (1994) has derived a Laplace expansion of the posterior expectation of positive functions of the parameter.

In practice, it is not always easy to assess *a priori* the number of components. If the number of components in the model is too small then the posterior distribution converges towards the corresponding Kullback–Leibler projection and the usual asymptotic properties for misspecified models can be applied. If the number of components is too large, that is, if the model is overspecified (or overfitted), then the model is not identifiable, nor is it regular. The non-identifiability has been detailed in Section 4.2.2.4. The irregularity of the model comes from the fact that for all $(\eta, \theta) \in \Theta^*$, which is the subset of parameter space leading to the same mixture distribution as $(\eta^*, \theta^*)$, the Fisher information matrix is degenerate. Hence in this case the results described in Theorem 4.2 do not apply and a new theory needs to be devised. This is presented in the following subsection.

### 4.3.3   Boundary parameters: overfitted mixtures

In overfitted mixtures the parameter is not identifiable and the maximum likelihood estimator at best converges to the set $\Omega^*$ of parameter values leading to the same mixing distribution $H^* = \sum_{g=1}^{G^*} \eta_g^* \delta_{(\theta_g^*)}$. Moreover, it has long been known that the minimax rate of estimation for the mixing distribution $H$, when $G^* < G$ or when $H^*$ is very close to a distribution with a number of components smaller than $G$ (i.e. living near the boundary of the parameter space), is much slower than $1/\sqrt{n}$. For instance, Chen (1995) showed that the minimax rate of convergence in the Wasserstein distance of the mixing distribution is bounded from above by $n^{-1/4}$, and some refinements have been obtained more recently, in particular in Ho & Nguyen (2016) where it is shown that the minimax rate of convergence deteriorates when the discrepancy between the true number of components and that of the model increases. Hence, even for the estimation of the mixing distribution $H$, the irregularity and non-identifiability associated with overspecifying the model ($G > G^*$) are problematic.

Interestingly, when looking at pointwise convergence, Rousseau & Mengersen (2011) showed that the shape of the prior distribution on the weight parameter $\eta$ was influential and had a regularizing effect on the posterior distribution. Then consistent convergence towards a single configuration of the parameter space $\Omega^*$ is achievable, with posterior concentration rate of order $n^{-1/2}$. We recall their result in this section.

Consider $H^* = \sum_{g=1}^{G^*} \eta_g^* \delta_{(\theta_g^*)}$ the true mixing distribution with $G^* < G$ and denote $f_0(\cdot) = \sum_{g=1}^{G^*} \eta_g^* f(\cdot|\theta_g^*)$. We first introduce some notation to better describe

$$\Omega^* = \left\{ (\eta, \theta) \in \mathcal{V}_G \times \Theta^G; \sum_{g=1}^{G} \eta_g \delta_{(\theta_g)} = H^* \right\}.$$

Following Liu & Shao (2004), let $\mathbf{t} = (t_g)_{g=0}^{G^*}$ with $0 = t_0 < t_1 < \ldots < t_{G^*} \le G$ be a partition of $\{1, \ldots, G\}$. For all $(\eta, \theta) \in \Omega^*$, there exists a $\mathbf{t}$ as defined above such that, up to a permutation of the labels,

$$\forall g = 1, \ldots, G^*, \quad \theta_{t_{g-1}+1} = \ldots = \theta_{t_g} = \theta_g^*, \quad \eta(g) = \sum_{j=t_{g-1}+1}^{t_g} \eta_j = \eta_g^*,$$

$$\eta_{t_{G^*}+1} = \ldots = \eta_G = 0.$$

The set $I_g = \{t_{g-1}+1, \ldots, t_g\}$ represents the cluster of $\{1, \ldots, G\}$ sharing the same parameter as $\theta_g^*$. We can then define the following reparameterization of $(\eta, \theta)$, up to a permutation of the labels:

$$\phi_{\mathbf{t}} = \left( (\theta_g)_{g=1}^{t_{G^*}}, (s_g)_{g=1}^{G^*-1}, (\eta_g)_{g=t_{G^*}+1}^{G} \right) \in \mathbb{R}^{dt_{G^*}+G^*+G-t_{G^*}-1}, \quad s_g = \eta(g) - \eta_g^*, \; g = 1, \ldots, G^*,$$

and

$$\psi_{\mathbf{t}} = \left( (q_g)_{g=1}^{t_{G^*}}, \theta_{t_{G^*}+1}, \ldots, \theta_G \right), \quad q_j = \frac{\eta_j}{\eta(g)}, \quad \text{when } j \in I_g.$$

Note that $f_0$ corresponds to

$$\phi_{\mathbf{t}}^* = (\theta_1^*, \ldots, \theta_1^*, \theta_2^*, \ldots, \theta_2^*, \ldots, \theta_{G^*}^*, \ldots, \theta_{G^*}^*, 0, \ldots, 0, \ldots, 0),$$

where $\theta_g^*$ is repeated $t_g - t_{g-1}$ times in the above vector, for any $\psi_{\mathbf{t}}$.

Then we parameterize $(\eta, \theta)$ as $(\phi_{\mathbf{t}}, \psi_{\mathbf{t}})$, so that $p_{\eta,\theta} = p_{(\phi_{\mathbf{t}}, \psi_{\mathbf{t}})}$, and we denote $p'_{(\phi_{\mathbf{t}}^*, \psi_{\mathbf{t}})}$ and $p''_{(\phi_{\mathbf{t}}^*, \psi_{\mathbf{t}})}$ the first and second derivatives of $p_{(\phi_{\mathbf{t}}, \psi_{\mathbf{t}})}$ with respect to $\phi_{\mathbf{t}}$ and computed at $(\eta^*, \theta^*) = (\phi_{\mathbf{t}}^*, \psi_{\mathbf{t}})$. Denote by $D^{(\ell)} f(\cdot|\theta)$ the array whose components are

$$\frac{\partial^\ell f(\cdot|\theta)}{\partial \theta_{i_1} \ldots \partial \theta_{i_\ell}}, \quad 2 \le \ell \le 3,$$

and by $\nabla f(\cdot|\theta)$ the gradient of $f(\cdot|\theta)$ with respect to $\theta$, and define, for all $g \le G^*$,

$$\underline{f}(\cdot|\theta_g^*, \delta) \le f(\cdot|\theta) \le \bar{f}(\cdot|\theta_g^*, \delta), \quad \forall |\theta - \theta_g^*| \le \delta.$$

We consider the following assumptions, in the spirit of conditions **C1**–**C3**.

**D1** $L_1$ *consistency.* There exists $\delta_n \le (\log n)^q/\sqrt{n}$, for some $q \ge 0$ such that

$$\lim_{M \to +\infty} \limsup_n E^* \left\{ P \left( \|f_0 - f_{\eta,\theta}\| \ge M\delta_n|\mathbf{y} \right) \right\} = 0.$$

**D2** *Regularity.* The model $\theta \in \Theta \to f(\cdot|\theta)$ is three times differentiable and regular in the sense that, for all $\theta \in \Theta$, the Fisher information matrix associated with the model $f_\theta$ is positive definite at $\theta$. For all $g \le G^*$, there exists $\delta > 0$ such that $\underline{f}(\cdot|\theta_g^*, \delta) = \underline{f}(\cdot|\theta_g^*)$ and $\bar{f}(\cdot|\theta_g^*, \delta) = \bar{f}(\cdot|\theta_g^*)$ exist and satisfy

$$\underline{f}(\cdot|\theta_g^*) \le f(\cdot|\theta) \le \bar{f}(\cdot|\theta_g^*), \quad \forall |\theta - \theta_g^*| \le \delta,$$

and

$$F_0\left(\frac{\bar{f}(\cdot|\theta_g^*)^3}{\underline{f}(\cdot|\theta_g^*)^3}\right) < +\infty, \quad F_0\left(\frac{\sup_{|\theta-\theta_g^*|\leq\delta}|\nabla f(\cdot|\theta)|^3}{\underline{f}(\cdot|\theta_g^*)^3}\right) < +\infty, \quad F_0\left(\frac{|\nabla f(\cdot|\theta_g^*)|^4}{f_0^4}\right) < +\infty,$$

$$F_0\left(\frac{\sup_{|\theta-\theta_g^*|\leq\delta}|D^2 f(\cdot|\theta_g^*)|^2}{\underline{f}(\cdot|\theta_g^*)^2}\right) < +\infty, \quad F_0\left(\frac{\sup_{|\theta-\theta_g^*|\leq\delta}|D^3 f(\cdot|\theta)|}{\underline{f}(\cdot|\theta_g^*)}\right) < +\infty.$$

Assume also that $\theta_g^* \in \text{int}(\Theta)$, the interior of $\Theta$, for all $g = 1, \ldots, G^*$.

**D3** *Integrability.* There exists $\Theta_0 \subset \Theta$ satisfying $\text{Leb}(\Theta_0) > 0$ (where $\text{Leb}(\cdot)$ stands for the Lebesgue integral), and for all $g \leq G^*$,

$$d(\theta_g^*, \Theta_0) = \inf_{\theta \in \Theta_0} |\theta - \theta_g^*| > 0,$$

such that, for all $\theta \in \Theta_0$,

$$F_0\left(\frac{f(\cdot|\theta)^4}{f_0^4}\right) < +\infty, \quad F_0\left(\frac{f(\cdot|\theta)^3}{\underline{f}(\cdot|\theta_g^*)^3}\right) < +\infty, \quad \forall g \leq G^*.$$

**D4** *Stronger identifiability.* For all $\mathbf{t}$ partitions of $\{1, \ldots, G\}$ as defined above, let $(\eta, \theta) \in \mathcal{V}_G \times \Theta^G$ and write $(\eta, \theta)$ as $(\phi_{\mathbf{t}}, \psi_{\mathbf{t}})$; then

$$(\phi_{\mathbf{t}} - \phi_{\mathbf{t}}^*)^\top p'_{\phi_{\mathbf{t}}^*, \psi_{\mathbf{t}}} + \frac{1}{2}(\phi_{\mathbf{t}} - \phi_{\mathbf{t}}^*)^\top p''_{\phi_{\mathbf{t}}^*, \psi_{\mathbf{t}}}(\phi_{\mathbf{t}} - \phi_{\mathbf{t}}^*) = 0 \Leftrightarrow$$

$$\forall g \leq G^*, s_g = 0 \quad \& \quad \forall j \in I_g \quad q_j(\theta_j - \theta_g^*) = 0, \quad \forall g \geq t_{G^*} + 1 \quad \eta_g = 0.$$

Assuming also that if $\theta \notin \{\theta_1, \ldots, \theta_G\}$ then for all functions $h_\theta$ which are linear combinations of derivatives of $f(\cdot|\theta)$ of order less than or equal to 2 with respect to $\theta$, and all functions $h_1$ which are also linear combinations of derivatives of the $f(\cdot|\theta_g)$, $g = 1, \ldots, G$, and its derivatives of order less than or equal to 2, we have $h_\theta + h_1 = 0$, if and only if $h_\theta = h_1 = 0$.

In Rousseau & Mengersen (2011) an extension to the non-compact case is presented which we do not repeat here.

**D5** *Prior.* The prior density with respect to the Lebesgue measure on $\Theta$ is continuous and positive, and the prior on $(\eta_1, \ldots, \eta_G)$ satisfies

$$\eta_1^{e_1-1} \cdots \eta_G^{e_G-1} \lesssim p(\eta) \lesssim \eta_1^{e_1-1} \cdots \eta_G^{e_G-1},$$

for some $0 < e_1, \ldots, e_G$.

As discussed in Rousseau & Mengersen (2011), condition **D4** is a stronger version of condition **C3**, since it also involves the second derivatives of the component's densities $f(\cdot|\theta)$. It is, however, satisfied in many common statistical models, including location and location–scale mixtures of Gaussian, Poisson mixtures, exponential and Student $t$ mixtures if the degrees of freedom vary in a compact subset of $[1, +\infty)$.

The following result shows that there is a phase transition behaviour of the posterior distribution depending on the positions of the hyperparameters $e_1, \ldots, e_G$ with respect to $r$, where $r$ is the dimension of $\theta_g$ (i.e. $\Theta \subset \mathbb{R}^r$).

**Theorem 4.3** *Let $\mathfrak{S}(G)$ be the set of permutations of $\{1, \ldots, G\}$, $e_0^{\max} = \max(e_g, g \leq G)$ and $e_0^{\min} = \min(e_g, g \leq G)$. Under assumptions **D1**–**D5**, the posterior distribution satisfies the following conditions.*

(a) *If $e_0^{\max} < r/2$ and $\rho = (rG^* + G^* - 1 + e_0^{\max}(G - G^*))/(r/2 - e_0^{\max})$, then*

$$\lim_{M \to +\infty} \limsup_n \mathrm{E}^* \left\{ \mathrm{P} \left[ \min_{\mathfrak{s} \in \mathfrak{S}(G)} \sum_{g=G^*+1}^{G} \eta_{\mathfrak{s}(g)} > Mn^{-1/2}(\log n)^{q(1+\rho)} \middle| \mathbf{y} \right] \right\} = 0.$$

(b) *If $e_0^{\min} > r/2$ and $\rho' = (rG^* + G^* - 1 + r(r - G^*)/2)/((e_0^{\min} - r/2)(G - G^*))$, then*

$$\lim_{\epsilon \to 0} \limsup_n \mathrm{E}^* \left\{ \mathrm{P} \left[ \min_{\mathfrak{s} \in \mathfrak{S}(G)} \sum_{g=G^*+1}^{G} \eta_{\mathfrak{s}(g)} < \epsilon(\log n)^{-q(1+\rho')} \middle| \mathbf{y} \right] \right\} = 0.$$

This theorem has interesting consequences. First, note that the case where $e_0^{\max} < r/2$ leads to a posterior distribution on the parameters which is much easier to interpret. Indeed, in that case a consequence of the proof and of the concentration result of Theorem 4.3 is that the posterior distribution on the parameters $(\mathfrak{s}^*(\theta_g), \mathfrak{s}^*(\eta_g), g \leq G^*)$ concentrates around $(\eta^*, \theta^*)$ at the rate $n^{-1/2}(\log n)^{q(1+\rho)}$ ($\mathfrak{s}^*(\cdot)$ is the permutation of the labels such that $\sum_{g \geq G^*+1} \eta_{\mathfrak{s}(g)} \leq Mn^{-1/2}(\log n)^{q(1+\rho)}$ and which minimizes the distance between $(\theta_{\mathfrak{s}(g)}, \eta_{\mathfrak{s}(g)}, g \leq G^*)$ and $(\theta^*, \eta^*)$ among all permutations $\mathfrak{s}$ of the $G^*$ *actual* components). Hence, in practice, the posterior distribution on the weights is a strong indicator of the components that can be considered as actual components and of those that can be considered as superfluous components. The second configuration is also interpretable, since when $e_0^{\min} > r/2$, the superfluous components are merging. This, however, leads to a less stable behaviour since there are multiple possible merging configurations. Note that there exists no Bernstein–von Mises type of result (or even asymptotic normality of the posterior distribution on the parameters of the non-negligible components), when $G > G^*$.

The posterior concentration result of Theorem 4.3 is pointwise in $H^*$. Since the minimax estimation rate of $H$ in Wasserstein distance over the sets of distributions with at most $G$ supporting points is strictly slower than $n^{-1/2}(\log n)^{q(1+\rho)}$, it is clear that Theorem 4.3 cannot be extended to a uniform result over the class of all mixing distributions with at most $G$ supporting points. Whether it can be made uniform over the class of mixing distributions having strictly less than $G$ supporting points remains, however, an open question. Choosing $e_0^{\max} < r/2$ acts as an informative prior and might lead to a suboptimal rate in a minimax sense, when the whole space of mixing distribution having at most $G$ support points is considered. In particular, it is to be expected that, if the mixing distribution $H = \sum_{g=1}^{G} \eta_g \delta_{(\theta_g)}$ is such that some of the weights $\eta_g$ are close to zero but non-zero, they might be underestimated with such a prior distribution.

This result has led to various extensions, both methodological and theoretical, as explained in Section 4.2.2.4. Interestingly, it cannot easily be extended to hidden Markov models. In the case of finite state space hidden Markov models, the hidden states $z_i$ are modelled as a Markov chain with transition matrix $\xi = (\xi_{jg})_{j,g \leq G}$. The same type of non-identifiability and irregularity occurs when the true latent variables live in a state space with $G^* < G$ points; however, in this case, there is an extra difficulty. This difficulty comes from the fact that in any neighbourhood of the true parameter $(\theta_j^*, j \leq G^*, \xi)$ there exist parameters $(\theta_j, j \leq G, \xi)$ for which $\xi$ is either non-ergodic or close to being so. The behaviour of the likelihood at such parameter values is not well understood and in a simple model with two components with known parameters (i.e. only $\xi$ is unknown) Gassiat & Keribin (2000) show that the likelihood computed at the maximum likelihood estimator has a pathological behaviour. For the posterior distribution to avoid the regions where $\xi$ is non-ergodic, one needs to consider priors which penalize for small values of $\sum_{g=1}^{G} \min_{j \leq G} \xi_{jg}$. This rules out priors favouring small values of $\xi_{jg}$ in a symmetrical way (in the labels), that is, priors where the rows of the transition matrix $\xi$ are independent Dirichlet distributions with parameters $e_{jg}$ which are either all equal to $e_0$ or equal to $e_0$ if $j \neq g$ and equal to $e_0'$ if $j = g$, with $e_0'$ small. The asymptotic behaviour of the posterior distribution for finite state space hidden Markov models has been studied in Gassiat & Rousseau (2014) and in van Havre et al. (2016).

### 4.3.4 Asymptotic behaviour of posterior estimates of the number of components

As presented in Chapter 1, the number of components can also be estimated directly using a hierarchical prior where one first selects $G$ from a prior $P_G$ and then, given $G$, considers a prior distribution on the parameters $(\eta, \theta, \psi) \in \mathcal{V}_G \times \Theta^G \times \Xi$:

$$G \sim P_G, \quad (\eta, \theta, \psi)|G \sim P_{|G}.$$

A Bayesian estimate of $G$ is then the posterior mode (i.e. $\hat{G} = \mathrm{argmax}_G p(G|\mathbf{y})$), where

$$p(G|\mathbf{y}) \propto p_G(G) p(\mathbf{y}|G), \quad p(\mathbf{y}|G) = \int_{\mathcal{V}_G \times \Theta^G} p(\mathbf{y}|\theta, \eta, \psi) dp_{|G}(\eta, \theta, \psi|G).$$

In the context of regular models, the logarithm of the marginal likelihood $p(\mathbf{y}|G)$ can be approximated using the Bayesian information criterion (BIC) formula and leads to consistent tests or model selection procedures; see also Chapter 7 for further material concerning the choice of $G$.

   In the context of mixtures, however, the models are irregular when they are overspecified. It has, however, been proved in Chambaz & Rousseau (2008) that marginal likelihood approaches can be used to consistently estimate the number of components. Two approaches are considered. The first is to consider the global posterior mode described above, and the second is to consider $\tilde{G}$ defined as

$$\tilde{G} = \min\{G; p(\mathbf{y}|G) \geq p(\mathbf{y}|G+1)\}.$$

Both approaches are consistent under the same type of assumptions as **D1–D2**. Using approaches borrowed from algebraic statistics, Yamazaki & Watanabe (2003) developed the equivalent of the BIC formula for some mixture models, with an implicit expression for the penalty term. Recently, an algorithm has been proposed in Drton & Plummer (2017) for computing this new BIC-type formula.

## 4.4 Concluding Remarks

This chapter has focused on the impact of prior distributions in the context of mixture models. Choosing a prior distribution complexified by the fact that improper priors typically lead to improper posteriors is a difficult task. There exists a vast literature on this subject in the context of weak informative priors, as described in Section 4.2, from conjugate priors to partial Jeffreys priors and data-dependent priors. Interestingly, if the model is well specified (i.e. the correct number of components is known), the usual asymptotic theory holds and the prior becomes non-influential in the definition of the posterior distribution. However, if the model is ill specified, then the prior remains asymptotically influential on the inference, as explained in Section 4.3.3. With a carefully chosen prior, a well-behaved posterior can still be obtained. The results presented here are all for finite mixture models. Infinite mixtures have also received a great amount of attention, both from a modelling or practical point of view (see Chapter 6) as well as from a theoretical point of view (see Ghosal & van der Vaart, 2017, and the literature therein).

# Bibliography

BERGER, J. O. (1985). *Statistical Decision Theory and Bayesian Analysis*. New York: Springer-Verlag, 2nd ed.

BERNARDO, J. M. & GIRÓN, F. J. (1988). A Bayesian analysis of simple mixture problems. In *Bayesian Statistics 3*, J. M. Bernardo, M. H. DeGroot, D. V. Lindley & A. F. M. Smith, eds. Oxford: Oxford University Press, pp. 67–78.

BERNARDO, J. M. & SMITH, A. F. M. (1994). *Bayesian Theory*. New York: John Wiley.

CELEUX, G., HURN, M. A. & ROBERT, C. (2000). Computational and inferential difficulties with mixtures posterior distribution. *Journal of the American Statistical Association* **95**, 957–979.

CHAMBAZ, A. & ROUSSEAU, J. (2008). Bounds for Bayesian order identification with application to mixtures. *Annals of Statistics* **36**, 938–962.

CHEN, J. (1995). Optimal rate of convergence for finite mixture models. *Annals of Statistics* **23**, 221–233.

CRAWFORD, S. (1994). An application of the Laplace method to finite mixture distributions. *Journal of the American Statistical Association* **89**, 259–267.

DIEBOLT, J. & ROBERT, C. P. (1994). Estimation of finite mixture distributions by Bayesian sampling. *Journal of the Royal Statistical Society, Series B* **56**, 363–375.

DRTON, M. & PLUMMER, M. (2017). A Bayesian information criterion for singular models. *Journal of the Royal Statistical Society, Series B* **79**, 323–380.

DURANTE, D., DUNSON, D. B. & VOGELSTEIN, J. T. (2017). Nonparametric Bayes modeling of populations of networks. *Journal of the American Statistical Association* **112**, 1516–1530.

FIGUEIREDO, M. A. T. & JAIN, A. K. (2002). Unsupervised learning of finite mixture models. *IEEE Transactions on Pattern Analysis and Machine Intelligence* **24**, 381–396.

FRÜHWIRTH-SCHNATTER, S. (2001). Markov chain Monte Carlo estimation of classical and dynamic switching and mixture models. *Journal of the American Statistical Association* **96**, 194–209.

FRÜHWIRTH-SCHNATTER, S. (2006). *Finite Mixture and Markov Switching Models*. New York: Springer-Verlag.

GASSIAT, E. & KERIBIN, C. (2000). The likelihood ratio test for the number of components in a mixture with Markov regime. *ESAIM: Probability and Statistics* **4**, 25–52.

GASSIAT, E. & ROUSSEAU, J. (2014). About the posterior distribution in hidden Markov models with unknown number of states. *Bernoulli* **20**, 2039–2075.

GEWEKE, J. (2007). Interpretation and inference in mixture models: Simple MCMC works. *Comput. Statist. Data Analysis* **51**, 3529–3550.

GHOSAL, S., GHOSH, J. K. & VAN DER VAART, A. W. (2000). Convergence rates of posterior distributions. *Annals of Statistics* **28**, 500–531.

GHOSAL, S. & VAN DER VAART, A. (2017). *Fundamentals of Nonparametric Bayesian Inference*, vol. 44 of *Cambridge Series in Statistical and Probabilistic Mathematics*. Cambridge: Cambridge University Press.

GRAZIAN, C. & ROBERT, C. P. (2018). Jeffreys priors for mixture estimation: properties and alternatives. *Computational Statistics and Data Analysis* **121**, 149–163.

GRUET, M., PHILIPPE, A. & ROBERT, C. (1999). MCMC control spreadsheets for exponential mixture estimation. *Journal of Computational and Graphical Statistics* **8**, 298–317.

HO, N. & NGUYEN, X. (2016). Convergence rates of parameter estimation for some weakly identifiable finite mixtures. *Annals of Statistics* **44**, 2726–2755.

JASRA, A., HOLMES, C. C. & STEPHENS, D. A. (2005). Markov chain Monte Carlo methods and the label switching problem in Bayesian mixture modeling. *Statistical Science* **20**, 50–67.

JEFFREYS, H. (1939). *Theory of Probability*. Oxford: Clarendon Press.

KAMARY, K., LEE, J. & ROBERT, C. P. (2018). Non-informative reparameterisations for location-scale mixtures. *Journal of Computational and Graphical Statistics* (to appear). doi:10.1080/10618600.2018.1438900.

LIU, X. & SHAO, Y. (2004). Asymptotics for likelihood ratio tests under loss of identifiability. *Annals of Statistics* **31**, 807–832.

MALSINER-WALLI, G., FRÜHWIRTH-SCHNATTER, S. & GRÜN, B. (2016). Model-based clustering based on sparse finite Gaussian mixtures. *Statistics and Computing* **26**, 303–324.

MALSINER-WALLI, G., FRÜHWIRTH-SCHNATTER, S. & GRÜN, B. (2017). Identifying mixtures of mixtures using Bayesian estimation. *Journal of Computational and Graphical Statistics* **26**, 285–295.

MARIN, J.-M. & ROBERT, C. (2008). Approximating the marginal likelihood in mixture models. *Bulletin of the Indian Chapter of ISBA* **V**, 2–7.

MENGERSEN, K. L. & ROBERT, C. (1996). Testing for mixtures: A Bayesian entropic approach (with discussion). In *Bayesian Statistics 5*, J. O. Berger, J. M. Bernardo, A. P. Dawid, D. V. Lindley & A. F. M. Smith, eds. Oxford University Press, Oxford, pp. 255–276.

PAPASTAMOULIS, P. (2013). label.switching: *Relabelling MCMC outputs of mixture models*. R package version 1.2.

PETRALIA, F., RAO, V. & DUNSON, D. B. (2012). Repulsive mixtures. In *Advances in Neural Information Processing Systems 25*, F. C. N. Pereira, C. J. C. Burges, L. Bottou & K. Q. Weinberger, eds. Red Hook, NY: Curran Associates, Inc., pp. 1889–1897.

RAFTERY, A. E. (1996). Hypothesis testing and model selection. In *Markov Chain Monte Carlo in Practice*, W. Gilks, D. Spiegelhalter & S. Richardson, eds. London: Chapman & Hall, pp. 163–188.

RICHARDSON, S. & GREEN, P. J. (1997). On Bayesian analysis of mixtures with an unknown number of components. *Journal of the Royal Statistical Society, Series B* **59**, 731–792.

ROBERT, C. & TITTERINGTON, M. (1998). Reparameterisation strategies for hidden Markov models and Bayesian approaches to maximum likelihood estimation. *Statistics and Computing* **8**, 145–158.

ROBERT, C. P. (2007). *The Bayesian Choice: From Decision-Theoretic Foundations to Computational Implementation*. New York: Springer.

RODRIGUEZ, C. & WALKER, S. (2014). Label switching in Bayesian mixture models: Deterministic relabeling strategies. *Journal of Computational and Graphical Statistics* **21**, 23–45.

ROEDER, K. & WASSERMAN, L. (1997). Practical Bayesian density estimation using mixtures of normals. *Journal of the American Statistical Association* **92**, 894–902.

ROSSI, P. & McCULLOCH, R. (2010). Bayesm: Bayesian inference for marketing/micro-econometrics. *R package version* **2**, 357–365.

ROUSSEAU, J. & MENGERSEN, K. (2011). Asymptotic behaviour of the posterior distribution in overfitted mixture models. *Journal of the Royal Statistical Society, Series B* **73**, 689–710.

RUBIO, F. & STEEL, M. (2014). Inference in two-piece location-scale models with Jeffreys priors. *Bayesian Analysis* **9**, 1–22.

STEPHENS, M. (2000). Dealing with label switching in mixture models. *Journal of the Royal Statistical Society, Series B* **62**, 795–809.

TITTERINGTON, D., SMITH, A. F. M. & MAKOV, U. (1985). *Statistical Analysis of Finite Mixture Distributions*. New York: John Wiley.

VAN DER VAART, A. (1998). *Asymptotic Statistics*. Cambridge: Cambridge University Press.

VAN HAVRE, Z., ROUSSEAU, J., WHITE, N. & MENGERSEN, K. L. (2016). Overfitting hidden Markov models with an unknown number of states. Preprint, arXiv:1602.02466.

VAN HAVRE, Z., WHITE, N., ROUSSEAU, J. & MENGERSEN, K. (2015). Overfitting Bayesian mixture models with an unknown number of components. *PLoS ONE* **10**, e0131738.

WASSERMAN, L. (2000). Asymptotic inference for mixture models using data dependent priors. *Journal of the Royal Statistical Society, Series B* **62**, 159–180.

YAMAZAKI, K. & WATANABE, S. (2003). Singularities in mixture models and upper bounds of stochastic complexity. *International Journal of Neural Networks* **16**, 1029–1038.

YANG, Y. & DUNSON, D. (2014). Minimax optimal Bayesian aggregation. Preprint, arXiv:1403.1345.

# 5

# Computational Solutions for Bayesian Inference in Mixture Models

**Gilles Celeux, Kaniav Kamary, Gertraud Malsiner-Walli, Jean-Michel Marin, and Christian P. Robert**

*INRIA Saclay, France; INRIA Saclay, France; Vienna University of Economics and Business, Austria; Université de Montpellier, France; Université Paris-Dauphine, France and University of Warwick, UK*

## CONTENTS

This chapter surveys the most standard Monte Carlo methods available for simulating from a posterior distribution associated with a mixture and conducts some experiments about the robustness of the Gibbs sampler in high-dimensional Gaussian settings.

## 5.1 Introduction

It may sound paradoxical that a statistical model that is written as a sum of Gaussian densities poses a significant computational challenge to Bayesian (as well as non-Bayesian) statisticians, but this is nonetheless the case. Estimating (and more globally running a Bayesian analysis on) the parameters of a mixture model has long been deemed impossible, except for the most basic cases, as illustrated by the approximations found in the literature of the 1980s (Smith & Makov, 1978; Titterington et al., 1985; Bernardo & Girón, 1988;

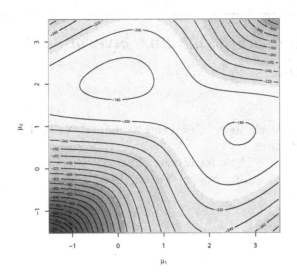

**FIGURE 5.1**
Posterior density surface for 100 observations from the Gaussian mixture $\frac{3}{10}\mathcal{N}(\mu_1, 1) +$ $\frac{7}{10}\mathcal{N}(\mu_2, 1)$, clearly identifying a modal region near the true value of the parameter $(\mu_1, \mu_2) = (0, \frac{5}{2})$ used to simulate the data, plus a secondary mode associated with the inversion of the data allocation to the two components. The figure was produced by computing the likelihood and prior functions over a $100 \times 100$ grid over $(-1, 3) \times (-1, 3)$. The level sets in the image are expressed on a logarithmic scale.

Crawford et al., 1992). Before the introduction of Markov chain Monte Carlo (MCMC) methods to the Bayesian community, there was no satisfactory way to bypass this problem, and it is no surprise that mixture models were among the first applications of Gibbs sampling to appear (Diebolt & Robert, 1990b; Gelman & King, 1990; West, 1992). The reason for this computational challenge is the combinatorically explosive nature of the development of the likelihood function, which contains $G^n$ terms when using $G$ components over $n$ observations. As detailed in other chapters, such as Chapter 1, the natural interpretation of the likelihood is to contemplate all possible partitions of the $n$-observation sample $\mathbf{y} = (y_1, \ldots, y_n)$. While the likelihood function can be computed in $\mathcal{O}(G \times n)$ time, being expressed as

$$p(\mathbf{y}|\theta) = \prod_{i=1}^{n} \sum_{g=1}^{G} \eta_g f(y_i|\theta_g),$$

where $\theta = (\theta_1, \ldots, \theta_G, \eta_1, \ldots, \eta_G)$, there is no structure in this function that allows for its exploration in an efficient way. Indeed, as demonstrated for instance by Chapter 2, the variations of this function are not readily available and require completion steps as in the EM algorithm. Given a specific value of $\theta$, one can compute $p(\mathbf{y}|\theta)$, but this numerical value does not provide useful information on the shape of the likelihood in a neighbourhood of $\theta$. The value of the gradient also is available in $\mathcal{O}(G \times n)$ time, but does not help much in this regard. (Several formulations of the Fisher information matrix, e.g. for Gaussian mixtures through special functions are available; see, for example, Behboodian, 1972, and Cappé et al., 2004.)

Computational advances have thus been essential to Bayesian inference on mixtures, while this problem has retrospectively fed new advances in Bayesian computational methodology.[1] In Section 5.2 we cover some of the proposed solutions, from the original data augmentation of Tanner & Wong (1987) that pre-dated the Gibbs sampler of Gelman & King (1990) and Diebolt & Robert (1990b) to specially designed algorithms, to the subtleties of label switching (Stephens, 2000).

As stressed by simulation experiments in Section 5.4, there nonetheless remain major difficulties in running Bayesian inference on mixtures of moderate to large dimensions. First of all, among all the algorithms that will be reviewed only Gibbs sampling seems to scale to high dimensions. Second, the impact of the prior distribution remains noticeable for sample sizes that would seem high enough in most settings, while larger sample sizes see the occurrence of extremely peaked posterior distributions that are a massive challenge for exploratory algorithms such as MCMC methods. Section 5.5 is specifically devoted to Gibbs sampling for high-dimensional Gaussian mixtures and a new prior distribution is introduced that seems to scale appropriately to high dimensions.

## 5.2 Algorithms for Posterior Sampling

### 5.2.1 A computational problem? Which computational problem?

When considering a mixture model from a Bayesian perspective (see, for example, Chapter 4), the associated posterior distribution

$$p(\theta|\mathbf{y}) \propto p(\theta) \prod_{i=1}^{n} \sum_{g=1}^{G} \eta_g f(y_i|\theta_g),$$

based on the prior distribution $p(\theta)$, is available in closed form, up to the normalizing constant, because the number of terms to compute is of order $\mathcal{O}(G \times n)$ in dimension 1 and of order $\mathcal{O}(G \times n \times d^2)$ in dimension $d$. This means that two different values of the parameter can be compared through their (non-normalized) posterior values. See, for instance, Figure 5.1 which displays the posterior density surface in the case of the univariate Gaussian mixture

$$\frac{3}{10} \mathcal{N}(\mu_1, 1) + \frac{7}{10} \mathcal{N}(\mu_2, 1),$$

clearly identifying a modal region near the true value of the parameter $(\mu_1, \mu_2) = \left(0, \frac{5}{2}\right)$ actually used to simulate the data. However, this does not mean that a probabilistic interpretation of $p(\theta|\mathbf{y})$ is immediately manageable: deriving posterior means, posterior variances, or simply identifying regions of high posterior density value remains a major difficulty when considering only this function. Since the dimension of the parameter space grows quite rapidly with $G$ and $d$, being for instance $3G - 1$ for a unidimensional Gaussian mixture against $G - 1 + dG + (d(d + 1)/2)G$ for a $d$-dimensional Gaussian mixture,[2] numerical integration cannot be considered as a viable alternative. Laplace approximations (see, for example, Rue et al., 2009) are incompatible with the multimodal nature of the posterior distribution. The only practical solution is thus to produce a simulation technique that approximates outcomes from the posterior. In principle, since the posterior density can be

---

[1]To some extent, the same is true for the pair consisting of maximum likelihood estimation and the EM algorithm, as discussed in Chapter 2.

[2]This $\mathcal{O}(d^2)$ magnitude of the parameter space explains why we *in fine* deem Bayesian inference for generic mixtures in large dimensions to present quite an challenge.

computed up to a constant, MCMC algorithms should operate smoothly in this setting. As we will see in this chapter, this is not always the case.

## 5.2.2 Gibbs sampling

Prior to 1989 – or, more precisely, prior to the publication by Tanner & Wong (1987) of their data augmentation paper, which can be considered as a precursor to Gelfand and Smith's (1990) Gibbs sampling paper – there was no manageable way to handle mixtures from a Bayesian perspective. As an illustration on a univariate Gaussian mixture with two components, Diebolt & Robert (1990a) studied a "grey code" implementation[3] for exploring the collection of all partitions of a sample of size $n$ into two groups and managed to reach $n = 30$ within a week of computation (in 1990s standards).

Consider, therefore, a mixture of $G$ components

$$\sum_{g=1}^{G} \eta_g \, f(y|\theta_g), \quad y \in \mathbb{R}^d, \tag{5.1}$$

where, for simplicity's sake, $f(\cdot|\theta)$ belongs to an exponential family

$$f(y|\theta) = h(y) \, \exp\{\theta \cdot y - \psi(\theta)\}$$

over the set $\mathbb{R}^d$, and where $(\theta_1, \ldots, \theta_G)$ is distributed from a product of conjugate priors

$$p(\theta_g|\alpha_g, \lambda_g) \propto \exp\{\lambda_g(\theta_g \cdot \alpha_g - \psi(\theta_g))\},$$

with hyperparameters $\lambda_g > 0$ and $\alpha_g \in \mathbb{R}^d$ $(g = 1, \ldots, G)$, while $(\eta_1, \ldots, \eta_G)$ follows the usual Dirichlet prior,

$$(\eta_1, \ldots, \eta_G) \sim \mathcal{D}(e_1, \ldots, e_G).$$

These prior choices are only made for convenience' sake, with hyperparameters requiring inputs from the modeller. Most obviously, alternative priors can be proposed at the cost of increased computational complexity. Given a sample $(y_1, \ldots, y_n)$ from (5.1), then as already explained in Chapter 1, we can associate to every observation an indicator random variable $z_i \in \{1, \ldots, G\}$ that indicates which component of the mixture is associated with $y_i$, that is, which term in the mixture was used to generate $y_i$. The demarginalization (or *completion*) of model (5.1) is then

$$z_i \sim \mathcal{M}(1, \eta_1, \ldots, \eta_G), \quad y_i|z_i \sim f(y_i|\theta_{z_i}).$$

Thus, considering $x_i = (y_i, z_i)$ (instead of $y_i$) entirely eliminates the mixture structure from the model since the likelihood of the completed model (the so-called complete-data likelihood function) is given by

$$
\begin{aligned}
L_c(\theta|x_1, \ldots, x_n) &\propto \prod_{i=1}^{n} \eta_{z_i} \, f(y_i|\theta_{z_i}) \\
&= \prod_{g=1}^{G} \prod_{i; z_i = g} \eta_g \, f(y_i|\theta_g).
\end{aligned}
$$

This latent structure is also exploited in the original implementation of the EM algorithm, as discussed in Chapter 2. Both steps of the Gibbs sampler (Robert & Casella, 2004) are then

---

[3]That is, an optimized algorithm that minimizes computing costs.

**Algorithm 5.1** Mixture posterior simulation

Set hyperparameters $(\alpha_g, e_g, \lambda_g)_g$. Repeat the following steps $T$ times, after a suitable burn-in phase:

1. Simulate $z_i$ $(i = 1, \ldots, n)$ from

$$\mathrm{P}(z_i = g | \theta_g, \eta_g, y_i) \propto \eta_g \, f(y_i | \theta_g) \quad (g = 1, \ldots, G),$$

   and compute the statistics

$$n_g = \sum_{i=1}^{n} \mathbb{I}(z_i = g), \quad n_g \overline{y}_g = \sum_{i=1}^{n} \mathbb{I}(z_i = g) y_i .$$

2. Generate $(g = 1, \ldots, G)$

$$\theta_g | \mathbf{z}, \mathbf{y} \sim p \left( \theta_g \,\middle|\, \frac{\lambda_g \alpha_g + n_g \overline{y}_g}{\lambda_g + n_g}, \lambda_g + n_g \right),$$

$$(\eta_1, \ldots, \eta_G) | \mathbf{z}, \mathbf{y} \sim \mathcal{D}_G(e_1 + n_1, \ldots, e_G + n_G) .$$

provided in Algorithm 5.1, with a straightforward simulation of all components' indices in parallel in step 1 and a simulation of the parameters of the mixture exploiting the conjugate nature of the prior against the complete-data likelihood function in step 2. Some implementations of this algorithm for various distributions from the exponential family can be found in the reference book of Frühwirth-Schnatter (2006).

*Illustration*

As an illustration, consider the setting of a univariate Gaussian mixture with two components with equal and known variance $\sigma^2$ and fixed weights $(\eta, 1 - \eta)$:

$$\eta \mathcal{N}(\mu_1, \sigma^2) + (1 - \eta) \mathcal{N}(\mu_2, \sigma^2) . \tag{5.2}$$

The only parameters of the model are thus $\mu_1$ and $\mu_2$. We assume in addition a normal $\mathcal{N}(0, 10\sigma^2)$ prior distribution on both means $\mu_1$ and $\mu_2$. Generating directly i.i.d. samples of $(\mu_1, \mu_2)$ distributed according to the posterior distribution associated with an observed sample $\mathbf{y} = (y_1, \ldots, y_n)$ from (5.2) quickly become impossible, as discussed for instance in Diebolt & Robert (1994) and Celeux et al. (2000), because of a combinatoric explosion in the number of calculations, which grow as $\mathcal{O}(2^n)$. The posterior is indeed solely interpretable as a well-defined mixture of standard distributions that involves that number of components.

As explained above, a natural completion of $(\mu_1, \mu_2)$ is to introduce the (unobserved) component indicators $z_i$ of the observations $y_i$, in a similar way as for the EM algorithm, namely,

$$\mathrm{P}(z_i = 1) = 1 - \mathrm{P}(z_i = 2) = \eta \quad \text{and} \quad y_i | z_i = g \sim \mathcal{N}(\mu_g, \sigma^2).$$

The completed distribution with $\mathbf{z} = (z_1, \ldots, z_n)$ is thus

$$p(\mu_1, \mu_2, \mathbf{z} | \mathbf{y}) \propto \exp\{-(\mu_1^2 + \mu_2^2)/20\sigma^2\} \prod_{i; z_i = 1} \eta \exp\{-(y_i - \mu_1)^2/2\sigma^2\}$$

$$\times \prod_{i; z_i = 2} (1 - \eta) \exp\{-(y_i - \mu_2)^2/2\sigma^2\} .$$

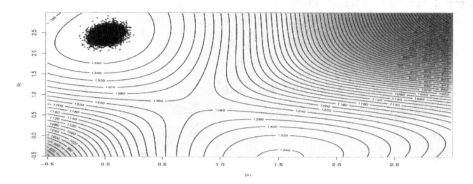

**FIGURE 5.2**
Gibbs sample of 5000 points for the mixture posterior against the posterior surface. The
level sets in the image are expressed on a logarithmic scale. (Reproduced with permission
from Robert & Casella, 2004.)

Since $\mu_1$ and $\mu_2$ are independent, given $(\mathbf{z}, \mathbf{y})$, the conditional distributions are $(g = 1, 2)$

$$\mu_g|\mathbf{y} \sim \mathcal{N}\left(\sum_{i;z_i=g} y_i / (0.1 + n_g), \sigma^2 / (0.1 + n_g)\right),$$

where $n_g$ denotes the number of $z_i$ equal to $g$ and 0.1=1/10 represents the prior precision.
Similarly, the conditional distribution of $\mathbf{z}$ given $(\mu_1, \mu_2)$ is a product of binomials, with

$$P(z_i = 1|y_i, \mu_1, \mu_2)$$
$$= \frac{\eta \exp\{-(y_i - \mu_1)^2/2\sigma^2\}}{\eta \exp\{-(y_i - \mu_1)^2/2\sigma^2\} + (1 - \eta) \exp\{-(y_i - \mu_2)^2/2\sigma^2\}}.$$

Figure 5.2 illustrates the behaviour of the Gibbs sampler in that setting, with a simulated
data set of 5000 points from the $0.7\mathcal{N}(0, 1) + 0.3\mathcal{N}(2.5, 1)$ distribution. The representation of
the MCMC sample after 15,000 iterations is quite in agreement with the posterior surface,
represented via a grid on the $(\mu_1, \mu_2)$ space and some contours; while it may appear to be
too concentrated around one mode, the second mode represented on this graph is much
lower since there is a difference of at least 50 in log-posterior units. However, the Gibbs
sampler may also fail to converge, as described in Diebolt & Robert (1994) and illustrated
in Figure 5.3. When initialized at the secondary mode of the likelihood, the magnitude of
the moves around this mode may be too limited to allow for exploration of further modes
(in a realistic number of iterations).

*Label switching*

Unsurprisingly, given the strong entropy of the local modes demonstrated in the illustrative example, Gibbs sampling performs very poorly in terms of label switching (already
discussed in Chapter 1) in that it rarely switches between equivalent modes of the posterior
distribution. One (if not the only) reason for this behaviour is that, due to the allocation of
the observations to the various components, that is, by completing the unknown parameters
with the unknown (and possibly artificial) latent variables, the Gibbs sampler faces enormous difficulties in switching between equivalent modes, because this amounts to changing
the values of most latent variables to a permuted version all at once. It can actually be

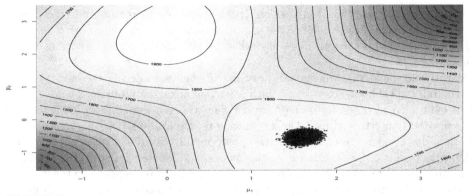

**FIGURE 5.3**
Gibbs sample for the two-mean mixture model, when initialized close to the second and lower mode, for true values $\mu_1 = 0$ and $\mu_2 = 2.5$, over the log likelihood surface. (Reproduced with permission from Robert & Casella, 2004.)

shown that the probability of seeing a switch goes down to zero as the sample size $n$ goes to infinity. This difficulty increases with dimension, in the sense that the likelihood function increasingly peaks with the dimension.

Some (e.g. Geweke, 2007) would argue the Gibbs sampler does very well by naturally selecting a mode and sticking to it. This certainly favours estimating the parameters of the different components. However, there is no guarantee that the Gibbs sampler will remain in the same mode over iterations.

The problem can alleviated to a certain degree by enhancing the Gibbs sampler (or any other simulation-based technique) to switch between equivalent modes. A simple, but efficient, solution to obtain a sampler that explores all symmetric modes of the posterior distribution is to enforce balanced label switching by concluding each MCMC draw by a random permutation of the labels. Let $\mathfrak{S}(G)$ denote the set of $G!$ permutations of $\{1, \ldots, G\}$. Assume that $(\eta_1, \ldots, \eta_G)$ follows the symmetric Dirichlet distribution $\mathcal{D}_G(e_0)$ (which corresponds to the $\mathcal{D}(e_0, \ldots, e_0)$ distribution and is invariant to permuting the labels by definition) and assume that also the prior on $\theta_g$ is invariant in this respect, that is, $p(\theta_{\mathfrak{s}(1)}, \ldots, \theta_{\mathfrak{s}(G)}) = p(\theta_1, \ldots, \theta_G)$ for all permutations $\mathfrak{s} \in \mathfrak{S}(G)$. Then, for any given posterior draw $\theta = (\eta_1, \ldots, \eta_G, \theta_1, \ldots, \theta_G)$, jumping between the equivalent modes of the posterior distribution can be achieved by defining the permuted draw $\mathfrak{s}(\theta) := (\eta_{\mathfrak{s}(1)}, \ldots, \eta_{\mathfrak{s}(G)}, \theta_{\mathfrak{s}(1)}, \ldots, \theta_{\mathfrak{s}(G)})$ for some permutation $\mathfrak{s} \in \mathfrak{S}(G)$.

This idea underlies the random permutation sampler introduced by Frühwirth-Schnatter (2001), where each of the $T$ sweeps of Algorithm 5.1 is concluded by such a permutation of the labels, based on randomly selecting one of the $G!$ permutations $\mathfrak{s} \in \mathfrak{S}(G)$; see also Frühwirth-Schnatter (2006, Section 3.5.6). Admittedly, this method works well only, if $G! \ll T$, as the expected number of draws from each modal region is equal to $T/G!$. Geweke (2007) suggests considering *all* of the $G!$ permutations in $\mathfrak{S}(G)$ for each of the $T$ posterior draws, leading to a completely balanced sample of $T \times G!$ draws; however, the resulting sample size can be enormous, if $G$ is large.

As argued in Chapter 1, the difficulty in exploring all modes of the posterior distribution in the parameter space is not a primary concern provided the space of mixture distributions (that are impervious to label-switching) is correctly explored. Since this is a space that is much more complex than the Euclidean parameter space, checking proper convergence is an issue. Once again, this difficulty is more acute in larger dimensions and it is compounded by the fact that secondary modes of the posterior become so "sticky" that a standard Gibbs sampler cannot escape their (fatal) attraction. It may therefore be reasonably argued, as in Celeux et al. (2000), that off-the-shelf Gibbs sampling does not necessarily work for mixture estimation, which would not be an issue in itself were alternative generic solutions readily available!

Label switching, however, still matters very much when considering the statistical evidence (or marginal likelihood) associated with a particular mixture model; see also Chapter 7. This evidence can be easily derived from Bayes' theorem by Chib's (1995) method, which can also be reinterpreted as a Dickey–Savage representation (Dickey, 1968), except that the Rao–Blackwell representation of the posterior distribution of the parameters is highly dependent on a proper mixing over the modes. As detailed in Neal (1999) and expanded in Frühwirth-Schnatter (2004), a perfect symmetry must be imposed on the posterior sample for the method to be numerically correct. In the most usual setting when this perfect symmetry fails, it must be imposed in the estimate, as proposed in Berkhof et al. (2003) and Lee & Robert (2016); see also Section 7.2.3 of Chapter 7 for more details.

### 5.2.3    Metropolis–Hastings schemes

As detailed at the beginning of this chapter, computing the likelihood function at a given parameter value $\theta$ is not a computational challenge provided (i) the component densities are available in closed-form[4] and (ii) the sample size remains manageable, for example, fits within a single computer memory. This property means that an arbitrary proposal can be used to devise a Metropolis–Hastings algorithm (Robert & Casella, 2004; Lee et al., 2008) associated with a mixture model, from a simple random walk to more elaborate solutions such as Langevin and Hamiltonian Monte Carlo.

The difficulty with this Metropolis–Hastings approach is to figure out an efficient way of implementing the simulating principle, which does not provide guidance on the choice of the proposal distribution. Parameters are set in different spaces, from the $G$-dimensional simplex of $\mathbb{R}^G$ to real vector spaces. A random walk is thus delicate to calibrate in such a context and it is often preferable to settle for a Gibbs sampler that updates one group of parameters at a time, for instance the weights, then the variances, then the means in the case of a location–scale mixture. This solution, however, shares some negative features with the original Gibbs sampler in the sense that it may prove hard to explore the entire parameter space (even without mentioning the label switching problem). Still, the implementation of the unidimensional location-parameterization of Kamary et al. (2018) relies on this blockwise version and manages to handle a reasonably large number of components, if not a larger number of dimensions. Indeed, when the dimension $d$ of the observables increases, manipulating $d \times d$ matrices gets particularly cumbersome, and we know of no generic solution to devise an automatic random walk Metropolis–Hastings approach in this case.

Among the strategies proposed to increase the efficiency of a Metropolis–Hastings algorithm in dimension one, let us single out the following:

1. Deriving independent proposals based on sequential schemes starting, for instance, from maximum likelihood and other classical estimates, since those are usually fast to derive, and followed by mixture proposals based on subsamples, at least in low-dimensional models.

2. An overparameterization of the weights $\eta_g$ defined as $\eta_g = \alpha_g / \sum_j \alpha_j$, with a natural extension of the Dirichlet prior $\mathcal{D}(e_0, \dots, e_0)$ into a product of $d$ gamma priors on the $\alpha_g$, where $\alpha_g \sim \mathcal{G}(e_0, 1)$. This representation avoids restricted simulations over the simplex

---

[4]Take, for example, a mixture of $\alpha$-stable distributions as a counter-example.

of $\mathbb{R}^G$, and adding an extra parameter means the corresponding Markov chain mixes better.

3. The inclusion of the original Gibbs steps between random exploration proposals, in order to thoroughly survey the neighbourhood of the current value. Gibbs sampling indeed shares to some extent the same property as the EM algorithm of shooting rapidly towards a local mode when started at random. In this respect, it eliminates the random walk poor behaviour of a standard Metropolis–Hastings algorithm. Cumulating this efficient if myopic exploration of the Gibbs sampler with the potential for mode jumping associated with the Metropolis–Hastings algorithm may produce the best of two worlds, in the sense that mixture proposals often achieve higher performance than both of their components (Tierney, 1994).

4. Other mixings of different types of proposals, using for instance reparameterization, overparameterization, or underparameterization. One important example is given by Rousseau & Mengersen (2011). In this paper, already discussed in Chapters 1 and 4, the authors consider mixtures with "too many" components and demonstrate that a Bayesian analysis of such overfitted models manages to eliminate the superfluous components. While this is an asymptotic result and while it imposes some constraints on the choice of the prior distribution on the weights, it nonetheless produces a theoretical warranty that working with more components than needed is not ultimately damaging to inference, while allowing in most situations for a better mixing behaviour of the MCMC sampler.

5. Deriving nonparametric estimates based, for instance, on Dirichlet process mixtures returns a form of clustering that can be exploited to build a componentwise proposal.

6. Approximate solutions such as nested sampling (see below), variational Bayes (Zobay, 2014), or expectation–propagation (EP; Minka, 2001; Titterington, 2011) may lead to independent proposals that contribute to a larger degree of mobility over the posterior surface.

7. Further sequential, tempering, and reversible jump solutions as discussed below.

At this stage, while the above has exhibited a medley of potential fixes to the Metropolis–Hasting woes, it remains urgent to warn the reader that no generic implementation is to be found so far (in the sense of generic software able to handle a wide enough array of cases). The calibration stage of those solutions remains a challenging issue that hinders and in some cases prevents an MCMC resolution of the computational problem. This is almost invariably the case when the dimension of the model gets into double digits; see Section 5.5.

## 5.2.4 Reversible jump MCMC

It may seem inappropriate to include a section or even a paragraph on reversible jump MCMC (Green, 1995) in this chapter since we are not directly concerned with estimating the number of components. However, this approach to $G$ variable or unknown environments is sometimes advanced as a possible means to explore better the parameter space by creating passageways through spaces of larger (and possibly smaller) dimensions and numbers of components. As discussed in Chopin & Robert (2010), this strategy is similar to bridge sampling. Once again, calibration of the method remains a major endeavour (Richardson & Green, 1997), especially in multidimensional settings, and we thus abstain from describing this solution any further.

### 5.2.5　Sequential Monte Carlo

Sequential Monte Carlo methods (see Del Moral et al., 2006) approach posterior simulation by a sequence of approximations, each both closer to the distribution of interest and borrowing strength from the previous approximation. They therefore apply even in settings where the data are static and entirely available from the start. They also go under the names of particle systems and particle filters.

Without getting into a full description of the way particle systems operate, let us recall here that this is a particular type of iterated importance sampling where, at each iteration $t$ of the procedure, a weighted sample $(\theta_{1t}, \ldots, \theta_{Nt})$ of size $N$ is produced, with weights $\omega_{it}$ targeting a distribution $\pi_t$. The temporal and temporary target $\pi_t$ may well be supported by a space other than the support of the original target $\pi$. For instance, in Everitt et al. (2016), the $\pi_t$ are the posterior distributions of mixtures with a smaller number of components, while in Chopin (2002) they are posterior distributions of mixtures with a smaller number of observations.[5] The sample or *particle system* at time $t$ is instrumental in designing the importance proposal for iteration $t+1$, using for instance MCMC-like proposals for simulating new values. When the number of iterations is large enough and the temporary targets $\pi_t$ are too different, the importance weights necessarily deteriorate down to zero (by basic martingale theory) and particle systems include optional resampling steps to select particles at random based on the largest weights, along with a linear increase in the number of particles as $t$ grows. The construction of the sequence of temporary targets $\pi_t$ is open and intended to facilitate the exploration of intricate and highly variable distributions, although its calibration is delicate and may jeopardize convergence.

In the particular setting of Bayesian inference, and in the case of mixtures, a natural sequence can be associated with subsample posteriors, that is, posteriors constructed with only a fraction of the original data, as proposed for instance in Chopin (2002, 2004). The starting target $\pi_0$ may for instance correspond to the true prior or to a posterior with a minimal sample size (Robert, 2007). A more generic solution is to replace the likelihood with small powers of the likelihood in order to flatten out the posterior and hence facilitate the exploration of the parameter space. A common version of the proposal is then to use a random walk, which can be calibrated in terms of the parameterization of choice and of the scale based on the previous particle system. The rate of increase of the powers of the likelihood can also be calibrated in terms of the degeneracy rate in the importance weights. This setting is studied by Everitt et al. (2016) who point out the many connections with reversible jump MCMC.

### 5.2.6　Nested sampling

While nested sampling is a latecomer to the analysis of mixtures (Skilling, 2004), one of the first examples of the use of this evidence estimation technique is a mixture example. We cannot recall the basics and background of this technique here and simply remind the reader that it involves simulating a sequence of particles over subsets of the form

$$\{\theta; \mathrm{L}(\theta|\mathbf{y}) \geq \alpha\},$$

where L is the likelihood function and $\alpha$ a bound updated at each iteration of the algorithm, by finding the lowest likelihood in the current sample and replacing the argument with a new value with a higher likelihood. For a mixture model this approach offers the advantage of using solely the numerical value of the likelihood at a given parameter value, rather

---

[5]One could equally conceive of the sequence of targets as being a sequence of posterior distributions of mixtures with a smaller number of dimensions or with smaller correlation structure, for instance borrowing from variational Bayes.

than exploiting more advanced features of this function. The resulting drawback is that the method is myopic, resorting to the prior or other local approximations for proposals. As the number of components, hence the dimension of the parameter, increases, it becomes more and more difficult to find moves that lead to higher values of the likelihood. In addition, the multimodality of the target distribution implies that there are more and more parts of the space that are not explored by the method (Chopin & Robert, 2010; Marin & Robert, 2010). While dimension (of the mixture model as well as of the parameter space) is certainly an issue and presumably a curse (Buchner, 2014), the MultiNest version of the algorithm manages dimensions up to 20 (Feroz et al., 2013), which remains a small number when considering multivariate mixtures.[6]

## 5.3 Bayesian Inference in the Model-Based Clustering Context

In addition to being a probabilistic model *per se*, finite mixture models provide a well-known probabilistic approach to clustering. In the model-based clustering setting each cluster is associated with a mixture component. Usually clustering is relevant and useful for data analysis when the number of observations is large, involving, say, several hundred observations, and so is the number of variables, with, say, several dozen variables. Moreover, choosing the unknown number $G$ of mixture components corresponding to the data clusters is a sensitive and critical task in this setting. Thus efficient Bayesian inference for model-based clustering requires MCMC algorithms working well and automatically in large dimensions with potentially numerous observations and smart strategies to derive a relevant number of clusters; see Chapter 7 for a detailed discussion.

Malsiner-Walli et al. (2016) devoted considerable efforts to assessing relevant Bayesian procedures in a model-based clustering context and we refer the reader to this paper for detailed coverage. In this section, we only summarize their inference strategy, which consists primarily of choosing relevant prior distributions. As an illustration, their approach is implemented in the following section in a realistic if specific case with 50 variables and a relatively large number of observations.

We recall that the goal of the approach is to cluster $n$ individuals described by $d$ quantitative variables. In Malsiner-Walli et al. (2016), each cluster is associated with a multivariate Gaussian distribution, resulting formally in a multivariate Gaussian mixture sample with $G$ components,

$$\sum_{g=1}^{G} \eta_g \mathcal{N}_d \left( \mu_g, \Sigma_g \right),$$

where $G$ is typically unknown.

*Choosing priors for the mixing proportions*

Malsiner-Walli et al.'s (2016) strategy involves starting the analysis with an overfitted mixture, that is, a mixture with a number of components $G$ most likely beyond the supposed (if unknown) number of relevant clusters. Assuming a symmetric Dirichlet prior $\mathcal{D}_G (e_0)$ on the mixing proportions, they argue, based upon asymptotic results established in Rousseau & Mengersen (2011) and as observed in Frühwirth-Schnatter (2011), that if $e_0 < r/2$, $r$ being the dimension of the component-specific parameter $\theta_g$, then the posterior expectation of the mixing proportions converges to zero for superfluous components. But if $e_0 > r/2$

---

[6] A bivariate Gaussian mixture with four components involves more than 20 parameters.

then the posterior density handles overfitting by defining at least two identical components, each with non-negligible weights. Thus Malsiner-Walli et al. (2016) favour small values of $e_0$ to allow emptying of superfluous components. More precisely, they consider a hierarchical prior distribution, namely a symmetric Dirichlet prior $\mathcal{D}_G(e_0)$ on the component weight distribution $(\eta_1, \ldots, \eta_G)$ and a gamma hyperprior for $e_0$:

$$\eta_1, \ldots, \eta_G | e_0 \sim \mathcal{D}_G(e_0), \quad e_0 \sim \mathcal{G}(a, aG).$$

Based on numerical experiments on simulated data, they recommend setting $a = 10$. An alternative choice is to fix the hyperparameter $e_0$ at a given small value.

*Choosing the priors for the component means and covariance matrices*

Following Frühwirth-Schnatter (2011), Malsiner-Walli et al. (2016) recommend putting a shrinkage prior, namely the normal-gamma prior, on component means. This prior is designed to handle high-dimensional mixtures, where not all variables contribute to the clustering structure, but a number of irrelevant variables are expected to be present without knowing *a priori* which variables this could be. For any such variable $y_{il}$ in dimension $l$, the components means $\mu_{l1}, \ldots, \mu_{lG}$ are pulled towards a common value $b_{0l}$, due to a local, dimension-specific shrinkage parameter $\lambda_l$. More specifically, the following hierarchical prior based on a multivariate normal distribution for the component means $\mu_g$ is chosen:

$$\mu_g \sim \mathcal{N}_d(b_0, B_0), \quad g = 1, \ldots, G,$$
$$b_0 \sim \mathcal{N}_d(m_0, M_0), \quad B_0 = \Lambda R_0 \Lambda, \quad R_0 = \mathrm{Diag}(R_1^2, \ldots, R_d^2),$$
$$\Lambda = \mathrm{Diag}(\sqrt{\lambda_1}, \ldots, \sqrt{\lambda_d}), \quad \lambda_l \sim \mathcal{G}(\nu_1, \nu_2), \quad l = 1, \ldots, d,$$

where $R_l$ is the range of $y_{il}$ (across $i = 1, \ldots, n$). Malsiner-Walli et al. (2016) suggest setting the hyperparameters $\nu_1$ and $\nu_2$ to 0.5 to allow for a sufficient shrinkage of the prior variance of the component means. For $b_0$, they specify an improper and empirical prior where $m_0 = \mathrm{median}(y)$ and $M_0^{-1} = 0$.

A standard conjugate hierarchical prior is considered on the component covariance matrices $\Sigma_g$:[7]

$$\Sigma_g^{-1} \sim \mathcal{W}(c_0, C_0), \quad C_0 \sim \mathcal{W}(g_0, G_0),$$
$$c_0 = 2.5 + \frac{d-1}{2}, \quad g_0 = 0.5 + \frac{d-1}{2}, \quad G_0 = \frac{100 g_0}{c_0} \mathrm{Diag}(1/R_1^2, \ldots, 1/R_d^2). \tag{5.3}$$

Under such prior distributions, an MCMC sampler is detailed in (Malsiner-Walli et al., 2016, Appendix 1). The point process representation of the MCMC draws introduced in Frühwirth-Schnatter (2006) and recalled in Chapter 1 is exploited to study the posterior distribution of the component-specific parameters, regardless of potential label switching. This is achieved through a $G$-centroid cluster analysis based on Mahalanobis distance as detailed in (Malsiner-Walli et al., 2016, Appendix 2).

Based on the simulation experiments conducted in Section 5.4, we conclude that prior (5.3) is problematic for high-dimensional mixtures with large values of $d$. In Section 5.5 a modification of prior (5.3) is presented that works well also for very high-dimensional mixtures.

A final remark is that in a model-based clustering context the means of the various

---

[7]Unlike the other chapters, here we use the same parameterizations of the Wishart and inverse Wishart distribution as employed in Frühwirth-Schnatter (2006): $\Sigma_g^{-1} \sim \mathcal{W}(c_0, C_0)$ if and only if $\Sigma_g \sim \mathcal{W}^{-1}(c_0, C_0)$, with $\mathrm{E}(\Sigma_g^{-1}) = c_0 C_0^{-1}$ and $\mathrm{E}(\Sigma_g) = C_0/(c_0 - (d+1)/2)$. In this parameterization, the standard gamma and inverse gamma distribution are recovered when $d = 1$.

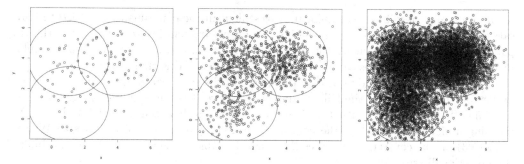

**FIGURE 5.4**
Scatter plots of simulated data sets (first two dimensions), with $n = 100$ (left), $n = 1000$ (middle) and $n = 10,000$ (right) and circles denoting the 95% probability regions for each component of mixture (5.4).

clusters are expected to be distinct. It is thus advisable to initialize the MCMC algorithm with a $k$-means algorithm in $\mathbb{R}^d$ in order to make sure that the Markov chain mixes properly and to avoid being stuck in a slow convergence area. Obviously, in large or even moderate dimensions, there is a clear need to perform several runs of the MCMC algorithm from different random positions to ensure that the MCMC algorithm has reached its stationary distribution.

## 5.4  Simulation Studies

In this section, we study the specific case of a multivariate Gaussian mixture model with three components in dimension $d = 50$. We simulate observations $\mathbf{y} = (y_1, \ldots, y_n)$ from $y \in \mathbb{R}^d$ such that

$$y \sim \sum_{g=1}^{3} \eta_g \mathcal{N}_d(\mu_g, \Sigma_g) \tag{5.4}$$

with $(\eta_1, \eta_2, \eta_3) = (0.35, 0.2, 0.45)$, $\mu_1 = (1, 4, \ldots, 1, 4)^\top$, $\mu_2 = (1, 1, \ldots, 1)^\top$, $\mu_3 = (4, 4, \ldots, 4)^\top$ and $\Sigma_g = \tau I_d$ $(g = 1, 2, 3)$. The parameter $\tau$ is chosen in the simulations to calibrate the overlap between the components of the mixture. Examples of data sets are shown in Figure 5.4 for varying $n$.

The simulated distribution is homoscedastic and the true covariance matrices are isotropic. However, we refrained from including this information in the prior distributions to check the performance of the proposed samplers. A direct implication of this omission is that the three covariance matrices involve $3 \left( \frac{d \times (d-1)}{2} + d \right) = 3825$ parameters instead of a single parameter. Furthermore, taking into account the three component means and the mixture weights, the dimension of the parameter space increases to 3977.

It would be illuminating to fit a mixture model to the simulated data sets using the various approaches outlined in Section 5.2 and to compare the computational performance of the different algorithms. However, our first observation is that the Gibbs sampler appears to be the sole manageable answer to generate parameter samples from the posterior distribution in such a high-dimensional experiment. Alternative methods such as Metropolis–Hastings schemes and sequential Monte Carlo samplers turned out to be extremely delicate to calibrate and, despite some significant investment in the experiment, we failed to find a satisfying calibration of the corresponding tuning parameters in order to recover the true parameters used for simulating the data. For this reason, investigation of the sampling methods is limited to Gibbs sampling for the remainder of this chapter.

A second interesting finding is that the choice of the scale parameter $\tau$ in (5.4) played a major role in the ability of the Gibbs sampler to recover the true parameters. In particular for simulations associated with $\tau \geq 5$ and a sample size between $n = 100$ and $n = 500$, meaning a significant overlap, the choice of the hyperparameters of the inverse Wishart prior (5.3) on the covariance matrices has a considerable impact on the shape and bulk of the posterior distribution.

Using the Matlab library bayesf associated with Frühwirth-Schnatter (2006), we observe poor performance of default choices, namely, either the conditionally conjugate prior proposed by Bensmail et al. (1997) or the hierarchical independence prior introduced by Richardson & Green (1997) for univariate Gaussian mixture and extended by Stephens (1997) to the multivariate case; see (Frühwirth-Schnatter, 2006, Section 6.3.2) for details on these priors. For such priors, we indeed found that inference led to a posterior distribution that is concentrated quite far from the likelihood. The reason for the discrepancy is that the posterior mass accumulates in regions of the parameter space where some mixture weights are close to zero. Thanks to the assistance of Sylvia Frühwirth-Schnatter, we managed to fix the issue for the hierarchical independence prior. Note that this prior distribution happens to be quite similar to the one introduced by Malsiner-Walli et al. (2016) and described in the previous section. We found through trial and error that we had to drastically increase the value of $c_0$ in the inverse Wishart prior (5.3) to, for example, $c_0 = 50 + \frac{d-1}{2}$, instead of the default choice, $c_0 = 2.5 + \frac{d-1}{2}$. The central conclusion of this experiment remains the fact that in high-dimensional settings prior distributions are immensely delicate to calibrate. A new, fully automatic way to choose the hyperparameters of (5.3) in a high-dimensional setting will be discussed in Section 5.5.

On the other hand, experiments for $\tau \leq 5$ indicate that concentration difficulties tend to vanish. In the following experiment, we chose the value $\tau = 1$ which produced well-separated clusters for illustration.

### 5.4.1   Known number of components

Recall that the proposal of Malsiner-Walli et al. (2016) has been designed to estimate both the number of components and the Gaussian expectations and covariances. However, the proposed methodology can easily be implemented to estimate the parameter values, while the number of components is supposed to be known. The only modification is to adjust the prior hyperparameters by setting $G = 3$ (which is the true value) and specify a fixed value $e_0 = 0.01$.[8]

In order to evaluate whether the Gibbs sampler is able to recover the true component parameters, we simulate two data sets from (5.4), with $n = 100$ and $n = 1000$ observations, respectively. For these two data sets, we repeat running the Gibbs sampler $M = 10$ times with $T = 10,000$ iterations after a burn-in period of 1000 iterations. The posterior estimates are computed after a reordering process based on the $G$-centroid clustering of $T$ simulated component means in the point process representation.

In order to check the sensitivity of the Gibbs sampling to the choice of the initial values

---

[8]Although the later value actually assumes that the specified mixture model is overfitting (which is not the case for these investigations), this prior setting still worked well in our simulations. Obviously, this observation remains conditional on these simulations.

**TABLE 5.1**
Estimated number of components $\hat{G}$ for a data set of $n$ observations drawn from the three-component mixture (5.4), averaged number across $M = 30$ independent replicates of Gibbs sampling with $T$ iterations

| $T$ | $n$ | $\min(\hat{G})$ | $\max(\hat{G})$ |
|---|---|---|---|
| $10^3$ | 100 | 10 | 10 |
| $10^4$ | 100 | 10 | 10 |
| $10^3$ | 500 | 10 | 10 |
| $10^4$ | 500 | 10 | 10 |
| $10^4$ | 1000 | 10 | 10 |
| $10^3$ | 10,000 | 3 | 3 |

for the MCMC algorithm, thus indicating potential issues with convergence, we compared three different methods of initialization:

(a) Initialization 1 determines initial values by performing $k$-means clustering of the data,

(b) Initialization 2 considers maximum likelihood estimates of the component parameters computed by the Rmixmod[9] package as initial values.

(c) Initialization 3 allocates a random value to each parameter, simulated from the corresponding prior distribution.

Both for $n = 100$ and $n = 1000$, the posterior estimates obtained by implementing the Gibbs sampler based on the three different initialization methods explained above are all very similar, as shown by our experiments where ten repeated calls to the Gibbs sampler showed no visible discrepancy between the three posterior estimates. This observed stability of the resulting estimations in terms of initial values is quite reassuring from the point of view of the convergence of the Gibbs sampler. (However, we have to acknowledge that it is restricted to a specific mixture model.) Furthermore, the componentwise parameter estimates associated with all three initialization methods are close to the corresponding true values.

### 5.4.2 Unknown number of components

In this section we consider the joint estimation of both the number of components and the parameter values for various data sets simulated from (5.4) with different numbers of observations $n$. We implement the same Gibbs sampler as used in Section 5.4.1, but with $e_0 = 0.0001$ and maximum number of components $G = 10$. According to Malsiner-Walli et al. (2016), this prior setting empties the redundant components and the unknown number of components can be estimated by the most frequent number of non-empty clusters; see also Chapter 7, Section 7.4.5. For both data sets, we run $M = 30$ independent Gibbs samplers with $T = 10,000$ iterations after a burn-in period of 1000, using an initial clustering of the data points into 10 groups obtained through $k$-means clustering.

As shown in Table 5.1, when $n$ is large ($10^4$), the method of Malsiner-Walli et al. (2016) always manages to pinpoint the true value of the number of mixture components (which is equal to 3) in all replications of our Gibbs sampler. However, for smaller data sets with $n$ ranging from 100 to 1000, the estimation of $G$ produces an overfit of the true value. Even when the number of iterations was increased to $T = 10^4$, the superfluous components did

[9]https://cran.r-project.org/web/packages/Rmixmod/

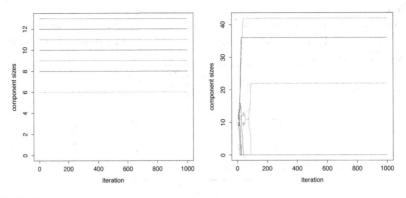

**FIGURE 5.5**
Trace plot of the number of observations assigned to the 10 components during MCMC sampling (each line corresponds to a component), under the prior used in Malsiner-Walli et al. (2016) with $e_0 = 0.01$ fixed (left) and under the prior setting according to the determinant criterion (right).

not get emptied during MCMC sampling, mainly because the Gibbs sampler got stuck in the initial classification. This hints at a prior-likelihood conflict in this high-dimensional setting which cannot be resolved for small data sets. Therefore, the following section proposes an idea for enforcing convergence of the Gibbs sampler for high-dimensional mixtures through the specification of a "suitable" prior on the component covariance matrices.

## 5.5  Gibbs Sampling for High-Dimensional Mixtures

As reported in Section 5.4.2, the approach by Malsiner-Walli et al. (2016) failed when estimating the number of components for high-dimensional data with $d = 50$ as the Gibbs sampler did not converge to a smaller number of non-empty clusters. When starting with an initial classification consisting of ten (overfitting) data clusters allocated to the different components, no merging of the components to a smaller number of non-empty components took place during MCMC sampling. As can be seen on the left in Figure 5.5, the number of observations assigned to the various groups is constant during MCMC sampling and corresponds to the starting classification.

The main reason why Gibbs sampling gets "stuck" lies in the small amount of overlapping probability between the component densities of a mixture model in high-dimensional spaces. In the starting configuration, the data points are partitioned into many small data clusters. Due to the large distances between component means in the high-dimensional space, the resulting component densities are rather isolated and almost no overlapping takes place, even for neighbouring components, as can be seen in Table 5.2 where the overlapping probability for two components of the mixture model (5.4) is reported.

As a consequence, in the classification step of the sampler (see step 1 of Algorithm 5.1), where an observation is assigned to the different components according to the evaluated component densities, barely any movement can be observed due to the missing overlap: once an observation is assigned to a component, it is extremely unlikely that it is allocated to another component in the next iteration.

**TABLE 5.2**

Overlap of components 2 and 3 in mixture (5.4) for increasing dimension $d$

| Dimension $d$ | Overlap |
|:---:|:---:|
| 2 | 0.034 |
| 4 | 0.003 |
| 50 | $2.48 \cdot 10^{-6}$ |
| 100 | $7.34 \cdot 10^{-51}$ |
| 200 | $7.21 \cdot 10^{-100}$ |

To overcome this curse of dimensionality, a promising idea is to encourage *a priori* "flat" component densities towards achieving a stronger overlapping of them also in higher dimensions. To this end, we propose to specify the prior on the component covariances $\Sigma_g \sim \mathcal{W}^{-1}(c_0, C_0)$ such that *a priori* the ratio of the volume $|\Sigma_g|$ with respect to the total data spread $|\mathrm{Cov}(y)|$ is kept constant across the dimensions. Using $\mathrm{E}(\Sigma_g) = C_0/(c_0 - 1)$, the specification of the scale matrix $C_0$ determines the prior expectation of $\Sigma_g$. In the following subsection, guidance is given on how to select $C_0$ in order to obtain a constant ratio $|\Sigma_g|/|\mathrm{Cov}(y)|$.

### 5.5.1 Determinant coefficient of determination

Consider the usual inverse Whishart prior $\Sigma_g^{-1} \sim \mathcal{W}(c_0, C_0)$, where $c_0 = \nu + (d-1)/2$, with $\nu > 0$ being fixed. In this subsection, we discuss the choice of $C_0$ for high-dimensional mixtures.

We define $C_0 = \phi S_y$, with $S_y$ being the empirical covariance matrix of the data, as suggested in Bensmail et al. (1997), among others, and exploit the variance decomposition of a multivariate mixture of normals, outlined in Frühwirth-Schnatter (2006):

$$\mathrm{Cov}(y) = \sum_{g=1}^{G} \eta_g \Sigma_g + \sum_{g=1}^{G} \eta_g (\mu_g - \mathrm{E}(y))(\mu_g - \mathrm{E}(y))^{\top}. \tag{5.5}$$

Thus the total variance $\mathrm{Cov}(y)$ of a mixture distribution with $G$ components arises from two sources of variability, namely the within-group heterogeneity, generated by the component variances $\Sigma_g$, and between-group heterogeneity, generated by the spread of the component means $\mu_g$ around the overall mixture mean $\mathrm{E}(y)$.

To measure how much variability may be attributed to unobserved between-group heterogeneity, Frühwirth-Schnatter (2006) considers the following two coefficients of determination derived from (5.5), namely the trace criterion $R_{tr}^2$ and the determinant criterion $R_{det}^2$:

$$R_{tr}^2 = 1 - \frac{\mathrm{tr}(\sum_{g=1}^{G} \eta_g \Sigma_g)}{\mathrm{tr}(\mathrm{Cov}(y))}, \tag{5.6}$$

$$R_{det}^2 = 1 - \frac{|\sum_{g=1}^{G} \eta_g \Sigma_g|}{|\mathrm{Cov}(y)|}. \tag{5.7}$$

Frühwirth-Schnatter (2006, p. 170) suggests choosing $\phi$ in $C_0 = \phi S_y$ such that a certain amount of explained heterogeneity according to the trace criterion $R_{tr}^2$ is achieved:

$$\phi_{tr} = (1 - \mathrm{E}(R_{tr}^2))(c_0 - (d+1)/2). \tag{5.8}$$

For instance, if $c_0 = 2.5 + (d-1)/2$, then choosing 50% expected explained heterogeneity

**TABLE 5.3**

$\phi_{det}$ for selected values of $d$ and $R^2_{det}$ with $c_0 = 2.5 + (d-1)/2$

|  | $d=2$ | $d=4$ | $d=50$ | $d=100$ | $d=150$ | $d=200$ |
|---|---|---|---|---|---|---|
| $R^2_{det} = 0.50$ | 1.225 | 1.831 | 11.09 | 20.55 | 29.90 | 39.22 |
| $R^2_{det} = 0.67$ | 0.995 | 1.651 | 11.00 | 20.46 | 29.82 | 39.14 |
| $R^2_{det} = 0.75$ | 0.866 | 1.540 | 10.94 | 20.41 | 29.77 | 39.08 |
| $R^2_{det} = 0.90$ | 0.548 | 1.225 | 10.74 | 20.22 | 29.59 | 38.90 |

yields $\phi_{tr} = 0.75$, while choosing a higher percentage of explained heterogeneity such as $R^2_{tr} = 2/3$ yields $\phi_{tr} = 0.5$. Obviously, the scaling factor $\phi_{tr}$ in (5.8) is independent of the dimension $d$ of the data. However, the experiments in Section 5.4 show that this criterion yields poor clustering solutions in high dimensions such as $d = 50$.

In this section we show that choosing $\phi$ according to the determinant criterion $R^2_{det}$ given in (5.7) yields a scale matrix $C_0$ in the prior of the covariance matrix $\Sigma_g$ which increases as $d$ increases. The determinant criterion $R^2_{det}$ also measures the volume of the corresponding matrices which leads to a more sensible choice in a model-based clustering context.

If we substitute in (5.7) the term $\sum \eta_g \Sigma_g$ by the prior expected value $\Sigma_{\tilde{g}}$, with $\tilde{g}$ being an arbitrary component, and estimate $\text{Cov}(y)$ through $S_y$, then we obtain

$$R^2_{det} = 1 - \frac{1}{|\Sigma_{\tilde{g}}^{-1} S_y|} = 1 - \frac{\phi^d_{det}}{|W|},$$

where $W \sim \mathcal{W}(c_0, I)$. Taking expectations,

$$E(R^2_{det}) = 1 - \phi^d_{det} E(|W^{-1}|),$$

and using

$$E(1/|W|^a) = \frac{\Gamma_d(c_0 - a)}{\Gamma_d(c_0)},$$

where

$$\Gamma_d(c) = \pi^{d(d-1)/4} \prod_{j=1}^{d} \Gamma\left(\frac{2c+1-j}{2}\right)$$

is the generalized gamma function, we obtain a scaling factor $\phi_{det}$ which depends on the dimension $d$ of the data:

$$\phi_{det} = (1 - R^2_{det})^{1/d} \frac{\Gamma_d(c_0)}{\Gamma_d(c_0 - 1)}. \tag{5.9}$$

Hence, the modified prior on $\Sigma_g^{-1}$ reads

$$\Sigma_g^{-1} \sim \mathcal{W}(c_0, \phi_{det} S_y), \tag{5.10}$$

where $c_0 = 2.5 + (d-1)/2$, $S_y$ is the empirical covariance matrix of the data, and $\phi_{det}$ is given by (5.9). Table 5.3 reports $\phi_{det}$ for selected values of $d$ and $R^2_{det}$. Note that for this prior $C_0 = \phi_{det} S_y$ is set to this fixed value, that is, no hyperprior on $C_0$ is specified.

**TABLE 5.4**

Clustering results of the simulation study involving an increasing number $d$ of variables and $n$ of observations, based on 100 simulated data sets for each combination of $n$ and $d$. The prior on $\Sigma_k^{-1}$ is chosen as in (5.10) with $C_0 = \phi_{det} S_y$ and $R_{det}^2 = 0.50$. The most frequently visited number of clusters $\tilde{G}$ (with frequencies in parentheses), the posterior mean of the number of clusters $\hat{G}$ and the adjusted Rand index $r_a$ are averaged across the 100 simulated data sets

| | | $n = 100$ | | | $n = 1000$ | | | $n = 10,000$ | | |
|---|---|---|---|---|---|---|---|---|---|---|
| $d$ | $\phi_{det}$ | $\tilde{G}$ | $\hat{G}$ | $r_a$ | $\tilde{G}$ | $\hat{G}$ | $r_a$ | $\tilde{G}$ | $\hat{G}$ | $r_a$ |
| 2 | 1.23 | 2(93) | 2.0 | 0.45 | 3(100) | 3.0 | 0.73 | 3(98) | 3.0 | 0.75 |
| 4 | 1.83 | 2(70) | 2.3 | 0.69 | 3(100) | 3.0 | 0.93 | 3(100) | 3.0 | 0.93 |
| 50 | 11.09 | 3(100) | 3.0 | 1.00 | 3(100) | 3.0 | 1.00 | 3(100) | 3.0 | 1.00 |
| 100 | 20.55 | 3(100) | 3.0 | 1.00 | 3(100) | 3.0 | 1.00 | 3(100) | 3.0 | 1.00 |
| 150 | 29.90 | 3(85) | 3.5 | 0.99 | 3(100) | 3.0 | 1.00 | 3(100) | 3.0 | 1.00 |
| 200 | 39.22 | 3(31) | 10.7 | 0.77 | 3(97) | 3.1 | 0.99 | 3(100) | 3.0 | 1.00 |

## 5.5.2 Simulation study using the determinant criterion

In order to evaluate whether the proposed determinant criterion for selecting $C_0$ enables the Gibbs sampler to converge to the true number of components, a simulation study is performed. One hundred data sets are sampled from the three-component mixture given in (5.4), with dimensionality $d$ varying from 2 to 200.

A sparse finite mixture model is fitted to the data sets using the Gibbs sampler and priors as described in Malsiner-Walli et al. (2016); however, the following prior modifications are made. The scale parameter of the Wishart prior on $\Sigma_k^{-1}$ is set to $C_0 = \phi_{det} S_y$ as in (5.10), where $\phi_{det}$ is selected according (5.9) with $R_{det}^2 = 0.5$; see also the first row in Table 5.3. This corresponds to a prior proportion of heterogeneity explained by the component means of $R^2 \approx 0.67$. The Dirichlet parameter $e_0$ is fixed to 0.01 and we increase the maximum number $G$ of components to $G = 30$.

The shrinkage factor $\lambda_l$ in the prior specification (5.3) is fixed to 1, since all variables $y_{il}$ are relevant for clustering in the present context. By design, for each single variable $y_{il}$, two of the means $\mu_{l1}, \mu_{l2}, \mu_{l3}$ are different. Furthermore, all pairs of variables $(y_{il}, y_{i,l+2})$ exhibit three well-separated component means located at (1,1), (4,4), (1,4) or (4,1) in their bivariate marginal distribution $p(y_{il}, y_{i,l+2})$. Hence, none of the variables is irrelevant in the sense that $\mu_{l1} \approx \mu_{l2} \approx \mu_{l3}$.

The estimated number of components is reported in Table 5.4, based on a Gibbs sampling with $T = 8000$ iterations after discarding the first 800 iterations as burn-in for each of the 100 data sets. For $n = 1000$ and $n = 10,000$, the resulting clustering is almost perfect, even for very high dimensions such as $d = 200$. The Gibbs sampler converges to the true number of components, as can be seen in the trace plot on the right in Figure 5.5, by leaving empty all specified components but three. For small data sets with $n = 100$, the criterion leads to an underfitting solution ($\hat{G} = 2$) for small $d$. However, as can be seen in the scatter plot on the left in Figure 5.4, a small data set of only 100 observations may not contain enough information for estimating three groups. Also, if $d > n$ the mixture model is not well defined, and the Gibbs sampler gets stuck again. However, for $d = 50$ and $d = 100$ the approach works well also for small data sets with $n = 100$.

## 5.6   Concluding Remarks

This chapter discussed practical Bayesian inference for finite mixtures using sampling-based methods. While these methods work well for moderately sized mixtures, Gibbs sampling turned out to be the only operational method for handling high-dimensional mixtures. Based on a simulation study, aiming to estimate Gaussian mixture models with 50 variables, we were unable to tune commonly used algorithms such as the Metropolis–Hastings algorithm or sequential Monte Carlo, due to the curse of dimensionality. Gibbs sampling turned out to be the exception in this collection of algorithms. However, we also found that the Gibbs sampler may get stuck when initialized with many small data clusters under previously published priors.

Hence, we consider that calibrating prior parameters in high-dimensional spaces remains a delicate issue. For Gaussian mixtures, we examined the role of the determinant criterion introduced by Frühwirth-Schnatter (2006) for incorporating the prior expected proportion of heterogeneity explained by the component means (and thereby determining the hetero-geneity covered by the components themselves). This led us to a new choice of the scaling matrix in the inverse Wishart prior. The resulting prior, in combination with Gibbs sampling, worked quite well, when estimating the number of components for Gaussian mixtures, even for very high-dimensional data sets with 200 variables. Starting with a strongly over-fitted mixture, the Gibbs sampler was able to converge to the true model.

Many other computational issues deserve further investigation, in particular for mixture models for high-dimensional data. In addition to the "big $p$" (in our notation "big $d$") and the "big $n$" problem, the "big $G$" problem is also relevant. Are we able to estimate several tens data clusters or more, a case that arises in high-energy physics (see Frühwirth et al., 2016) or genomics (see Rau & Maugis-Rabusseau, 2018)?

# Bibliography

BEHBOODIAN, J. (1972). Information matrix for a mixture of two normal distributions. *Journal of Statistical Computation and Simulation* **1**, 295–314.

BENSMAIL, H., CELEUX, G., RAFTERY, A. & ROBERT, C. (1997). Inference in model-based cluster analysis. *Statistics and Computing* **7**, 1–10.

BERKHOF, J., VAN MECHELEN, I. & GELMAN, A. (2003). A Bayesian approach to the selection and testing of mixture models. *Statistica Sinica* **13**, 423–442.

BERNARDO, J. M. & GIRÓN, F. J. (1988). A Bayesian analysis of simple mixture problems. In *Bayesian Statistics 3*, J. M. Bernardo, M. H. DeGroot, D. V. Lindley & A. F. M. Smith, eds. Oxford: Oxford University Press, pp. 67–78.

BUCHNER, J. (2014). A statistical test for nested sampling algorithms. Preprint, arXiv:1407.5459.

CAPPÉ, O., MOULINES, E. & RYDÉN, T. (2004). *Hidden Markov Models*. New York: Springer-Verlag.

CELEUX, G., HURN, M. & ROBERT, C. (2000). Computational and inferential difficulties with mixture posterior distributions. *Journal of the American Statistical Association* **95**, 957–979.

CHIB, S. (1995). Marginal likelihood from the Gibbs output. *Journal of the American Statistical Association* **90**, 1313–1321.

CHOPIN, N. (2002). A sequential particle filter method for static models. *Biometrika* **89**, 539–552.

CHOPIN, N. (2004). Central limit theorem for sequential Monte Carlo methods and its application to Bayesian inference. *Annals of Statistics* **32**, 2385–2411.

CHOPIN, N. & ROBERT, C. (2010). Properties of nested sampling. *Biometrika* **97**, 741–755.

CRAWFORD, S., DEGROOT, M. H., KADANE, J. & SMALL, M. J. (1992). Modelling lake-chemistry distributions: Approximate Bayesian methods for estimating a finite-mixture model. *Technometrics* **34**, 441–453.

DEL MORAL, P., DOUCET, A. & JASRA, A. (2006). Sequential Monte Carlo samplers. *Journal of the Royal Statistical Society, Series B* **68**, 411–436.

DICKEY, J. M. (1968). Three multidimensional integral identities with Bayesian applications. *Annals of Statistics* **39**, 1615–1627.

DIEBOLT, J. & ROBERT, C. (1990a). Bayesian estimation of finite mixture distributions, Part I: Theoretical aspects. Tech. Rep. 110, LSTA, Université Paris VI, Paris.

DIEBOLT, J. & ROBERT, C. (1990b). Estimation des paramètres d'un mélange par échantillonnage bayésien. *Notes aux Comptes-Rendus de l'Académie des Sciences I* **311**, 653–658.

DIEBOLT, J. & ROBERT, C. (1994). Estimation of finite mixture distributions by Bayesian sampling. *Journal of the Royal Statistical Society, Series B* **56**, 363–375.

EVERITT, R. G., CULLIFORD, R., MEDINA-AGUAYO, F. & WILSON, D. J. (2016). Sequential Bayesian inference for mixture models and the coalescent using sequential Monte Carlo samplers with transformations. Preprint, arXiv:1612.06468.

FEROZ, F., HOBSON, M. P., CAMERON, E. & PETTITT, A. N. (2013). Importance nested sampling and the MultiNest algorithm. Preprint, arXiv:1306.2144.

FRÜHWIRTH, R., ECKSTEIN, K. & FRÜHWIRTH-SCHNATTER, S. (2016). Vertex finding by sparse model-based clustering. *Journal of Physics: Conference Series* **762**, 012055.

FRÜHWIRTH-SCHNATTER, S. (2001). Markov chain Monte Carlo estimation of classical and dynamic switching and mixture models. *Journal of the American Statistical Association* **96**, 194–209.

FRÜHWIRTH-SCHNATTER, S. (2004). Estimating marginal likelihoods for mixture and Markov switching models using bridge sampling techniques. *Econometrics Journal* **7**, 143–167.

FRÜHWIRTH-SCHNATTER, S. (2006). *Finite Mixture and Markov Switching Models*. New York: Springer-Verlag.

FRÜHWIRTH-SCHNATTER, S. (2011). Dealing with label switching under model uncertainty. In *Mixture Estimation and Applications*, K. Mengersen, C. P. Robert & D. Titterington, eds., chap. 10. Chichester: Wiley, pp. 213–239.

GELFAND, A. & SMITH, A. F. M. (1990). Sampling based approaches to calculating marginal densities. *Journal of the American Statistical Association* **85**, 398–409.

GELMAN, A. & KING, G. (1990). Estimating incumbency advantage without bias. *American Journal of Political Science* **34**, 1142–1164.

GEWEKE, J. (2007). Interpretation and inference in mixture models: Simple MCMC works. *Computational Statistics & Data Analysis* **51**, 3529–3550.

GREEN, P. (1995). Reversible jump MCMC computation and Bayesian model determination. *Biometrika* **82**, 711–732.

KAMARY, K., LEE, J. E. & ROBERT, C. P. (2018). Weakly informative reparameterisations for location-scale mixtures. *Computational Statistics & Data Analysis* (to appear). doi:10.1080/10618600.2018.1438900.

LEE, J. E. & ROBERT, C. P. (2016). Importance sampling schemes for evidence approximation in mixture models. *Bayesian Analysis* **11**, 573–597.

LEE, K., MARIN, J.-M., MENGERSEN, K. L. & ROBERT, C. (2008). Bayesian inference on mixtures of distributions. In *Platinum Jubilee of the Indian Statistical Institute*, N. N. Sastry, ed. Indian Statistical Institute, Bangalore.

MALSINER-WALLI, G., FRÜHWIRTH-SCHNATTER, S. & GRÜN, B. (2016). Model-based clustering based on sparse finite Gaussian mixtures. *Statistics and Computing* **26**, 303–324.

MARIN, J.-M. & ROBERT, C. P. (2010). Importance sampling methods for Bayesian discrimination between embedded models. In *Frontiers of Statistical Decision Making and Bayesian Analysis*, M.-H. Chen, D. K. Dey, P. Müller, D. Sun & K. Ye, eds. New York: Springer-Verlag.

MINKA, T. (2001). Expectation propagation for approximate Bayesian inference. In *UAI '01: Proceedings of the 17th Conference in Uncertainty in Artificial Intelligence*, D. K. Jack S. Breese, ed. University of Washington, Seattle.

NEAL, R. (1999). Erroneous results in "Marginal likelihood from the Gibbs output". Tech. rep., University of Toronto.

RAU, A. & MAUGIS-RABUSSEAU, C. (2018). Transformation and model choice for RNA-seq co-expression analysis. *Briefings in Bioinformatics* **19**, 425–436.

RICHARDSON, S. & GREEN, P. J. (1997). On Bayesian analysis of mixtures with an unknown number of components (with discussion). *Journal of the Royal Statistical Society, Series B* **59**, 731–792.

ROBERT, C. (2007). *The Bayesian Choice*. New York: Springer-Verlag.

ROBERT, C. & CASELLA, G. (2004). *Monte Carlo Statistical Methods*. New York: Springer-Verlag, 2nd ed.

ROUSSEAU, J. & MENGERSEN, K. (2011). Asymptotic behaviour of the posterior distribution in overfitted mixture models. *Journal of the Royal Statistical Society, Series B* **73**, 689–710.

RUE, H., MARTINO, S. & CHOPIN, N. (2009). Approximate Bayesian inference for latent Gaussian models using integrated nested Laplace approximations (with discussion). *Journal of the Royal Statistical Society, Series B* **71**, 319–392.

SKILLING, J. (2004). Nested sampling. In *Bayesian Inference and Maximum Entropy Methods: 24th International Workshop*, R. Fisher, R. Preuss & U. von Toussiant, eds., vol. 735. AIP Conference Proceedings.

SMITH, A. F. M. & MAKOV, U. (1978). A quasi-Bayes sequential procedure for mixtures. *Journal of the Royal Statistical Society, Series B* **40**, 106–112.

STEPHENS, M. (1997). *Bayesian Methods for Mixtures of Normal Distributions*. Unpublished D.Phil. thesis, University of Oxford.

STEPHENS, M. (2000). Bayesian analysis of mixture models with an unknown number of components – an alternative to reversible jump methods. *Annals of Statistics* **28**, 40–74.

TANNER, M. & WONG, W. (1987). The calculation of posterior distributions by data augmentation. *Journal of the American Statistical Association* **82**, 528–550.

TIERNEY, L. (1994). Markov chains for exploring posterior distributions (with discussion). *Annals of Statistics* **22**, 1701–1786.

TITTERINGTON, D. (2011). The EM algorithm, variational approximations and expectation propagation for mixtures. In *Mixtures: Estimation and Applications*, K. Mengersen, C. Robert & D. Titterington, eds., chap. 1. New York: John Wiley, pp. 1–21.

TITTERINGTON, D., SMITH, A. F. M. & MAKOV, U. (1985). *Statistical Analysis of Finite Mixture Distributions*. New York: John Wiley.

WEST, M. (1992). Modelling with mixtures. In *Bayesian Statistics 4*, J. M. Bernardo, J. O. Berger, A. P. Dawid & A. F. M. Smith, eds. Oxford: Oxford University Press.

ZOBAY, O. (2014). Variational Bayesian inference with Gaussian-mixture approximations. *Electronic Journal of Statistics* **8**, 355–389.

# 6

## Bayesian Nonparametric Mixture Models

**Peter Müller**

*UT Austin, USA*

## CONTENTS

We review the use of Bayesian nonparametric priors for inference in mixtures. Interpreting a mixture model as an expectation with respect to a mixing measure, it becomes natural to complete the model with a prior probability model on the unknown mixing measure. Prior models on random probability measures, like the mixing measure here, are known as Bayesian nonparametric models. We review some commonly used models, including in particular the Dirichlet process prior, normalized random measures with independent increments, and the determinantal point process and variations. Many applications of such models include inference on the implied partition of the experimental units, that is, a clustering of the data. This gives rise to predictive distributions that again take the form of a mixture model.

## 6.1 Introduction

Inference for mixture models and closely related hierarchical models is one of the big success stories of Bayesian inference. This is particularly true for applications in biostatistics. For example, popular Bayesian models for inference on patient subpopulations are variations of the following model. Let $y_i$ denote a response for the $i$th patient. We assume

$$y_i \sim \sum_{g=1}^{G} \eta_g p(y_i \mid \mu_g), \qquad (6.1)$$

$i = 1, \ldots, n$, including possibly $G = \infty$. The component-specific model $p(y_i \mid \mu_g)$ could be, for example, a survival model with parameters $\mu_g$, possibly including a regression on patient covariates. Let $\eta = (\eta_1, \ldots, \eta_G)$ denote the weights. The model is completed with a prior probability model $p(\eta)$ and $p(\mu_1, \ldots, \mu_G)$, typically assuming independence of the $\mu_g$, perhaps conditional on some hyperparameters. Inference in this model becomes a problem of Bayesian nonparametric inference when (6.1) is interpreted as an expectation with respect to a random mixing measure, as follows.

*Random discrete mixing measure*

Let $H = \sum \eta_g \delta_{\mu_g}$ denote a (discrete) random probability measure with atoms at $\mu_g$. The mixture model (6.1) can then be written as $y_i \mid H \sim \int p(y_i \mid \theta) \, dH(\theta)$. Equivalently, we can introduce latent variables $\theta_i$ and replace the integral with respect to $H$ by a hierarchical model

$$y_i \mid \theta \sim p(y_i \mid \theta_i), \quad \theta_i \mid H \sim H. \qquad (6.2)$$

In model (6.2) it becomes natural to consider a Bayesian nonparametric (BNP) prior $p(H)$ on $H$. BNP refers to prior probability models on infinite-dimensional parameter spaces, for example, random distributions such as $H$ in (6.2). See, for example, Hjort et al. (2010), Phadia (2013), Ghoshal & van der Vaart (2017), and Müller et al. (2015) for recent reviews of Bayesian nonparametric inference. The most widely used BNP prior $p(H)$ is the Dirichlet process (DP) prior (Ferguson, 1973). We briefly introduce the DP model to serve as a running example in this introduction, but defer more details to Section 6.2. A DP prior is indexed with two parameters, $\alpha > 0$ and a probability distribution $H_0$, and defines a prior for $H = \sum \eta_g \delta_{\mu_g}$ by assuming

$$\eta_g = v_g \prod_{\ell < g} (1 - v_\ell), \quad \text{with } v_g \sim \mathcal{B}e(1, \alpha), \text{ i.i.d. and } \mu_g \sim H_0, \text{ i.i.d.} \qquad (6.3)$$

This is known as the stick-breaking representation (Sethuraman, 1994). We write $H \sim \mathcal{DP}(\alpha, H_0)$. Model (6.2) with a DP prior $p(H) = \mathcal{DP}(\alpha, H_0)$ is known as a DP mixture. It is perhaps the most widely used BNP model; see also Chapter 17 for applications in finance. It was first introduced in Lo (1984), Escobar (1988, 1994), and Escobar & West (1995).

*Latent partitions*

Closely related to inference for a discrete mixing measure $H$ is the problem of inference on a random partition of the experimental units $[n] = \{1, \ldots, n\}$. The nature of (6.1) as a random partition of $[n]$ is evident if we rewrite (6.1) as a hierarchical model,

$$y_i \mid \theta_i \sim p(y_i \mid \theta_i), \quad \mathrm{P}(\theta_i = \mu_g \mid \eta) = \eta_g, \qquad (6.4)$$

with latent variables $\theta = (\theta_1, \ldots, \theta_n)$. Let $\{\theta_1^\star, \ldots, \theta_{G_+}^\star\} \subseteq \{\mu_g, \ g = 1, \ldots, G\}$ denote the $G_+ \leq \min(n, G)$ distinct values among the $\theta_i$. Then $C_k = \{i : \ \theta_i = \theta_k^\star\}$, $k = 1, \ldots, G_+$, defines a partition $\mathcal{C}_n = \{C_1, \ldots, C_{G_+}\}$ of $[n]$ into $G_+$ clusters. We use $n_k = |C_k|$ to denote the cardinality of the clusters. When $n$ is not understood from the context we add a subindex $n$ and write $G_{+,n}$ and $n_{nk}$. Sometimes it will be convenient to introduce cluster membership indicators $z_i = k$ if $i \in C_k$ and use $\mathbf{z} = (z_1, \ldots, z_n)$ to denote the partition. Note that

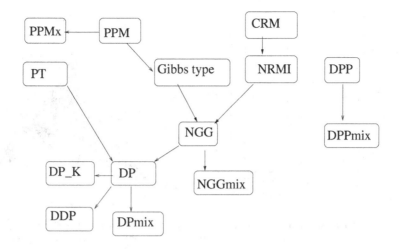

**FIGURE 6.1**
The diagram shows how BNP models in the upcoming discussion are related to each other. An arrow indicates that one model is a special case of the other (or constructed using the other model). The diagram is far from complete, including only models that are introduced in this chapter.

alternatively we could have introduced clusters $\tilde{C}_g = \{i : \theta_i = \mu_g\}$, $g = 1, \ldots, G$, allowing for possibly empty clusters. However, to avoid ambiguous terminology we will throughout use a convention of indexing only occupied clusters $C_k$.

Being equivalent model statements, inference on a random partition $\mathcal{C}_n$ is also implied by (6.2). To see this, note that the discrete nature of $H$ implies many ties among the $\theta_i$. If we use the arrangement of ties to define clusters we are back to a prior on the random partition $\mathcal{C}_n$ as in (6.4). That is, let $\theta_k^\star$, $k = 1, \ldots, G_+$, denote the $G_+ \leq n$ unique values and define $C_k = \{i : \theta_i = \theta_k^\star\}$, as before. In other words, any prior $p(H)$ in (6.2) implies a prior $p(\mathcal{C}_n)$. In fact, one can show that any exchangeable random partition $p(\mathcal{C}_n)$ can be introduced this way (Pitman, 2006). Here, exchangeability refers to invariance of $p(\mathcal{C}_n)$ under any permutation of the indices $i = 1, \ldots, n$, for any $n$. And we assume that the random partition $p(\mathcal{C}_n)$ is consistent across $n$ in the sense that the restriction of a random partition $p(\mathcal{C}_{n+1})$ of $[n+1]$ to $[n]$ gives $p(\mathcal{C}_n)$. That is, $p(\mathcal{C}_n) = \sum_{z_{n+1}} p(\mathcal{C}_{n+1})$.

In this chapter we review some of the popular BNP mixture models, starting with the widely used DP mixture models in Section 6.2 and some generalizations of the DP prior in Section 6.3. We then discuss inference that exploits the posterior distribution on the implied random partition and variations of these models in Section 6.4. Finally, in Section 6.5, we review the use of repulsive priors on the mixing measure, using in particular the determinantal point process (DPP).

In the following discussion we will run into a proliferation of BNP models and acronyms. Figure 6.1 summarizes how these models are related to each other. The diagram is far from complete, including only models that are introduced or at least mentioned in this chapter. For similar diagrams covering more BNP models see, for example, (Phadia, 2013, Figure 1.1) and (Müller et al., 2015, Figure 1).

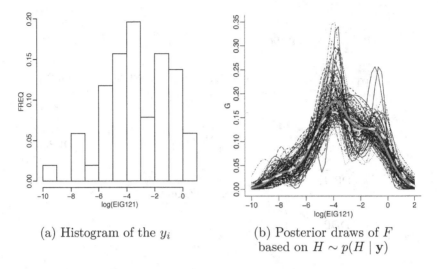

(a) Histogram of the $y_i$         (b) Posterior draws of $F$
                                        based on $H \sim p(H \mid \mathbf{y})$

**FIGURE 6.2**
Example 6.1. (a) Data and (b) posterior inference for $F$. Panel (b) shows 96 posterior draws of the mixture model $F$ based on $H \sim p(H \mid \mathbf{y})$ (thin black curves) and the posterior mean $\mathrm{E}(F \mid \mathbf{y})$ (thick gray curve). For comparison the figure also shows a kernel density estimate (dashed thick white line).

## 6.2 Dirichlet Process Mixtures

### 6.2.1 The Dirichlet process prior

The original definition of the DP is due to Ferguson (1973). A random distribution $H$ on $\Theta$ is said to follow a DP prior with baseline probability measure $H_0$ and mass parameter $\alpha$, denoted $H \sim \mathcal{DP}(\alpha, H_0)$, if for any partition $\{A_1, \dots, A_k\}$ of $\Theta$,

$$(H(A_1), \dots, H(A_k)) \sim \mathcal{D}(\alpha H_0(A_1), \dots, \alpha H_0(A_k)).$$

In the introduction we already saw an alternative defining property of the DP, known as the stick-breaking construction. Implicit in the stick-breaking construction is the fact that $H$ is discrete, even if $H_0$ is a continuous distribution.

*Pólya urn*

For later reference we state yet another defining property of the DP. Let $\theta_1, \theta_2, \dots$ be an i.i.d. sequence such that

$$\theta_i \mid H \sim H \quad \text{and} \quad H \sim \mathcal{DP}(\alpha, H_0), \tag{6.5}$$

as in (6.2). As before, let $\{\theta_1^\star, \dots, \theta_{G_{+,n}}^\star\}$ denote the unique values, let $n_{nk} = \sum_{i=1}^{n} \mathbb{I}(\theta_i = \theta_k^\star)$ be the number of $\theta_i$, $i \leq n$, equal to $\theta_k^\star$, and let $z_i = k$ if $\theta_i = \theta_k^\star$. Blackwell & MacQueen (1973) showed that the joint distribution of the $z_i$ can be characterized in terms of the predictive probability function,

$$\mathrm{P}(z_i = k \mid z_1, \dots, z_{i-1}) \propto \begin{cases} n_{i-1,k}, & k = 1, \dots, G_{+,i-1}, \\ \alpha, & k = G_{+,i-1} + 1. \end{cases} \tag{6.6}$$

This is known as the Pólya urn. For later reference note that (6.6) implies $p(\theta_n \mid \theta_1, \ldots, \theta_{n-1}) \propto \sum_{k=1}^{G_{+}, n-1} n_{n-1,k} \delta_{\theta_k^{\star}} + \alpha H_0$. Let $\theta_{-i} = (\theta_{\ell}, \ \ell \neq i)$ and let $G_{+}^{-i}$ and $n_j^{-i}$ denote the number of unique values and the multiplicities in $\theta_{-i}$. Since (6.5) is symmetric in the indices, the same expression as (6.6) must hold for $p(\theta_i \mid \theta_{-i})$:

$$p(\theta_i \mid \theta_{-i}) \propto \sum_{k=1}^{G_{+}^{-i}} n_k^{-i} \delta_{\theta_k^{\star}} + \alpha H_0. \tag{6.7}$$

*Posterior distributions*

One of the reasons for the wide use of the DP prior is computational ease and, closely related, conjugacy with respect to i.i.d. sampling. If $\theta = (\theta_1, \ldots, \theta_n)$ is an i.i.d. sample with $\theta_i \mid H \sim H$ and $H \sim \mathcal{DP}(\alpha, H_0)$ then

$$H \mid \theta \sim \mathcal{DP}(\alpha + n, H_1)$$

with $H_1 \propto \alpha H_0 + \sum_{i=1}^{n} \delta_{\theta_i}$. The posterior mean $\mathrm{E}(H \mid \theta) = H_1$ can be interpreted as a weighted average between the baseline measure $H_0$ and the empirical distribution of the $\theta_i$.

*Dirichlet process mixtures*

In many applications the discrete nature of $H$ is awkward. This motivates the DP mixture (DPM) model. For example, consider a density estimation problem $y_i \sim F$, i.i.d., $i = 1, \ldots, n$. To proceed with Bayesian inference on the unknown distribution $F$ we complete the model with a prior on $F$. In most applications it would be unreasonable to directly use a DP prior, because of the implied discrete nature of a DP random measure. Instead we use a convolution of a discrete DP random measure with a continuous kernel, that is, $F(y) = \int p(y \mid \theta) \, dH(\theta)$ and a hyperprior $H \sim \mathcal{DP}(\alpha, H_0)$. This is, of course, exactly the mixture model (6.2) with a DP prior on $H$,

$$y_i \mid \theta \sim p(y_i \mid \theta_i), \quad \theta_i \mid H \sim H, \quad H \sim \mathcal{DP}(\alpha, H_0). \tag{6.8}$$

**Example 6.1 (Gene expression data)** *Figure 6.2a shows measurements $y_i$ corresponding to EIG121 gene expression for $n = 51$ uterine cancer patients. We assume $y_i \sim F$ with a DPM prior (6.8). Figure 6.2b shows posterior inference on $F$ (as a pdf). Inference is based on 500 iterations of a Gibbs sampler, using an approximation with a finite DP prior.*

In Example 6.1 we used a DPM model for density estimation for an unknown distribution $F$ in an application that called for a continuous distribution. In some cases the DPM model is useful also for discrete distributions, when inference includes extrapolation beyond the support of the data, or, more generally, borrowing of strength across different parts of the sample space. For example, in the following problem $y_i = 0$ is censored, by the nature of the experiment, and we use a DPM of Poisson distributions for inference on $F$, including $F(0)$. An honest description of uncertainties on $F(0)$ is critical for this example.

**Example 6.2 (T-cell receptors)** *Guindani et al. (2014) consider data on counts of distinct T-cell receptors. The diversity of T-cell receptor types is an important characteristic of the immune system. A common summary of the diversity is the clonal-size distribution. The clonal-size distribution is the table of frequencies $\widehat{F}_y$ of counts $y = 1, 2, \ldots, n$. For example, $\widehat{F}_2 = 11$ means that there were 11 distinct T-cell receptors that were observed twice, etc. Table 6.1 shows the observed frequencies for one of the experiments considered in Guindani et al. (2014).*

**TABLE 6.1**
Example 6.2. Frequencies $\widehat{F}_y$ of counts $y_i = 1, 2, \ldots$; $\widehat{F}_0$ is censored

| counts $y$ | 0 | 1 | 2 | 3 | 4 | $\geq 5$ |
|---|---|---|---|---|---|---|
| frequencies $\widehat{F}_y$ | – | 37 | 11 | 5 | 2 | 0 |

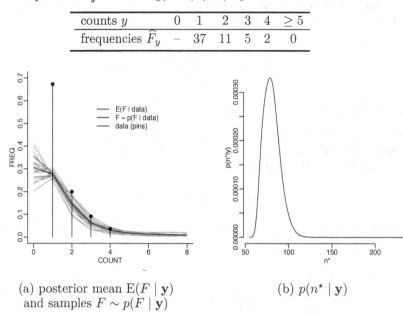

(a) posterior mean $E(F \mid \mathbf{y})$
and samples $F \sim p(F \mid \mathbf{y})$

(b) $p(n^\star \mid \mathbf{y})$

**FIGURE 6.3**
Example 6.2. (a) Posterior mean $E(F \mid \mathbf{y})$ (solid line) of the clonal size distribution $F$, and posterior draws $F \sim p(F \mid \mathbf{y})$ (dotted thin lines). Only the probabilities at the integer values $y = 1, 2, \ldots$ are meaningful. The points are connected only for a better display. The pins show the empirical frequencies $\widehat{F}_y$ (standardized for $y \geq 1$; whereas $F$ is standardized for $y \geq 0$). (b) The implied posterior $p(n^\star \mid \mathbf{y})$ for the total number of distinct T-cell receptors.

*Consider a model $y_i \sim F$, with Poisson kernels, $y_i \sim \mathcal{P}(\theta_i)$ and $\theta_i \sim H$ with $H \sim \mathcal{DP}(\alpha, H_0)$. Importantly, we do not include the constraint $y_i \geq 1$. Instead we assume that T-cell receptor counts are generated as $y_i \sim F$, $i = 1, \ldots, n^\star$, for $n^\star \geq n$, with $y_i = 0$ for $i = n + 1, \ldots, n^\star$. Without loss of generality we assume that the observed non-zero counts are the first $n$ counts. The last $n^\star - n$ counts are censored. The total number of T-cell receptors $n^\star$ becomes another model parameter. Posterior inference is implemented using a Gibbs sampler, with one of the transition probabilities sampling $n^\star$ conditional on currently imputed $\theta_i$, $i = 1, \ldots, n$. Figure 6.3a shows inference on the estimated distribution $F(y) = \int \mathcal{P}(y \mid \theta) \, dH(\theta)$ together with posterior draws $F \sim p(F \mid \mathbf{y})$. Figure Figure 6.3b shows the implied posterior distribution $p(n^\star \mid \mathbf{y})$.*

### 6.2.2 Posterior simulation in Dirichlet process mixture models

Posterior inference is usually implemented as posterior MCMC simulation, which becomes particularly easy when the sampling model $p(y_i \mid \theta_i)$ and the base measure $H_0$ of the DP prior are conjugate. Recall the definition of the cluster membership indicators $z_i = k$ if $i \in C_k$ and the unique values $\{\theta_1^\star, \ldots, \theta_{G_+}^\star\}$, and let $\mathbf{y}_k^\star = (y_i; \ i \in C_k)$ denote the data arranged by clusters. Also, let $\theta_{-i}$ denote $\theta$ with the $i$th element removed, and similarly for $\mathbf{z}_{-i}$ and $G_+^{-i}$.

*Posterior MCMC simulations*

Escobar (1988) proposed the first posterior Gibbs sampler for the DPM model (6.2), based on transition probabilities that update $\theta_i$ by draws from the complete conditional posterior $p(\theta_i \mid \theta_{-i}, \mathbf{y})$. However, this Gibbs sampler can be argued to suffer from a slowly mixing Markov chain. Bush & MacEachern (1996) proposed a variation using two types of transition probabilities. One updates $z_i$ by draws from the complete conditional posterior probability $p(z_i \mid \mathbf{z}_{-i}, \mathbf{y})$, after marginalizing with respect to $\theta$. The other type of transition probability generates $\theta_k^\star$ from $p(\theta_k^\star \mid \mathbf{z}, \mathbf{y})$. We first discuss the latter. This will help us to establish notation that we will need for the other transition probability.

One transition probability is defined as a Gibbs step, generating $\theta_k^\star$ from the complete conditional posterior

$$p(\theta_k^\star \mid \mathbf{z}, \mathbf{y}) \propto H_0(\theta_k^\star) \prod_{i \in C_k} p(y_i \mid \theta_k^\star). \tag{6.9}$$

Here we used that *a priori* $p(\theta_k^\star) = H_0(\theta_k^\star)$. This follows from the stick-breaking definition of the DP random measure. The posterior $p(\theta_k^\star \mid \mathbf{z}, \mathbf{y})$ is simply the posterior on $\theta_k^\star$ in a parametric model with prior $H_0(\theta_k^\star)$ and sampling model $p(y_i \mid \theta_k^\star)$, restricted to data $y_i$, $i \in C_k$.

The other type of transition probabilities updates $z_i$ by draws from the complete conditional posterior distribution $P(z_i = k \mid \mathbf{z}_{-i}, \mathbf{y})$, which is derived as follows. First consider $p(\theta_i \mid \theta_{-i}, \mathbf{y})$. The prior given in (6.7) is multiplied with the sampling distribution to get

$$p(\theta_i \mid \theta_{-i}, \mathbf{y}) \propto \sum_{k=1}^{G_+^{-i}} n_k^{-i} p(y_i \mid \theta_k^{\star,-i}) \delta_{\theta_k^{\star,-i}}(\theta_i) + \alpha p(y_i \mid \theta_i) H_0(\theta_i).$$

Recall that $G_+^{-i}$ is the number of unique values after excluding $\theta_i$, and similar for $n_k^{-i}$ and $\theta_k^{\star,-i}$. Recognizing that $\theta_i = \theta_k^{\star,-i}$ implies $z_i = k$, we can write the same conditional as a joint distribution for $(\theta_i, z_i)$. Finally, marginalizing with respect to $\theta_k^\star$ using (6.9) replaces $p(y_i \mid \theta_k^{\star,-i})$ by $p(y_i \mid z_i = k, \mathbf{y}_k^{\star,-i}) = \int p(y_i \mid \theta_k^{\star,-i}) \, dp(\theta_k^{\star,-i} \mid \mathbf{y}_k^{\star,-i})$, and we eventually find

$$P(z_i = k \mid \mathbf{z}_{-i}, \mathbf{y}) \propto \begin{cases} n_k^{-i} p(y_i \mid z_i = k, \mathbf{y}_k^{\star,-i}), & \text{for } k = 1, \ldots, G_+^{-i}, \\ \alpha \, h_0(y_i), & \text{for } k = G_+^{-i} + 1, \end{cases} \tag{6.10}$$

where $h_0(y_i) = \int p(y_i \mid \theta) \, dH_0(\theta)$. Iterating over draws from (6.9) and (6.10) defines a widely used posterior MCMC scheme for DPM models.

Of course, (6.10) is only of practical use if $h_0$ is easily evaluated, that is, when $p(y \mid \theta)$ and $H_0(\theta)$ form a conjugate pair. Several alternative posterior simulation methods have been proposed for the more general case. A good discussion appears in Neal (2000). In particular, Algorithm 8 in Neal (2000) provides an easily implemented posterior MCMC scheme for general DPM models. A common characteristic of (6.10) and Algorithm 8 is the use of marginal probabilities $p(z_i \mid \mathbf{z}_{-i}, \mathbf{y})$, marginalizing over the unknown $H$.

*Finite Dirichlet process*

Alternatively, Ishwaran & James (2001) introduced an approximate DPM model by truncating the stick-breaking representation (6.3) of the DP after $G$ terms by setting $v_G = 1$. Recall that $v_g$ are the beta-distributed fractions in (6.3). Let $H \sim \mathcal{DP}_G$ denote the finite DP truncated after $G$ terms, let $v = (v_1, \ldots, v_{G-1})$, $\mu = (\mu_1, \ldots, \mu_G)$ and define cluster membership indicators as $z_i = g$ if $\theta_i = \mu_g$ (allowing here for possibly empty clusters with $n_g = 0$). The model is known as the truncated DP. We write $H \sim \mathcal{DP}_G$. It is straightforward to implement a Gibbs sampler for $p(v, \mu, \mathbf{z} \mid \mathbf{y})$.

### 6.2.3   Dependent mixtures – the dependent Dirichlet process model

An attractive feature of the DPM model is the easy generalization to multiple related mixture models. Such models arise frequently in applications when inference for related populations, cases, etc. is required. Generically, let $x_i$ denote some covariate for the $i$th observation. For example, $x_i \in \{0, 1\}$ might record standard care versus experimental therapy for patients $i = 1, \ldots, n$ in a clinical trial. For the moment assume that $x_i$ is binary and consider the model

$$F_x(y_i) = p(y_i \mid x_i = x, H_x) = \int p(y_i \mid \theta) \, dH_x(\theta), \qquad (6.11)$$

where $\{H_0, H_1\}$ are two mixing measures, for example, with marginal DP prior, $H_x \sim \mathcal{DP}(\alpha, H_0)$. In many applications one would want to complete the model by specifying $p(H_0, H_1)$ such that $H_0$ and $H_1$ are dependent. Let

$$H_x = \sum_g \eta_g \delta_{\mu_{xg}}. \qquad (6.12)$$

The weights are already indexed by a single index $g$, common across $x$, in anticipation of the upcoming construction. A very elegant construction to achieve the desired dependence is by defining $p(\mu_{0g}, \mu_{1g})$ as a bivariate dependent distribution, but independent across $g$. In the general case of $\{H_x;\ x \in X\}$ with arbitrary index set $X$ the bivariate distribution $p(\mu_{0g}, \mu_{1g})$ is replaced by a suitable stochastic process indexed by $x$; for example, a Gaussian process for $x \in \mathbb{R}$. This is the construction of the dependent DP (DDP) of MacEachern (1999). In one of several variations of the model the weights are common across $x$, as we already anticipated above in (6.12).

**Example 6.3 (Oral cancer)**   *We use a dependent mixture of normal models as in (6.11) and (6.12) with a bivariate normal prior on $(\mu_{0g}, \mu_{1g})$ to model log survival times for $n = 80$ patients with cancer of the mouth (Klein & Moeschberger, 2003, Section 1.11). The covariate $x_i \in \{0, 1\}$ is an indicator for aneuploid ($x = 1$, abnormal number of chromosomes) versus diploid ($x = 0$) tumors. Figure 6.4 shows the posterior estimated distributions $\mathrm{E}(F_x \mid \mathbf{y})$, $x = 0, 1$. The estimates are plotted as survival functions on the absolute scale of the survival times. The thin lines show posterior simulations $F_x \sim p(F_x \mid \mathbf{y})$.*

Inference under the DP, the DDP, and many variations (and many other BNP models) is implemented in the R package DPpackage@DPpackage (Jara, 2007; Jara et al., 2011). Other public domain implementations include functions for some DP models in the R package bayesm (Rossi et al., 2005; Rossi & McCulloch, 2008). Also the Bayesian regression software by Karabatsos (2014) includes a wide variety of BNP regression problems in a menu-driven package.

## 6.3   Normalized Generalized Gamma Process Mixtures

### 6.3.1   NRMI construction

Yet another defining property of the DP is the construction as a normalized gamma process (Ferguson, 1973). The gamma process is a particular example of a much wider class of models known as completely random measures (CRMs; Kingman, 1993, Chapter 8). Consider any non-intersecting measurable subsets $A_1, \ldots, A_k$ of the desired sample space $X$. The defining

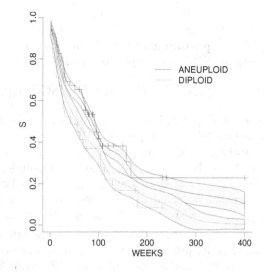

**FIGURE 6.4**
Estimated survival function $E(F_x \mid \mathbf{y})$ by tumor type (solid black and dashed curves).
The bands around the estimated survival functions show pointwise $\pm 1.0$ posterior standard
deviation bounds. The piecewise constant lines plot the Kaplan–Meier estimates.

property of a CRM $\tilde{H}$ is that $\tilde{H}(A_j)$ be mutually independent. The gamma process is a
CRM with $\tilde{H}(A_j) \sim \mathcal{G}(\alpha H_0(A_j), 1)$, for a probability measure $H_0$ and $\alpha > 0$. Normalizing
$\tilde{H}$ by $H(A_j) = \tilde{H}(A_j)/\tilde{H}(X)$ defines a DP prior with base measure $\alpha H_0$.

*Completely random measures and the normalized generalized gamma process*

Replacing the gamma process by any other CRM defines alternative BNP priors for random probability measures. Such priors are known as normalized random measures with
independent increments (NRMI) and were first described in Regazzini et al. (2003); they
include a large number of BNP priors. A recent review of NRMIs appears in Lijoi & Prünster (2010). Besides the DP prior, another interesting example is the normalized generalized
gamma process (NGG), discussed in Lijoi et al. (2007). We write $H \sim \mathrm{NGG}(\alpha, \kappa, \gamma, H_0)$.
The NGG is indexed by a total mass parameter $\alpha > 0$, two more scalar parameters $\kappa \geq 0$
and $\gamma \in [0, 1)$, and a base probability measure $H_0$. In fact, the DP is a special case of the
NGG with $\kappa = 1$ and $\gamma = 0$.

As for any CRM, a realization from the generalized gamma process (before normalization) can be generated using the following constructive definition (Kingman, 1993, Section
8.2). Assume we wish to generate a random measure $\tilde{H}$ on a measurable space $X$, for ex-
ample $\mathbb{R}^d$. We set up a Poisson process over $\mathbb{R}^+ \times X$ with Poisson intensity $\nu(\tilde{\eta}, \mu)$. The
choice of $\nu(\cdot)$ determines different CRMs. The arguments are already labeled in anticipation
of the next step. For the generalized gamma process we use

$$\nu(\tilde{\eta}, \mu) = \nu_1(\tilde{\eta})\,\alpha H_0(\mu), \quad \text{with } \nu_1(\tilde{\eta}) = e^{-\kappa\tilde{\eta}}\tilde{\eta}^{-(1+\gamma)}/\Gamma(1-\gamma). \tag{6.13}$$

Let $(\tilde{\eta}_g, \mu_g)$, $g = 1, \ldots$, denote a realization of this Poisson process. Then $\tilde{H} \propto \sum \tilde{\eta}_g \delta_{\mu_g}$ is a
realization of the desired CRM. Here, $\tilde{H}$ is still the (non-normalized) CRM. The normalized
measure $H$ rescales the weights $\tilde{\eta}_g$ to unit total mass.

*NGG mixtures*

Barrios et al. (2013) discuss mixture models with an NGG prior on the mixing measure, similar to (6.8), but with an NGG prior replacing the DP prior:

$$y_i \mid \theta \sim p(y_i \mid \theta_i), \quad \theta_i \mid H \sim H, \quad H \sim \text{NGG}(\alpha, \kappa, \gamma, H_0). \tag{6.14}$$

The discussion in Barrios et al. (2013) is more general, allowing for any other NRMI, but the NGG is a sufficiently rich model for most purposes. However, in comparison to the DP prior the additional flexibility of the NGG is important for modeling. This is extensively discussed in De Blasi et al. (2014) and Barrios et al. (2013). For example, consider two clusters $k$ and $\ell$ with cluster sizes $n_k > n_\ell$. As before, let $z_i$ denote a latent cluster membership indicator in an equivalent hierarchical model version of (6.14), let $\mathbf{z}_{-i} = (z_j, \ j \neq i)$, and define $n_k^{-i} = \sum_{j \neq i} \mathbb{I}(z_j = k)$ to be cluster sizes without the $i$th unit. Then *a priori* $\text{P}(z_i = k \mid \mathbf{z}_{-i})/\text{P}(z_i = \ell \mid \mathbf{z}_{-i}) = (n_k^{-i} - \gamma)/(n_\ell^{-i} - \gamma)$. The implication for data analysis is that cluster sizes under NGG priors with $\gamma > 0$ tend to be more concentrated, with few large clusters including most experimental units. Perhaps more importantly, the implied prior on the number of clusters, $G_+$, is more flexible under the NGG prior, in the sense that for matching prior means, hyperprior parameters can be chosen to allow for substantially more prior variance for $G_+$. This allows the number of clusters to be *a posteriori* adjusted as needed for the data. Under the DP prior, $p(G_+)$ is centered around approximately $\alpha \log(n)$. That is, prior centering determines $\alpha$, leaving no more flexibility to inflate prior variance.

### 6.3.2 Posterior simulation for normalized generalized gamma process mixtures

Most importantly, posterior inference under (6.14) is still easily implemented. Barrios et al. (2013), Favaro & Teh (2013), and Argiento et al. (2010) outline specific MCMC algorithms. These are based on a representation of the posterior distribution for NRMIs discussed in James et al. (2009), under independent sampling, $\theta_i \sim H$ as in (6.14), and an NRMI prior for $H$. Details of the general result are not needed for the upcoming algorithm for NGG mixtures. We only outline the setup, and give specific details for the NGG mixture. The representation involves a model augmentation of the posterior $p(\theta, \tilde{H} \mid \mathbf{y})$ under (6.14) with a latent variable, $u$, using

$$p(u \mid \theta, \tilde{H}, \mathbf{y}) = p(u \mid G_+) \propto u^{n-1}(u+\kappa)^{G_+ + \gamma - n} e^{-\alpha(u+\kappa)^\gamma / \gamma}. \tag{6.15}$$

For the following description of the algorithm it is convenient to distinguish atoms of $H$ that are matched with currently imputed $\theta_i$ versus unmatched atoms. Also posterior inference is most easily discussed for the random measure $\tilde{H}$, before normalization. As before, let $\{\theta_k^\star, \ k = 1, \ldots, G_+\}$ denote the unique $\theta_i$s. Then

$$\tilde{H} = \sum_{k=1}^{G_+} \eta_k^\star \delta_{\theta_k^\star} + \tilde{H}_C, \quad \text{with } \tilde{H}_C = \sum_{g=1}^{\infty} \tilde{\eta}_g \delta_{\mu_g}. \tag{6.16}$$

Note that the split of $\tilde{H}$ in (6.16) implicitly is a function of $\theta$ (to identify the unique $\theta_k^\star$) and can only be used when conditioning on $\theta$. In the MCMC implementation we approximate $\tilde{H}_C$ by using the $G$ terms with largest $\tilde{\eta}_g$ only. This is possible since the algorithm for generating $\tilde{H}_C$ samples the $\tilde{\eta}_g$ in decreasing order; see below. We can therefore assume that the weights are indexed by decreasing order, $\tilde{\eta}_g \geq \tilde{\eta}_{g+1}$. Let $\tilde{\eta} = (\tilde{\eta}_1, \ldots, \tilde{\eta}_G)$, $\mu = (\mu_1, \ldots, \mu_G)$. Finally, let $\eta^\star = (\eta_1^\star, \ldots, \eta_{G_+}^\star)$ and $\boldsymbol{m} = (n_1, \ldots, n_{G_+})$.

*An MCMC scheme for normalized generalized gamma process mixture models*

We describe the particular Gibbs sampling implementation that is proposed in Barrios et al. (2013). The following steps define transition probabilities of an MCMC scheme for the posterior distribution $p(\theta, \tilde{H}, u \mid \mathbf{y})$ under (6.14), augmented with (6.15). The algorithm includes five transition probabilities. Let $[a \mid b, c]$ indicate sampling from the conditional distribution of parameter $a$ given $b, c$. All distributions are complete conditional posterior distributions, with the absence of any variables in the conditioning set indicating conditional indpendence. In some cases dependence on $\theta$ is only indirectly through $\mathbf{z}$, or even just $G_+$ or $n_j$. The five steps are: (i) $[u \mid G_+]$; (ii) $[\eta_k^\star \mid u, n_k]$; (iii) $[\theta_k^\star \mid \mathbf{z}, \mathbf{y}]$; (iv) $[\eta, \mu \mid u]$; (v) $[\theta_i \mid \tilde{H}]$.

In step (i) we generate $u$ from (6.15). Favaro & Teh (2013) recommend instead sampling $v = \log(u)$, as the complete conditional distribution $p(v \mid G_+)$ turns out to be log concave, allowing for easier random variable generation. In step (ii) we update $\eta_k^\star$, $k = 1, \ldots, G_+$, by generating from the complete conditional posterior which under the NGG simplifies to

$$\eta_k^\star \mid u, n_j \sim \mathcal{G}(n_j - \gamma, \kappa + u).$$

In step (iii) we draw from the complete conditional posterior distribution for $\theta_k^\star$. This step is identical to (6.9).

Step (iv) updates $\tilde{H}_C$. James et al. (2009) show that, conditional on $u$, the random $\tilde{H}_C$ is again a CRM with Poisson intensity $\nu^\star(\tilde{\eta}, \mu)$, that is, replacing the original $\nu(\cdot, \cdot)$ by an updated Poisson intensity $\nu^\star(\cdot, \cdot)$. In the case of the NGG this simpifies to

$$\tilde{H}_C \sim \mathrm{NGG}(\alpha, \kappa + u, \gamma, H_0).$$

We can use the following easy algorithm to generate $\tilde{H}_C = \sum \tilde{\eta}_g \delta_{\mu_g}$. Ferguson & Klass (1972) introduce a clever scheme to generate the weights $\tilde{\eta}_g$ in decreasing order. The construction requires a function $N(v) = \int_v^\infty \nu_1(\tilde{\eta}) d\tilde{\eta}$, using the factor $\nu_1(\tilde{\eta})$ from definition (6.13), with $\kappa^\star = \kappa + u$ in place of $\kappa$. That is, $N(v) = \frac{\alpha}{\Gamma(1-\gamma)} \int_v^\infty e^{-(\kappa+u)\tilde{\eta}} \tilde{\eta}^{-(1+\gamma)} d\tilde{\eta}$. Next let $\xi_1, \xi_2, \ldots$ denote a realization from a unit-rate Poisson process, that is, $\xi_g - \xi_{g-1} \sim \mathcal{E}(1)$ are i.i.d. exponential draws (starting with $\xi_0 = 0$). Then

$$\tilde{\eta}_g = N^{-1}(\xi_g).$$

In words, plot the function $N(v)$ against $v \geq 0$, mark the $\xi_g$ on the vertical axis, and then use $N^{-1}(\cdot)$ to map $\xi_1, \xi_2, \ldots,$ to the horizontal $v$-axis. The construction delivers $\tilde{\eta}_g$, already ordered by decreasing size. The construction of $\tilde{H}_C$ is completed by generating the locations $\mu_g \sim H_0$, i.i.d.

Finally, in step (v) we resample $\theta$ by generating $p(\theta_i \mid \tilde{H}, y_i) \propto \tilde{H}(\theta_i) p(y_i \mid \theta_i)$. Write $\tilde{H} = \sum_{\ell=1}^\infty \tilde{w}_\ell \delta_{m_\ell}$, using a single running index $\ell$ for all terms in (6.16). Then $\mathrm{P}(\theta_i = m_\ell \mid y_i) \propto \tilde{w}_\ell \, p(y_i \mid m_\ell)$.

The described MCMC scheme is implemented in the R package BNPdensity, which is available in the CRAN package repository http://cran.r-project.org/.

An alternative MCMC scheme for posterior inference under model (6.14) is described in Favaro & Teh (2013) and also in Argiento et al. (2010). Favaro & Teh (2013) describe what can be characterized as a modified version of the Pólya urn. Recall that the Pólya urn (6.6) defines the marginal distribution of $(\theta_1, \ldots, \theta_n)$ under the DP prior, after marginalizing with respect to $H$. Similarly, Favaro & Teh (2013) describe a method for sampling $p(\theta_1, \ldots, \theta_n \mid u, \mathbf{y})$, marginalizing with respect to $H$. Generating $u \mid \theta$ proceeds as in step (i) above. Additionally, they describe the complete conditional posterior distributions for the NGG hyperparameters. This allows model (6.14) to be augmented with a hyperprior on the NGG parameters.

## 6.4    Bayesian Nonparametric Mixtures with Random Partitions

Recall that we started the discussion by observing that a mixture model (6.1) can naturally be thought of as a mixture with respect to a mixing measure, as in (6.2), and we proceeded by assuming BNP priors on the mixing measure.

There is another feature of hierarchical models like (6.2) with a discrete BNP prior $p(H)$ that naturally leads to a mixture model. Consider the posterior predictive distribution $F_{n+1}(y_{n+1}) = p(y_{n+1} \mid \mathbf{y})$. With an argument similar to (6.10) for $i = n + 1$, but without conditioning on $y_{n+1}$ and with instead an additional convolution with $p(y_{n+1} \mid z_{n+1} = k, \mathbf{y})$, we find

$$F_{n+1}(y_{n+1}) \equiv p(y_{n+1} \mid \mathbf{y}) \propto \sum_{k=1}^{G_+} n_k p(y_{n+1} \mid \mathbf{y}_k^\star) + \alpha h_0(y_{n+1}). \qquad (6.17)$$

Let $F(y) = \int p(y \mid \theta) \, dH(\theta)$, as earlier, and let $\overline{F} = \mathrm{E}\,(F \mid \mathbf{y})$ denote the posterior expectation. Then $F_{n+1} = \overline{F}$. That is, the posterior predictive distribution (6.17) coincides with the posterior expectation of the random probability measure. This is easily seen by considering $\mathrm{P}(y_{n+1} \leq c \mid \mathbf{y}) = \mathrm{E}(\mathrm{P}(y_{n+1} \leq c \mid F, \mathbf{y}) \mid \mathbf{y}) = \mathrm{E}(F(c) \mid \mathbf{y})$. Here, we overload notation to let $F(c)$ indicate the cdf under the probabilty measure $F$.

In the outlined construction the nature of $F_{n+1}$ as a mixture model arises from the implied random partition $p(\mathcal{C}_n)$ under i.i.d. sampling $\theta_i \sim H$ from the discrete random probability measure $H = \sum_g \eta_g \delta_{\mu_g}$. In that case the mixture model $F_{n+1}$ is just another manifestation of the assumed mixture model $F(y) = \sum \eta_g p(y \mid \theta_g)$, and does not introduce fundamentally new structure. In fact, any exchangeable random partition $p(\mathcal{C}_n)$ can be argued to arise from such a construction. See, for example, Lee et al. (2013b) for a review. The attraction of exchangeable random partitions is coherence and mathematical tractability.

However, if the inference goal is a posterior predictive distribution in the form of a mixture model, as in (6.17), then the same form can be achieved with any underlying random partition model $p(\mathcal{C}_n)$, including possibly non-exchangeable random partitions.

### 6.4.1    Locally weighted mixtures

An attractive general framework for random partitions are the product partition models (PPMs; Hartigan, 1990) which take the form

$$p(\mathcal{C}_n = \{C_1, \dots, C_{G_+}\}) \propto \prod_{j=1}^{G_+} c(C_j)$$

for some functions $c(C_j)$, which are known as the *cohesion* functions. The cohesion functions are restricted to be non-negative functions of $C_j$, but in principle any such function is valid. If $c(C)$ is only a function of the size of $C$, then the resulting model for $\mathcal{C}_n$ is invariant under permutations of the indices. If additionally $p(\mathcal{C}_n) = \sum_{z_{n+1}} p(\mathcal{C}_{n+1})$, then we are back to exchangeable partitions. For example, with $c(C) = \alpha \times (|C| - 1)!$ the PPM reduces to the Pólya urn (6.6). More general, it can be shown that the family of all PPM models with cohesion function $c(C) = c(|C|)$ that depend on $C$ only indirectly through the cardinality and that define exchangeable random partitions coincides with the family of so-called Gibbs-type priors. See, for example, De Blasi et al. (2014) for a discussion. A subset of the Gibbs-type priors, in turn, are the NGG models that we discussed before (De Blasi et al., 2014). Another notable Gibbs-type prior is the Pitman–Yor model, of which incidentally the DP is again a special case (De Blasi et al., 2014).

However, abandoning the restriction to exchangeable partitions allows for other interesting variations of PPM models. Consider, for example, the problem of clustering patients in a clinical trial. Usually important baseline covariates $x_i$ are available for each patient, and it might be desireable to favor clusters of patients who are more homogeneous with respect to these baseline covariates. Let $\mathbf{x}_k^\star = (x_i; \; i \in C_k)$ denote the baseline covariates arranged by cluster, and similarly for $\mathbf{y}_k^\star$. Müller et al. (2011) introduce the PPMx model by replacing the cohesion function $c(C_k)$ in a PPM with $c(C_k)g(\mathbf{x}_k^\star)$. Here $g(\mathbf{x}_k^\star)$ is termed a similarity function. It is any function that penalizes for lack of heterogeneity of $\mathbf{x}_k^\star$. For example, for categorical covariates $g(\mathbf{x}_k^\star) = 1/m_k$ could be related to the number $m_k$ of distinct values $x_i$, $i \in C_k$. The implied posterior predictive distribution is a locally weighted mixture. Let $\mathbf{x} = (x_1, \ldots, x_n)$ and $\mathbf{z} = (z_1, \ldots, z_n)$, and write $y = y_{n+1}$, $x = x_{n+1}$ and $z = z_{n+1}$ for the variables of a future patient. Then

$$\eta_k(x) \equiv \mathrm{P}(z = k \mid x, \theta, \mathbf{x}, \mathbf{z}) \propto \frac{g(\mathbf{x}_k^\star \cup x)c(C_k \cup \{n+1\})}{g(\mathbf{x}_k^\star)c(C_k)}$$

for $k = 1, \ldots, G_+, G_+ + 1$, with the understanding that the denominator evaluates as 1 for $k = G_+ + 1$. And the posterior predictive distribution for a future observation $y = y_{n+1}$ with covariates $x = x_{n+1}$, still conditional on $\theta$ and $\mathbf{z}$, becomes

$$p(y \mid x, \mathbf{x}, \mathbf{z}) = \sum_{k=1}^{G_+} \eta_k(x)p(y \mid x, \mathbf{x}_k^\star, \mathbf{y}_k^\star) + \eta_{G_+ + 1}(x)h_0(y \mid x), \qquad (6.18)$$

where $p(y \mid x, \mathbf{x}_k^\star, \mathbf{y}_k^\star) = \int p(y \mid x, \theta_k^\star) \, dp(\theta_k^\star \mid \mathbf{x}_k^\star, \mathbf{y}_k^\star)$ and $h_0(y \mid x) = \int p(y \mid x, \theta) \, dH_0(\theta)$. The form of the predictive distribution is a mixture over the clusters of a random partition, similar to $F_{n+1}$ in (6.17), but now as a locally weighted mixture with weights $\eta_k(x)$ that are indexed by the covariate $x$. Marginalizing with respect to $\theta$ and $\mathbf{z}$, the posterior predictive distribution $p(y_{n+1} \mid x_{n+1}, \mathbf{y})$ adds additional posterior averaging with respect to $p(\theta, \mathbf{z} \mid \mathbf{y})$, that is, $F_{n+1}(y \mid x, \mathbf{y}) = \mathrm{E}(p(y \mid x, \theta, \mathbf{x}, \mathbf{z}) \mid \mathbf{y})$. The form of (6.18) as a mixture with covariate dependent weights defines a mixture of experts model as discussed in Chapter 12.

### 6.4.2   Conditional regression

An alternative, natural way to introduce similar covariate dependent mixtures, again as a posterior predictive distribution $p(y_{n+1} \mid \mathbf{y}, \mathbf{x}, x_{n+1})$ in a BNP model, is the following construction. We proceed as if the pairs $(x_i, y_i)$ were independent random samples from a joint distribution $(x_i, y_i) \sim F$, complete the model with a DPM on $F$, and report inference on $F$. This is introduced in Müller et al. (1996) and Park & Dunson (2010). The construction is easiest when both $x_i$ and $y_i$ are continuous. Consider a DPM of normal kernels, mixing with respect to location and scale. Write the DPM as a hierarchical model as in (6.8), using the kernel $p(x, y \mid \theta = (\mu, \Sigma)) = \mathcal{N}(\mu, \Sigma)$:

$$(x_i, y_i \mid \theta_i) \overset{\mathrm{ind}}{\sim} \mathcal{N}(\mu_i, \Sigma_i), \quad \theta_i \sim H \text{ and } H \sim \mathcal{DP}(\alpha, H_0).$$

As before, let $\theta_k^\star = (\mu_k^\star, \Sigma_k^\star)$, $k = 1, \ldots, G_+$, denote the unique values of $\theta_i$, $i = 1, \ldots, n$, with multiplicities $n_k$. Let $f(y \mid x, \theta_k^\star)$ denote the conditional normal density in $y$ given $x$ that is implied by the multivariate normal $\mathcal{N}(\mu_k^\star, \Sigma_k^\star)$, and let $\eta(x \mid \theta_k^\star)$ denote the marginal normal density in $x$ under $\mathcal{N}(\mu_k^\star, \Sigma_k^\star)$. Similarly, let $f_0(y \mid x)$ and $\eta_0(x)$ denote the implied conditional and marginal when $\theta^\star$ is generated from $H_0$, that is, $f_0(y \mid x) = \int f(y \mid x, \theta) \, dH_0(\theta)$ and $\eta_0(x) = \int \eta(x \mid \theta) \, dH_0(\theta)$. The posterior predictive distribution for a future observation $y = y_{n+1}$ with covariates $x = x_{n+1}$ takes the form

$$p(y \mid x, \theta^\star) \propto \alpha\, \eta_0(x) f_0(y \mid x) + \sum_{k=1}^{G_+} n_k\, \eta(x \mid \theta_k^\star)\, f(y \mid x, \theta_k^\star),$$

similar to (6.18). The construction introduced an additional – one could argue, inappropriate – factor in the likelihood by including $x_i$ in the hypothetical multivariate response $(x_i, y_i)$. The approach only works easily when $x_i$ and $y_i$ are both continuous. In the case of mixed data types it is more natural to separate the two, as in the PPMx model.

*Dependent Dirichlet process mixtures*

Finally, recall the construction of DDP model. This provides yet another approach to construct covariate dependent mixture models, now on the basis of a BNP prior $p(H_x, \ x \in X)$ on the mixing measure in a mixture, now indexed with the covariate $x$. This construction is not much used.

## 6.5   Repulsive Mixtures (Determinantal Point Process)

In many applications investigators seek to interpret the components of a mixture model, or equivalently, the clusters in a random partition model, as scientifically meaningful structure. For example, researchers have recognized that breast cancer is highly heterogeneous and different subgroups involve different disease mechanisms. Depending on the nature of the experiment and the data, a mixture model such as (6.1) could be used to recover such latent subgroups. For example, investigators might identify molecular drivers based on gene expression profiles of different tumor subtypes. In such applications, maximizing the diversity of inferred molecular drivers helps identify and interpret these distinct mechanisms.

However, the use of independent priors for component-specific parameters $\mu_g$ in (6.1) and, similarly, $\theta_g^\star$ in (6.2) does not provide any prior regularization to favor such diversity. To the contrary, the independent prior gives rise to concerns about overfitting which generates redundant mixture components with similar parameters, leading to unnecessarily complex models and poor interpretability. In particular, this compromises the interpretation of the mixture components as biologically meaningful structure. Rousseau & Mengersen (2011) argue that concerns are partially mitigated with carefully chosen priors which lead to redundant structure being asymptotically removed; see also Chapter 4. Alternatively, Petralia et al. (2012) proposed a class of repulsive priors for mixture components. The proposed repulsive prior is based on a distance metric in which small distances are penalized. However, posterior computations are complex and do not easily scale to higher-dimensional problems.

*Determinantal Point Process*

Xu et al. (2016) argue for the use of a determinantal point process (DPP) to build an alternative prior that imposes the desired repulsive prior regularization across mixture components. The DPP is a point process that generates random configurations $\{\mu_1, \ldots, \mu_G\}$ in a way that highly favors very diverse atoms $\mu_g$. The cardinality $G$ is part of the random configuration. The use of DPP priors for statistical inference in mixtures was first proposed in aff (????). Xu et al. (2016) generalize the setup by discussing more general inference for

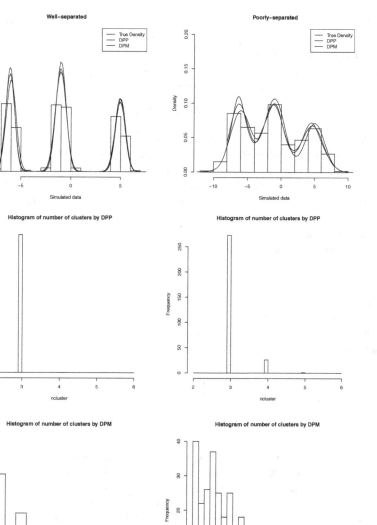

**FIGURE 6.5**

The top panel plots show the histograms of two simulated data sets with true density (solid), estimated density by DPP prior (dashed) and DPM (dotted). The middle and bottom panels present the histograms of the estimated number of clusters by DPP prior and DPM, respectively.

latent structures and by developing an efficient transdimensional posterior MCMC scheme that allows inference across $G$.

The DPP defines a point process on $C \subseteq \mathbb{R}^d$. We first define a point process for a finite state space, $S = \{\omega_1, \dots, \omega_R\}$, for example a grid in $\mathbb{R}^d$ (in that case $R$ would be the number of grid cells). Let $C$ denote an $(R \times R)$ positive semidefinite matrix, constructed, for example, as $C_{ij} = C(\omega_i, \omega_j)$ with a covariance function $C(\omega_i, \omega_j)$. Let $C_A$ denote the submatrix of rows and columns indicated by $A \subseteq S$. We define a random point configuration $X = \{\mu_1, \ldots, \mu_G\}$. The $\mu_g$ will later become the component-specific parameters, when we use the DPP to construct a prior probability model for a mixture. We define

$$\mathrm{P}(X = A) \propto \det(C_A) \tag{6.19}$$

as a probability distribution on the $2^n$ possible point configurations $X \subset S$. This defines a subclass of DPPs known as L-ensembles. It is easy to see why the DPP defines a repulsive point process if one interprets the determinant as the volume of a parallelotope spanned by the column vectors of $C_A$. Equal or similar column vectors span less volume than very diverse ones. A good review of DPP models for finite state spaces, including the derivation of the normalizing constant in (6.19), appears in Kulesza & Taskar (2012).

For a continuous state space $C \subseteq \mathbb{R}^d$, we define an L-ensemble by a density $f(X)$ with respect to the unit-rate Poisson process as

$$f(X) \propto \det(C_X), \tag{6.20}$$

for $X = \{\mu_1, \ldots, \mu_G\}$. As before, $C_X$ is a $(G \times G)$ matrix with $(i, j)$th entry defined by a continuous covariance function $C(\mu_i, \mu_j)$. We write $X \sim \mathrm{DPP}(C)$, or $X \sim \mathrm{DPP}(C, \phi, \tau)$ when $\phi$ and $\tau$ are included as unknown hyperparameters in the definition of the covariance function $C(\mu_i, \mu_j)$.

*Determinantal point process mixtures*

Xu et al. (2016) propose the DPP mixture model as the sampling model (6.1) with prior

$$(\mu_1, \ldots, \mu_G) \sim \mathrm{DPP}(C), \quad \eta \mid G, \delta \sim \mathcal{D}(\delta, \ldots, \delta).$$

The number of terms $G$ in the mixture is part of the random point configuration. Posterior inference is implemented as transdimensional MCMC. Essentially, the density (6.20) with respect to the unit-rate Poisson process can be used in a reversible jump MCMC as if it were a density with respect to the Lebesgue measure. In the Metropolis–Hastings acceptance probability the density $f(X)$ is replaced by $f(X)/G!$ (Xu et al., 2016).

Figure 6.5 summarizes a small simulation study using the proposed DPP mixture model, with a normal kernel $y_i \mid \mu_g \sim \mathcal{N}(\mu_g, \sigma^2)$ and an additional gamma prior on the common precision parameter, $1/\sigma^2 \sim \mathcal{G}(a_0, b_0)$. We used a squared exponential covariance function. The implemented model also includes hyperparameters for the covariance function. See Xu et al. (2016) for details about inference for the covariance function hyperparameters. The top panel shows the simulation truth, density estimates under the DPP model and under an equivalent DPM model for two simulated data sets. The middle and bottom panels show the posterior distribution of the number $G$ of clusters under the DPP prior and DPM, respectively. Both the DPP and DPM models lead to very similar density estimates. However, inference under the DPP reports far fewer clusters, simply because of its repulsive feature.

## 6.6   Concluding Remarks

Starting from an interpretation of a generic mixture model as an expectation with respect to a random mixing measure, we discussed BNP approaches to inference in mixture models.

In some cases the BNP nature of a mixture model is only one of perspective. For example, inference under a finite mixture model can be meaningfully seen as inference under a BNP model when the focus is on inference for the implied (finite) discrete mixing measure or when the finite mixing measure is constructed as an approximation of an infinite discrete BNP prior. This is the case, for example, for the popular finite DP prior.

We reviewed BNP priors for the mixing measure in mixtures of a parametric kernel with respect to a nonparametric model. Taking this perspective, we excluded in particular a discussion of mixtures of BNP models, that is, mixtures of nonparametric models with respect to parametric (hyper)priors. A common example of such models are mixtures of Pólya tree models which are used to mitigate the lack of smoothness of density estimates under a Pólya tree prior (Hanson & Johnson, 2002).

Another related large class of BNP models generalizes the random partition models that we briefly described in this chapter. A partition is a family of non-overlapping subsets of $[n]$. More general feature allocation models generate instead possibly overlapping subsets, also without the requirement that the union be $[n]$. One of the most widely used feature allocation models is the Indian buffet process; see, for example, Broderick et al. (2013) for a good review of random partitions and feature allocation models.

Finally, another interesting class of BNP models in mixtures that we did not discuss in this review are hierarchical extensions of the DPM. The hierarchical DP of Teh et al. (2006) defines multiple DPM models, with a hierarchical hyperprior that allows the mixtures to share atoms across submodels. Similar models are discussed in Rodríguez et al. (2008) and Lee et al. (2013a), with different notions of hierarchical extensions and ways of linking the submodels; see also Chapter 17 for further BNP models and their application in finance.

# Bibliography

(????).

ARGIENTO, R., GUGLIELMI, A. & PIEVATOLO, A. (2010). Bayesian density estimation and model selection using nonparametric hierarchical mixtures. *Computational Statistics & Data Analysis* **54**, 816–832.

BARRIOS, E., NIETO-BARAJAS, L. E. & PRÜNSTER, I. (2013). A study of normalized random measures mixture models. *Statistical Science* **28**, 313–334.

BLACKWELL, D. & MACQUEEN, J. B. (1973). Ferguson distributions via Pólya urn schemes. *Annals of Statistics* **1**, 353–355.

BRODERICK, T., JORDAN, M. I. & PITMAN, J. (2013). Cluster and feature modeling from combinatorial stochastic processes. *Statistical Science* **28**, 289–312.

BUSH, C. A. & MACEACHERN, S. N. (1996). A semiparametric Bayesian model for randomised block designs. *Biometrika* **83**, 275–285.

DE BLASI, P., FAVARO, S., LIJOI, A., MENA, R., PRÜNSTER, I. & RUGGIERO, M. (2014). Are Gibbs-type priors the most natural generalization of the Dirichlet process? *IEEE Transactions on Pattern Analysis and Machine Intelligence* **37**, 212–229.

ESCOBAR, M. D. (1988). *Estimating the Means of Several Normal Populations by Nonparametric Estimation of the Distributions of the Means*. Unpublished doctoral thesis, Department of Statistics, Yale University.

ESCOBAR, M. D. (1994). Estimating normal means with a Dirichlet process prior. *Journal of the American Statistical Association* **89**, 268–277.

ESCOBAR, M. D. & WEST, M. (1995). Bayesian density estimation and inference using mixtures. *Journal of the American Statistical Association* **90**, 577–588.

FAVARO, S. & TEH, Y. W. (2013). MCMC for normalized random measure mixture models. *Statistical Science* **28**, 335–359.

FERGUSON, T. S. (1973). A Bayesian analysis of some nonparametric problems. *Annals of Statistics* **1**, 209–230.

FERGUSON, T. S. & KLASS, M. J. (1972). A representation of independent increment processes without Gaussian components. *Annals of Mathematical Statistics* **43**, 1634–1643.

GHOSHAL, S. & VAN DER VAART, A. (2017). *Fundamentals of Nonparametric Bayesian Inference*. Cambridge: Cambridge University Press.

GUINDANI, M., SEPÚLVEDA, N., PAULINO, C. D. & MÜLLER, P. (2014). A Bayesian semi-parametric approach for the differential analysis of sequence counts data. *Applied Statistics* **63**, 385–404.

HANSON, T. & JOHNSON, W. O. (2002). Modeling regression error with a mixture of Polya trees. *Journal of the American Statistical Association* **97**, 1020–1033.

HARTIGAN, J. A. (1990). Partition models. *Communications in Statistics: Theory and Methods* **19**, 2745–2756.

HJORT, N. L., HOLMES, C., MÜLLER, P. & WALKER, S. (2010). *Bayesian Nonparametrics*. Cambridge: Cambridge University Press.

ISHWARAN, H. & JAMES, L. F. (2001). Gibbs sampling methods for stick-breaking priors. *Journal of the American Statistical Association* **96**, 161–173.

JAMES, L. F., LIJOI, A. & PRÜNSTER, I. (2009). Posterior analysis for normalized random measures with independent increments. *Scandinavian Journal of Statistics* **36**, 76–97.

JARA, A. (2007). Applied Bayesian non- and semi-parametric inference using DPpackage. *Rnews* **7**, 17–26.

JARA, A., HANSON, T. E., QUINTANA, F. A., MÜLLER, P. & ROSNER, G. L. (2011). DPpackage: Bayesian semi- and nonparametric modeling in R. *Journal of Statistical Software* **40**, 1–30.

KARABATSOS, G. (2014). Software user's manual for Bayesian regression: Nonparametric and parametric models. Tech. rep., University of Illinois-Chicago.

KINGMAN, J. F. C. (1993). *Poisson Processes*. Oxford: Oxford University Press.

KLEIN, J. P. & MOESCHBERGER, M. L. (2003). *Survival Analysis: Techniques for Censored and Truncated Data*. New York: Springer-Verlag.

KULESZA, A. & TASKAR, B. (2012). Determinantal point processes for machine learning. Preprint, arXiv:1207.6083.

LEE, J., MÜLLER, P., ZHU, Y. & JI, Y. (2013a). A nonparametric Bayesian model for local clustering with application to proteomics. *Journal of the American Statistical Association* **108**, 775–788.

LEE, J., QUINTANA, F., MÜLLER, P. & TRIPPA, L. (2013b). Defining predictive probability functions for species sampling models. *Statistical Science* **28**, 209–222.

LIJOI, A., MENA, R. H. & PRÜNSTER, I. (2007). Controlling the reinforcement in Bayesian non-parametric mixture models. *Journal of the Royal Statistical Society, Series B* **69**, 715–740.

LIJOI, A. & PRÜNSTER, I. (2010). Models beyond the Dirichlet process. In *Bayesian Nonparametrics*, N. L. Hjort, C. Holmes, P. Müller & S. G. Walker, eds. Cambridge: Cambridge University Press, pp. 80–136.

LO, A. Y. (1984). On a class of Bayesian nonparametric estimates I: Density estimates. *Annals of Statistics* **12**, 351–357.

MACEACHERN, S. (1999). Dependent nonparametric processes. In *ASA Proceedings of the Section on Bayesian Statistical Science*. American Statistical Association, Alexandria, VA.

MÜLLER, P., ERKANLI, A. & WEST, M. (1996). Bayesian curve fitting using multivariate normal mixtures. *Biometrika* **83**, 67–79.

MÜLLER, P., QUINTANA, F., JARA, A. & HANSON, T. (2015). *Bayesian Nonparametric Data Analysis*. Cham: Springer-Verlag.

MÜLLER, P., QUINTANA, F. & ROSNER, G. (2011). A product partition model with regression on covariates. *Journal of Computational and Graphical Statistics* **20**, 260–278.

NEAL, R. M. (2000). Markov chain sampling methods for Dirichlet process mixture models. *Journal of Computational and Graphical Statistics* **9**, 249–265.

PARK, J.-H. & DUNSON, D. (2010). Bayesian generalized product partition models. *Statistica Sinica* **20**, 1203–1226.

PETRALIA, F., RAO, V. & DUNSON, D. B. (2012). Repulsive mixtures. In *Advances in Neural Information Processing Systems 25*, F. C. N. Pereira, C. J. C. Burges, L. Bottou & K. Q. Weinberger, eds. Red Hook, NY: Curran Associates, Inc., pp. 1889–1897.

PHADIA, E. G. (2013). *Prior Processes and Their Applications*. Heidelberg: Springer-Verlag.

PITMAN, J. (2006). *Combinatorial Stochastic Processes*, vol. 1875 of *Lecture Notes in Mathematics*. Berlin: Springer-Verlag. Lectures from the 32nd Summer School on Probability Theory held in Saint-Flour, July 7–24, 2002, with a foreword by Jean Picard.

REGAZZINI, E., LIJOI, A. & PRÜNSTER, I. (2003). Distributional results for means of normalized random measures with independent increments. *Annals of Statistics* **31**, 560–585.

RODRÍGUEZ, A., DUNSON, D. B. & GELFAND, A. E. (2008). The nested Dirichlet process, with discussion. *Journal of the American Statistical Association* **103**, 1131–1144.

ROSSI, P., ALLENBY, G. & MCCULLOCH, R. (2005). *Bayesian Statistics and Marketing*. New York: John Wiley.

ROSSI, P. & MCCULLOCH, R. (2008). *bayesm: Bayesian inference for marketing/microeconometrics*. R package version 2.2-2.

ROUSSEAU, J. & MENGERSEN, K. (2011). Asymptotic behaviour of the posterior distribution in overfitted mixture models. *Journal of the Royal Statistical Society, Series B* **73**, 689–710.

SETHURAMAN, J. (1994). A constructive definition of Dirichlet priors. *Statistica Sinica* **4**, 639–650.

TEH, Y. W., JORDAN, M. I., BEAL, M. J. & BLEI, D. M. (2006). Sharing clusters among related groups: Hierarchical Dirichlet processes. *Journal of the American Statistical Association* **101**, 1566–1581.

XU, Y., MÜLLER, P. & TELESCA, D. (2016). Bayesian inference for latent biologic structure with determinantal point processes (DPP). *Biometrics* **72**, 955–964.

# 7

# Model Selection for Mixture Models – Perspectives and Strategies

**Gilles Celeux, Sylvia Frühwirth-Schnatter and Christian P. Robert**

*INRIA Saclay, France; Vienna University of Economics and Business, Austria; Université Paris-Dauphine, France and University of Warwick, UK*

## CONTENTS

## 7.1   Introduction

Determining the number $G$ of components in a finite mixture distribution defined as

$$y \sim \sum_{g=1}^{G} \eta_g f_g(y|\theta_g), \tag{7.1}$$

is an important and difficult issue. This is a most important question, because statistical inference about the resulting model is highly sensitive to the value of $G$. Selecting an erroneous value of $G$ may produce a poor density estimate. This is also a most difficult question from a theoretical perspective as it relates to unidentifiability issues of the mixture model, as discussed already in Chapter 4. This is a most relevant question from a practical viewpoint since the meaning of the number of components $G$ is strongly related to the modelling purpose of a mixture distribution.

From this perspective, the famous quote from Box (1976), "All models are wrong, but some are useful" is particularly relevant for mixture models since they may be viewed as a semi-parametric tool when addressing the general purpose of density estimation or as a model-based clustering tool when concerned with unsupervised classification; see also Chapter 1. Thus, it is highly desirable and ultimately profitable to take into account the grand modelling purpose of the statistical analysis when selecting a proper value of $G$, and we distinguish in this chapter between selecting $G$ as a density estimation problem in Section 7.2 and selecting $G$ in a model-based clustering framework in Section 7.3.

Both sections will discuss frequentist as well as Bayesian approaches. At a foundational level, the Bayesian approach is often characterized as being highly directive, once the prior distribution has been chosen (see, for example, Robert, 2007). While the impact of the prior on the evaluation of the number of components in a mixture model or of the number of clusters in a sample from a mixture distribution cannot be denied, there exist competing ways of assessing these quantities, some borrowing from point estimation and others from hypothesis testing or model choice, which implies that the solution produced will strongly depend on the perspective adopted. We present here some of the Bayesian solutions to the different interpretations of picking the "right" number of components in a mixture, before concluding on the ill-posed nature of the question.

As already mentioned in Chapter 1, there exists an intrinsic and foundational difference between frequentist and Bayesian inferences: only Bayesians can truly *estimate* $G$, that is, treat $G$ as an additional unknown parameter that can be estimated simultaneously with the other model parameters $\theta = (\eta_1, \ldots, \eta_G, \theta_1, \ldots, \theta_G)$ defining the mixture distribution (7.1). Nevertheless, Bayesians very often rely on model selection perspectives for $G$, meaning that Bayesian inference is carried out for a range of values of $G$, from 1, say, to a pre-specified maximum value $G_{\max}$, given a sample $\mathbf{y} = (y_1, \ldots, y_n)$ from (7.1). Each value of $G$ thus corresponds to a potential model $\mathcal{M}_G$, and those models are compared via Bayesian model selection. A typical choice for conducting this comparison is through the values of the marginal likelihood $p(\mathbf{y}|G)$,

$$p(\mathbf{y}|G) = \int p(\mathbf{y}|\theta, G) p(\theta|G) \mathrm{d}\theta, \tag{7.2}$$

separately for each mixture model $\mathcal{M}_G$, with $p(\theta|G)$ being a prior distribution for all unknown parameters $\theta$ in a mixture model with $G$ components.

However, cross-model Bayesian inference on $G$ is far more attractive, at least conceptually, as it relies on one-sweep algorithms, namely computational procedures that yield

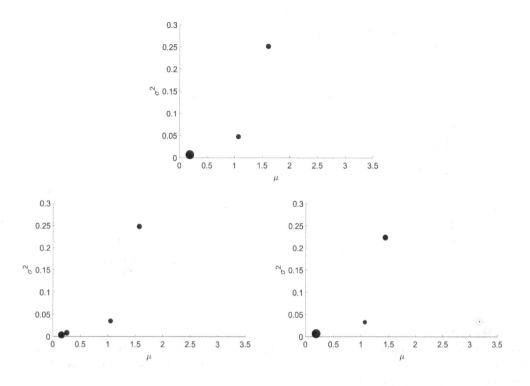

**FIGURE 7.1**

Point process representation of the estimated mixture parameters $(\hat{\mu}_g, \hat{\sigma}_g^2)$ for three mixture distributions fitted to the enzyme data using a Bayesian framework under the prior of Richardson & Green (1997). The size of each point corresponds to the mixture weight $\hat{\eta}_g$. Top: $G = 3$. Bottom: $G = 4$ with $\eta \sim \mathcal{D}_4(4)$ (left) and $\eta \sim \mathcal{D}_4(0.5)$ (right; the very small fourth component is marked by a circle).

estimators of $G$ jointly with the unknown model parameters. Section 7.4 reviews such one-sweep Bayesian methods for cross-model inference on $G$, ranging from well-known methods such as reversible jump Markov chain Monte Carlo (MCMC) to more recent ideas involving sparse finite mixtures relying on overfitting in combination with a prior on the weight distribution that forces sparsity.

## 7.2 Selecting $G$ as a Density Estimation Problem

When the estimation of the data distribution is the main purpose of the mixture modelling, it is generally assumed that this distribution truly is a finite mixture distribution. One inference issue is then to find the true number of mixture components, $G$, that is, the *order* of the mixture behind the observations. This assumption is supposed to produce well-grounded tests and model selection criteria.

The *true order* of a finite mixture model is the smallest value of $G$ such that the components of the mixture in (7.1) are all distinct and the mixing proportions are all positive (that is, $\theta_g \neq \theta_{g'}$, $g \neq g'$ and $\eta_g > 0$). This definition attempts to deal with the ambiguity (or

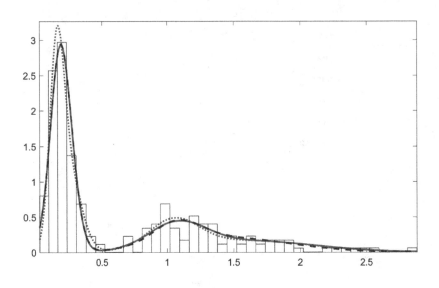

**FIGURE 7.2**
Histogram of the enzyme data together with three fitted mixture distributions: $G = 3$ (solid line); $G = 4$ and $\eta \sim \mathcal{D}_4 (4)$ (dotted line); $G = 4$ and $\eta \sim \mathcal{D}_4 (0.5)$ (dashed line). The dashed and solid lines are nearly identical.

non-identifiability) due to overfitting, discussed in Section 1.3 of Chapter 1 and Section 4.2.2 of Chapter 4: a mixture with $G$ components can equally be defined as a (non-identifiable) mixture with $G + 1$ components where the additional component either has a mixing proportion $\eta_{G+1}$ equal to zero or the parameter $\theta_{G+1}$ is identical to the parameter $\theta_g$ of some other component $g \in \{1, \ldots, G\}$. These identifiability issues impact both frequentist and Bayesian methods for selecting $G$. Hence, the order $G$ is a poorly defined quantity and in practical mixture analysis it is often difficult to decide what order $G$ describes the data best.

By way of illustration, a mixture of normal distributions $\mathcal{N}(\mu_g, \sigma_g^2)$ with $G = 3$ components is fitted within a Bayesian framework to the enzyme data studied in Richardson & Green (1997), using the same prior as Richardson & Green, in particular a uniform prior on the weight distribution $\eta = (\eta_1, \ldots, \eta_G)$. In addition, mixtures with $G = 4$ components are fitted, but with different symmetric Dirichlet priors for $\eta$, namely $\eta \sim \mathcal{D}_4 (4)$ and $\eta \sim \mathcal{D}_4 (0.5)$. As discussed in Section 4.2.2 above, the first prior favours overlapping components, whereas the second prior favours small components, should the mixture be overfitting.

Full conditional Gibbs sampling is applied for posterior inference. All three mixture models are identified by $k$-means clustering in the point process representation of the posterior draws of $(\mu_g, \sigma_g)$. The estimated component parameters $(\hat{\mu}_g, \hat{\sigma}_g^2, \hat{\eta}_g)$ are visualized through a point process representation in Figure 7.1. Obviously, the parameters for the four-component mixture are quite different and emerge in quite different ways than the components of the three-component mixture. The component $(\hat{\mu}_g, \hat{\sigma}_g^2, \eta_g) = (0.19, 0.007, 0.61)$ is split into the two components $(\hat{\mu}_g, \hat{\sigma}_g^2) = (0.16, 0.003)$ and $(\hat{\mu}_{g'}, \hat{\sigma}_{g'}^2) = (0.26, 0.008)$ with weights $0.38 + 0.23 = 0.61$ under the prior $\eta \sim \mathcal{D}_4 (4)$. Under the prior $\eta \sim \mathcal{D}_4 (0.5)$, the variance of the two components with the larger means is reduced and a fourth tiny component with weight 0.012 and a large mean is added.

Figure 7.2 shows the density of these three mixture distributions together with a histogram of the data. The density of $G = 4$ under the prior $\eta \sim \mathcal{D}_4 (0.5)$ is nearly identical to the density of $G = 3$ with the tiny fourth component capturing the largest observations. The density of $G = 4$ under the prior $\eta \sim \mathcal{D}_4 (4)$ is also very similar to the density of $G = 3$, but tries to capture the skewness in the large, well-separated cluster with the smallest observations. Clearly, it is not easy to decide which of these three densities describes the data best.

### 7.2.1   Testing the order of a finite mixture through likelihood ratio tests

From a frequentist perspective, a natural approach to the determination of the order of a mixture distribution is to rely on the likelihood ratio test associated with the hypotheses of $G$ ($H_0$) versus $G+1$ ($H_A$) non-empty components. However, as a consequence of the above-mentioned identifiability problem, regularity conditions ensuring a standard asymptotic distribution for the maximum likelihood (ML) estimates do not hold; see Section 4.3.3. When one component is superfluous ($H_0$), the parameter $\theta_{G+1}$ under the alternative hypothesis $H_A$ lies on the boundary of the parameter space. Moreover, the remainder term appearing within a series expansion of the likelihood ratio test statistic is not uniformly bounded under $H_A$. Therefore, its distribution remains unknown.

Many attempts have been made to modify the likelihood ratio test in this setting; see, for example, the references in McLachlan & Peel (2000) and Frühwirth-Schnatter (2006). Here, we wish to mention the seminal works of Dacunha-Castelle & Gassiat (1997, 1999), which make use of a locally conic parameterization to deal with non-identifiability. This research has been updated and extended to ensure a consistent estimation of $G$ with penalized ML when $G$ is bounded for independent and dependent finite mixtures (Gassiat, 2002). Note that this boundary on $G$ has been relaxed in the paper of Gassiat & van Handel (2013) for a mixture of translated distributions. Moreover, an early reference that deals explicitly with testing $G$ against $G + 1$ in Markov switching models (see Chapter 13) is Hansen (1992).

Adopting a different perspective, McLachlan (1987) proposed using a parametric bootstrap test to select the number of components in a normal mixture. This approach can be extended without difficulty to other mixture distributions. To test the null hypothesis that $G = G_0$ against the alternative that $G = G_1$ at the level $\alpha$, McLachlan (1987) suggests the following procedure: draw $B$ bootstrap samples from a mixture model of order $G_0$ with the parameters being equal to the maximum likelihood estimator (MLE) $\hat{\theta}_{G_0}$ and compute the log likelihood ratio statistic (LRS) of $G = G_0$ versus $G = G_1$ for all bootstrap samples. If the LRS computed on the original sample is smaller than the $1 - \alpha$ quantile of the distribution of the bootstrapped LRSs, then the hypothesis $G = G_0$ is not rejected. It must be pointed out that this bootstrap test is *biased* since the $p$-value is computed from a bootstrap sample where the parameter value $\theta_{G_0}$ has been estimated from the *whole* observed sample. One way to address this bias is to resort to *double bootstrapping*: first, $B$ bootstrap samples are used to compute an estimate $\hat{\theta}_{G_0}^b$ for each bootstrap sample $b = 1, \ldots, B$, while a second bootstrap layer produces an LRS for each bootstrap sample $b$ of the first bootstrap layer. Unfortunately, this double bootstrap procedure is extremely computer-intensive.

As far as we know, technical difficulties aside, statistical tests are rarely used to estimate the order of a mixture. There are several reasons for this. First, the mixture models under comparison are not necessarily embedded. And second, the proposed tests are numerically difficult to implement and slow. Hence, other procedures such as optimizing penalized log likelihood or resorting to Bayesian methods are preferable.

## 7.2.2  Information criteria for order selection

Various information criteria for selecting the order of a mixture distribution are discussed in this section, including the Akaike (AIC) and Bayesian (BIC) information criteria (Section 7.2.2.1), the slope heuristic (Section 7.2.2.2), the deviance information criterion (DIC) (Section 7.2.2.3), and the minimum message length (Section 7.2.2.4) and we refer to the literature for additional criteria such as the approximate weight of evidence (AWE) criterion (Banfield & Raftery, 1993). Information criteria are based on penalizing the log likelihood function of a mixture model $\mathcal{M}_G$ with $G$ components, $\ell_o(\theta; G) = \log L_o(\theta; G)$, where

$$L_o(\theta; G) = \prod_{i=1}^{n} \left[ \sum_{g=1}^{G} \eta_g f_g(y_i \mid \theta_g) \right] \tag{7.3}$$

is also known as the observed-data likelihood. The penalty is proportional to the number of free parameters in $\mathcal{M}_G$, denoted by $v_G$, and the various criteria differ in the choice of the corresponding proportionality factor. The number $v_G$ increases linearly in $G$ and quantifies the complexity of the model. For a multivariate mixture of Gaussian distributions with unconstrained covariance matrices generating observations of dimension $r$, for instance, $v_G = G(1 + r + r(r+1)/2) - 1$.

### 7.2.2.1  AIC and BIC

Let $\hat{\theta}_G$ be the MLE corresponding to the observed-data likelihood $L_o(\theta; G)$, defined in (7.3). The AIC (Akaike, 1974) and BIC (Schwarz, 1978) are popular model selection criteria for solving the bias–variance dilemma for choosing a parsimonious model. $\mathrm{AIC}(G)$ is defined as

$$\mathrm{AIC}(G) = -2\,\ell_o(\hat{\theta}_G; G) + 2\,v_G, \tag{7.4}$$

whereas $\mathrm{BIC}(G)$ is defined as

$$\mathrm{BIC}(G) = -2\,\ell_o(\hat{\theta}_G; G) + v_G \log(n). \tag{7.5}$$

Both criteria are asymptotic criteria and assume that the sampling pdf is within the model collection. On the one hand, the AIC aims to minimize the Kullback–Leibler divergence between model $\mathcal{M}_G$ and the sampling pdf. On the other hand, the BIC approximates the marginal likelihood of model $\mathcal{M}_G$, defined in (7.2), by ignoring the impact of the prior.

In some settings and under proper regularity conditions, the BIC can be shown to be consistent, meaning it eventually picks the true order of the mixture, while the AIC is expected to have a good predictive behaviour and happens to be minimax optimal, that is, to minimize the maximum risk among all estimators, in some regular situations (Yang, 2005). However, in a mixture setting both penalized log likelihood criteria face the same difficulties as the likelihood ratio test due to the identifiability problems mentioned in Section 7.2.1.

Under proper regularity conditions, the BIC enjoys the following asymptotic properties.

(a) The BIC is consistent: if there exists $G^*$ such that the true distribution $p_0$ generating the data is equal to $p(\cdot|G^*)$, then, for $n$ large enough, BIC selects $G^*$.

(b) Even if such a $G^*$ does not exist, good behaviour of the BIC can be expected, if $p_0$ is close to $p(\cdot|G^*)$ for the value $G^*$ selected by the BIC.

Unfortunately, the regularity conditions that validate the above Laplace approximation require the model parameters to be identifiable. As seen above, this is not true in general for most mixture models. However, the BIC has been shown to be consistent when the pdfs of the mixture components are bounded (Keribin, 2002). This is, for example, the case for a Gaussian mixture model with equal covariance matrices. In practice, there is no reason to think that the BIC is not consistent for selecting the number of mixture components when the mixture model is used to estimate a density (see, for instance, Roeder & Wasserman, 1997; Fraley & Raftery, 2002).

For singular models for which the Fisher information matrix is not everywhere invertible, Drton & Plummer (2017) proposed the so-called sBIC criterion. This criterion makes use of the Watanabe (2009) marginal likelihood approximation of a singular model. It is the solution of a fixed point equation approximating the weighted average of the log marginal likelihoods of the models in competition. The sBIC criterion is proven to be consistent. It coincides with the BIC criterion when the model is regular. But, while the usual BIC is in fact not Bayesian, the sBIC is connected to the large-sample behaviour of the log marginal likelihood (Drton & Plummer, 2017).

However, the BIC does not lead to a prediction of the observations that is asymptotically optimal; see Yang (2005) and Drton & Plummer (2017) for further discussion on the comparative properties of the AIC and BIC. In contrast to the BIC criterion, the AIC is known to suffer from a marked tendency to overestimate the true value of $G$ (see, for instance, Celeux & Soromenho (1996) for illustrations). However, a modification of AIC, the so-called AIC3 criterion, proposed in Bozdogan (1987), which replaces the penalty $2v_G$ with $3v_G$, provides a good assessment of $G$ when the latent class model is used to estimate the density of categorical data (Nadif & Govaert, 1998). Nevertheless, the theoretical reasons for this interesting behaviour of the AIC3 (in this particular context) remain for the most part mysterious.

Finally, when the BIC is used to select the number of a mixture components for real data, it has a marked tendency to choose a large number of components or even to choose the highest proposed number of components. The reason for this behaviour is once more related to the fact that the penalty of the BIC is independent of the data, apart from the sample size $n$. When the bias in the mixture model does not vanish when the number of components increases, the BIC always increases by adding new mixture components. In a model-based clustering context, this under-penalization tendency is often counterbalanced by the entropy of the mixture, added to the BIC in the ICLbic criterion (see Section 7.3.2.1), which could lead to a compromise between the fit of a mixture model and its ability to produce a sensible clustering of the data. But there are many situations where the entropy of the mixture is not enough for counterbalancing this tendency and, moreover, the ICLbic is not really relevant when the modelling purpose is not related to clustering.

### 7.2.2.2 The Slope Heuristics

The so-called slope heuristics (Birgé & Massart, 2001, 2007), are a data-driven method to calibrate a penalized criterion that is known up to a multiplicative constant $\kappa$. It has been successfully applied to many situations, and particularly to mixture models when using the observed-data log likelihood; see Baudry et al. (2012). As shown by Baudry (2015), it can be extended without difficulty to other contrasts including the conditional classification log likelihood, which will be defined in Section 7.3.2.1. Roughly speaking, as with the AIC and BIC, the penalty function pen($G$) is assumed to be proportional to the number of free parameters $v_G$ (i.e. the model dimension), pen($G$) $\propto \kappa v_G$.

The penalty is calibrated using the data-driven slope estimation (DDSE) procedure, available in the R package capushe (Baudry et al., 2012). The method assumes a linear relation between the observed-data log likelihood and the penalty. It is important to note that this assumption must and may easily be verified in practice via a simple plot. Then the DDSE procedure directly estimates the slope of the expected linear relationship between the contrast (here the observed-data log likelihood, but other contrasts such as the conditional classification likelihood are possible) and the model dimension $v_G$ which is a function of the number $G$ of components. The estimated slope $\kappa$ defines a minimal penalty $\kappa v_G$ below which smaller penalties give rise to the selection of more complex models, while higher penalties should select models with reasonable complexity. Arguments are provided in Birgé & Massart (2007) and Baudry et al. (2012) that the optimal (oracle) penalty is approximately twice the minimal penalty. Thus, by setting the penalty to be $2\kappa v_G$, the slope heuristics criterion is defined as

$$\mathrm{SH}(G) = -\ell_o(\hat{\theta}_G; G) + 2\kappa v_G,$$

when considering mixture models in a density estimation framework. For more details about the rationale and the implementation of the slope heuristics, see Baudry et al. (2012).

The slope heuristics method relies on the assumption that the bias of the fitted models decreases as their complexity increases and becomes almost constant for the most complex model. In the mixture model framework, this requires the family of models to be roughly nested. More discussion, technical developments and illustrations are given in Baudry (2015).

The ability of the slope heuristics method, which is not based on asymptotic arguments, to detect the stationarity of the model family bias (namely the fact that the bias becomes almost constant) is of prime relevance. It leads this criterion to propose more parsimonious models than the BIC or even the integrated complete-data likelihood criterion (to be discussed in Section 7.3.2.1). Many illustrations of this practical behaviour can be exhibited in various domains of application of mixture models; see, for instance, a clustering use of the slope heuristics to choose the number of components of a multivariate Poisson mixture with RNASeq transcriptome data (Rau et al., 2015) or in a model-based clustering approach for comparing bike sharing systems (Bouveyron et al., 2015).

### 7.2.2.3   DIC

In recent years, the deviance information criterion introduced by Spiegelhalter et al. (2002) has become a popular criterion for Bayesian model selection because it is easily computed from posterior draws, using MCMC methods. Like other penalized log likelihood criteria, the DIC involves a trade-off between goodness of fit and model complexity, measured in terms of the so-called effective number of parameters. However, the use of the DIC to choose the order $G$ of a mixture model is not without issues, as discussed by De Iorio & Robert (2002) and Celeux et al. (2006).

To apply the DIC in a mixture context, several decisions have to be made. As for any latent variable model, a first difficulty arises in the choice of the appropriate likelihood function. Should the DIC be based on the *observed-data* log likelihood $\log p(\mathbf{y}|\theta, G)$, the *complete-data* log likelihood $\log p(\mathbf{y}, \mathbf{z}|\theta, G)$ or the *conditional* log likelihood $\log p(\mathbf{y}|\mathbf{z}, \theta, G)$, where $\mathbf{z} = (z_1, \ldots, z_n)$ are the latent allocations generating the data (see also Section 7.3.1)? Second, the calculation of the DIC requires an estimate $\hat{\theta}_G$ of the unknown parameter $\theta$ which may suffer from label switching, making the DIC (which is based on averaging over MCMC draws) unstable. Finally, if the definition of the DIC involves either the complete-data or conditional likelihood, the difficulty that $\mathbf{z}$ is unobserved must be dealt with, either by integrating against the posterior $p(\mathbf{z}|\mathbf{y}, G)$ or by using a plug-in estimator of $\mathbf{z}$ in which case once again the label switching problem must be addressed to avoid instability.

In an attempt to calibrate these difficulties, Celeux et al. (2006) investigate in total eight different DIC criteria. $\mathrm{DIC}_2$, for instance, focuses on the marginal distribution of the data and considers the allocations $\mathbf{z}$ as nuisance parameters. Consequently, it is based on the observed-data likelihood:

$$\mathrm{DIC}_2(G) = -4\mathrm{E}_\theta \left( \log p(\mathbf{y}|\theta, G)|\mathbf{y} \right) + 2\log p(\mathbf{y}|\hat{\theta}_G, G),$$

where the posterior mode estimator $\hat{\theta}_G$ (which is invariant to label switching) is obtained from the observed-data posterior $p(\theta|\mathbf{y}, G)$ and $\mathrm{E}_\theta$ is the expectation with respect to the posterior $p(\theta|\mathbf{y}, G)$.

Based on several simulation studies, Celeux et al. (2006) recommend using $\mathrm{DIC}_4$ which is based on computing first DIC for the complete-data likelihood function and then integrating over $\mathbf{z}$ with respect to the posterior $p(\mathbf{z}|\mathbf{y}, G)$. This yields

$$\mathrm{DIC}_4(G) = -4\mathrm{E}_{\theta, \mathbf{z}} \left( \log p(\mathbf{y}, \mathbf{z}|\theta, G)|\mathbf{y} \right) + 2\mathrm{E}_\mathbf{z} \left( \log p(\mathbf{y}, \mathbf{z}|\hat{\theta}_G(\mathbf{z}))|\mathbf{y} \right),$$

where $\hat{\theta}_G(\mathbf{z})$ is the complete-data posterior mode which must be computed for each draw from the posterior $p(\mathbf{z}|\mathbf{y}, G)$. This is straightforward if the complete-data posterior $p(\theta_g|\mathbf{y}, \mathbf{z})$ is available in closed form. If this is not the case, Celeux et al. (2006) instead use the posterior mode estimator $\hat{\theta}_G$ of the observed-data posterior $p(\theta|\mathbf{y})$. This leads to an approximation of $\mathrm{DIC}_4(G)$, called $\mathrm{DIC}_{4a}(G)$, which is shown to be a criterion that penalizes $\mathrm{DIC}_2(G)$ by the expected entropy, defined in (7.16):

$$\mathrm{DIC}_{4a}(G) = \mathrm{DIC}_2(G) + 2\mathrm{E}_\theta \left( \mathrm{ENT}(\theta; G)|\mathbf{y} \right).$$

Both $\mathrm{DIC}_2(G)$ and $\mathrm{DIC}_{4a}(G)$ are easily estimated from (MCMC) draws from the posterior $p(\theta|\mathbf{y}, G)$ by substituting all expectations $\mathrm{E}_\bullet(\cdot|\mathbf{y})$ by an average over the corresponding draws. Note that label switching is not a problem here, because both $\log p(\mathbf{y}|\theta, G)$ and $\mathrm{ENT}(\theta; G)$ are invariant to the labelling of the groups.

However, in practical mixture modelling, the DIC turns out to be very unstable, as shown by Celeux et al. (2006) for the galaxy data (Roeder, 1990). A similar behaviour was observed by Frühwirth-Schnatter & Pyne (2010) who fitted skew-normal mixtures to Alzheimer disease data under various prior assumptions. While the marginal likelihood selected $G = 2$ with high confidence for all priors, $\mathrm{DIC}_{4a}(G)$ selected $G = 1$, regardless of the chosen prior, whereas the number of components selected by $\mathrm{DIC}_2(G)$ ranged from 2 to 4, depending on the prior.

### 7.2.2.4 The minimum message length

Assuming that the form of the mixture models is fixed (e.g. Gaussian mixture models with free covariance matrices or Gaussian mixture models with a common covariance matrix), several authors have proposed dealing with the estimation of the mixture parameters and $G$ in a single algorithm with the minimum message length (MML) criterion (see, for instance, Rissanen, 2012; Wallace & Freeman, 1987). Considering the MML criterion in a Bayesian perspective and choosing Jeffreys' non-informative prior $p(\theta)$ for the mixture parameter, Figueiredo & Jain (2002) propose minimizing the criterion

$$\mathrm{MML}(\theta; G) = -\log p(\mathbf{y}|\theta, G) - \log p(\theta|G) + \frac{1}{2}\log |I(\theta)| + \frac{v_G}{2}(1 - \log(12)),$$

where $I(\theta)$ is the expected Fisher information matrix which is approximated by the complete-data Fisher information matrix $I_C(\theta)$.

As we know, for instance from Section 4.2.2 above, Jeffreys' non-informative prior does not work for mixtures. Figueiredo & Jain (2002) circumvent this difficulty by only considering the parameters of the components whose proportion is non-zero, namely the components $g$ such that $\hat{\eta}_g > 0$.

Assuming, for instance, that the mixture model considered arises from the general Gaussian mixture family with free covariance matrices, this approach leads to minimizing the criterion

$$\text{MML}(\theta; G) = -\log p(\mathbf{y}|\theta, G) + \frac{G^\star}{2} \log \frac{n}{12}$$

$$+ \frac{\dim(\theta_g)}{2} \sum_{g:\hat{\eta}_g > 0} \{\log(n \cdot \dim(\theta_g)/12) + G^\star(\dim(\theta_g) + 1)\}, \quad (7.6)$$

with $G^\star = \text{card}\{g|\hat{\eta}_g > 0\}$. In this Bayesian context, the approach of Figueiredo & Jain (2002) involves optimizing iteratively the criterion (7.6), starting from a large number of components $G_{\max}$, and cancelling the components $g$ such that, at iteration $s$,

$$\sum_{i=1}^{n} \hat{\tau}_{ig}^{(s)} < \frac{\dim(\theta_g^{(s)})}{2}, \quad (7.7)$$

where $\hat{\tau}_{ig}^{(s)}$ are the elements of the fuzzy classification matrix defined in (7.18). Thus, the chosen number of components $G^\star$ is the number of components remaining at the convergence of the iterative algorithm. This iterative algorithm could be the EM algorithm, but Figueiredo & Jain (2002) argue that with EM, for large $G$, it can happen that no component has enough initial support, as the criterion for cancellation defined in (7.7) is fulfilled for *all* $G$ components. Thus, they prefer to make use of the componentwise EM algorithm of Celeux et al. (2001), which updates the $\eta_g$ and the $\theta_g$ sequentially: update $\eta_1$ and $\theta_1$, recompute $\tau_{i1}$ for $i = 1, \ldots, n$, update $\eta_2$ and $\theta_2$, recompute $\tau_{i2}$ for $i = 1, \ldots, n$, and so on.

Zeng & Cheung (2014) use exactly the same approach with the completed-data or the classification likelihood instead of the observed-data likelihood. Thus, roughly speaking, the procedure of Figueiredo & Jain (2002) is expected to provide a similar number of components to the BIC, while the procedure of Zeng & Cheung (2014) is expected to provide a similar number of clusters to the ICLbic presented in Section 7.3.2.1.

## 7.2.3 Bayesian model choice based on marginal likelihoods

From a Bayesian testing perspective, selecting the number of components can be interpreted as a model selection problem, given the probability of each model within a collection of all models corresponding to the different numbers of components (Berger, 1985). The standard Bayesian tool for making this model choice is based on the marginal likelihood (also called *evidence*) of the data $p(\mathbf{y}|G)$ for each model $\mathcal{M}_G$, defined in (7.2), which naturally penalizes models with more components (and more parameters) (Berger & Jefferys, 1992).

While the BIC is often considered as one case of information criterion, it is important to recall (see Section 7.2.2.1) that it was first introduced by Schwartz (1965) as an approximation to the marginal likelihood $p(\mathbf{y}|G)$. Since this approximation does not depend on the choice of the prior $p(\theta|G)$, it is not of direct appeal for a Bayesian evaluation of the number of components, especially when considering that the marginal likelihood itself can be approximated by simulation-based methods, as discussed in this section.

### 7.2.3.1 Chib's method, limitations and extensions

The reference estimator for evidence approximation is Chib's (1995) representation of the marginal likelihood of model $\mathcal{M}_G$ as[1]

$$p(\mathbf{y}|G) = \frac{p(\mathbf{y}|\theta^o, G)p(\theta^o|G)}{p(\theta^o|\mathbf{y}, G)}, \quad (7.8)$$

---

[1]This was earlier called *the candidate's formula* by Julian Besag (1989).

which holds for *any* choice of the plug-in value $\theta^o$. While the posterior $p(\theta^o|\mathbf{y}, G)$ is not available in closed form for mixtures, a Gibbs sampling decomposition allows for a Rao–Blackwellized approximation of this density (Robert & Casella, 2004) that furthermore converges at a parametric speed, as already noticed in Gelfand & Smith (1990):

$$\hat{p}(\theta^o|\mathbf{y}, G) = \frac{1}{M} \sum_{m=1}^{M} p(\theta^o|\mathbf{y}, \mathbf{z}^{(m)}, G),$$

where $\mathbf{z}^{(m)}, m = 1, \ldots, M$, are the posterior draws for the latent allocations $\mathbf{z} = (z_1, \ldots, z_n)$, introduced earlier in Chapter 1; see Chapter 5 for a review of posterior sampling methods.

However, for mixtures, the convergence of this estimate is very much hindered by the fact that it requires perfect symmetry in the Gibbs sampler, that is, complete label switching within the simulated Markov chain. When the completed chain $(z_1^{(m)}, \ldots, z_n^{(m)})$ remains instead concentrated around one single or a subset of the modes of the posterior distribution, the approximation of $\log \hat{p}(\theta^o|\mathbf{y}, G)$ based on Chib's representation fails, in that it is usually off by a numerical factor of order $O(\log G!)$. Furthermore, this order cannot be used as a reliable correction, as noted by Neal (1999) and Frühwirth-Schnatter (2006).

A straightforward method of handling Markov chains that are not perfectly mixing (which is the usual setting) is found in Berkhof et al. (2003) (see also Frühwirth-Schnatter, 2006, Section 5.5.5; Lee et al., 2009) and can be interpreted as a form of Rao–Blackwellization. The proposed correction is to estimate $\hat{p}(\theta^o|\mathbf{y}, G)$ as an average computed over all possible permutations of the labels, thus forcing the label switching and the exchangeability of the labels to occur in a "perfect" manner. The new approximation can be expressed as

$$\tilde{p}(\theta^o|\mathbf{y}, G) = \frac{1}{MG!} \sum_{\mathfrak{s} \in \mathfrak{S}(G)} \sum_{m=1}^{M} p(\theta^o|\mathbf{y}, \mathfrak{s}(\mathbf{z}^{(m)}), G),$$

where $\mathfrak{S}(G)$ traditionally denotes the set of the $G!$ permutations of $\{1, \ldots, G\}$ and where $\mathfrak{s}$ is one of those permutations. Note that the above correction can also be rewritten as

$$\tilde{p}(\theta^o|\mathbf{y}, G) = \frac{1}{MG!} \sum_{\mathfrak{s} \in \mathfrak{S}(G)} \sum_{m=1}^{M} p(\mathfrak{s}(\theta^o)|\mathbf{y}, \mathbf{z}^{(m)}, G), \tag{7.9}$$

as this may induce some computational savings. Further savings can be found in the importance sampling approach of Lee & Robert (2016), who reduce the number of permutations to be considered.

While Chib's representation has often been advocated as a reference method for computing the evidence, other methods abound, among them nested sampling (Skilling, 2007; Chopin & Robert, 2010), reversible jump MCMC (Green, 1995; Richardson & Green, 1997), particle filtering (Chopin, 2002), bridge sampling (Frühwirth-Schnatter, 2004) and path sampling (Gelman & Meng, 1998). Some of these methods are discussed next.

### 7.2.3.2 Sampling-based approximations

If $G$ is moderate, sampling-based techniques are particularly useful for estimating the marginal likelihood of finite mixture models; see Frühwirth-Schnatter (2004) and Lee & Robert (2016). Frühwirth-Schnatter (2004) considered three such estimation techniques, namely importance sampling, reciprocal importance sampling, and bridge sampling.

For sampling-based techniques, one selects an importance density $q_G(\theta)$ which is easy to sample from and provides a rough approximation to the posterior density $p(\theta|\mathbf{y}, G)$. Given a suitable importance density $q_G(\theta)$, an importance sampling approximation to the marginal likelihood is based on rewriting (7.2) as

$$p(\mathbf{y}|G) = \int \frac{p(\mathbf{y}|\theta, G)p(\theta|G)}{q_G(\theta)} q_G(\theta)\mathrm{d}\theta.$$

Based on a sample $\theta^{(l)} \sim q_G(\theta)$, $l = 1, \ldots, L$, from the importance density $q_G(\theta)$, the importance sampling estimator of the marginal likelihood is given by

$$\hat{p}_{IS}(\mathbf{y}|G) = \frac{1}{L} \sum_{l=1}^{L} \frac{p(\mathbf{y}|\theta^{(l)}, G)p(\theta^{(l)}|G)}{q_G(\theta^{(l)})}. \tag{7.10}$$

Gelfand & Dey (1994) introduced reciprocal importance sampling, which is based on the observation that (7.8) can be written as

$$\frac{1}{p(\mathbf{y}|G)} = \frac{p(\theta|\mathbf{y}, G)}{p(\mathbf{y}|\theta, G)p(\theta|G)}.$$

Integrating both sides of this equation with respect to the importance density $q_G(\theta)$ yields

$$\frac{1}{p(\mathbf{y}|G)} = \int \frac{q_G(\theta)}{p(\mathbf{y}|\theta, G)p(\theta|G)} p(\theta|\mathbf{y}, G).$$

This leads to the reciprocal importance sampling estimator of the marginal likelihood, where the inverse of the ratio appearing in (7.10) is evaluated at the MCMC draws $\theta^{(m)}$, $m = 1, \ldots, M$, and no draws from the importance density $q_G(\theta)$ are required:

$$\hat{p}_{RI}(\mathbf{y}|G) = \left( \frac{1}{M} \sum_{m=1}^{M} \frac{q_G(\theta^{(m)})}{p(\mathbf{y}|\theta^{(m)}, G)p(\theta^{(m)}|G)} \right)^{-1}.$$

These two estimators are special cases of bridge sampling (Meng & Wong, 1996):

$$p(\mathbf{y}|G) = \frac{\mathrm{E}_{q_G(\theta)}(\alpha(\theta)p(\mathbf{y}|\theta, G)p(\theta|G))}{\mathrm{E}_{p(\theta|\mathbf{y}, G)}(\alpha(\theta)q_G(\theta))},$$

with specific functions $\alpha(\theta)$. The (formally) optimal choice for $\alpha(\theta)$ yields the bridge sampling estimator $\hat{p}_{BS}(\mathbf{y}|G)$ and combines draws $\theta^{(l)}$, $l = 1, \ldots, L$, from the importance density with MCMC draws $\theta^{(m)}$, $m = 1, \ldots, M$. Using $\hat{p}_{IS}(\mathbf{y}|G)$ as a starting value for $\hat{p}_{BS,0}(\mathbf{y}|G)$, the following recursion is applied until convergence to estimate $\hat{p}_{BS}(\mathbf{y}|G) = \lim_{t\to\infty} \hat{p}_{BS,t}(\mathbf{y}|G)$:

$$\hat{p}_{BS,t}(\mathbf{y}|G) = \frac{L^{-1} \displaystyle\sum_{l=1}^{L} \frac{p(\mathbf{y}|\theta^{(l)}, G)p(\theta^{(l)}|G)}{Lq_G(\theta^{(l)}) + Mp(\mathbf{y}|\theta^{(l)}, G)p(\theta^{(l)}|G)/\hat{p}_{BS,t-1}(\mathbf{y}|G)}}{M^{-1} \displaystyle\sum_{m=1}^{M} \frac{q_G(\theta^{(m)})}{Lq_G(\theta^{(m)}) + Mp(\mathbf{y}|\theta^{(m)}, G)p(\theta^{(m)}|G)/\hat{p}_{BS,t-1}(\mathbf{y}|G)}}. \tag{7.11}$$

The reliability of these estimators depends on several factors. First, as shown by Frühwirth-Schnatter (2004), the tail behaviour of $q_G(\theta)$ compared to the mixture posterior $p(\theta|\mathbf{y}, G)$ is relevant. Whereas the bridge sampling estimator $\hat{p}_{BS}(\mathbf{y}|G)$ is fairly robust to the tail behaviour of $q_G(\theta)$, $\hat{p}_{IS}(\mathbf{y}|G)$ is sensitive if $q_G(\theta)$ has lighter tails than $p(\theta|\mathbf{y}, G)$, and $\hat{p}_{RI}(\mathbf{y}|G)$ is sensitive if $q_G(\theta)$ has fatter tails than $p(\theta|\mathbf{y}, G)$. Second, as pointed out by Lee & Robert

(2016), for any of these methods it is essential that the importance density $q_G(\theta)$ exhibits the same kind of multimodality as the mixture posterior $p(\theta|\mathbf{y}, G)$ and all modes of the posterior density are covered by the importance density also for increasing values of $G$. Otherwise, sampling-based estimators of the marginal likelihood are prone to be biased for the same reason Chib's estimator is biased, as discussed in Section 7.2.3.1. A particularly stable estimator is obtained when bridge sampling is combined with a perfectly symmetric importance density $q_G(\theta)$. Before the various estimators are illustrated for three well-known data sets (Richardson & Green, 1997), we turn to the choice of appropriate importance densities.

*Importance densities for mixture analysis*

As manual tuning of the importance density $q_G(\theta)$ for each model under consideration is rather tedious, methods for choosing sensible importance densities in an unsupervised manner are needed. DiCiccio et al. (1997), for instance, suggested various methods to construct Gaussian importance densities from the MCMC output. However, the multimodality of the mixture posterior density with $G!$ equivalent modes evidently rules out such a simple choice. Frühwirth-Schnatter (1995) is one of the earliest references that used Rao–Blackwellization to construct an unsupervised importance density from the MCMC output to compute marginal likelihoods via sampling-based approaches and applied this idea to model selection for linear Gaussian state space models. Frühwirth-Schnatter (2004) extends this idea to finite mixture and Markov switching models where the complete-data posterior $p(\theta|\mathbf{y}, \mathbf{z})$ is available in closed form. Lee & Robert (2016) discuss importance sampling schemes based on (nearly) perfectly symmetric importance densities.

For a mixture distribution, where the component-specific parameters $\theta_g$ can be sampled in one block from the complete-data posterior $p(\theta_g|\mathbf{z}, \mathbf{y})$, Rao–Blackwellization yields the importance density

$$q_G(\theta) = \frac{1}{S} \sum_{s=1}^{S} p(\eta|\mathbf{z}^{(s)}) \prod_{g=1}^{G} p(\theta_g|\mathbf{z}^{(s)}, \mathbf{y}), \qquad (7.12)$$

where $\mathbf{z}^{(s)}$ are the posterior draws for the latent allocations. The construction of this importance density is fully automatic and it is sufficient to store the moments of these conditional densities (rather than the allocations $\mathbf{z}$ themselves) during MCMC sampling for later evaluation. This method can be extended to cases where sampling $\theta_g$ from $p(\theta_g|\mathbf{z}, \mathbf{y})$ requires two (or even more) blocks such as for Gaussian mixtures where $\theta_g = (\mu_g, \sigma_g^2)$ is sampled in two steps from $p(\mu_g|\sigma_g^2, \mathbf{z}, \mathbf{y})$ and $p(\sigma_g^2|\mu_g, \mathbf{z}, \mathbf{y})$.

Concerning the number of components in (7.12), on the one hand $S$ should be small for computational reasons, because $q_G(\theta)$ has to be evaluated for each of the $S$ components numerous times (e.g. $L$ times for the importance sampling estimator (7.10)). On the other hand, as mentioned above, it is essential that $q_G(\theta)$ covers all symmetric modes of the mixture posterior, and this will require a dramatically increasing number of components $S$ as $G$ increases. Hence, any of these estimators is limited to moderate values of $G$, say up to $G = 6$.

Various strategies are available to ensure multimodality in the construction of the importance density. Frühwirth-Schnatter (2004) chooses $S = M$ and relies on random permutation Gibbs sampling (Frühwirth-Schnatter, 2001) by applying a randomly selected permutation $\mathfrak{s}_m \in \mathfrak{S}(G)$ at the end of the $m$th MCMC sweep to define a permutation $\mathbf{z}^{(s)} = \mathfrak{s}_m(\mathbf{z}^{(m)})$ of the posterior draw $\mathbf{z}^{(m)}$ of the allocation vector. The random permutations $\mathfrak{s}_1, \ldots, \mathfrak{s}_M$ guarantee multimodality of $q_G(\theta)$ in (7.12); however, as discussed above, it is important to ensure good mixing of the underlying permutation sampler over all $G!$ equivalent posterior modes. Only if $S$ is large compared to $G!$ are all symmetric modes visited by random permutation sampling. Choosing, for instance, $S = S_0 G!$ ensures that each mode is visited on average $S_0$ times.

As an alternative to random permutation sampling, approaches exploiting full permutations have been suggested; see, for example, Frühwirth-Schnatter (2004). Importance sampling schemes exploiting full permutation were discussed in full detail in Lee & Robert (2016). The definition of a fully symmetric importance density $q_G(\theta)$ is related to the correction for Chib's estimator discussed earlier in (7.9):

$$q_G(\theta) = \frac{1}{S_0 G!} \sum_{\mathfrak{s} \in \mathfrak{S}(G)} \sum_{s=1}^{S_0} p(\eta | \mathfrak{s}(\mathbf{z}^{(s)})) \prod_{g=1}^{G} p(\theta_g | \mathfrak{s}(\mathbf{z}^{(s)}), \mathbf{y}). \qquad (7.13)$$

This construction, which has $S = S_0 G!$ components, is based on a small number $S_0$ of particles $\mathbf{z}^{(s)}$, as $q_G(\theta)$ needs to be only a rough approximation to the mixture posterior $p(\theta | \mathbf{y}, G)$ and estimators such as bridge sampling will be robust to the tail behaviour of $q_G(\theta)$. In (7.13), all symmetric modes are visited exactly $S_0$ times. The moments of the $S_0$ conditional densities need to be stored for only one of the $G!$ permutations and, again, this construction can be extended to the case where the components of $\theta_g$ are sampled in more than one block. Lee & Robert (2016) discuss strategies for reducing the computational burden associated with evaluating $q_G(\theta)$.

Frühwirth-Schnatter (2006, p. 146) and Lee & Robert (2016) discuss a simplified version of (7.13) where the random sequence $\mathbf{z}^{(s)}$, $s = 1, \ldots, S_0$, is substituted by a single optimal partition $\mathbf{z}^*$ such as the maximum *a posteriori* (MAP) estimator:

$$q_G(\theta) = \frac{1}{G!} \sum_{\mathfrak{s} \in \mathfrak{S}(G)} p(\theta | \mathfrak{s}(\mathbf{z}^*), \mathbf{y}).$$

In MATLAB, the bayesf package (Frühwirth-Schnatter, 2018) allows one to estimate $\hat{p}_{BS}(\mathbf{y}|G)$, $\hat{p}_{IS}(\mathbf{y}|G)$ and $\hat{p}_{RI}(\mathbf{y}|G)$ with the importance density being constructed either as in (7.12) using random permutation sampling or as in (7.13) using full permutation sampling.

*Example: Marginal likelihoods for the data sets in Richardson & Green (1997)*

By way of illustration, marginal likelihoods are computed for mixtures of $G$ univariate normal distributions $\mathcal{N}(\mu_g, \sigma_g^2)$ for $G = 2, \ldots, 6$ for the acidity data, the enzyme data and the galaxy data studied by Richardson & Green (1997) in the framework of reversible jump MCMC (see Section 7.4.2 for a short description of this one-sweep method). We use the same priors as Richardson & Green, namely the symmetric Dirichlet prior $\eta \sim \mathcal{D}_G(1)$, the normal prior $\mu_g \sim \mathcal{N}(m, R^2)$, the inverse gamma prior $\sigma_g^2 \sim \mathcal{IG}(2, C_0)$ and the gamma prior $C_0 \sim \mathcal{G}(0.2, 10/R^2)$, where $m$ and $R$ are the midpoint and the length of the observation interval. For a given $G$, full conditional Gibbs sampling is performed for $M = 12,000$ draws after a burn-in of 2000, by iteratively sampling from $p(\sigma_g^2 | \mu_g, C_0, \mathbf{z}, \mathbf{y})$, $p(\mu_g | \sigma_g^2, \mathbf{z}, \mathbf{y})$, $p(C_0 | \sigma_1^2, \ldots, \sigma_G^2)$, $p(\eta | \mathbf{z})$ and $p(\mathbf{z} | \theta, \mathbf{y})$.

A fully symmetric importance density $q_{G,F}(\theta)$ is constructed from (7.13), where $S_0 = 100$ components are selected for each mode. For comparison, an importance density $q_{G,R}(\theta)$ is constructed from (7.12) with $S = S_0 G!$, ensuring that for random permutation sampling each mode is visited on average $S_0$ times. However, unlike $q_{G,F}(\theta)$, the importance density $q_{G,R}(\theta)$ is not fully symmetric. Ignoring the dependence between $\mu_g$ and $\sigma_g^2$, the component densities are constructed from conditionally independent densities, given the $s$th draw of $(\mathbf{z}, \theta_1, \ldots, \theta_G, C_0)$:

$$p(\mu_g, \sigma_g^2 | \mathbf{z}^{(s)}, \theta_g^{(s)}, C_0^{(s)}, \mathbf{y}) = p(\mu_g | \sigma_g^{2,(s)}, \mathbf{z}^{(s)}, \mathbf{y}) p(\sigma_g^2 | \mu_g^{(s)}, C_0^{(s)}, \mathbf{z}^{(s)}, \mathbf{y}).$$

Prior evaluation is based on the marginal prior $p(\sigma_1^2, \ldots, \sigma_G^2)$, where $C_0$ is integrated out.

This yields in total six estimators, $\hat{p}_{BS,F}(\mathbf{y}|G)$, $\hat{p}_{IS,F}(\mathbf{y}|G)$ and $\hat{p}_{RI,F}(\mathbf{y}|G)$ for full permutation sampling and $\hat{p}_{BS,R}(\mathbf{y}|G)$, $\hat{p}_{IS,R}(\mathbf{y}|G)$ and $\hat{p}_{RI,R}(\mathbf{y}|G)$ for random permutation sampling, for each $G = 2, \ldots, 6$. Results are visualized in Figure 7.3, by plotting the six estimators $\log \hat{p}_{\bullet}(\mathbf{y}|G)$ as well as $\log \hat{p}_{\bullet}(\mathbf{y}|G) \pm 3\,\mathrm{SE}$ over $G$ for all three data sets. For each estimator, the standard errors SE are computed as in Frühwirth-Schnatter (2004). Good estimators should be unbiased with small standard errors and the order in which the six estimators are arranged (which is the same for all $G$s) is related to this quality measure.

There is a striking difference in the reliability of the six estimators, in particular as $G$ increases. Reciprocal importance sampling is particularly unreliable and the estimated values of $\log \hat{p}_{RI,R}(\mathbf{y}|G)$ under $q_{G,R}(\theta)$ tend to be extremely biased for $G \geq 4$, even if the bias is reduced to a certain extent by choosing the fully symmetric importance density $q_{G,F}(\theta)$. Also the two other estimators $\log \hat{p}_{IS,R}(\mathbf{y}|G)$ and $\log \hat{p}_{BS,R}(\mathbf{y}|G)$ tend to be biased under $q_{G,R}(\theta)$, and bridge sampling is more sensitive than importance sampling to choosing an importance density that is not fully symmetric.

Unlike for reciprocal importance sampling, the bias disappears for both bridge sampling and importance sampling under the fully symmetric importance density $q_{G,F}(\theta)$, and $\log \hat{p}_{IS,F}(\mathbf{y}|G)$ and $\log \hat{p}_{BS,F}(\mathbf{y}|G)$ yield more or less identical results. However, due to the robustness of bridge sampling with respect to the tail behaviour of $q_{G,F}(\theta)$, we find that the standard errors of $\log \hat{p}_{BS,F}(\mathbf{y}|G)$ are often considerably smaller than the standard errors of $\log \hat{p}_{IS,F}(\mathbf{y}|G)$, in particular for the enzyme data.

Based on $\log \hat{p}_{BS,F}(\mathbf{y}|G)$, marginal likelihood evaluation yields the following results for the three data sets. For the acidity data, $\log \hat{p}_{BS,F}(\mathbf{y}|G = 3) = -198.2$ and $\log \hat{p}_{BS,F}(\mathbf{y}|G = 4) = -198.3$ are more less the same, with the log odds of $G = 3$ over $G = 4$ being equal to 0.1. Also for the enzyme data, with $\log \hat{p}_{BS,F}(\mathbf{y}|G = 3) = -74.2$ and $\log \hat{p}_{BS,F}(\mathbf{y}|G = 4) = -74.3$, the log odds of $G = 3$ over $G = 4$ are equal to 0.1. Finally, for the galaxy data, $\log \hat{p}_{BS,F}(\mathbf{y}|G = 5) = \log \hat{p}_{BS,F}(\mathbf{y}|G = 6) = -225.9$. Hence, under the prior $p(\theta|G)$ employed by Richardson & Green (1997), for all three data sets no clear distinction can be made between two values of $G$ based on the marginal likelihood. However, if the marginal likelihoods are combined with a prior on the number of components such as $G - 1 \sim \mathcal{P}(1)$ (Nobile, 2004), then the log posterior odds, being equal to 1.5 for the acidity and the enzyme data and 1.8 for the galaxy data, yield evidence for the smaller of the two values of $G$ for all three data sets.

## 7.3   Selecting $G$ in the Framework of Model-Based Clustering

Assuming that the data stem from one of the models under comparison is most often unrealistic and can be misleading when using the AIC or BIC. Now a common feature of standard penalized likelihood criteria is that they abstain from taking the modelling purpose into account, except when inference is about estimating the data density. In particular, misspecification can lead to overestimating the complexity of a model in practical situations. Taking the modelling purpose into account when selecting a model leads to alternative model selection criteria that favor useful and parsimonious models. This viewpoint is particularly relevant when considering a mixture model for model-based clustering; see Chapter 8 for a review of this important application of mixture models.

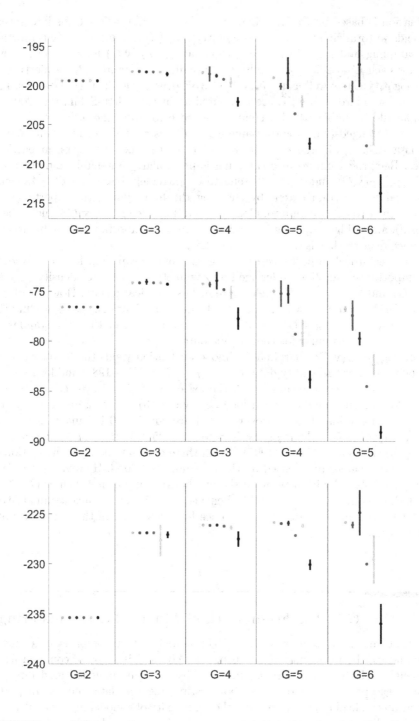

**FIGURE 7.3**
Marginal likelihood estimation for the benchmarks in Richardson & Green (1997): the acidity data (top), the enzyme data (middle) and the galaxy data (bottom) over $G = 2, \ldots, G = 6$. For each $G$, six estimators $\log \hat{p}_{\bullet}(\mathbf{y}|G)$ are given together with $\log \hat{p}_{\bullet}(\mathbf{y}|G) \pm 3\,\mathrm{SE}$ in the order $\log \hat{p}_{BS,F}(\mathbf{y}|G)$, $\log \hat{p}_{IS,F}(\mathbf{y}|G)$, $\log \hat{p}_{IS,R}(\mathbf{y}|G)$, $\log \hat{p}_{BS,R}(\mathbf{y}|G)$, $\log \hat{p}_{RI,F}(\mathbf{y}|G)$ and $\log \hat{p}_{RI,R}(\mathbf{y}|G)$ from left to right.

### 7.3.1 Mixtures as partition models

Clustering arises in a natural way when an i.i.d. sample is drawn from the finite mixture distribution (7.1) with weights $\eta = (\eta_1, \ldots, \eta_G)$. As explained in Chapter 1, Section 1.1.3, each observation $y_i$ can be associated with the component, indexed by $z_i$, that generated this data point:

$$z_i | \eta \sim \mathcal{M}(1, \eta_1, \ldots, \eta_G), \tag{7.14}$$
$$y_i | z_i \sim f_{z_i}(y_i | \theta_{z_i}).$$

Let $\mathbf{z} = (z_1, \ldots, z_n)$ be the collection of all component indicators that were used to generate the $n$ data points $\mathbf{y} = (y_1, \ldots, y_n)$. Obviously, $\mathbf{z}$ defines a partition of the data. A cluster $C_g = \{i : z_i = g\}$ is thus defined as a subset of the data indices $\{1, \ldots, n\}$, containing all observations with identical allocation variables $z_i$. Hence, the indicators $\mathbf{z}$ define a partition $\mathcal{C} = \{C_1, \ldots, C_{G_+}\}$ of the $n$ data points, where $y_i$ and $y_j$ belong to the same cluster if and only if $z_i = z_j$. The partition $\mathcal{C}$ contains $G_+ = |\mathcal{C}|$ clusters, where $|\mathcal{C}|$ is the cardinality of $\mathcal{C}$. In a Bayesian context, finite mixture models imply *random partitions* over the lattice

$$\mathcal{S}_G^n = \{(z_1, \ldots, z_n) : z_i \in \{1, \ldots, G\}, i = 1, \ldots, n\},$$

as will be discussed in detail in Section 7.3.3.

In model-based clustering, a finite mixture model is applied to recover the (latent) allocation indicators $\mathbf{z}$ from the data and to estimate a suitable partition of the data. A useful quantity in this respect is the so-called fuzzy classification matrix $\tau$. The elements $\tau_{ig}$, with $i = 1, \ldots, n$ and $g = 1, \ldots, G$, of $\tau$ are equal to the conditional probability that observation $y_i$ arises from component $g$ in a mixture model of order $G$ given $y_i$:

$$\tau_{ig} = \mathrm{P}(z_i = g | y_i, \theta) = \mathrm{P}(z_{ig} = 1 | y_i, \theta) = \frac{\eta_g f_g(y_i | \theta_g)}{\sum_{j=1}^{G} \eta_j f_j(y_i | \theta_j)}, \tag{7.15}$$

where $z_{ig} = \mathbb{I}(z_i = g)$. The entropy $\mathrm{ENT}(\theta; G)$ corresponding to a fuzzy classification matrix $\tau$ is defined as

$$\mathrm{ENT}(\theta; G) = - \sum_{g=1}^{G} \sum_{i=1}^{n} \tau_{ig} \log \tau_{ig} \geq 0. \tag{7.16}$$

Both $\tau$ and $\mathrm{ENT}(\theta; G)$ are data-driven measures of the ability of a $G$-component mixture model to provide a relevant partition of the data. If the mixture components are well separated for a given $\theta$, then the classification matrix $\tau$ tends to define a clear partition of the data set $\mathbf{y} = (y_1, \ldots, y_n)$, with $\tau_{ig}$ being close to 1 for one component and close to 0 for all other components. In this case, $\mathrm{ENT}(\theta; G)$ is close to 0. On the other hand, if the mixture components are poorly separated, then $\mathrm{ENT}(\theta; G)$ takes values larger than zero. The maximum value $\mathrm{ENT}(\theta; G)$ can take is $n \log G$, which is the entropy of the uniform distribution which assigns $y_i$ to all $G$ clusters with the same probability $\tau_{ig} \equiv 1/G$.

In a Bayesian context, the fuzzy classification matrix is instrumental for joint estimation of the parameter $\theta$ and $\mathbf{z}$ within Gibbs sampling using data augmentation (see, for example, Robert & Casella, 2004). In a frequentist framework, the estimated classification matrix $\hat{\tau}$, given a suitable estimate $\hat{\theta}_G$ of the mixture parameters $\theta$ (e.g. the MLE), can be used to derive an estimator $\hat{\mathbf{z}}$ of the partition of the data; see also Chapter 8, Section 8.2.4. As will be discussed in Section 7.3.2, the entropy of the estimated classification matrix $\hat{\tau}$ plays an important role in defining information criteria for choosing $G$ in a clustering context.

## 7.3.2   Classification-based information criteria

As discussed in Section 7.2.2.1 within the framework of density estimation, the BIC enjoys several desirable properties; however, within cluster analysis it shows a tendency to overestimate $G$; see, for instance, Celeux & Soromenho (1996). The BIC does not take the clustering purposes for assessing $G$ into account, regardless of the separation of the clusters. To overcome this limitation, an attractive possibility is to select $G$ so that the resulting mixture model leads to the clustering of the data with the largest evidence. This is the purpose of various classification-based information criteria such as the integrated complete-data likelihood criterion that are discussed in this subsection.

In a classification context, it is useful to state a simple relation linking the log of the observed-data density $p(\mathbf{y}|\theta)$ and the complete-data density $p(\mathbf{y},\mathbf{z}|\theta)$. The observed-data log likelihood of $\theta$ for a sample $\mathbf{y}$, denoted by $\ell_o(\theta;G)$, is given by

$$\ell_o(\theta;G) = \sum_{i=1}^{n} \log \left[ \sum_{g=1}^{G} \eta_g f_g(y_i \mid \theta_g) \right],$$

whereas the complete-data log likelihood of $\theta$ for the complete sample $(\mathbf{y},\mathbf{z})$, denoted by $\ell_c(\theta;G)$, reads

$$\ell_c(\theta,\mathbf{z};G) = \sum_{i=1}^{n} \sum_{g=1}^{G} z_{ig} \log(\eta_g f_g(y_i \mid \theta_g)),$$

where $z_{ig} = \mathbb{I}(z_i = g)$, $g = 1,\dots,G$. These log likelihoods are linked in the following way:

$$\ell_c(\theta,\mathbf{z};G) = \ell_o(\theta;G) - \mathrm{EC}(\theta,\mathbf{z};G), \tag{7.17}$$

where

$$\mathrm{EC}(\theta,\mathbf{z};G) = - \sum_{g=1}^{G} \sum_{i=1}^{n} z_{ig} \log \tau_{ig} \geq 0.$$

Since $\mathrm{E}(z_{ig}|\theta,y_i) = \mathrm{P}(z_{ig}=1|\theta,y_i) = \tau_{ig}$, we obtain that the expectation of $\mathrm{EC}(\theta,\mathbf{z};G)$ with respect to the conditional distribution $p(\mathbf{z}|\mathbf{y},\theta)$ for a given $\theta$ is equal the entropy $\mathrm{ENT}(\theta;G)$ defined in (7.16). Hence, the entropy can be regarded as a penalty for the observed-data likelihood in cases where the resulting clusters are not well separated.

### 7.3.2.1   The integrated complete-data likelihood criterion

The integrated (complete-data) likelihood related to the complete data $(\mathbf{y},\mathbf{z})$ is

$$p(\mathbf{y},\mathbf{z} \mid G) = \int_{\Theta_G} p(\mathbf{y},\mathbf{z} \mid G,\theta) p(\theta \mid G) \mathrm{d}\theta,$$

where

$$p(\mathbf{y},\mathbf{z} \mid G,\theta) = \prod_{i=1}^{n} p(y_i, z_i \mid G,\theta) = \prod_{i=1}^{n} \prod_{g=1}^{G} \eta_g^{z_{ig}} \left[ f_g(y_i \mid \theta_g) \right]^{z_{ig}}.$$

This integrated complete-data likelihood (ICL) takes the missing data $\mathbf{z}$ into account and can be expected to be relevant for choosing $G$ in a clustering context. However, computing the ICL is challenging for various reasons. First, computing the ICL involves an integration in high dimensions. Second, the labels $\mathbf{z}$ are unobserved (missing) data. To approximate the ICL, a BIC-like approximation is possible (Biernacki et al., 2000):

$$\log p(\mathbf{y},\mathbf{z} \mid G) \approx \log p(\mathbf{y},\mathbf{z} \mid G,\hat{\theta}^{\mathbf{z}}) - \frac{\upsilon_G}{2} \log n,$$

where

$$\hat{\theta}^{\mathbf{z}} = \arg\max_{\theta} p(\mathbf{y}, \mathbf{z} \mid G, \theta),$$

and $\upsilon_G$ is the number of free parameters of the mixture model $\mathcal{M}_G$. Note that this approximation involves the complete-data likelihood, $\mathrm{L}_c(\theta, \mathbf{z}; G) = p(\mathbf{y}, \mathbf{z} \mid G, \theta)$; however, $\mathbf{z}$ and, consequently, $\hat{\theta}^{\mathbf{z}}$ are unknown. First, approximating $\hat{\theta}^{\mathbf{z}} \approx \hat{\theta}_G$, with $\hat{\theta}_G$ being the MLE of the $G$-component mixture parameter $\theta$, is expected to be valid for well-separated components. Second, given $\hat{\theta}_G$, the missing data $\mathbf{z}$ are imputed using the MAP estimator $\hat{\mathbf{z}} = \mathrm{MAP}(\hat{\theta}_G)$ defined by

$$\hat{z}_{ig} = \left\{ \begin{array}{ll} 1, & \text{if } \mathrm{argmax}_l \tau_{il}(\hat{\theta}_G) = g, \\ 0, & \text{otherwise.} \end{array} \right.$$

This leads to the criterion

$$\mathrm{ICLbic}(G) = \log p(\mathbf{y}, \hat{\mathbf{z}} \mid G, \hat{\theta}_G) - \frac{\upsilon_G}{2} \log n.$$

Exploiting (7.17), one obtains that the ICLbic criterion takes the form of a BIC criterion, penalized by the estimated entropy

$$\mathrm{ENT}(\hat{\theta}_G; G) = - \sum_{g=1}^{G} \sum_{i=1}^{n} \hat{\tau}_{ig} \log \hat{\tau}_{ig} \geq 0,$$

with $\hat{\tau}_{ig}$ denoting the conditional probability that $y_i$ arises from the $g$th mixture component ($i = 1, \ldots, n$, $g = 1, \ldots, G$) under the parameter $\hat{\theta}_G$; see (7.15).

Because of this additional entropy term, the ICLbic criterion favours values of $G$ giving rise to partitions of the data with the highest evidence. In practice, the ICLbic appears to provide a stable and reliable estimation of $G$ for real data sets and also for simulated data sets from mixtures when the components do not overlap too much. However, it should be noted that the ICLbic, which is not concerned with discovering the true number of mixture components, can underestimate the number of components for simulated data arising from mixtures with poorly separated components.

*An illustrative comparison of the BIC and ICLbic*

Obviously, in many situations where the mixture components are well separated, the BIC and ICLbic select the same number of mixture components. But the following small numerical example aims to illustrate a situation where these two criteria give different answers.

We start from a benchmark (genuine) data set known as the *Old Faithful Geyser*. Each of the 272 observations consists of two measurements: the duration of the eruption and the waiting time before the next eruption of the Old Faithful Geyser, in Yellowstone National Park, USA. We consider a bivariate Gaussian mixture model with component densities $\mathcal{N}(\mu_k, \Sigma_k)$ with unconstrained covariance matrices $\Sigma_k$.

For this data set, Figure 7.4 shows that the ICLbic selects with a large evidence $G = 2$, while the BIC slightly prefers $G = 3$ to $G = 2$. The BIC solution with $G = 3$ components appears to model deviations from normality in one of the two obvious clusters, rather than a relevant additional cluster.

### 7.3.2.2 The conditional classification likelihood

In a model-based clustering context where a cluster is associated with a mixture component, it is sensible in view of (7.17) to maximize the conditional expectation of the complete-data log likelihood (Baudry, 2015),

$$\log \mathrm{L}_{cc}(\theta; G) = \mathrm{E}_{\mathbf{z}}(\ell_c(\theta, \mathbf{z}; G)) = \ell_o(\theta; G) - \mathrm{ENT}(\theta; G),$$

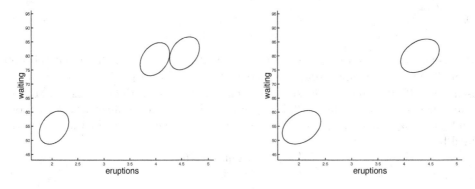

**FIGURE 7.4**
Cluster ellipses for the Old Faithful Geyser data: (left) the BIC solution; (right) the ICLbic solution.

rather than the observed-data log likelihood function $\ell_o(\theta; G)$. This can be done through an EM-type algorithm where the M step at iteration $s+1$ involves finding

$$\theta^{(s+1)} \in \underset{\theta \in \Theta_G}{\mathrm{argmax}} \left( \ell_o(\theta; G) + \sum_{i=1}^{n} \sum_{g=1}^{G} \tau_{ig}^{(s)} \log \tau_{ig} \right), \qquad (7.18)$$

where the $\tau_{ig}$ are defined as in (7.15) and

$$\tau_{ig}^{(s)} = \frac{\eta_g^{(s)} f_g(y_i \mid \theta_g^{(s)})}{\sum_{j=1}^{G} \eta_j^{(s)} f_j(y_i \mid \theta_j^{(s)})}.$$

This M step can be performed by using an adaptation of the so-called Bayesian expectation maximization (BEM) of Lange (1999). The resulting algorithm inherits the fundamental property of EM to increase the criterion $\log L_{cc}(\theta)$, which does not depend on **z**, at each iteration.

In this context, Baudry (2015) considered choosing $G$ from a penalized criterion of the form

$$L_{cc}\text{-}ICL(G) = -\log L_{cc}(\widehat{\theta}_G^{\mathrm{MLccE}}; G) + \mathrm{pen}(G),$$

where $\widehat{\theta}_G^{\mathrm{MLccE}} = \arg\max_\theta \log L_{cc}(\theta; G)$. Under standard regularity conditions and assuming that pen : $\{1, \ldots, G_{\max}\} \to \mathbb{R}^+$ satisfies

$$\begin{cases} \mathrm{pen}(G) = o_{\mathbb{P}}(n), & \text{as } n \to \infty, \\ (\mathrm{pen}(G) - \mathrm{pen}(G')) \xrightarrow[n \to \infty]{\mathbb{P}} \infty, & \text{if } G' < G, \end{cases}$$

Baudry (2015) proved that

$$\mathbb{P}[\widehat{G} \neq G_0] \xrightarrow[n \to \infty]{} 0,$$

where $\widehat{G} = \min \arg\min_G L_{cc}\text{-}ICL(G)$ and $G_0$ is the minimum number of components such that the bias of the models is stationary for $G \geq G_0$,

$$G_0 = \min \underset{G}{\mathrm{argmax}} \, E_{p_0} \left[ \ell_c(\theta_G^0) \right],$$

with

$$\theta_G^0 = \underset{\theta \in \Theta_G}{\mathrm{argmin}} \left\{ d_{\mathrm{KL}}\left(p_0, p(\,.\,; \theta)\right) + E_{p_0}\left[EC(\theta; G)\right] \right\},$$

$d_{\mathrm{KL}}\big(p_0, p(\,.\,;\theta)\big)$ being the Kullback–Leibler distance between the true distribution $p_0$ of the data and the mixture distribution with parameter $\theta$. Moreover, Baudry (2015) deduces that, by analogy with the BIC, an interesting identification criterion to be minimized is

$$\mathrm{L}_{cc}\text{-}\mathrm{ICL}(G) = -\log \mathrm{L}_{cc}(\widehat{\theta}_G^{\mathrm{MLccE}}; G) + \frac{\upsilon_G}{2}\log n.$$

The criterion ICLbic can thus be viewed as an approximation of $\mathrm{L}_{cc}$-ICL. Therefore, the criterion $\mathrm{L}_{cc}$-ICL underlies a notion of class that is a compromise between the "mixture component" and the "cluster" points of view.

### 7.3.2.3   Exact derivation of the ICL

Like the BIC, the ICL has been defined in a Bayesian framework, but its asymptotic approximations ICLbic and $\mathrm{L}_{cc}$-ICL are not intrinsically Bayesian, since they do not depend on the associated prior distribution. However, if the mixture components belong to the exponential family, it is possible to get closed-form expressions for the ICL (see Biernacki et al., 2010, or Bertoletti et al., 2015). With such closed-form expressions, it is possible to compute the ICL values by replacing the missing labels $\mathbf{z}$ with their most probable values using the MAP operator after estimating the parameter $\hat{\theta}_G$ as the posterior mode or the MLE (see Biernacki et al., 2010). An alternative is to optimize the exact ICL in $\mathbf{z}$. The limitations of approaches based on exact ICL computing are twofold.

*Choosing non-informative prior distributions*

Except for categorical data which involve mixtures of multivariate discrete distributions, there is no proper consensual non-informative prior distribution for other classes of mixture models such as Gaussian or Poisson mixture models (see Chapter 4). It is obviously possible to choose exchangeable weakly informative hyperparameters with conjugate prior distributions for the parameters of the mixture components. However, the posterior distribution and thus the resulting ICL values will inevitably depend on these hyperparameters. For the latent class model on categorical data, deriving the exact ICL is easier, since the non-informative conjugate Dirichlet prior distributions $\mathcal{D}_G(e_0)$ are proper for the weight distribution of the mixture. Following the recommendation of Frühwirth-Schnatter (2011), it has been demonstrated that choosing $e_0 = 4$ is expected to provide a stable selection of $G$ (see, for instance, Keribin et al., 2015). Numerical experiments on simulated data proved that exact ICL computed with plug-in estimates $\hat{\theta}_G$ of the parameter could provide different and more reliable estimation of $G$ than the ICLbic for small sample sizes. Thus, when conjugate non-informative prior distributions are available, deriving a non-asymptotic approximation of ICL can be feasible.

*Optimizing the exact ICL*

Several authors have considered the direct optimization of the exact ICL in $\mathbf{z}$ without estimating $\theta$. Bertoletti et al. (2015), Côme & Latouche (2015) and Wyse et al. (2017) have proposed greedy algorithms, while Tessier et al. (2006) proposed using evolutionary optimization algorithms. At this point, it is important to remark that the optimization problem has to be solved in a search space with about $O(G_{\max}^n)$ elements, where $G_{\max}$ is the maximum number of components allowed. This means that the optimization problem becomes quite formidable for $n$ large. In addition, the proposed greedy algorithms are highly sensitive to the numerous local optima and have only been experimented with for moderate sample sizes. This is the reason why evolutionary algorithms are expected to be useful but they need to be calibrated (to choose the tuning parameters) and are expensive in computing time.

### 7.3.3   Bayesian clustering

In the context of Bayesian clustering (see Lau & Green, 2007, for an excellent review), where the allocation indicator $\mathbf{z} = (z_1, \ldots, z_n)$ is regarded as a latent variable, a finite mixture model implies *random partitions* over the lattice $\mathcal{S}_G^n$. Hence, for a given order $G$ of the mixture distribution (7.1), both the prior density $p(\mathbf{z}|G)$ and the posterior density $p(\mathbf{z}|G, \mathbf{y})$ are discrete distributions over the lattice $\mathcal{S}_G^n$. Although this induces a change of prior modelling, Lau & Green (2007) discuss Bayesian nonparametric (BNP; see Chapter 6) methods to estimate the number of clusters. We discuss the BNP perspective further in Section 7.4.4 and refer to Chapter 6 for a comprehensive treatment.

For a finite mixture model, the Dirichlet prior $\eta \sim \mathcal{D}(e_1, \ldots, e_G)$ on the weight distribution strongly determines what the prior distribution $p(\mathbf{z}|G)$ looks like. To preserve symmetry with respect to relabelling, typically the symmetric Dirichlet prior $\mathcal{D}_G(e_0)$ is employed, where $e_1 = \ldots = e_G = e_0$. The corresponding prior $p(\mathbf{z}|G) = \int \prod_{i=1}^n p(z_i|\eta) d\eta$ is given by

$$p(\mathbf{z}|G) = \frac{\Gamma(Ge_0)}{\Gamma(n + Ge_0)\Gamma(e_0)^{G_+}} \prod_{g:n_g>0} \Gamma(n_g + e_0), \qquad (7.19)$$

where $n_g = \sum_{i=1}^n \mathbb{I}(z_i = g)$ is the number of observations in cluster $g$ and $G_+$ is defined as the number of non-empty clusters,

$$G_+ = G - \sum_{g=1}^G \mathbb{I}(n_g = 0). \qquad (7.20)$$

As mentioned earlier, in model-based clustering interest lies in estimating the number of clusters $G_+$ in the $n$ data points rather than the number of components $G$ of the mixture distribution (7.1), and it is important to distinguish between both quantities. Only a few papers make this clear distinction between the number of mixture components $G$ and the number of data cluster $G_+$ for finite mixture models (Nobile, 2004; Malsiner-Walli et al., 2017; Miller & Harrison, 2018; Frühwirth-Schnatter & Malsiner-Walli, 2018).

A common criticism concerning the application of finite mixture models in a clustering context is that the number of components $G$ needs to be known *a priori*. However, what is yet not commonly understood is (a) that the really relevant question is whether or not the *number of clusters* $G_+$ in the data is known *a priori* and (b) that even a finite mixture with a fixed value of $G$ can imply a random prior distribution on $G_+$. By way of further illustration, let $n_g = \sum_{i=1}^n \mathbb{I}(z_i = g)$ be the number of observations generated by the components $g = 1, \ldots, G$. Then (7.14) implies that $n_1, \ldots, n_G$ follow a multinomial distribution:

$$n_1, \ldots, n_G|\eta \sim \mathcal{M}(n, \eta_1, \ldots, \eta_G). \qquad (7.21)$$

Depending on the weights $\eta = (\eta_1, \ldots, \eta_G)$ appearing in the mixture distribution (7.1), multinomial sampling according to (7.21) may lead to partitions with $n_g$ being zero, leading to so-called "empty components". In this case, fewer than $G$ mixture components were used to generate the $n$ data points which contain $G_+$ non-empty clusters, where $G_+$ is defined as in (7.20).

In a Bayesian framework towards finite mixture modelling, the Dirichlet prior $\eta \sim \mathcal{D}_G(e_0)$ on the component weights controls whether, *a priori*, $G_+$ is equal to $G$ and no empty components occur. In particular, if $e_0$ is close to 0, then $G_+$ is a random variable taking *a priori* values smaller than $G$ with high probability. Exploiting the difference between $G_+$ and $G$ in an overfitting mixture with a prior on the weight distribution that strongly shrinks redundant component weights towards 0 is a cornerstone of the concept of

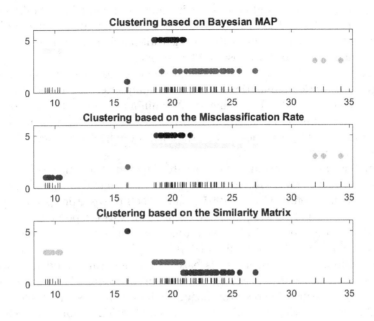

**FIGURE 7.5**
Bayesian clustering of the galaxy data (Roeder, 1990), assuming a Gaussian mixture with $G = 5$ components. The data are indicated through a rug plot. Partitions resulting from the MAP estimator (top), the minimum risk estimator (middle) and minimizing Binder's loss function.

sparse finite mixtures (Malsiner-Walli et al., 2016) which will be discussed in Section 7.4.5 as a one-sweep method to determine $G_+$ for a fixed $G$.

In Bayesian clustering (rather than Bayesian mixture estimation), the main object of interest is the (marginal) posterior of the allocations $\mathbf{z}$, rather than the (marginal) posterior distribution of the mixture parameters $\theta$. Depending on the mixture under investigation, the integrated likelihood $p(\mathbf{y}|\mathbf{z}, G)$ for $G$ known may be available in closed form, in particular, if the component densities $f_g(y|\theta_g)$ come from exponential families and a conditionally conjugate prior $p(\theta_g)$ is employed for $\theta_g$. As noted, for instance, by Casella et al. (2004), for many mixture models it is then possible to derive an explicit form for the marginal posterior $p(\mathbf{z}|\mathbf{y}, G)$ of the indicators $\mathbf{z}$, where dependence on the parameter $\theta$ is integrated out. By Bayes' theorem, the marginal posterior $p(\mathbf{z}|\mathbf{y}, G)$ is given by

$$p(\mathbf{z}|\mathbf{y}, G) \propto p(\mathbf{y}|\mathbf{z}, G)p(\mathbf{z}|G), \qquad (7.22)$$

where the integrated prior $p(\mathbf{z}|G)$ is given by (7.19) and the integrated likelihood $p(\mathbf{y}|\mathbf{z}, G)$ takes the form

$$p(\mathbf{y}|\mathbf{z}, G) = \int p(\mathbf{y}|\mathbf{z}, \theta_1, \ldots, \theta_G, \eta, G)p(\theta_1, \ldots, \theta_G, \eta|G)\mathrm{d}(\theta_1, \ldots, \theta_G, \eta). \qquad (7.23)$$

To explore the posterior of the allocations, efficient methods to sample from the posterior $p(\mathbf{z}|\mathbf{y}, G)$ are needed, and some of these methods will be discussed in Section 7.4.3. This exploration is quite a computational challenge, as the size of the lattice $\mathcal{S}_G^n$ increases rapidly with both the number $n$ of observations and the number $G$ of components and is given by the Bell number. For $n = 10$ and $G = 3$, for instance, there are 59,049 different allocations $\mathbf{z}$, whereas for $n = 100$ and $G = 3$ the number of different allocations is of the order of $5 \cdot 10^{47}$. This means that it is impossible to visit all possible partitions $\mathcal{C}$ during posterior sampling and many partitions are visited at best once.

This large set of partitions raises the question of how to summarize the posterior $p(\mathbf{z}|\mathbf{y}, G)$, given posterior simulations. Common summaries are based on deriving point estimators $\hat{\mathbf{z}}$, such as the MAP estimator, the minimum risk estimator or the partition minimizing Binder's loss function (Binder, 1978), see Section 8.3.2 for more details. However, these estimators (even if they differ) do not fully reflect the uncertainty in assigning observations to clusters.

By way of illustration, a mixture of univariate Gaussian distributions is used for Bayesian clustering of the galaxy data (Roeder, 1990), assuming that $G = 5$ is fixed. Prior specification follows Richardson & Green (1997), and 12,000 draws from $p(\mathbf{z}|\mathbf{y}, G)$ are obtained using full conditional Gibbs sampling. In Figure 7.5, various point estimators $\hat{\mathbf{z}}$ derived from the posterior draws of $\mathbf{z}$ are displayed, together with a rug plot of the data. While the MAP estimator and the estimator minimizing Binder's loss function are invariant to label switching, the minimum risk estimator is based on an identified model. Label switching is resolved by applying $k$-means clustering to the point process representation of the MCMC draws of $(\mu_g, \sigma_g)$. Classification over the various point estimators $\hat{\mathbf{z}}$ is stable for observations in the two clusters capturing the tails, but the classification for observations in the centre of the distribution tends to be rather different.

To quantify such uncertainty, Wade & Gharhamani (2018) develop not only appropriate point estimates, but also credible sets to summarize the posterior distribution of the partitions based on decision- and information-theoretic techniques.

### 7.3.4    Selecting $G$ under model misspecification

Mixture models are a very popular tool for model-based clustering, in both the frequentist and Bayesian frameworks. However, success in identifying meaningful clusters in the data very much hinges on specifying sensible component densities, and Bayesian inferences towards estimating the number of clusters are sensitive to misspecifications of the component densities, as are most penalized likelihood criteria discussed in the previous subsections. Most commonly, a finite mixture model with (multivariate) Gaussian component densities is fitted to the data to identify homogeneous data clusters within a heterogeneous population:

$$y \sim \sum_{g=1}^{G} \eta_g \mathcal{N}(\mu_g, \Sigma_g). \tag{7.24}$$

Similarly to the likelihood approach, Bayesian cluster analysis has to address several issues. First, as discussed above, even if we fit a correctly specified mixture model (7.1) to data generated by this model, an estimate of the number of components $G$ will not necessarily be a good estimator of the number of clusters $G_+$ in the data, and a more reliable estimate is obtained when exploring the partitions.

However, problems with the interpretation of $G_+$ might nevertheless occur, in particular if the component density is misspecified and several components have to be merged to address this misspecification. A typical example is fitting the multivariate Gaussian mixture distribution (7.24) to data such as the *Old Faithful Geyser* data. As shown in Figure 7.4, more than one Gaussian component is needed to capture departure from normality such as skewness and excess kurtosis for one of the two clusters. As discussed before, the BIC is particularly sensitive to this kind of misspecification, and classification-based information criteria such as the ICL criterion introduced in Section 7.3.2.1 are more robust in this respect.

In both Bayesian and frequentist frameworks, misspecification has been resolved by choosing more flexible distributions for the components densities. Many papers demonstrate the usefulness of mixtures of parametric non-Gaussian component densities in this context (see Frühwirth-Schnatter & Pyne, 2010, and Lee & McLachlan, 2013, among many others), and Chapter 10 also addresses this problem. Unsurprisingly, the estimated $G_+$ of such a non-Gaussian mixture often provides a much better estimator of the number of clusters than does the Gaussian mixture. With respect to inference, the Bayesian framework offers a slight advantage, as MCMC methods are able to deal with non-standard component densities in a more flexible way than the EM algorithm.

In higher dimensions it might be difficult to choose an appropriate parametric distribution for characterizing a data cluster, and mixture models with more flexible (not necessarily parametric) cluster densities turn out to be useful. The mixture of Gaussian mixtures approach, for instance, exploits the ability of normal mixtures to accurately approximate a wide class of probability distributions, and models the non-Gaussian cluster distributions themselves by Gaussian mixtures. This introduces a hierarchical framework where in the upper level a non-Gaussian mixture is fitted as in (7.1), whereas at a lower level each component density $f_g(y|\theta_g)$ itself is described by a mixture of $H_g$ Gaussian distributions. On the upper level, $G_+$ is identified as the number of such clusters, whereas the number of subcomponents $H_g$ in each cluster reflects the quality of the semi-parametric mixture approximation.

Two different approaches are available to "estimate" the number of clusters in such a framework. Any such approach has to deal with the following additional identifiability problems for this type of mixtures: the observed-data likelihood ascertains this model just as one big mixture of Gaussian distributions with $\tilde{G} = H_1 + \ldots + H_G$ components, and it does not change when we exchange subcomponents between clusters on the lower level, even though this leads to different cluster distributions on the upper level of the mixture of mixtures model. Hence, a mixture of mixtures model is not identifiable in the absence of additional information, and this is most naturally dealt with within a Bayesian framework.

Within the Bayesian approach, it is common to estimate the hierarchical mixture of mixtures model directly by including such prior information; see, in particular, Malsiner-Walli et al. (2017) who consider a random-effects prior to introduce prior dependence among the $H_g$ means of the subcomponent Gaussian mixture defining $f_g(y|\theta_g)$. A different approach which is prevalent in the frequentist literature employs a two-step procedure and tries to create meaningful clusters after having fitted a Gaussian mixture as in (7.24) with $G = G_{\max}$. The clusters are determined by successively merging components according to some criterion such as the entropy of the resulting partition (Baudry et al., 2010); see Chapter 8, Section 8.2.2 for additional approaches and further details.

## 7.4 One-Sweep Methods for Cross-model Inference on $G$

From a Bayesian perspective, inference methods that treat $G$ or $G_+$ as an unknown parameter to be estimated jointly with the component-specific parameters $\theta$ are preferable to processing $G$ as a model index and relying on testing principles. Several such approaches are reviewed in this section.

### 7.4.1   Overfitting mixtures

Rousseau & Mengersen (2011) examine the issue of an overfitting mixture, that is, the estimation of a mixture model with $G$ components when the true distribution behind the data has fewer than $G$, say $G_0$, components. This setting complicates even further the non-identifiability of the mixture model, since there are $\binom{G}{G_0}$ ways of picking $G_0$ components out of the $G$ (while cancelling the others); see also Chapter 4, Section 4.2.2.

Rousseau & Mengersen (2011) show that the posterior distribution on the parameters of the overfitted mixture has a much more stable behaviour than the likelihood function when the prior on the weights of the mixture is sufficiently concentrated on the boundaries of the parameter space, that is, with many weights being close to zero. In fact, the central result of Rousseau & Mengersen (2011) is that, if the dimension $r$ of the component parameters is larger than twice the hyperparameter $e_0$ of a symmetric Dirichlet prior $\mathcal{D}_G(e_0)$ on the weights, then the sum of the weights of the extra $G - G_0$ components asymptotically concentrates at zero. This result has the additional appeal of validating less informative priors as asymptotically consistent. In practice, it means that selecting a Dirichlet $\mathcal{D}_G(1/2)$ and an arbitrary prior on the component parameters should see superfluous components vanish as the sample size grows to be large enough, even though the impact of the choice of $e_0$ can be perceived for finite sample sizes.

### 7.4.2   Reversible jump MCMC

Reversible jump MCMC (RJMCMC; Green, 1995) was exploited by Richardson & Green (1997) to select the number of components $G$ for univariate mixtures of Gaussian distributions. As briefly discussed in Chapter 1, Section 1.4.3, this simulation method is based on creating a Markov chain that moves over a space of variable dimensions, namely between the parameter spaces of finite mixtures with different numbers of components, while retaining the fundamental detailed balance property that ensures the correct stationary (posterior) distribution.

The intuition behind the RJMCMC method is to create bijections between pairs of parameter spaces by creating auxiliary variables that equate the dimensions of the augmented spaces and to keep the same bijection for a move and its reverse. When designing a RJMCMC algorithm, those pairwise moves have to be carefully selected in order to reach sufficiently probable regions in the new parameter space. Richardson & Green (1997) discuss at length their *split-and-merge* moves which split (or aggregate) one (or two) components of the current mixture, with better performance than the basic *birth-and-death* moves, but performance may deteriorate as the number of components increases. The design of suitable proposals for higher-dimensional mixtures is quite a challenge, as demonstrated by Dellaportas & Papageorgiou (2006) and Zhang et al. (2004) for multivariate normal mixtures. In an attempt to extend RJMCMC methods to hidden Markov models, Cappé et al. (2002) had to face acceptance rates as low as 1%. RJMCMC is a natural extension of the traditional Metropolis–Hastings algorithm, but calibrating it is often perceived as too great an obstacle to its implementation, and it is not competitive with within-model simulations in the case of a small number of values of $G$ in competition.

### 7.4.3   Allocation sampling

As discussed in Section 7.3.3, the main object of interest in Bayesian clustering is the marginal posterior of the allocations, that is, $p(\mathbf{z}|\mathbf{y}, G)$ (if $G$ is known) or $p(\mathbf{z}|\mathbf{y})$ (if $G$ is unknown). Hence, Bayesian clustering has to rely on efficient methods to sample from the posterior $p(\mathbf{z}|\mathbf{y}, G)$ (or $p(\mathbf{z}|\mathbf{y})$).

While full conditional Gibbs sampling from the joint distribution $p(\theta, \mathbf{z} | \mathbf{y}, G)$ will yield draws from the (marginal) posterior $p(\mathbf{z} | \mathbf{y}, G)$, several authors considered alternative algorithms of "allocation sampling". Early Bayesian clustering approaches without parameter estimation are based on sampling from the marginal posterior distribution $p(\mathbf{z} | \mathbf{y}, G)$, defined earlier in (7.22), for known $G$. Chen & Liu (1996) were among the first to show how sampling of the allocations from $p(\mathbf{z} | \mathbf{y}, G)$ (for a fixed $G$) becomes feasible through MCMC methods, using either single-move Gibbs sampling or the Metropolis–Hastings algorithm; see Frühwirth-Schnatter (2006, Section 3.4) and Marin et al. (2005) for more details.

We want to stress here the following issue. Although these MCMC samplers operate in the marginal space of the allocations $\mathbf{z}$, neither the integrated likelihood $p(\mathbf{y} | \mathbf{z}, G)$, defined earlier in (7.23), nor the prior $p(\mathbf{z} | G)$, given in (7.19), can be (properly) defined without specifying a prior distribution $p(\theta_1, \ldots, \theta_G, \eta | G)$ for the unknown parameters of a mixture model with $G$ components. This problem is closely related to the problem discussed in Section 7.3.2.3 of having to choose priors for the exact ICL criterion. As discussed in Section 4.2.2, the choice of such a prior is not obvious and may have considerable impact on posterior inference.

These early sampling algorithms focus on computational aspects and do not explicitly account for the problem that the number $G_+$ of clusters in the sampled partitions $\mathbf{z}$ might differ from $G$, taking the identity of $G$ and $G_+$ more or less for granted. Still, as discussed above and again in Section 7.4.5, whether this applies or not very much depends on the choice of the hyperparameter $e_0$ in the Dirichlet prior $\mathcal{D}_G(e_0)$ on the weights.

Nobile & Fearnside (2007) address the problem of an unknown number of components $G$ in the context of allocation sampling. For a given $G$, they employ the usual Dirichlet prior $\eta | G \sim \mathcal{D}(e_1, \ldots, e_G)$ on the weight distribution, but treat $G$ as an unknown parameter, associated with a prior $p(G)$ (e.g. $G - 1 \sim \mathcal{P}(1)$), as justified by Nobile (2004). An MCMC sampler is developed that draws from the joint posterior $p(\mathbf{z}, G | \mathbf{y})$, by calling either Gibbs or Metropolis–Hastings moves based on the conditional distribution of $p(\mathbf{z} | \mathbf{y}, G)$ for a given $G$ and by running RJMCMC type moves for switching values of $G$. Based on these posterior draws, $\hat{G}_+$ is estimated from the posterior draws of the number of non-empty clusters $G_+$. Several post-processing strategies are discussed for solving the label switching problem that is inherent in this sampler and for estimating $\hat{\mathbf{z}}$.

### 7.4.4 Bayesian nonparametric methods

A quite different approach of selecting the number $G_+$ of clusters exists outside the framework of finite mixture models and relies on Bayesian nonparametric approaches based on mixture models with countably infinite number of components, as discussed in Chapter 6 in full detail.

For Dirichlet process (DP) mixtures (Müller & Mitra, 2013), for instance, the discrete mixing distribution in the finite mixture (7.1) is substituted by a random distribution $H \sim DP(\alpha, H_0)$, drawn from a DP prior with precision parameter $\alpha$ and base measure $H_0$. As a draw $H$ from a DP is almost surely discrete, the corresponding model has a representation as an infinite mixture,

$$y \sim \sum_{g=1}^{\infty} \eta_g f_g(y | \theta_g), \tag{7.25}$$

with i.i.d. atoms $\theta_g \overset{iid}{\sim} H_0$ drawn from the base measure $H_0$ and weights $\eta_g$ obeying the stick-breaking representation

$$\eta_g = v_g \prod_{j=1}^{g-1} (1 - v_j), \quad g = 1, 2, \ldots, \tag{7.26}$$

with $v_g \sim \mathcal{B}e(1, \alpha)$ (Sethuraman, 1994).

As DP priors induce ties among the observations, such an approach automatically induces a random partition (or clustering) $\mathcal{C}$ of the data with a corresponding random cardinality $G_+$ (see Section 6.4). Since there are infinitely many components in (7.25) (i.e. $G = \infty$), there is no risk of confusing $G$ and $G_+$ as for finite mixtures. For a DP prior with precision parameter $\alpha$, the prior distribution over the partitions $\mathcal{C}$ is given by

$$p(\mathcal{C}|\alpha, G_+) = \alpha^{G_+} \frac{\Gamma(\alpha)}{\Gamma(n + \alpha)} \prod_{g:n_g>0} \Gamma(n_g), \tag{7.27}$$

where $n_g$ and $G_+$ are defined as in (7.19). Another defining property of the DP prior is the structure of the prior predictive distribution $p(z_i|\mathbf{z}_{-i})$, where $\mathbf{z}_{-i}$ denotes all indicators excluding $z_i$. Let $G_+^{-i}$ be the number of non-empty clusters implied by $\mathbf{z}_{-i}$ and let $n_g^{-i}$, $g = 1, \ldots, G_+^{-i}$, be the corresponding cluster sizes. Then the probability that $z_i$ is assigned to an existing cluster $g$ is given by

$$\mathrm{P}(z_i = g|\mathbf{z}_{-i}, n_g^{-i} > 0) = \frac{n_g^{-i}}{n - 1 + \alpha}, \tag{7.28}$$

whereas the prior probability that $z_i$ creates a new cluster (indexed by $G_+^{-i} + 1$) is equal to

$$\mathrm{P}(z_i = G_+^{-i} + 1|\mathbf{z}_{-i}) = \frac{\alpha}{n - 1 + \alpha}. \tag{7.29}$$

Given this strong focus on BNP mixtures as random partition models, it is not surprising that the main interest in posterior inference is again in the draws from the posterior $p(\mathbf{z}|\mathbf{y})$ of the allocations which are exploited in various ways to choose an appropriate partition $\hat{\mathbf{z}}$ of the data and to estimate the number of clusters $G_+$.

Lau & Green (2007) compare BNP methods to estimate the number of clusters with the outcome associated with finite mixtures. They also show in detail how to derive a single (optimal) point estimate $\hat{\mathbf{z}}$ from the posterior $p(\mathbf{z}|\mathbf{y})$, with the number of distinct clusters $\hat{G}_+$ in $\hat{\mathbf{z}}$ being an estimator of $G_+$ in this framework. To derive a partition of the data, Molitor et al. (2010) cluster the data using the pairwise association matrix as a distance measure which is obtained by aggregating over all partitions obtained during MCMC sampling, using partitioning around medoids. The optimal number of clusters is determined by maximizing an associated clustering score; see also Liverani et al. (2013).

A well-known limitation of DP priors is that *a priori* the cluster sizes are expected to be geometrically ordered, with one big cluster, geometrically smaller clusters, and many singleton clusters (Müller & Mitra, 2013). This initiated the investigation of alternative BNP mixtures and their usefulness for clustering. A popular BNP two-parameter mixture is obtained from the Pitman–Yor process (PYP) prior $PY(\beta, \alpha)$ with $\beta \in [0, 1)$, $\alpha > -\beta$ (Pitman & Yor, 1997), with a stick-breaking representation as in (7.26) with $v_g \sim \mathcal{B}e(1 - \beta, \alpha + k\beta)$. The DP prior occurs as a special case when $\beta = 0$. PYP mixtures are known to be more useful than the DP mixture for data with many significant, but small clusters.

For a DP as well as a PYP mixture, the prior expected number of data clusters $G_+$ increases as the number $n$ of observations increases, where for the DP process $G_+ \sim \alpha \log(n)$ (Korwar & Hollander, 1973) and $G_+ \sim n^\beta$ obeys a power law for PYP mixtures. As will be discussed in the next subsection, finite mixtures are quite different in this respect.

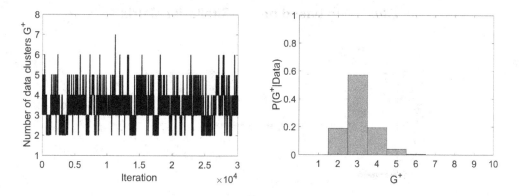

**FIGURE 7.6**
Sparse finite mixture modelling of the enzyme data: (left) 30,000 posterior draws of the number of data clusters $G_+$; (right) posterior distribution $p(G_+|\mathbf{y})$.

### 7.4.5 Sparse finite mixtures for model-based clustering

Inspired by the important insights of Rousseau & Mengersen (2011), Malsiner-Walli et al. (2016) introduced the concept of sparse finite mixture models for model-based clustering as an alternative to infinite mixtures, following ideas presented earlier in Frühwirth-Schnatter (2011). A similar approach is pursued by van Havre et al. (2015).

While remaining within the framework of finite mixtures, sparse finite mixture models provide a semi-parametric Bayesian approach in so far as the number $G_+$ of non-empty mixture components used to generate the data is not assumed to be known in advance, but random, as already discussed in Section 7.3.3. The basic idea of sparse finite mixture modelling is to deliberately specify an *overfitting* finite mixture model with too many components $G$. Sparse finite mixtures stay within the common finite mixture framework by assuming a symmetric Dirichlet prior $\eta \sim \mathcal{D}_G(e_0)$ on the weight distribution; however, the hyperparameter $e_0$ of this prior is selected such that superfluous components are emptied automatically during MCMC sampling and sparse solutions with regard to the number $G_+$ of clusters are induced through the prior on the weight distribution. This proposal leads to a simple Bayesian framework where a straightforward MCMC sampling procedure is applied to jointly estimate the unknown number of non-empty data clusters $G_+$ together with the remaining parameters.

As discussed in Section 7.3.3, for such a mixture model, the number $G$ of components does not reflect the number of data clusters, as many components will remain unused. Following Nobile (2004), Malsiner-Walli et al. (2016) derive the posterior distribution $\mathrm{P}(G_+ = g|\mathbf{y})$, $g = 1, \ldots, G$, of the number $G_+$ of data clusters from the MCMC output of the allocations $\mathbf{z}$. Therefore, for each iteration $m$ of MCMC sampling, all components $g$ to which some observations been assigned are identified from $\mathbf{z}^{(m)}$ and the corresponding number of non-empty components is considered:

$$G_+^{(m)} = G - \sum_{g=1}^{G} \mathbb{I}(n_g^{(m)} = 0),$$

where, for $g = 1, \ldots, G$, $n_g^{(m)} = \sum_{i=1}^{n} \mathbb{I}(z_i^{(m)} = g)$ is the number of observations allocated to component $g$, and $\mathbb{I}(\cdot)$ denotes the indicator function. The posterior distribution $\mathrm{P}(G_+ = g|\mathbf{y})$, $g = 1, \ldots, G$, is then estimated by the corresponding relative frequency.

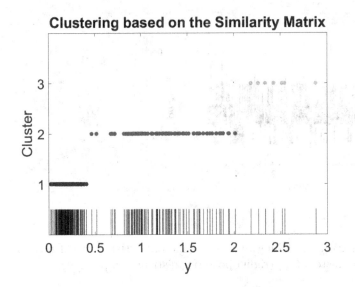

**FIGURE 7.7**
Sparse finite mixture modelling of the enzyme data, displayed as a rug plot. Partition $\hat{\mathbf{z}}$ optimizing Binder's loss function. The number of clusters in this partition is equal to three.

The number of clusters $G_+$ can be derived as a point estimator from this distribution, for example, the posterior mode estimator $\tilde{G}_+$ that maximizes the (estimated) posterior distribution $\mathrm{P}(G_+ = g|\mathbf{y})$. This happens to be the most frequent number of clusters visited during MCMC sampling. The posterior mode estimator appears to be sensible in the present context when adding very small clusters hardly changes the marginal likelihood. This makes the posterior distribution $\mathrm{P}(G_+ = g|\mathbf{y})$ extremely right-skewed, and other point estimators such as the posterior mean are extremely sensitive to prior choices, as noted by Nobile (2004). However, under a framework where sparse finite mixtures are employed for density estimation, very small components might be important and other estimators of $G_+$ might be better justified.

An alternative way to summarize clustering based on sparse finite mixtures is by exploring the posterior draws of the partitions $\mathbf{z}$ and determining some optimal partition, such as the partition $\hat{\mathbf{z}}$ minimizing Binder's loss function. This can be done without the need to resolve label switching or to stratify the draws with respect to $G_+$. The cardinality $\hat{G}_+$ of such an optimal partition $\hat{\mathbf{z}}$ is yet another estimator of the number of clusters. The posterior mode estimator $\tilde{G}_+$ and $\hat{G}_+$ do not necessarily coincide, and differences in these estimators reflect uncertainty in the posterior distribution over the partition space. As discussed by Frühwirth-Schnatter et al. (2018), the approach of Wade & Gharhamani (2018) to quantifying such uncertainty can be applied immediately to sparse finite mixture models.

The appropriate choice of the hyperparameter $e_0$ is important for the application of the sparse finite mixture approach in a clustering context. While in a density estimation framework the asymptotic criterion of Rousseau & Mengersen (2011) suggests the choice $e_0 < r/2$, with $r$ being the dimension of $\theta_g$, this rule is not necessarily a sensible choice for selecting the number of clusters $G_+$ in a data set of finite size $n$, as demonstrated for a broad range of mixture models in Malsiner-Walli et al. (2016) and Frühwirth-Schnatter & Malsiner-Walli (2018). Indeed, these papers show that values of $e_0 \ll r/2$ much smaller than the asymptotic criterion of Rousseau & Mengersen (2011) are needed to identify the right number of clusters, and recommend choosing either very small fixed values such as $e_0 = 0.001$ or applying a hyperprior with $e_0 \sim \mathcal{G}(a_e, b_e)$ such that $\mathrm{E}(e_0) = a_e/b_e$ is very small (e.g. $e_0 \sim \mathcal{G}(1, 200)$).

Under the provision that $G_+$ underestimates $G$, this approach constitutes a simple and generic strategy for model selection without making use of model selection criteria, RJMCMC, or marginal likelihoods. Applications include Gaussian mixtures as well as mixtures of Gaussian mixtures (Malsiner-Walli et al., 2017) and sparse mixtures for discrete-valued data (Frühwirth-Schnatter & Malsiner-Walli, 2018). By way of further illustration, the enzyme data (shown earlier in Figure 7.2) are reanalysed using sparse finite mixtures, taking the prior of Richardson & Green (1997) as base measure. The maximum number of data clusters is chosen as $G = 10$ and the hierarchical sparse Dirichlet prior $\eta \sim \mathcal{D}_G(e_0)$, $e_0 \sim \mathcal{G}(1, 200)$ is applied.

Figure 7.6 shows 30,000 posterior draws of the number of data clusters $G_+$ as well as the corresponding posterior distribution $p(G_+|\mathbf{y})$. The posterior mode estimator yields three clusters with $\mathrm{P}(G_+ = 3|\mathbf{y}) = 0.57$. Also two clusters are supported with $\mathrm{P}(G_+ = 2|\mathbf{y}) = 0.19$, which is not unexpected in the light of Figure 7.2, showing two (albeit non-Gaussian) data clusters. Due to this misspecification of the component densities the four-cluster solution is equally supported with $\mathrm{P}(G_+ = 4|\mathbf{y}) = 0.19$. Finally, Figure 7.7 shows the partition $\hat{\mathbf{z}}$ optimizing Binder's loss function together with a rug plot of the data. The number of clusters in this partition is equal to three, supporting the choice based on the posterior mode. The resulting clustering nicely captures the three distinct groups of data points.

*Relation to BNP methods*

The concept of sparse finite mixtures is related in various ways to DP mixtures, discussed in Section 7.4.4. If the weight distribution follows the Dirichlet prior $\eta \sim \mathcal{D}_G(\alpha/G)$ and the base measure $H_0$ serves as prior for the component parameters (i.e. $\theta_g \sim H_0$), then, as shown by Green & Richardson (2001), the finite mixture in (7.1) converges to a DP mixture with mixing distribution $H \sim DP(\alpha, H_0)$ as $G$ increases. This relationship has mainly been exploited to obtain a finite mixture approximation to the DP mixture. In this sense, the sparse finite Gaussian mixture introduced in Malsiner-Walli et al. (2016) could be seen as an approximation to a DP mixture. Nevertheless, as argued by Malsiner-Walli et al. (2017), it makes sense to stay within the framework of finite mixtures and to consider $G$ as a second parameter which is held fixed at a finite value, as this provides a two-parameter alternative to DP mixtures with related properties.

Representations similar to BNP mixtures exist also for finite mixture models under the symmetric prior $\eta \sim \mathcal{D}_G(e_0)$, but are not commonly known, although they shed further light on the relation between the two model classes. First of all, a stick-breaking representation of the weights $\eta_1, \eta_2, \ldots, \eta_G$ as in (7.26) in terms of a sequence of independently (albeit not identically) distributed random variables exists also for finite mixtures, with $v_g \sim \mathcal{B}e(e_0, (G - g)e_0)$, $g = 1, \ldots, G - 1$, $v_G = 1$; see, for example, Frühwirth-Schnatter (2011).

Second, as already discussed in Section 7.3.1, finite mixture models can be regarded as random partition models and the prior distribution over all random partitions $\mathcal{C}$ of $n$ observations can be derived from the joint (marginal) prior $p(\mathbf{z}|G)$ given in (7.19) (see, for example, Malsiner-Walli et al. (2017)):

$$p(\mathcal{C}|e_0, G_+) = \frac{G!}{(G - G_+)!} \frac{\Gamma(Ge_0)}{\Gamma(n + Ge_0)\Gamma(e_0)^{G_+}} \prod_{g:n_g>0} \Gamma(n_g + e_0).$$

This prior takes the form of a product partition model as for DP mixtures (see (7.27)) and is invariant to permuting the cluster labels.

Finally, as for BNP mixtures, it is possible to derive the prior predictive distribution $p(z_i|\mathbf{z}_{-i})$, where $\mathbf{z}_{-i}$ denotes all indicators excluding $z_i$. Let $G_+^{-i}$ be the number of non-empty clusters implied by $\mathbf{z}_{-i}$, and let $n_g^{-i}$, $g = 1, \ldots, G_+^{-i}$, be the corresponding cluster sizes. Then the probability that $z_i$ is assigned to an existing cluster $g$ is given by

$$P(z_i = g|\mathbf{z}_{-i}, n_g^{-i} > 0) = \frac{n_g^{-i} + e_0}{n - 1 + e_0 G},$$

which is closely related to (7.28), in particular if $e_0 = \alpha/G$ and $G$ increases. However, the prior probability that $z_i$ creates a new cluster with $z_i \in I = \{g : n_g^{-i} = 0\}$ is equal to

$$P(z_i \in I|\mathbf{z}_{-i}) = \frac{e_0(G - G_+^{-i})}{n - 1 + e_0 G}, \tag{7.30}$$

and is quite different from (7.29). In particular, for $e_0$ independent of $G$, this probability not only depends on $e_0$, but also increases with $G$. Hence a sparse finite mixture model can be regarded as a two-parameter model, where both $e_0$ and $G$ influence the prior expected number of data clusters $G_+$, which is determined for a DP mixture solely by $\alpha$. Furthermore, the prior probability (7.30) of creating new clusters decreases as the number $G_+^{-i}$ of non-empty clusters increases, as opposed to DP mixtures where this probability is constant and to PYP mixtures where this probability increases. Hence, sparse finite mixtures are useful for clustering data that arise from a moderate number of clusters that does not increase as the number of data points $n$ increases.

## 7.5   Concluding Remarks

The issue of selecting the number of mixture components has always been contentious, both in frequentist and Bayesian terms, and this chapter has reflected on this issue by presenting a wide variety of solutions and analyses. The main reason for the difficulty in estimating the order $G$ of a mixture model is that it is a poorly defined quantity, even when setting aside identifiability and label switching aspects. Indeed, when considering a single sample of size $n$ truly generated from a finite mixture model, there is always a positive probability that the observations in that sample are generated from a subset of the components of the mixture of size $G_+$ rather than from all components $G$. As shown by the asymptotic results in Chapter 4, the issue goes away as the sample size $n$ goes to infinity (provided $G$ remains fixed), but this does not bring a resolution to the quandary of whether or not $G$ is estimable. In our opinion, inference should primarily bear on the number of data clusters $G_+$, since the conditional posterior distribution of $G$ given $G_+$ mostly depends on the prior modelling and very little on the data. Without concluding like Larry Wasserman (on his now defunct *Normal Deviate* blog) that "mixtures, like tequila, are inherently evil and should be avoided at all costs", we must acknowledge that the multifaceted uses of mixture models imply that the estimation of a quantity such as the number of mixture components should be impacted by the purpose of modelling via finite mixtures, as for instance through the prior distributions in a Bayesian setting.

# Bibliography

AKAIKE, H. (1974). A new look at the statistical model identification. *IEEE Transactions on Automatic Control* **19**, 716–723.

BANFIELD, J. D. & RAFTERY, A. E. (1993). Model-based Gaussian and non-Gaussian clustering. *Biometrics* **49**, 803–821.

BAUDRY, J.-P. (2015). Estimation and model selection for model-based clustering with the conditional classification likelihood. *Electronic Journal of Statistics* **9**, 1041–1077.

BAUDRY, J.-P., MAUGIS, C. & MICHEL, B. (2012). Slope heuristics: Overview and implementation. *Statistics and Computing* **22**, 455–470.

BAUDRY, J.-P., RAFTERY, A. E., CELEUX, G., LO, K. & GOTTARDO, R. (2010). Combining mixture components for clustering. *Journal of Computational and Graphical Statistics* **19**, 332–353.

BERGER, J. (1985). *Statistical Decision Theory and Bayesian Analysis*. New York: Springer-Verlag, 2nd ed.

BERGER, J. & JEFFERYS, W. (1992). Sharpening Ockham's razor on a Bayesian strop. *American Statistician* **80**, 64–72.

BERKHOF, J., VAN MECHELEN, I. & GELMAN, A. (2003). A Bayesian approach to the selection and testing of mixture models. *Statistica Sinica* **13**, 423–442.

BERTOLETTI, M., FRIEL, N. & RASTELLI, R. (2015). Choosing the number of components in a finite mixture using an exact integrated completed likelihood criterion. *Metron* **73**, 177–199.

BESAG, J. (1989). A candidate's formula: A curious result in Bayesian prediction. *Biometrika* **76**, 183.

BIERNACKI, C., CELEUX, G. & GOVAERT, G. (2000). Assessing a mixture model for clustering with the integrated completed likelihood. *IEEE Transactions on Pattern Analysis and Machine Intelligence* **22**, 719–725.

BIERNACKI, C., CELEUX, G. & GOVAERT, G. (2010). Exact and Monte Carlo calculations of integrated likelihoods for the latent class model. *Journal of Statistical Planning Inference* **140**, 2991–3002.

BINDER, D. A. (1978). Bayesian cluster analysis. *Biometrika* **65**, 31–38.

BIRGÉ, L. & MASSART, P. (2001). Gaussian model selection. *Journal of the European Mathematical Society* **3**, 203–268.

BIRGÉ, L. & MASSART, P. (2007). Gausian model selection. *Probability Theory and Related Fields* **138**, 33–73.

BOUVEYRON, C., CÔME, E. & JACQUES, J. (2015). The discriminative functional mixture model for a comparative analysis of bike sharing systems. *Annals of Applied Statistics* **9**, 1726–1760.

BOX, G. E. P. (1976). An application of the Laplace method to finite mixture distributions. *Journal of the American Statistical Association* **71**, 791–799.

BOZDOGAN, H. (1987). Model selection and Akaike's information criterion (AIC): The general theory and its analytical extensions. *Psychometrika* **52**, 345–370.

CAPPÉ, O., ROBERT, C. P. & RYDÉN, T. (2002). Reversible jump MCMC converging to birth-and-death MCMC and more general continuous time samplers. *Journal of the Royal Statistical Society, Series B* **65**, 679–700.

CASELLA, G., ROBERT, C. P. & WELLS, M. T. (2004). Mixture models, latent variables and partitioned importance sampling. *Statistical Methodology* **1**, 1–18.

CELEUX, G., CHRÉTIEN, S., FORBES, F. & MKHADRI, A. (2001). A component-wise EM algorithm for mixtures. *Journal of Computational and Graphical Statistics* **10**, 697–712.

CELEUX, G., FORBES, F., ROBERT, C. P. & TITTERINGTON, D. M. (2006). Deviance information criteria for missing data models. *Bayesian Analysis* **1**, 651–674.

CELEUX, G. & SOROMENHO, G. (1996). An entropy criterion for assessing the number of clusters in a mixture model. *Journal of Classification* **13**, 195–212.

CHEN, R. & LIU, J. S. (1996). Predictive updating methods with application to Bayesian classification. *Journal of the Royal Statistical Society, Series B* **58**, 397–415.

CHIB, S. (1995). Marginal likelihood from the Gibbs output. *Journal of the American Statistical Association* **90**, 1313–1321.

CHOPIN, N. (2002). A sequential particle filter method for static models. *Biometrika* **89**, 539–552.

CHOPIN, N. & ROBERT, C. P. (2010). Properties of nested sampling. *Biometrika* **97**, 741–755.

CÔME, E. & LATOUCHE, P. (2015). Model selection and clustering in stochastic block models based on the exact integrated complete data likelihood. *Statistical Modelling* **6**, 564–589.

DACUNHA-CASTELLE, D. & GASSIAT, E. (1997). The estimation of the order of a mixture model. *Bernoulli* **3**, 279–299.

DACUNHA-CASTELLE, D. & GASSIAT, E. (1999). Testing the order of a model using locally conic parametrization: Population mixtures and stationary ARMA processes. *Annals of Statistics* **27**, 1178–1209.

DE IORIO, M. & ROBERT, C. P. (2002). Discussion of "Bayesian measures of model complexity and fit", by Spiegelhalter et al. *Journal of the Royal Statistical Society, Series B* **64**, 629–630.

DELLAPORTAS, P. & PAPAGEORGIOU, I. (2006). Multivariate mixtures of normals with unknown number of components. *Statistics and Computing* **16**, 57–68.

DICICCIO, T. J., KASS, R. E., RAFTERY, A. & WASSERMAN, L. (1997). Computing Bayes factors by combining simulation and asymptotic approximations. *Journal of the American Statistical Association* **92**, 903–915.

DRTON, M. & PLUMMER, M. (2017). A Bayesian information criterion for singular models. *Journal of the Royal Statistical Society, Series B* **79**, 1–18.

FIGUEIREDO, M. & JAIN, A. K. (2002). Unsupervised learning of finite mixture models. *IEEE Transactions on Pattern Analysis and Machine Intelligence* **24**, 381–396.

FRALEY, C. & RAFTERY, A. E. (2002). Model-based clustering, discriminant analysis, and density estimation. *Journal of the American Statistical Association* **97**, 611–631.

FRÜHWIRTH-SCHNATTER, S. (1995). Bayesian model discrimination and Bayes factors for linear Gaussian state space models. *Journal of the Royal Statistical Society, Series B* **57**, 237–246.

FRÜHWIRTH-SCHNATTER, S. (2001). Markov chain Monte Carlo estimation of classical and dynamic switching and mixture models. *Journal of the American Statistical Association* **96**, 194–209.

FRÜHWIRTH-SCHNATTER, S. (2004). Estimating marginal likelihoods for mixture and Markov switching models using bridge sampling techniques. *Econometrics Journal* **7**, 143–167.

FRÜHWIRTH-SCHNATTER, S. (2006). *Finite Mixture and Markov Switching Models*. New York: Springer-Verlag.

FRÜHWIRTH-SCHNATTER, S. (2011). Dealing with label switching under model uncertainty. In *Mixture Estimation and Applications*, K. Mengersen, C. P. Robert & D. Titterington, eds., chap. 10. Chichester: Wiley, pp. 193–218.

FRÜHWIRTH-SCHNATTER, S. (2018). *Applied Bayesian Mixture Modelling. Implementations in MATLAB using the package bayesf Version 4.0.* https://www.wu.ac.at/statmath/faculty-staff/faculty/sfruehwirthschnatter/.

FRÜHWIRTH-SCHNATTER, S., GRÜN, B. & MALSINER-WALLI, G. (2018). Contributed comment on article by Wade and Gharamani. *Bayesian Analysis* **13**, 601–603.

FRÜHWIRTH-SCHNATTER, S. & MALSINER-WALLI, G. (2018). From here to infinity – sparse finite versus Dirichlet process mixtures in model-based clustering. Preprint, arXiv:1706.07194.

FRÜHWIRTH-SCHNATTER, S. & PYNE, S. (2010). Bayesian inference for finite mixtures of univariate and multivariate skew normal and skew-$t$ distributions. *Biostatistics* **11**, 317–336.

GASSIAT, E. (2002). Likelihood ratio inequalities with applications to various mixtures. *Annales de l'Institut Henri Poincaré – Probabilités et Statistiques* **38**, 887–906.

GASSIAT, E. & VAN HANDEL, R. (2013). Consistent order estimation and minimal penalties. *IEEE Transactions on Information Theory* **59**, 1115–1128.

GELFAND, A. & DEY, D. (1994). Bayesian model choice: Asymptotics and exact calculations. *Journal of the Royal Statistical Society, Series B* **56**, 501–514.

GELFAND, A. & SMITH, A. (1990). Sampling based approaches to calculating marginal densities. *Journal of the American Statistical Association* **85**, 398–409.

GELMAN, A. & MENG, X. (1998). Simulating normalizing constants: From importance sampling to bridge sampling to path sampling. *Statistical Science* **13**, 163–185.

GREEN, P. J. (1995). Reversible jump MCMC computation and Bayesian model determination. *Biometrika* **82**, 711–732.

GREEN, P. J. & RICHARDSON, S. (2001). Modelling heterogeneity with and without the Dirichlet process. *Scandinavian Journal of Statistics* **28**, 355–375.

HANSEN, B. (1992). The likelihood ratio test under non-standard conditions: Testing the Markov switching model of GNP. *Journal of Applied Econometrics* **7**, S61–S82.

KERIBIN, C. (2002). Consistent estimation of the order of mixture models. *Sankhyā, Series A* **62**, 49–66.

KERIBIN, C., BRAULT, V., CELEUX, G. & GOVAERT, G. (2015). Estimation and selection for the latent block model on categorical data. *Statistics and Computing* **25**, 1201–1216.

KORWAR, R. M. & HOLLANDER, M. (1973). Contributions to the theory of Dirichlet processes. *Annals of Probability* **1**, 705–711.

LANGE, K. (1999). *Numerical Analysis for Statisticians*. New York: Springer-Verlag.

LAU, J. W. & GREEN, P. J. (2007). Bayesian model-based clustering procedures. *Journal of Computational and Graphical Statistics* **16**, 526–558.

LEE, J. & ROBERT, C. (2016). Importance sampling schemes for evidence approximation in mixture models. *Bayesian Analysis* **11**, 573–597.

LEE, K., MARIN, J.-M., MENGERSEN, K. & ROBERT, C. (2009). Bayesian inference on mixtures of distributions. In *Perspectives in Mathematical Sciences I: Probability and Statistics*, N. N. Sastry, M. Delampady & B. Rajeev, eds. Singapore: World Scientific, pp. 165–202.

LEE, S. X. & MCLACHLAN, G. J. (2013). EMMIXuskew: An R package for fitting mixtures of multivariate skew t-distributions via the EM algorithm. *Journal of Statistical Software* **55**, 1–22.

LIVERANI, S., HASTIE, D. I., PAPATHOMAS, M. & RICHARDSON, S. (2013). PReMiuM: An R package for profile regression mixture models using Dirichlet processes. Preprint, arXiv:1303.2836.

MALSINER-WALLI, G., FRÜHWIRTH-SCHNATTER, S. & GRÜN, B. (2016). Model-based clustering based on sparse finite Gaussian mixtures. *Statistics and Computing* **26**, 303–324.

MALSINER-WALLI, G., FRÜHWIRTH-SCHNATTER, S. & GRÜN, B. (2017). Identifying mixtures of mixtures using Bayesian estimation. *Journal of Computational and Graphical Statistics* **26**, 285–295.

MARIN, J.-M., MENGERSEN, K. & ROBERT, C. (2005). Bayesian modelling and inference on mixtures of distributions. In *Handbook of Statistics*, C. R. Rao & D. Dey, eds., vol. 25. New York: Springer-Verlag, pp. 459–507.

MCLACHLAN, G. (1987). On bootstrapping the likelihood ratio test statistic for the number of components in a normal mixture. *Applied Statistics* **36**, 318–324.

MCLACHLAN, G. & PEEL, D. (2000). *Finite Mixture Models.* New York: Wiley.

MENG, X. & WONG, W. (1996). Simulating ratios of normalizing constants via a simple identity: A theoretical exploration. *Statistica Sinica* **6**, 831–860.

MILLER, J. W. & HARRISON, M. T. (2018). Mixture models with a prior on the number of components. *Journal of the American Statistical Association* **113**, 340–356.

MOLITOR, J., PAPATHOMAS, M., JERRETT, M. & RICHARDSON, S. (2010). Bayesian profile regression with an application to the National Survey of Children's Health. *Biostatistics* **11**, 484–498.

MÜLLER, P. & MITRA, R. (2013). Bayesian nonparametric inference – why and how. *Bayesian Analysis* **8**, 269–360.

NADIF, M. & GOVAERT, G. (1998). Clustering for binary data and mixture models – choice of the model. *Applied Stochastic Models and Data Analysis* **13**, 269–278.

NEAL, R. (1999). Erroneous results in "Marginal likelihood from the Gibbs output". Tech. rep., University of Toronto.

NOBILE, A. (2004). On the posterior distribution of the number of components in a finite mixture. *Annals of Statistics* **32**, 2044–2073.

NOBILE, A. & FEARNSIDE, A. (2007). Bayesian finite mixtures with an unknown number of components: The allocation sampler. *Statistics and Computing* **17**, 147–162.

PITMAN, J. & YOR, M. (1997). The two-parameter Poisson-Dirichlet distribution derived from a stable subordinator. *Annals of Probability* **25**, 855–900.

RAU, A., MAUGIS-RABUSSEAU, C., MARTIN-MAGNIETTE, M.-L. & CELEUX, G. (2015). Co-expression analysis of high-throughput transcriptome sequencing data with Poisson mixture models. *Bioinformatics* **31**, 1420–1427.

RICHARDSON, S. & GREEN, P. (1997). On Bayesian analysis of mixtures with an unknown number of components (with discussion). *Journal of the Royal Statistical Society, Series B* **59**, 731–792.

RISSANEN, J. (2012). *Optimal Estimation of Parameters.* Cambrdige: Cambridge University Press.

ROBERT, C. P. (2007). *The Bayesian Choice.* New York: Springer.

ROBERT, C. P. & CASELLA, G. (2004). *Monte Carlo Statistical Methods.* New York: Springer-Verlag, 2nd ed.

ROEDER, K. (1990). Density estimation with confidence sets exemplified by superclusters and voids in galaxies. *Journal of the American Statistical Association* **85**, 617–624.

ROEDER, K. & WASSERMAN, L. (1997). Practical Bayesian density estimation using mixtures of normals. *Journal of the American Statistical Association* **92**, 894–902.

ROUSSEAU, J. & MENGERSEN, K. (2011). Asymptotic behaviour of the posterior distribution in overfitted mixture models. *Journal of the Royal Statistical Society, Series B* **73**, 689–710.

SCHWARTZ, L. (1965). On Bayes procedures. *Zeitschrift für Wahrscheinlichkeitstheorie und verwandte Gebiete* **4**, 10–26.

SCHWARZ, G. (1978). Estimating the dimension of a model. *Annals of Statistics* **6**, 461–464.

SETHURAMAN, J. (1994). A constructive definition of Dirichlet priors. *Statistica Sinica* **4**, 639–650.

SKILLING, J. (2007). Nested sampling for Bayesian computations. *Bayesian Analysis* **1**, 833–859.

SPIEGELHALTER, D. J., BEST, N. G., CARLIN, B. P. & VAN DER LINDE, A. (2002). Bayesian measures of model complexity and fit. *Journal of the Royal Statistical Society, Series B* **64**, 583–639.

TESSIER, D., SCHOENAUER, M., BIERNACKI, C., CELEUX, G. & GOVAERT, G. (2006). Evolutionary latent class clustering of qualitative data. Tech. Rep. RR-6082, INRIA.

VAN HAVRE, Z., WHITE, N., ROUSSEAU, J. & MENGERSEN, K. (2015). Overfitting Bayesian mixture models with an unknown number of components. *PLoS ONE* **10**, e0131739.

WADE, S. & GHARHAMANI, Z. (2018). Bayesian cluster analysis: Point estimation and credible balls (with discussion). *Bayesian Analysis* **13**, 559–626.

WALLACE, C. & FREEMAN, P. (1987). Estimation and inference via compact coding. *Journal of the Royal Statistical Society, Series B* **49**, 241–252.

WATANABE, S. (2009). *Algebraic Geometry and Statistical Learning Theory*. Cambridge: Cambridge University Press.

WYSE, J., FRIEL, N. & LATOUCHE, P. (2017). Inferring structure in bipartite networks using the latent block model and exact ICL. *Network Science* **5**, 45–69.

YANG, Y. (2005). Can the strengths of AIC and BIC be shared? A conflict between model identification and regression estimation. *Biometrika* **92**, 937–950.

ZENG, H. & CHEUNG, Y.-M. (2014). Learning a mixture model for clustering with the completed likelihood minimum message length criterion. *Pattern Recognition* **47**, 2011–2030.

ZHANG, Z., CHAN, K. L., WU, Y. & CHEN, C. (2004). Learning a multivariate Gaussian mixture model with the reversible jump MCMC algorithm. *Statistics and Computing* **14**, 343–355.

# Part II

# Mixture Modelling and Extensions

# 8

---

# Model-Based Clustering

## Bettina Grün

*Johannes Kepler University Linz, Austria*

## CONTENTS

Mixture models extend the toolbox of clustering methods available to the data analyst. They allow for an explicit definition of the cluster shapes and structure within a probabilistic framework and exploit estimation and inference techniques available for statistical models in general. In this chapter an introduction to cluster analysis is provided, model-based clustering is related to standard heuristic clustering methods and an overview of different ways to specify the cluster model is given. Post-processing methods to determine a suitable clustering, infer cluster distribution characteristics and validate the cluster solution are discussed. The versatility of the model-based clustering approach is illustrated by giving an overview of different areas of applications.

## 8.1   Introduction

Cluster analysis – also known as unsupervised learning – is used in multivariate statistics to uncover latent groups suspected in the data or to discover groups of homogeneous observations. The aim is thus often defined as partitioning the data such that the groups are as dissimilar as possible and that the observations within the same group are as similar as possible. The groups forming the partition are also referred to as clusters.

Cluster analysis can be used for different purposes. It can be employed (1) as an exploratory tool to detect structure in multivariate data sets such that the results allow the data to be summarized and represented in a simplified and shortened form, (2) to perform vector quantization and compress the data using suitable prototypes and prototype assignments and (3) to reveal a latent group structure which corresponds to unobserved heterogeneity. A standard statistical textbook on cluster analysis is, for example, Everitt et al. (2011).

Clustering is often referred to as an ill-posed problem which aims to reveal *interesting* structures in the data or to derive a *useful* grouping of the observations. However, specifying what is interesting or useful in a formal way is challenging. This complicates the specification of suitable criteria for selecting a clustering method or a final clustering solution. Hennig (2015) also emphasizes this point. He argues that the definition of the *true* clusters depends on the context and on the aim of clustering. Thus there does not exist a unique clustering solution given the data, but different aims of clustering imply different solutions, and analysts should in general be aware of the ambiguity inherent in cluster analysis and thus be transparent about their clustering aims when presenting the solutions obtained.

At the core of cluster analysis is the definition of what a cluster is. This can be achieved by defining the characteristics of the clusters which should emerge as output from the analysis. Often these characteristics can only be informally defined and are not directly useful for selecting a suitable clustering method. In addition, some notion of the total number of clusters suspected or the expected size of clusters might be needed to characterize the cluster problem. Furthermore, domain knowledge is important for deciding on a suitable solution, in the sense that the derived partition consists of interpretable clusters that have practical relevance. However, domain experts are often only able to assess the suitability of a solution once they are confronted with a grouping but are unable to provide clear characteristics of the desired clustering beforehand.

Model-based clustering can help in the application of cluster analysis by requiring the analyst to formulate the probabilistic model which is used to fit the data and thus making the aims and the cluster shapes aimed for more explicit than is generally the case if heuristic clustering methods are used. The use of mixture models for clustering is also discussed in the general textbooks on mixture models by McLachlan & Peel (2000a) and Frühwirth-Schnatter (2006). In addition, several review articles on model-based clustering are available, including Stahl & Sallis (2012) and McNicholas (2016b) as well as the monograph on mixture model-based classification by McNicholas (2016a).

In the following, heuristic clustering methods are described and their relation to Gaussian mixture modelling is elaborated. After discussing the specification of the clustering problem, strategies for selecting a suitable mixture model together with the appropriate clustering base are given in Section 8.2. The post-processing methods given a fitted model are outlined in Section 8.3 where it is discussed how to obtain an identified model, derive a partition of the data, characterize the cluster distributions, gain insights through suitable visualizations and validate the clustering. Model-based clustering has been used in a range of different applications from which also methodological advances have emerged. Some important areas are discussed in detail in Section 8.4.

### 8.1.1 Heuristic clustering

Heuristic clustering methods are based on the definition of similarities or dissimilarities between observations and groups of observations. These methods are used as input to the different cluster algorithms proposed. In general, two main clustering strategies are distinguished: hierarchical clustering and partitioning clustering.

In hierarchical clustering a nested sequence of partitions is constructed. This is accomplished using either a bottom-up (*agglomerative*) or a top-down (*divisive*) approach. In bottom-up or agglomerative clustering, each observation starts in its own cluster and at each step two clusters are merged until all observations are contained in a single cluster. By contrast, top-down or divisive clustering starts with a single cluster and at each step one cluster is split into two. A greedy search strategy is employed where at each single step the optimal one-step-ahead decision is made. However, this does not imply that any of the intermediate cluster solutions obtained is optimal.

In order to obtain an optimal partition of $n$ observations $\mathbf{y} = \{y_1, \ldots, y_n\}$ for a given number of clusters $G$, a partitioning cluster algorithm needs to be used. The $k$-means algorithm aims to find the partition $\mathcal{C} = \{C_1, \ldots, C_G\}$ which minimizes the within-cluster sum of squares weighted by inverse cluster sizes,

$$\arg\min_{\mathcal{C}} \sum_{g=1}^{G} \frac{1}{n_g} \sum_{i \in C_g} \sum_{j \in C_g} \|y_i - y_j\|^2,$$

where $n_g$ is the number of observations in group $g$, $C_g$ is the set of observations assigned to group $g$ by the partition $\mathcal{C}$ and $\|\cdot\|$ denotes the Euclidean distance. The solution to this objective function can also be obtained by solving the optimization problem

$$\arg\min_{\mathcal{C}} \sum_{g=1}^{G} \sum_{i \in C_g} \|y_i - \bar{y}_g\|^2,$$

where the cluster centroid $\mu_g$ is equal to $\bar{y}_g$, which is given by

$$\bar{y}_g = \frac{1}{n_g} \sum_{i \in C_g} y_i.$$

Note that if the partition is restricted to contain only non-empty elements, then $G$ is necessarily finite for a finite sample of size $n$, but if the partition $\mathcal{C}$ may also contain empty groups, then $G = \infty$ is also possible. This observation similarly holds also for the mixture models subsequently considered. For mixture models finite or infinite values for $G$ might be used to represent the theoretical data-generating process. However, the induced partitions will consist of a finite number of groups for finite samples. Also the number of components with observations assigned to them will be finite for finite samples.

Extensions of the $k$-means algorithm to dissimilarity measures other than the squared Euclidean distance have been proposed, leading to $k$-centroid cluster analysis (Leisch, 2006). These variants can also be used for multivariate data where variables are collected on different scale levels and which require specific dissimilarity measures. For example, for asymmetric binary data the Jaccard distance or Jaccard coefficient (Jaccard, 1912) has been argued to be a suitable dissimilarity measure because it disregards joint zeros (see, for example, Kaufman & Rousseeuw, 1990, p. 26).

### 8.1.2    From $k$-means to Gaussian mixture modelling

If finite mixtures of multivariate Gaussian distributions are used for model-based clustering, a probabilistic distribution is specified which is used as a data-generating process for the observed data. In particular, it is assumed that the data in each group or cluster are generated from a multivariate Gaussian distribution and the combined data stem from a convex combination of multivariate Gaussian distributions. This distribution used in Gaussian mixture modelling is given by

$$\sum_{g=1}^{G} \eta_g \phi(y|\mu_g, \Sigma_g),$$

where $\phi(\cdot|\mu, \Sigma)$ is the probability density function (pdf) of the multivariate Gaussian distribution with mean $\mu$ and covariance matrix $\Sigma$, corresponding to the cluster distribution, and $\eta_g$ are the cluster sizes with $\eta_g \geq 0, \sum_{g=1}^{G} \eta_g = 1$. The parameters $\theta$ in this model are the cluster sizes $\eta_g$, and the cluster-specific parameters consisting of the cluster means $\mu_g$ and the cluster covariance matrices $\Sigma_g$ for $g = 1, \ldots, G$.

This model class can be seen as a generalisation of $k$-means clustering. Define $\mathbf{z} = (z_1, \ldots, z_n)$, where $z_i \in \{1, \ldots, G\}$ is the cluster membership of observation $i$. The $k$-means clustering solution can also be obtained by maximizing the classification likelihood $p(\mathbf{y}|\mathbf{z}, \theta)$ of a finite mixture of multivariate Gaussian distributions with identical isotropic covariance matrices $\Sigma_g = \sigma^2 I$, where $I$ is the identity matrix, and equal weights $\eta_g = 1/G$ with respect to the mixture parameters $\theta$ and the cluster memberships $\mathbf{z}$:

$$\sum_{i=1}^{n} \frac{1}{G} \phi(y_i|\mu_{z_i}, \sigma^2 I).$$

This implies that quite specific cluster shapes are implicitly imposed when using $k$-means clustering, namely that the clusters are spherical with equal variability in every dimension. In such a situation it is important to define a suitable scaling for each of the dimensions because the $k$-means solution treats the variability in each of the dimensions equally and the obtained solution therefore is not invariant with respect to the scaling of the variables. A data-driven approach often used to achieve this is to standardize the data. This, however, is problematic if (1) extreme, outlying observations are present in the data and (2) the cluster structure is not equally strong in all dimensions, leading to different between-cluster dissimilarities in the different dimensions. The dissimilarities in dimensions containing strong cluster structure are attenuated by the standardization, and thus standardization deteriorates the quality of the resulting cluster solution. In addition the application of $k$-means favours cluster solutions where the clusters have the same size and volume. Thus making this relationship between Gaussian mixture models and $k$-means clustering explicit allows insights into the implicit assumptions imposed when $k$-means clustering is used. The notion that heuristic clustering methods impose fewer assumptions than model-based clustering is thus clearly in error. It is rather the case that users are often less aware of the assumptions implicitly imposed when using heuristic methods.

Extending the $k$-means approach to allow for arbitrary covariance matrices in the clusters leads to model-based clustering using finite mixtures where the classification likelihood $p(\mathbf{y}|\mathbf{z}, \theta)$ rather than the observed likelihood $p(\mathbf{y}|\theta)$ is maximized (Scott & Symons, 1971; Symons, 1981; McLachlan, 1982; Celeux & Govaert, 1992). In this case the parameters of the mixture model as well as the cluster memberships are estimated, implying that the number of quantities to be estimated grows with the number of observations. Under these conditions maximum likelihood estimates have been shown to be asymptotically biased (Bryant & Williamson, 1978).

The classification maximum likelihood approach can be implemented in two different ways: either excluding the cluster sizes $\eta = (\eta_1, \ldots, \eta_G)$ from the likelihood or including $\eta$ in the likelihood. The first case corresponds to implicitly assuming that these sizes are all equal, while the second case can be seen as adding a penalty term to the classification likelihood. Maximizing the classification likelihood instead of the observed likelihood has the advantage that the derived solution is potentially better suited for the clustering context and that iterative methods for model estimation such as the expectation-maximization (EM) algorithm converge faster. Biernacki et al. (2003) thus consider initializing the ordinary EM algorithm by first using several runs of the so-called classification EM, which implements an EM algorithm for maximizing the classification likelihood, and selecting the best solution detected (see Chapter 2 for a detailed discussion of the EM algorithm).

Applications of Gaussian mixture modelling are widespread, and this model class is often used as the basic model for clustering metric multivariate data if $k$-means clustering is not flexible enough. Recent exemplary applications of Gaussian mixture modelling are, among many others, Kim et al. (2014) in hydrology who cluster a multivariate data set of hydrochemical measurements from groundwater samples to separate anthropogenic and natural groundwater groups. Perera & Mo (2016) used Gaussian mixture models in ocean engineering to understand marine engine operating regions as part of a ship energy efficiency management plan. In environmental science, Skakun et al. (2017) used Gaussian mixture models to determine a data-driven classification method to distinguish winter crops from spring and summer crops.

Compared to $k$-means clustering, Gaussian mixture modelling has several advantages. The model explicitly allows for clusters of different size and different volume. In addition, the clusters are independent of the scaling used for the variables (except for potential numerical issues). This flexibility comes with several drawbacks. The likelihood is unbounded for the general Gaussian mixture model and spurious solutions might emerge. Assuming a one-to-one mapping between clusters and components in the mixture model might lead to counter-intuitive clustering solutions, because (a) several components form a single mode and (b) observations are not assigned to the cluster for which the component mean is closest in Euclidean space due to different component-specific covariance matrices. Finally, the fitted model might correspond to a semi-parametric approximation of the data distribution, rather than represent a useful clustering model.

Thus specification of a suitable model (see Section 8.2) and the use of appropriate post-processing techniques (Section 8.3) are essential to reach a cluster solution that is useful and meets the stated clustering aims.

### 8.1.3 Specifying the clustering problem

Hennig & Liao (2013) claim that there "are no unique 'true' or 'best' clusters in a data set", rather the user who applies clustering methods needs to define what a cluster is. In general, this involves specifying the characteristics of a cluster with regard to size and shape and how clusters are assumed to differ. This constitutes essential information for assessing which observations form a cluster and which belong to different clusters. These decisions need to be made regardless of whether a model-based approach or a heuristic approach to clustering is employed. However, using a model-based approach makes these decisions generally more explicit. The specified model clearly indicates what cluster distributions are considered. Furthermore, in a model-based approach model selection and evaluation are based on statistical inference methods. This allows the problem of choosing a suitable number of clusters to be recast as a statistical model selection problem and adds the possibility of rigorously assessing uncertainty.

Different notions of what defines a cluster have been proposed and common examples of

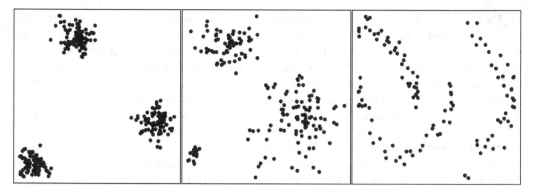

**FIGURE 8.1**
Illustration of clustering concepts: (left) compact clusters; (middle) density-based clusters; (right) connected clusters or clusters sharing a functional relationship.

such cluster characteristics are described in the following, consisting of compactness, density-based levels, connectedness and functional similarity. For illustration, two-dimensional scatter plots of data where a clear cluster structure regarding these notions is present are used.

*Compactness*

A cluster is characterized by points being close to one another. Separation between observations indicates that they stem from different clusters. In this case, the cluster centroid is a useful representative for all observations in the cluster. Often the notion of compactness is used to derive a cluster solution assuming that all clusters have similar levels of compactness. This implies that all clusters have a comparable volume and that the cluster centroids equally well represent observations in their clusters. The $k$-means algorithm minimizes the within-cluster sum of squares which can be seen as a measure of compactness and thus explicitly tries to address this clustering notion. In hierarchical clustering some linkage methods, that is, distance definitions between groups of observations, also lead to solutions with high compactness, such as complete linkage.

*Density-based levels*

Areas in the observation space where observations frequently occur are referred to as density clusters. This cluster concept is generally used for continuous data. In contrast to the notion of compactness, clusters might have different volumes and arbitrary shapes, that is, non-convex shapes are also possible. However, this also implies that the centroids – a concept which might even not be well defined for non-convex clusters – do not equally well represent their assigned observations. Furthermore, new observations might be hard to assign to a cluster if they occur in regions not identified as cluster regions, that is, assigning any (new) observations to one of the clusters might not be straightforward and unambiguous. Cluster methods which are based on an estimate of the data density and determine connected components in a density level set try to explicitly address this cluster notion (Azzalini & Menardi, 2014; Scrucca, 2016).

*Connectedness*

A cluster is defined by a friends-of-friends strategy. Observations are assigned to the same cluster if they are close to each other or if they are linked by other observations also assigned

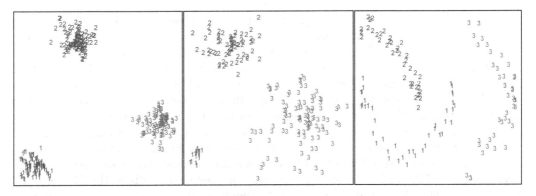

**FIGURE 8.2**
Illustration of clustering concepts including the true cluster membership using the same number for data points in the same cluster: (left) compact clusters; (middle) density-based clusters; (right) connected clusters or clusters sharing a functional relationship.

to the cluster. This implies that arbitrary cluster shapes are possible. However, no structure beyond that provided by the data themselves is imposed, indicating that solutions might be quite data-dependent and instable. Also characterizing a cluster through a centroid or a prototype might be impossible. Some linkage methods used in hierarchical clustering favour solutions of this kind, such as single-linkage where a *chaining phenomenon* occurs.

### Functional similarity

Observations are assigned to the same cluster if they share a common functional relationship between the variables in the different dimensions. For example, one variable might be seen as dependent and the others as independent, leading to a regression setting. Functional similarity then implies that the regression coefficients are similar within clusters, but differ across clusters. Network data represent another data setting where functional similarity might serve as notion for constructing clusters. In this case groups are formed, for example, by joining observations which have a similar linking behaviour to other observations.

### Illustration for artificial data sets

Figure 8.1 illustrates these different clustering concepts using three data sets containing two-dimensional metric data. The same data are also shown in Figure 8.2 where the information on the cluster membership of each observation is also included using the same number for data points in the same cluster. The scatter plot on the left visualizes a data set where the clusters are compact, that is, they have a similar level of compactness and shape and are very well separated. This is the easiest case for clustering where most of the clustering methods employed should be able to detect the correct cluster structure. The scatter plot in the middle shows a data set where clusters correspond to high density levels and can still be represented by a cluster centroid, but differ in their level of compactness. In contrast to $k$-means, model-based clustering allows these differences in compactness to be accounted for and the true cluster structure to be better extracted. The data set given in the scatter plot on the right illustrates the case where clusters can easily be identified using the concept of connectedness. In addition, these connected clusters might be identifiable by imposing cluster-specific functional relationships between the values on the $x$- and the $y$-axis and also by assuming that cluster sizes vary with the value of $x$.

If $k$-means is used to cluster the three data sets, good performance is expected for the first two, which are based on the compactness and density-based concepts for clusters. The $k$-means approach is expected to perform badly on the third data set and not to be able to detect the true cluster structure. The performance of $k$-means clustering on these data sets is investigated in Section 8.3.5. The quality of the true clustering solutions is visualized in Figure 8.3 using a silhouette plot based on the Euclidean distance. In addition, a silhouette-type plot is used to illustrate how model-based clustering methods perform on these three artificial data sets. The conditional probabilities of cluster memberships are determined based on suitable mixture models fitted using maximum likelihood estimation. They are then split by the true cluster memberships of the observations and visualized in Figure 8.5.

The application of standard model-based clustering methods generally does not ensure that compact, high-density or connected clusters or clusters with distinctly different functional relationships are obtained, and heuristic methods might be easier to tune to address any of these notions. However, this also implies that heuristic methods are more rigid in the solutions detected and might be more likely to come up with a specific clustering solution regardless of the inherent structure in the data. Thus solutions obtained using model-based clustering techniques which are agnostic of these cluster requirements can be assessed with regard to these characteristics to verify if such a structure might inherently be present in the data. This also emphasizes the need to specify a suitable mixture model as well as applying appropriate post-processing methods to ensure that the outcome of model-based clustering is a useful clustering solution.

## 8.2   Specifying the Model

For model-based clustering, finite as well as infinite mixture models have been used to fit a suitable distribution to the data and, in a subsequent step, to infer the cluster memberships from the fitted distributions and, potentially, also gain insights about typical characteristics of the cluster distributions. Fraley & Raftery (2002) point out that model-based clustering embeds the cluster problem in a probabilistic framework. The statistical framework allows statistical inference methods to be employed to obtain a suitable clustering of the data.

The starting point for model-based clustering is to define the distribution of a cluster and to decide how the cluster sizes are distributed. The definition of the cluster distribution can also be seen as the specification of the clustering kernel (see Frühwirth-Schnatter, 2011b). Depending on the available data and their characteristics, different kernels are suitable. In addition, the purpose of the cluster analysis and characteristics of the intended cluster solution also guide the choice of the clustering kernel. The requirement that a cluster distribution needs to be specified makes model-based clustering in general more transparent with respect to the targeted clustering solutions, while in applications of heuristic methods the characteristics of the targeted clustering solutions often remain implicit.

Heuristic clustering is based on notions of similarity and dissimilarity between observations and groups of observations. Model-based methods also use a notion of similarity between observations. Observations are "similar" if they are generated by the same cluster distribution. The dissimilarity between observations in different clusters is influenced by the differences allowed between the cluster distributions.

If the same parametric distribution is assumed for each of the clusters, differences between cluster distributions are determined by the specific parameter values only. In the case of mixtures of multivariate Gaussian distributions, for example, often a centroid-based approach is pursued. The cluster means are assumed to differ and characterize the cluster distributions, while the cluster covariance matrices represent nuisance parameters. The cluster covariance matrices might differ over clusters, but potentially could also be similar. If the covariance matrices are fixed to be identical over clusters, the dissimilarity between clusters is solely based on the cluster centroids. The conditional probabilities of cluster membership of an observation are then based on the Mahalanobis distance between the observation and the cluster centroid weighted by the cluster sizes.

## 8.2.1 Components corresponding to clusters

If mixture models are used for model-based clustering, the standard assumption is that each of the components in the mixture model corresponds to a cluster in the clustering solution. This implies that the component distribution specified in the mixture model is also the assumed cluster distribution.

For the specification of the mixture model it needs to be decided first if the clusters are all expected to follow a distribution from the same family of parametric distributions which differ only in the values of the parameter vectors or if different parametric distributions are assumed for the different clusters. The latter might only be possible in cases where a clear notion of the latent groups to be detected is available, for instance, if the presence of two groups corresponding to healthy and ill persons is assumed and some prior knowledge is available that the distributions of these two groups are structurally different. Otherwise – which is the usual case – the model specification is *a priori* the same for all of the components, that is, the same (parametric) model is imposed and components differ only in the specific values of the parameters.

The choice of the distribution for the components is governed by two aspects: (1) the data structure and (2) the suspected cluster distributions. In particular, the data structure can easily be verified and has led to a standard set of mixture models to be considered for certain kinds of data, for example, multivariate continuous data, multivariate categorical data, multivariate mixed data or longitudinal data (see also Chapter 11 for a discussion of mixture models of high-dimensional data).

*Multivariate continuous data*

For model-based clustering of multivariate continuous data of dimension $d$, the standard model is the finite mixture model of multivariate Gaussian distributions given by

$$y_i \sim \sum_{g=1}^{G} \eta_g \phi(y_i | \mu_g, \Sigma_g).$$

The advantages of Gaussian mixture modelling are that estimation methods are well established and that the component distributions are thoroughly understood, thus facilitating the interpretation of results. Drawbacks are that the likelihood is unbounded for arbitrary covariance matrices $\Sigma_g$ because of singular solutions leading to spurious results. Results are also strongly susceptible to extreme observations because of the light tails of the Gaussian distribution. When clustering data, it is often essential that the means of the cluster distributions are different and can be used as centroids to represent the clusters. However, this is not necessarily achieved when using this standard model class because there are no constraints – either implicitly or explicitly – imposed which would ensure that the fitted mixture model has these characteristics.

If a centroid-based approach to clustering is pursued, the covariance matrices of the components in the mixture model are seen as nuisance parameters which need to be suitably parameterized to allow for identification of the clusters. In particular, for high-dimensional data the specification of the covariance matrices is crucial because they add a substantial number of additional parameters to the model as the number of parameters of a single, unrestricted covariance matrix is $d(d+1)/2$ for data of dimension $d$.

To achieve a parsimonious parameterization, different restrictions on the covariance matrices have been proposed. Popular parameterizations, based on the spectral decomposition of the covariance matrix into shape, volume and orientation, have been proposed in Banfield & Raftery (1993) and are also discussed in Celeux & Govaert (1995) and Fraley & Raftery (2002). The spectral decomposition of the covariance matrix of the $g$th component is given by

$$\Sigma_g = \lambda_g D_g A_g D_g^\top,$$

where the positive scalar $\lambda_g$ corresponds to the volume, the $d \times d$ matrix $D_g$ to the orientation and the $d$-dimensional diagonal matrix $A_g$ to the shape. More parsimonious parameterizations can be achieved by restricting the components of the decomposition to be equal over components and by imposing certain values, for example, assuming $D_g$ to be the identity matrix (see Chapter 11, Section 11.4 for a detailed discussion).

An alternative approach for a parsimonious parameterization of the covariance matrices is to assume a latent structure leading, for example, to factor analysers as models for the covariance matrices (McLachlan & Peel, 2000b) or the unified latent Gaussian model approach which encompasses factor analysers and probabilistic principal component analysers as special cases (McNicholas & Murphy, 2008).

The use of mixtures of Gaussian distributions for clustering data imposes a prototypical shape on the clusters which implies symmetric and light-tailed distributions. However, in applications clusters are often assumed to have different shapes, for example, data clusters could exhibit skewed shapes and contain outlying observations. In order to account for skewness the extension to multivariate skew-normal distributions (Azzalini & Dalla Valle, 1996) has been considered, and for a more robust method the $t$ distribution is used for the components. For an application in the mixture context, see Frühwirth-Schnatter & Pyne (2010) and Lee & McLachlan (2013, 2014). Further approaches considered, for example, the use of shifted asymmetric Laplace distributions (Franczak et al., 2014) or multivariate normal inverse Gaussian distributions (O'Hagan et al., 2016). See also Chapter 10 for further examples.

*Multivariate categorical data*

In the context of multivariate binary data the latent class model has been developed as a useful tool for clustering (Goodman, 1974). This model can also be easily extended to the case of categorical data. Latent class models aim to explain the dependency structure present in multivariate data by introducing a discrete latent variable, (i.e. the cluster membership) such that, conditional on the latent variable, the observations in the different dimensions are independent. This assumption is also referred to as *conditional local independence*. The latent class model for multivariate binary data is given by

$$y_i \sim \sum_{g=1}^{G} \eta_g \left[ \prod_{j=1}^{d} \mathcal{B}er(y_{ij}|\pi_g^j) \right],$$

where $\mathcal{B}er(\cdot|\pi)$ is the density of the Bernoulli distribution with success parameter $\pi$ and $\pi_g^j = \mathrm{P}(y_{ij} = 1)$ is the success probability for the $j$th variable in the $g$th component (see also Chapter 11, Section 11.5).

*Multivariate mixed data*

The conditional local independence assumption can also be used for modelling mixed data. For groups of variables where a joint distribution allowing for dependencies might be hard to specify for the components, a product of the univariate distributions is used (Hunt & Jorgensen, 1999). Thus as long as a suitable parametric distribution is available in the univariate case, a mixture model can be specified and fitted.

Alternatively, an approach has been proposed which uses the following model for the components: the data in each cluster are assumed to be generated based on a latent variable which follows a multivariate Gaussian distribution where an arbitrary covariance matrix can be used. The latent variable is then mapped to the observed measurement scale, for example, a binary or ordinal variable (Cai et al., 2011; Browne & McNicholas, 2012; Gollini & Murphy, 2014); see also Chapter 11, Section 11.6.

*Multivariate data with special structures*

Special structures in multivariate data sets occur because of the variables having different roles or because of specific constraints restricting the values of the variables that can jointly be observed. If one variable represents a dependent variable and the others explanatory variables, this leads to a regression setting. If multivariate data occur only on a restricted support, rather than on the product of the support of each univariate variable, then this restriction has to be taken into account in model specification. Graph or network data also have a specific multivariate data structure.

In a regression setting one variable is identified as the dependent variable $y$ and all other variables act as potential independent variables $x$. The mixture model is then given by

$$y_i | x_i \sim \sum_{g=1}^{G} \eta_g f(y_i | x_i, \theta_g),$$

where $f(\cdot)$ represents a regression model and $\theta_g$ contains the regression parameters of component $g$. Clearly any regression model could be used for the components and identified with a cluster. The selection of the regression models for the components needs to ensure that any other inherent structure in the multivariate data set is captured. For example, repeated observations might be taken into account by including random effects.

Identifiability might be an issue in the context of mixtures of regressions (Follmann & Lambert, 1991; Hennig, 2000; Grün & Leisch, 2008); see also Chapter 12, Section 12.5. If the mixture model as specified is unidentifiable, this implies that neither the parameter estimates used to characterize the clusters nor the partitions derived can be uniquely determined, thus rendering the interpretation of results futile. If the non-identifiability leads to two or more isolated, clearly differing solutions which are observationally equivalent, the data do not allow us to distinguish between them. In this case domain knowledge might help in deciding which of these solutions represents a useful clustering and might allow the alternative solutions to be excluded.

Specific models might be useful if the multivariate data have a restricted support, for instance, when the observations or their squared values sum to one, in which case the multivariate data points have the unit simplex or the unit hypersphere as support. Clustering data on the unit hypersphere is sometimes also seen as clustering data only with respect to the directions implied, while neglecting the length information of the data points. Using a component distribution which has as support the unit hypersphere leads, for example, to mixtures of von Mises–Fisher distributions (Banerjee et al., 2005). This model has also been shown to represent the corresponding probabilistic model for spherical clustering where cosine similarity is used as similarity measure in $k$-centroid clustering, similar to how Gaussian mixtures are a probabilistic model for $k$-means clustering.

Analysing the topology of systems of interacting components is based on network data where the components correspond to nodes and their interactions to edges. Such graph or network data represent a specific data structure and are often stored using an adjacency matrix. If the data contain $n$ nodes, the adjacency matrix is of dimension $n \times n$ and the entries reflect whether there are edges between nodes together with the edge weights. In the simple case of a symmetric, unweighted graph, the adjacency matrix is a symmetric binary matrix. Clustering methods applied to network data aim to find community structures or to group nodes together which are similar in their interactions (Handcock et al., 2007; Newman & Leicht, 2007).

## 8.2.2 Combining components into clusters

While mixture models as considered in Section 8.2.1 are often able to capture the data distribution, the assumption that there is a one-to-one relationship between components and clusters might be violated. Some components might be too *similar* to form separate clusters and should rather be merged.

In this situation, a mixture of mixtures model can be useful where groups of components are combined to form a single cluster. The inference on clusters can be performed either simultaneously with model fitting or as a subsequent step, after having fitted a suitable model for approximating the data distribution. In the following, several strategies are discussed for multivariate continuous data, and some of these strategies might also be employed for other types of multivariate data.

Finite mixtures of Gaussian distributions are not only a suitable model for model-based clustering, but can also be applied for semi-parametric approximation of general distribution functions. A hierarchical mixture of mixtures of Gaussian distributions model can thus be used in a situation where a single Gaussian distribution is not flexible enough to capture the cluster distribution. In this hierarchical model the components on the upper level of the mixture correspond to clusters, while those on the lower level are only used for semi-parametric estimation of the cluster distribution.

In its hierarchical structure the model is given by

$$p(y_i|\theta) = \sum_{g=1}^{G} \eta_g f_g(y_i|\theta_g),$$

$$f_g(y_i|\theta_g) = \sum_{h=1}^{H_g} w_{gh}\phi(y_i|\mu_{gh}, \Sigma_{gh}).$$

While such a model is appealing as it involves only Gaussian distributions and thus is easy to understand and implement, identifiability of the model is an issue because the likelihood is invariant to assignment of components on the lower level to clusters on the upper level. In fact the density implied by this hierarchical representation is equivalent to

$$p(y_i|\theta) = \sum_{g=1}^{G}\sum_{h=1}^{H_g} \eta_g w_{gh}\phi(y_i|\mu_{gh}, \Sigma_{gh}). \tag{8.1}$$

That is to say that the hierarchical mixture of Gaussian mixtures model is equivalent to a mixture of Gaussian distributions with $\tilde{G} = \sum_{g=1}^{G} H_g$ components with weights $\eta_{gh} = \eta_g w_{gh}$ and component-specific parameter vectors $\mu_{gh}$ and $\Sigma_{gh}$. Clearly the components in equation (8.1) can be arbitrarily regrouped to form clusters without changing the overall density.

To achieve identifiability in a hierarchical mixture of mixtures model several approaches were proposed. They differ in that either identification is already aimed for during model fitting or that first a suitable semi-parametric approximation of the data distribution is obtained using mixtures of Gaussian distributions and then a second step is used for forming clusters by combining components.

*Direct inference*

In order to directly fit this kind of model in an identified way, strong identifiability constraints were imposed in a maximum likelihood framework. Bartolucci (2005) considers only the case of univariate data and specifies a mixture of Gaussian mixtures model where the mixtures on the upper level are restricted to being unimodal and the same for all clusters except for a mean shift. A similar restriction to having the same mixture distribution on the upper level except for a mean shift was considered in Di Zio et al. (2007) for multivariate data.

Within a Bayesian framework the identifiability issue present due to the invariance of the likelihood can be resolved by specifying informative priors which allow components from the same and different clusters to be automatically distinguished. Malsiner-Walli et al. (2017) consider this approach for finite mixtures, while Yerebakan et al. (2014) use an infinite mixture approach. Malsiner-Walli et al. (2017) proposed a prior structure which reflects the prior assumptions on the separateness of the clusters and the compactness of their shapes and which can be suitably adapted for different kinds of applications. Thus the clustering notions, that is, the cluster solutions aimed for, are explicitly included in the mixture model using informative priors in a Bayesian setting. Their prior also implies that the mixture on the lower level used for approximating the cluster distributions is allowed to contain components with different covariance matrices, but where the component means are pulled towards a common mean corresponding to the centre of the cluster. The prior structure employed by Yerebakan et al. (2014) is more rigid. In their approach, the covariance matrices on the lower level are assumed to be the same for the various Gaussian densities within a cluster and to be equal to a scaled version of a common covariance matrix $\Sigma$.

*Two-step procedures*

In the first step, mixtures of multivariate Gaussians are fitted as semi-parametric approximations of the data distribution. An issue with the use of multivariate Gaussian mixtures for the semi-parametric approximation is that the solutions might be ambiguous. Different Gaussian mixture models often allow one to similarly well approximate the data distribution, because the approximation might either use only a few components with complex component distributions or many components with simple component distributions. In case of Gaussian mixture models, the complexity of the component distribution is primarily governed by the structure of the covariance matrix.

In a second step, a merging approach is employed in order to form meaningful clusters from the Gaussian components of the fitted density. Different criteria were proposed for deciding on merging in a stepwise procedure such as closeness of the means (Li, 2005), the modality of the resulting clusters (Chan et al., 2008; Hennig, 2010), the entropy of the resulting partition (Baudry et al., 2010), the collocation of the observations (Molitor et al., 2010), the degree of overlap measured by misclassification probabilities (Melnykov, 2016) and the use of clustering cores (Scrucca, 2016).

This second step corresponds to another cluster analysis being performed where the estimated mixture components rather than the data points are the input. The mixture is assumed to be too fine-grained to represent a good cluster solution, and groups of components need to be formed to obtain the clusters. Note, in particular, that those approaches which only take the conditional probabilities of component memberships or collocations of observations into account can directly be used for any kind of mixture model to merge components into clusters regardless of the distributions used for the components of the mixture.

### 8.2.3   Selecting the clustering base

The variables included as clustering base are in general assumed to all equally contribute to the clustering solution. Each of the variables is assumed to be in line with and reflect the cluster structure. In general, the variables used for cluster analysis should be carefully selected, because choosing a different set of variables might change the meaning of the resulting cluster solution (Hennig, 2015). Even if efforts are made to select a suitable clustering base based on theoretical considerations driven by domain knowledge, the variables selected might not all be equally useful to cluster the data or contribute equally to a selected clustering solution. In fact, some of the variables included in the cluster base might turn out to either contain no cluster structure or be irrelevant for the clustering solution obtained. The inclusion of irrelevant variables has several drawbacks. In particular, their inclusion complicates model selection due to overfitting and makes the interpretation of the cluster solution a harder task than necessary. In the worst case, some variables might even perturb the cluster structure detected. Such variables are referred to as *masking* variables and, if included, deteriorate the cluster solution.

Alternatively, the aim could be to reduce the clustering base in order to eliminate redundant variables. This approach assumes that the clustering information contained in a subset of the variables is sufficient to characterise the cluster solution obtained using all variables. This task thus requires identification of the minimal set of variables necessary to identify the cluster solution. Using only the smaller set might then get rid of collinearity problems, facilitates the interpretation by providing a core set of essential variables, and reduces the number of variables that need to be collected for future analyses.

A range of variable selection methods for clustering have been proposed. They can be used in the heuristic and/or the model-based context and they can be applied before, during or after the cluster analysis to identify a suitable subset of variables (Gnanadesikan et al., 1995; Steinley & Brusco, 2008b); see also Chapter 11, Section 11.7.

*Prior filtering*

Prior filtering investigates the distribution of single variables and assesses how well suited they might be to reveal cluster structure in the data. Variables with a very homogeneous distribution clearly do not allow clusters to be extracted from the data. These variables can be excluded from the clustering base before performing the cluster analysis. Clearly for very high-dimensional data prior filtering is appealing because it might allow the dimensionality of the clustering problem to be substantially reduced.

In particular for continuous variables, indices to assess the *clusterability* of a variable have been developed. The aim of these clusterability indices is to determine whether a variable allows observations to be meaningfully clustered or if the observations exhibit a tendency to form clusters based on a specific variable. Steinley & Brusco (2008a), for instance, propose using a variance-to-range ratio for initial screening of each variable and excluding variables where these ratios are small.

*Variable selection*

Deciding on a suitable variable set during finite mixture model estimation is often complicated by the fact that the suitability of a variable set depends on the number of clusters and the specific component model used, for example, the specification of the covariance matrices in Gaussian mixture modelling. Thus the decision on the variable set needs to be made simultaneously while deciding on a suitable number of clusters and the component model.

Heuristic methods for exploring the model space with respect to different numbers of clusters and variable sets have been proposed in Raftery & Dean (2006), Maugis et al. (2009) and Dean & Raftery (2010) for Gaussian mixture models and for latent class analysis using maximum likelihood estimation and the Bayesian information criterion for model comparison.

Alternatively, within a Bayesian framework, Tadesse et al. (2005) propose using reversible jump Markov chain Monte Carlo (MCMC) methods in Gaussian mixture modelling to move between mixture models with different numbers of components while variable selection is accomplished by stochastic search through the model space. In the context of infinite mixtures, Kim et al. (2006) combine stochastic search for cluster-relevant variables with a Dirichlet process prior on the mixture weights to estimate the number of components. White et al. (2016) suggest using collapsed Gibbs sampling in the context of latent class analysis to perform inference on the number of clusters as well as the usefulness of the variables.

*Shrinkage methods*

A different approach to address the variable selection problem is to use shrinkage or penalization methods. In a frequentist setting these methods correspond to maximizing a penalized likelihood function, while in a Bayesian setting suitable priors are used to induce shrinkage. This approach aims to induce solutions which favour a homogeneous distribution over some variable or parameter in case the evidence for heterogeneity is insufficient. This prevents the overestimation of heterogeneity and provides insights into which variables or parameters are relevant for the cluster solution.

The shrinkage approaches pursued build on work in regression analysis for variable selection. In this context, for example, the lasso (least absolute shrinkage and selection operator; Tibshirani, 1996) was proposed to perform variable selection. For the lasso penalty the penalized likelihood estimate has exact zeros for some of the regression coefficients instead of small absolute values. Equivalent results are obtained for the maximum *a posteriori* (MAP) estimate if the lasso is used as prior for the regression coefficients. In the mixture context, the lasso penalty or prior is imposed on the parameter representing the difference between the cluster-specific value and the overall value of the parameter, thus inducing solutions where these differences are shrunk towards zero and the overall value of the parameter is the same over all clusters.

In Gaussian mixture models, shrinkage approaches have been proposed which only impose the penalty or shrinkage prior on the cluster means. This specification of the shrinkage reflects the assumption that the component means characterize the clusters and thus are different for variables relevant for clustering. This approach was employed in Pan & Shen (2007) using penalized maximum likelihood estimation. Resorting to a fully Bayesian approach, Malsiner-Walli et al. (2016) employed the normal-gamma prior for the differences in the component means of finite mixtures, and Yau & Holmes (2011) imposed a double exponential prior, the Bayesian lasso, on the differences in the component means while fitting infinite mixtures.

*Post-hoc selection*

This approach aims to arrive at a cluster solution for a subset of variables which is similar to the cluster solution obtained with all variables. This procedure is based on the assumption that the clustering base does not contain any masking variables, but the best cluster solution is obtained using all variables. However, some of the variables might be redundant because they either contain no or the same information on the cluster structure as other variables in the clustering base. In the post-hoc selection step, the aim is to identify these redundant variables and omit them. After omitting these variables, the cluster problem is recast as a lower-dimensional problem and results are easier to interpret. For example, Fraiman et al. (2008) propose two procedures to identify such variables, assuming that a "satisfactory" grouping is given.

### 8.2.4   Selecting the number of clusters

Selecting the number of clusters is quite a controversial topic, discussed in detail in Chapter 7 of this volume. For finite mixtures, a suitable number of components can be selected using different criteria. Information criteria such as the Akaike information criterion (AIC) or Bayesian information criterion (BIC) have been used in a model-based clustering context where it has been shown for the BIC that the number of components is consistently selected under certain conditions, in particular ensuring that the component densities remain bounded (Keribin, 2000).

If the model is fitted as part of the merging approach, there has been less controversy around methods to select a suitable model. In this case only a suitable semi-parametric approximation is required and the number of clusters is determined in the subsequent step.

In the case where components are assumed to correspond to clusters, the situation is more complicated. Information criteria developed for general model assessment usually aim to identify a solution which reflects the data-generating process well and thus select a model which represents a good semi-parametric approximation of the data distribution. The information criteria are ignorant of the final purpose of a mixture model in a clustering context. In cases where the component distribution does not exactly match the cluster distribution, the number of clusters tends to be overestimated using these criteria. This problem becomes more severe the larger the amount of data available. The larger the data set, the better the cluster distributions can be estimated and deviations from the imposed parametric distribution are more severely penalized.

In order to improve model selection performance when using mixture models of parametric distributions for model-based clustering, an alternative criterion has been proposed. This criterion measures not only the suitability of a model based on the goodness of fit for the data distribution, as indicated by the log likelihood, but also how well observations can be assigned to clusters. The latter is equivalent to aiming at well-separated cluster distributions where the conditional probabilities of component membership unambiguously allow the observations to be assigned to one of the components. Thus the entropy of the conditional probabilities of component membership is also taken into account, leading to the integrated completed-data likelihood information criterion (ICL; Biernacki et al., 2000),

$$\text{ICL}(G) = \sum_{i=1}^{n} \log f(y_i, \hat{z}_i | \hat{\theta}_G, G) - \frac{\upsilon_G}{2} \log n,$$

where $\hat{z}_i$ is the estimated cluster membership for observation $i = 1, \ldots, n$, $\hat{\theta}_G$ the estimated mixture parameters and $\upsilon_G$ the number of estimated parameters. In contrast to the AIC or BIC, the ICL aims to identify well separated clusters. This avoids overestimating the number of clusters by taking into account that the estimated model is used to assign observations to clusters and to obtain a suitable partition of the observations. We refer again to Chapter 7 for more details.

## 8.3 Post-processing the Fitted Model

### 8.3.1 Identifying the model

Mixture models suffer from generic non-identifiability issues due to label switching (Redner & Walker, 1984), that is, the mixture distribution is the same regardless of which label is assigned to which cluster. Only inference on label-invariant quantities can be easily accomplished while inference on quantities which depend on a unique labelling requires an identified mixture model (see also Chapter 4, Section 4.3 and Chapter 5, Section 5.2.2).

If only point estimates of mixture models are used, such as maximum likelihood estimates in a frequentist framework or MAP estimates in a Bayesian setting, a unique solution is easily obtained by imposing an ordering constraint. If the uncertainty of parameter estimation is to be quantified based on bootstrap techniques in frequentist estimation and MCMC samples in Bayesian estimation with symmetric priors, the situation is more complicated. Different methods for obtaining an identified model have been proposed, including (a) imposing an ordering constraint (Frühwirth-Schnatter, 2001), (b) applying label-invariant loss functions in cluster and relabelling algorithms (Stephens, 2000), (c) fixing the component membership of some observations (Chung et al., 2004), (d) relabelling with respect to the point estimate, for example, the MAP estimate (Marin et al., 2005), (e) clustering in the point process representation (Frühwirth-Schnatter, 2006, 2011a), and (f) probabilistic approaches taking the uncertainty of the relabelling into account (Sperrin et al., 2010). For an overview see also Jasra et al. (2005) and Papastamoulis (2016).

### 8.3.2 Determining a partition

Binder (1978) considered Bayesian cluster analysis as the task of determining a suitable partition, given a data set and assuming an underlying mixture model. Thus, in the Bayesian context, more focus has been given to develop methods for obtaining a suitable partition. In the frequentist setting a partition is in general determined using an identified model to classify observations separately and obtain a partition.

In the Bayesian context, Binder (1978) proposed obtaining the optimal partition by minimizing the expected loss given $p(\mathbf{z}|\mathbf{y})$ and suggested several different loss functions. This approach corresponds to minimizing

$$\mathrm{E}(R(\hat{\mathbf{z}}, \mathbf{z})|\mathbf{y})$$

with respect to $\hat{\mathbf{z}}$ given the loss $R(\cdot, \cdot)$ between two classifications of the data. One possible loss function is to use the 0–1 loss, where

$$R(\hat{\mathbf{z}}, \mathbf{z}) = \begin{cases} 0, & \text{if } \hat{\mathbf{z}} = \mathbf{z}, \\ 1, & \text{if } \hat{\mathbf{z}} \neq \mathbf{z}, \end{cases}$$

or a label-invariant version thereof. Using the 0–1 loss function results in the MAP estimate, hence to selecting the mode of $p(\mathbf{z}|\mathbf{y})$. The drawback of the 0–1 loss is that as long as two classifications are not the same they are assessed as equally different by assigning a loss of one. This drawback has led to a number of alternative loss functions being suggested. Depending on the loss function used, in general either an identified model or the posterior distribution of collocation is required to determine the loss minimizing classification. For more details, see also Frühwirth-Schnatter (2006, Section 7.1.7).

*Based on an identified model*

Given an identified model, the cluster memberships can be inferred from the latent allocation variables $z_i$ and conditional on the parameters $\theta$ of the mixture model for identically independently distributed observations. In this case the latent allocation variables are independent and group memberships can be inferred separately for each observation based on the conditional distributions of cluster memberships $\mathbf{z} = (z_1, \ldots, z_n)$ given the data $\mathbf{y} = (y_1, \ldots, y_n)$,

$$P(z_1 = g_1, \ldots, z_n = g_n | \mathbf{y}, \theta) \propto \prod_{i=1}^{n} \eta_{g_i} f_{g_i}(y_i | \theta_{g_i}),$$

where $g_i$ is the cluster to which observation $y_i$ is assigned. To determine a clustering solution $\hat{\mathbf{z}}$, different estimates can be used, based on the conditional probabilities of cluster membership. For instance, observations can be assigned to the cluster from which they most likely arise or they are assigned by drawing from the conditional probabilities of cluster membership.

Assigning the observations to the cluster from which they most likely arise results in the classification which is also obtained when minimizing the misclassification rate as loss function. This classification is obtained by setting

$$\hat{z}_i = \arg \max_g P(z_i = g | y_i, \theta)$$

for each observation $i$.

*Based on the collocation matrix*

Partitions have the advantage that they are label-invariant quantities. This implies that a final partition can be determined through procedures that do not require an identified model, but only rely on the estimated collocation of observations as input. This also implies that the only output required from model fitting is the information on the partitions, that is, collapsed sampling schemes can be used in an MCMC context.

The collocation matrix is a $n \times n$ matrix of values between 0 and 1, with 1s in the diagonal, where the $(i, j)$th entry represents the probability that observations $i$ and $j$ are assigned to the same component of the fitted mixture model, that is, $P(z_i = z_j | \mathbf{y})$. Several approaches for deriving a suitable partition from such a collocation matrix have been proposed in a Bayesian setting where draws from the posterior distribution of partitions are available. These suggestions include (a) minimizing a pairwise coincidence loss function (Binder, 1978; Lau & Green, 2007), (b) reformulating $P(z_i = z_j | \mathbf{y})$ as a dissimilarity matrix and using partitioning around medoids (PAM; Kaufman & Rousseeuw, 1990), a standard partitioning clustering technique (Molitor et al., 2010), (c) determining a partition which minimizes the squared distance as loss function (Fritsch & Ickstadt, 2009), and (d) minimizing the variation of information as loss function (Wade & Gharhamani, 2018). Of particular note is that the partition minimizing the posterior expected loss cannot in general be obtained directly, but iterative optimization techniques need to be employed. Due to the size of the partition space this is a computationally demanding task.

### 8.3.3 Characterizing clusters

The aim of cluster analysis is to determine groups in the data. The results of a cluster analysis, however, generally consist not only of a partition of the data, but also of a characterization of the groups. The clustering, together with the characterization of the clusters, allows one to summarize a multivariate data set and to give insights into its structure. The characterization of the clusters allows one to profile them and provides insights into the latent groups or types identified. This can be used to describe the groups or even assign them meaningful names.

In heuristic cluster analysis a prototype for each group or cluster is often determined. Comparing the differences between prototypes in heuristic clustering is not straightforward, because the clustering algorithm aims to maximize these differences.

In model-based clustering, the cluster distributions allow the clusters to be characterized. If parametric distributions are used for the clusters, the focus is typically on reporting and comparing the parameter estimates together with their associated uncertainty. Model-based methods thus allow differences between clusters to be assessed using a sound statistical framework. This facilitates identification of the variables which contribute most to the clustering, and assessment whether the resulting clustering is useful. The cluster distributions can be determined either based on an identified mixture model or conditional on the partition or clustering obtained for the data.

### 8.3.4 Validating clusters

General validation techniques have been developed for cluster analysis which can be applied regardless of the clustering method used. These validation methods can be divided into internal, external and stability-based methods. Furthermore, the assessment can be performed on the level of the global cluster solution or can be specific to a single cluster. A general overview of cluster validation methods is, for example, given in Halkidi et al. (2001), while Brock et al. (2008) provide an overview on internal and stability-based measures as well as biological ones in the context of bioinformatic applications.

*Internal measures*

The use of internal measures is appealing because they only require data which are already available. However, they rely on a notion of distance between observations. This notion is readily available when heuristic clustering methods are applied. For the application of model-based clustering methods no distance or dissimilarity measure between observations needs to be specified. This has to be done additionally in order to be able to calculate internal measures.

Most of the internal measures include information on within-cluster scattering (i.e. compactness) and between-cluster separation. Examples for these measures are silhouette width (Rousseeuw, 1987), the Dunn index (Dunn, 1974) and the Davies–Bouldin index (Davies & Bouldin, 1979). However, these measures are in general only useful for convex-shaped clusters and fail to provide suitable insights into the quality of a cluster solution in cases of arbitrarily shaped non-convex clusters and if noisy observations are present. It is also obvious that if the internal criterion coincides with the criterion minimized in the algorithm used for heuristic clustering, the solution obtained with this method should perform well. This thus seems to be an unfair comparison. Nevertheless, it might still be useful to compare different cluster results on this basis. This comparison provides insights into how much worse alternative solutions are which are derived using a different cluster criterion. In addition this investigation might also indicate what kind of cluster structure might naturally be contained in the data.

Also internal measures specifically developed for the mixture model context have been proposed. Celeux & Soromenho (1996) suggest assessing the suitability of a fitted mixture model to be used for clustering based on the entropy of the conditional probabilities of cluster memberships. Because the final partition of the data is derived from these conditional probabilities of cluster memberships, a mixture model is more suitable for clustering if observations can be unambiguously assigned to clusters. This measure also captures the loss of information incurred by using only the estimated partition as result and neglecting the uncertainty of cluster assignment.

### External measures

External measures relate the partition derived to some external structure (i.e. partition) imposed on the data. For instance, an additional categorical variable might be available which induces a partition of the data, but has not been used in the cluster analysis. Such a comparison assumes implicitly that the aim of the cluster analysis was to arrive at a partition of the data which is close to this partition. In general, using cluster analysis to extract a partition which corresponds to a partition induced by an observed categorical variable is questionable. If the target variable is observed, it would seem more natural to use a classification or supervised learning approach.

Standard methods for assessing classification accuracy (e.g. the misclassification rate) can be employed to compare the class labels to the cluster labels. However, this approach requires that a mapping from cluster labels to class labels is determined, which might eventually not be straightforward if the numbers of clusters and classes are different. One possible approach might be to choose the mapping which maximizes the classification accuracy criterion employed. As an alternative, label-invariant measures can be employed which determine the similarity between two partitions regardless of any labels assigned to each of the groups contained in the partitions. This is achieved by determining the numbers of pairs of observations which are in the same group for both partitions, in different groups for both partitions, and in different groups in one partition and in the same group for the other. Based on these numbers of pairs the Rand index (Rand, 1971) or adjusted Rand index (Hubert & Arabie, 1985), the Jaccard coefficient (Jaccard, 1912), the Fowlkes–Mallows index (Fowlkes & Mallows, 1983), among others, can be derived and used as validation measures. A further criterion used is *purity* (Zhong & Ghosh, 2003) which assesses the extent to which a cluster only contains observations from the same class, that is, this criterion does not penalize splitting classes into several clusters, but deteriorates if classes are merged into the same cluster.

### Stability measures

External measures compare two partitions. These measures thus can also be used to assess the stability of cluster solutions (see, for example, Hennig, 2007). The extent to which a cluster solution depends on a specific data set and how much it varies if a new data set is used can be assessed based on bootstrapping. Pairs of bootstrap samples are drawn from the data and clustered. These two cluster solutions each induce a partition of the original data set. The similarity between these partitions is determined using an external measure and can be used to assess stability. Alternatively, it is also of interest to assess stability of clustering solutions if different cluster algorithms are employed.

Dolnicar & Leisch (2010) point out that these stability assessments allow one to infer if cluster solutions can at least be constructed in a stable way, in cases where natural clusters (i.e. density clusters) are not contained in the data. They propose stability as a criterion to select a suitable clustering solution if no natural clusters are contained in the data set.

## 8.3.5   Visualizing cluster solutions

Suitable visualization methods allow one to assess the cluster quality, to illustrate the cluster shapes and to gain insight into the cluster distributions. Visualization is of particular importance in the clustering context as clustering is an exploratory data analysis tool. Assessing the quality of a cluster solution based on visualizations is also often necessary, because of the difficulty of formally defining the clustering problem in a way that ensures that the obtained solutions have the desired characteristics. Furthermore, input from domain experts to validate and optimize a cluster solution might be easier to obtain if they are able to assess a suggested solution using suitable visualizations.

*Assessing cluster quality*

Given different cluster validation indices, it might be easier to compare them and select a suitable solution dependent on their values using visualization methods. For instance, if a clear cluster structure is suspected in the data, an elbow or optimal value of the criterion might be visually discernible and used for model selection. In the context of merging components to clusters, Baudry et al. (2010) suggest plotting the entropy values against the number of clusters and selecting the solution where there is a break point in a piecewise linear fit.

An additional visualization technique based on an internal cluster validation index is the silhouette plot (Rousseeuw, 1987). The silhouette plot illustrates the quality of the cluster solution based on the silhouette values grouped by cluster. As an alternative, Leisch (2010) proposed the shadow plot which has the same structure but uses the shadow value instead of the silhouette value. The shadow value has the advantage that it is computationally less demanding to determine than the silhouette value. The drawback is that this index relies more on the cluster centroid being a good representative.

*Illustrating cluster shapes and separation*

Cluster shapes and separateness can be illustrated using scatter plots of the data points, at least for continuous data. However, in the case of high-dimensional data this might not be a feasible approach and lower-dimensional representations of the data might be more useful. The lower-dimensional representations could be based on either general techniques for dimension reduction such as principle component analysis or specific techniques for cluster analysis. In the context of Gaussian mixture modelling, Scrucca (2010) proposes determining the subspace which captures most of the cluster structure contained in the data. Alternatively, cluster-specific projections have been proposed which for a given cluster maximizes its distance to the other clusters (Hennig, 2004).

*Characterizing cluster prototypes*

Profile plots of the prototypes can help to quickly identify in which variables clusters differ and how they can be characterized (Dolnicar & Leisch, 2014). Profile plots use the information on the prototypes and visualize them. In the model-based context this information consists of characterizations of the cluster distributions, for example, the parameters in case of parametric distributions. Profile plots are based on conditional plots and separately visualize each of the cluster results, but allow for easy comparison. These plots can also be enhanced to include uncertainty information.

In the context of shrinkage priors imposed on the cluster means, Yau & Holmes (2011) and Malsiner-Walli et al. (2016) propose visualizing the posterior distribution of the shrinkage factors for each of the variables using boxplots. Such a visualization indicates the variables for which the clusters differ and thus can be used to characterize the clusters.

*Visualizing the example data sets*

The three data sets introduced in Section 8.1.3 and shown in Figures 8.1 and 8.2 are used to create silhouette plots and to visualize in a silhouette-type plot the cluster uncertainty inherent in the fitted mixture model.

The silhouette value for an observation $i$ is given by

$$s(i) = \frac{b(i) - a(i)}{\max\{a(i), b(i)\}},$$

where

$$a(i) = \frac{1}{n_{z_i}} \sum_{j \in C_{z_i}} \|y_i - y_j\|,$$

$$b(i) = \min_{g \in \{1, \ldots, G\} \setminus z_i} \frac{1}{n_g} \sum_{j \in C_g} \|y_i - y_j\|,$$

that is, $a(i)$ is the average Euclidean distance of observation $i$ to all observations $j$ assigned to the same cluster and $b(i)$ is the minimum average Euclidean distance of observation $i$ to observations $j$ which are assigned to a different cluster. This definition ensures that $s(i)$ takes values in $[-1, 1]$, where values close to 1 indicate "better" clustering solutions.

Rousseeuw (1987) suggests visualizing the silhouette values by grouping them by cluster and sorting them in decreasing order within cluster. The width of the values for each cluster indicates the size of the cluster and the distribution of silhouette values within a cluster reflects how well separated that cluster is from the other clusters in Euclidean space. The average silhouette value within a cluster indicated how compact and well separated this cluster is. The overall cluster solution can be assessed using the total average of silhouette values.

Figure 8.3 shows the silhouette values using Euclidean distance and the true clustering solution for the three artificial data sets. The top plot gives the silhouette plot in the case where compact clusters are present. The result indicates that these clusters are well separated in Euclidean space, and the plot also reflects the fact that the clusters are equally sized. For the case of density-based clusters the silhouette plot indicates that the clusters are of different size and the different levels of compactness of the clusters impact the silhouette values within clusters (middle plot). For the case of connected clusters which might be modelled using some functional relationship, the silhouette plot at the bottom indicates that centroid-based partitioning methods using the Euclidean distance might not be able to detect the true solution because most observations in the first cluster (i.e. the $U$-shaped cluster in Figures 8.1 and 8.2 on the right) are in Euclidean space closer on average to observations from a different cluster than their own.

If model-based clustering methods are used to fit the different data sets, one can use for the data set in Figure 8.3(a) a mixture of Gaussian distributions with identical spherical covariance matrices, for the data set in Figire 8.3(b) a mixture of Gaussian distributions with spherical covariance matrices differing in volume, and for the data set in Figure 8.3(c) a mixture of linear regression models with a horizontal line for the first cluster and polynomial regressions of degree 2 for the other two clusters in combination with a concomitant variable model (see also Section 8.4.2) based on the variable on the $x$-axis (i.e. the cluster sizes depend on the variable $x$ in the form of a multinomial logit model). The models obtained when fitted using the EM algorithm initialized in the true solution are shown in Figure 8.4. For the first two examples, the cluster means are indicated together with the 50% and 95% prediction ellipsoids (neglecting the uncertainty with which the parameters are estimated) for the fitted components given by circles using solid and dashed lines, respectively. For

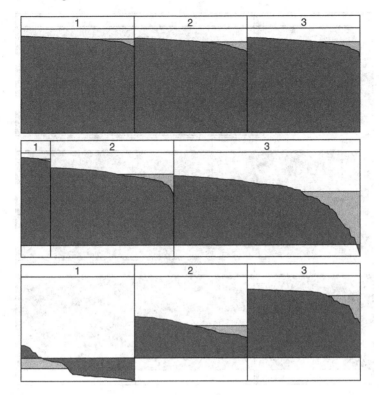

**FIGURE 8.3**
Silhouette plots based on the Euclidean distance and the true clustering for the three example data sets shown in Figure 8.1: (top) compact clusters; (middle) density-based clusters; (bottom) connected clusters or clusters sharing a functional relationship.

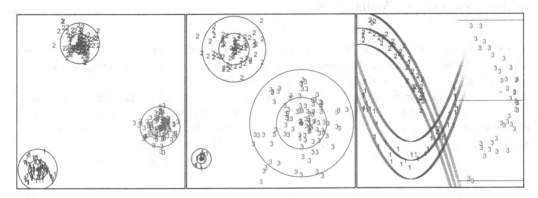

**FIGURE 8.4**
Plots of the example data sets together with the fitted mixture models: (left) a mixture of three Gaussian distributions with identical spherical covariance matrices; (middle) a mixture of three Gaussian distributions with different spherical covariance matrices; (right) a mixture of linear regression models with concomitant variables.

the third example, the fitted regression lines for each of the components are shown together with 95% pointwise prediction bands (neglecting the uncertainty with which the parameters are estimated) and using an alpha-shading (i.e. a transparency value) corresponding to the

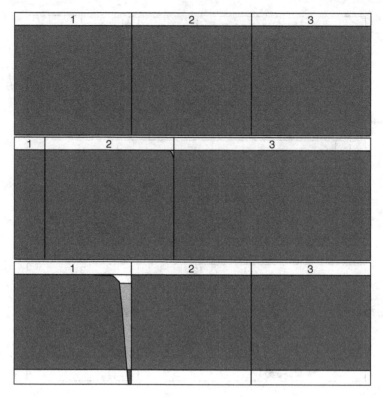

**FIGURE 8.5**
Plots of the conditional probabilities of cluster membership based on a finite mixture model and split by the true clustering for the three example data sets shown in Figure 8.1: (top) compact clusters; (middle) density-based clusters; (bottom) connected clusters or clusters sharing a functional relationship.

component size derived from the concomitant variable model. The points are numbered according to their component assignments based on the maximum conditional probabilities of cluster memberships obtained from the fitted mixture models.

The conditional probabilities of cluster memberships obtained for the three fitted mixture models split by the true cluster memberships are visualized in Figure 8.5. In this case clearly the fitted mixture models – even though not representing the true cluster generating mechanism – allow the cluster memberships to be identified very well.

## 8.4 Illustrative Applications

The areas of application are diverse, and specialized model-based clustering methods have been developed to meet the needs and challenges encountered in the different fields. Challenges are encountered due to specific data structures or specific desired cluster solution characteristics. Specific data structures are, for example, very high-dimensional data or the availability of time series panel data. Specific cluster solution characteristics are required if specific cluster shapes or the presence of very small clusters are suspected in the data.

### 8.4.1 Bioinformatics: Analysing gene expression data

In bioinformatics cluster methods have emerged as useful tools for analysing gene expression data, among other applications; for an introduction, see McLachlan et al. (2004). The aims of clustering in this area are (a) to reduce data dimensionality because of the large number of genes present in the data, (b) to verify if gene expression patterns differ between observed groups using an unsupervised learning approach (i.e. a clustering instead of a classification approach where the group labels are included in the analysis; Kebschull et al., 2014) to avoid overfitting, and (c) to detect latent groups potentially being present.

In contrast to other applications, analysing gene expression data poses specific problems where model-based clustering was adapted in suitable ways to provide better results than standard clustering methods. First of all, the data structure is different because in general the number of observations is small compared to the dimension of the data. Thus parsimonious Gaussian mixture models (McNicholas & Murphy, 2008) based on mixtures of factor analysers (McLachlan & Peel, 2000b; McLachlan et al., 2003; McNicholas & Murphy, 2010) have emerged as a useful model-based clustering method in this context. Furthermore, time-course data have led to the extension of mixtures of linear models to mixtures of linear mixed models using semi-parametric regression methods (Luan & Li, 2003, 2004; Celeux et al., 2005; Ng et al., 2006; Scharl et al., 2010; Grün et al., 2012).

In general, the distribution of the latent groups is suspected to be neither isotropic nor symmetric and to contain extreme observations, and thus a single Gaussian distribution is not able to capture the cluster distribution. In order to allow for more flexible shapes, mixtures of $t$ distributions have been considered to allow for heavier tails and skew distributions to account for asymmetry (Pyne et al., 2009; Frühwirth-Schnatter & Pyne, 2010; Franczak et al., 2014; Vrbik & McNicholas, 2014; Lee & McLachlan, 2014; O'Hagan et al., 2016); see also Chapter 10.

### 8.4.2 Marketing: Determining market segments

In market research, clustering methods are used for market segmentation; for an introduction see, for example, Wedel & Kamakura (2001) and Dolnicar et al. (2018). Market segmentation is a key instrument in strategic marketing. The aim of market segmentation is to divide the consumer or business market into subgroups. These subgroups can then be targeted separately, which provides competitive advantages. In order to be useful in practice, market segments need to fulfil certain criteria such as identifiability, substantiality, accessibility, stability, responsiveness, and actionability. Some of these criteria might even be seen as knock-out criteria such that clustering solutions that do not comply with them cannot be considered for implementing a successful market segmentation strategy.

As pointed out by Allenby & Rossi (1999), there is no clear consensus on how consumer heterogeneity is best modelled. While there exists agreement that consumers differ in their interests, preferences, etc., it is less clear if these heterogeneities are due to the presence of latent groups or because of continuous individual differences. While the presence of latent groups indicates the use of mixture models, continuous differences might be better captured by a random effects model. However, even if a random effects model is assumed to be better suited, it is still doubtful that consumer heterogeneity might be captured by a single Gaussian distribution. In general the random effects distribution is not known and the assumption of a Gaussian distribution (i.e. a symmetric unimodal distribution) will remain questionable. In this case, the random effects distribution could also be approximated by either a mixture distribution (see, for example, Aitkin, 1999) or the combination of a finite mixture with random effects within the components. The later model is also referred to as the heterogeneity model (Frühwirth-Schnatter et al., 2004a,b).

Dolnicar & Leisch (2010) also argue that density clusters rarely exist in consumer data and that the latent group model assuming homogeneity within the groups is hardly ever a good fit. They nevertheless defend the use of market segmentation and the extraction of subgroups. From a managerial point of view, grouping consumers into segments can still be beneficial and useful for targeting and positioning, even if these groups do not reflect natural clusters present in the data.

Specific extensions of mixture models developed for market segmentation are mixtures of regression models (Wedel & DeSarbo, 1995). These models have been useful for extracting market segments where the members have similar price sensitivity values or respond in a similar way to promotions. Previous to mixtures of regression models, a heuristic method referred to as clusterwise linear regression (Späth, 1979) was used to obtain clusters where members are similar in the regression coefficients. In clusterwise linear regression, an algorithmic approach to partitioning observations is pursued where linear regression models are fitted within each group to minimize a sum-of-squares criterion. Mixtures of regressions embed this approach in a probabilistic framework. Compared to clusterwise linear regression, mixtures of regressions have the advantage that they allow one to easily extend clustering of linear regression models to clustering of generalized linear models and even more general regression models.

Market segments are often defined using behavioural variables as a basis for segmentation, because the intention is to form groups of consumers who are similar in their behaviour. However, to ensure accessibility, these identified market segments also need to differ with respect to other variables (e.g. socio-demographic information). To improve the accessibility of the market segmentation solution obtained, concomitant variable models (Dayton & Macready, 1988) have been employed (Wedel & DeSarbo, 2002). While the cluster structure is still determined based on the behavioural variables only, the cluster memberships vary with the concomitant variables. This means that, for example, multinomial logit models are used to model the cluster memberships depending on the socio-demographic variables. This allows one to identify for which of the concomitant variables the cluster memberships are significantly different and to profile the segments (see Chapter 12 for a thorough review of such mixture of experts models).

### 8.4.3   Psychology and sociology: Revealing latent structures

Latent structure analysis has evolved in psychology to model dependency structures between observed variables; for a survey, see Andersen (1982). Latent class analysis assumes that the dependencies are caused through a latent variable which groups the observations; that is, conditional on the latent group, the variables are independent. For an introduction to latent class analysis, see Collins & Lanza (2010).

In the case of binary variables this implies that within each group the observations arise from a product of Bernoulli distributions and the differences in success probabilities between the groups lead to dependencies observed in the aggregate data (Goodman, 1974). In psychology, these latent groups are associated with different types of respondents, leading to different prototypical answers on dichotomous item batteries. Extension to ordinal variables also exist (Linzer & Lewis, 2011). In ordinal latent class analysis, the conditional independence assumption is retained, and for each dimension a different distribution for the ordinal variable is assumed for each group. An issue when applying this model class is identifiability. In general, these latent class models are not generically identifiable because only local identifiability can be ensured. Local identifiability implies that in a neighbourhood of a parameterization no other parameterization exists, implying the same mixture distribution. The lack of global identifiability in this case implies that there might exist a different parameterization of the same distribution somewhere else in the parameter space which is isolated from the other solution.

Areas of applications arise where behavioural variables, attitudes or values are collected on a dichotomous or categorical scale and the observed values are associated because of latent traits in the population which lead to jointly observing high or low values for a group of variables. Latent class analysis, for example, has been used for survey data on health and risk behaviour among youth to identify and characterize different risk groups in the population (Collins & Lanza, 2010); see Chapter 9 for further mixture models for binary, categorical and count data.

### 8.4.4 Economics and finance: Clustering time series

In economics, the model-based clustering approach has proven useful by accounting for unobserved heterogeneity in standard econometric models which, if neglected, might lead to biased results and thus to wrong conclusions being drawn from the data. For example, Alfó et al. (2008) investigate whether growth can be considered exogenous in the Solovian sense. They allow for heterogeneity between countries in order to obtain a grouping of countries based on their estimated model.

Specific applications of interest emerged for the analysis of time series data; for an overview see Frühwirth-Schnatter (2011b). For instance, mixtures of Markov switching models were developed to allow groups of time series to follow a different Markov switching model (Frühwirth-Schnatter & Kaufmann, 2008); see also Chapter 13. For an application to financial data consisting of time series on returns from 21 European stock markets see, for example, Dias et al. (2015) who identify three clusters in the data where the Markov switching differs. Mixtures of Markov chain models for categorical time series models are considered in Frühwirth-Schnatter et al. (2012, 2016). Frühwirth-Schnatter et al. (2012) analyse earnings development of young labour market entrants, and Frühwirth-Schnatter et al. (2016) investigate career paths of Austrian women after their first birth.

### 8.4.5 Medicine and biostatistics: Unobserved heterogeneity

Specific models developed and primarily used in medicine and biostatistics are those concerned with survival analysis. Model-based clustering has been employed in this area by extending the standard models for survival analysis to the mixture case such as mixtures of proportional hazard models (Rosen & Tanner, 1999).

Functional data are also often encountered in medical and biostatistical applications. Functional linear models relate a functional predictor with a scalar response by determining a smooth regression parameter function where the integral of the functional predictor times the parameter function over time equals the conditional mean of the scalar response. The generalization of this functional regression model to mixtures is investigated in Yao et al. (2011). Its application is illustrated on egg-laying data from a fertility study where the functional regression is used to relate fertility of the early life period to the total lifetime and to gain insights into two distinct mechanisms relating longevity and early fertility.

In the case of longitudinal data, growth mixture modelling has been considered (Muthén & Asparouhov, 2015; Muthén et al., 2002). Growth mixture modelling refers to the use of mixtures of generalized linear mixed models to longitudinal data which allow the intra-individual dependencies over time as well as differences between individuals to be captured.

A further model-based clustering application in health research uses Dirichlet process mixtures to assess the performance of health centres and classify them (Ohlssen et al., 2006; Zhao et al., 2015). The Dirichlet process clustering approach is used as a semi-parametric approach to approximate the latent heterogeneity distribution over health centres. The clustering structure obtained allows groups of particularly badly or well performing centres to be identified.

## 8.5   Concluding Remarks

Model-based clustering has emerged as a useful tool for performing cluster analysis. The flexibility of this approach stems from the fact that any statistical distribution or model can be used for the components. This allows one to come up with clustering techniques for any kind of data for which a statistical model is available. However, this flexibility also has drawbacks. The cluster structure aimed for might not necessarily be reflected in a model-based clustering solution. This is due to the difficulty of explicitly including the notions of the clustering aimed for in the specified model. While the specified model may capture the targeted cluster structure, nevertheless solutions are possible which are not useful in a clustering context. The selection of a suitable clustering method and deriving sensible clustering solutions thus remain ambiguous and might be perceived as lacking scientific rigour. Future developments in model-based clustering will thus hopefully address these issues and ameliorate these problems. Furthermore, new areas of application might arise and induce new methodological developments, pushing the boundaries of this modelling technique further.

# Bibliography

AITKIN, M. (1999). A general maximum likelihood analysis of variance components in generalized linear models. *Biometrics* **55**, 117–128.

ALFÓ, M., TROVATO, G. & WALDMANN, R. J. (2008). Testing for country heterogeneity in growth models using a finite mixture approach. *Journal of Applied Econometrics* **23**, 487–514.

ALLENBY, G. M. & ROSSI, P. E. (1999). Marketing models of consumer heterogeneity. *Journal of Econometrics* **89**, 57–78.

ANDERSEN, E. B. (1982). Latent structure analysis: A survey. *Scandinavian Journal of Statistics* **9**, 1–12.

AZZALINI, A. & DALLA VALLE, A. (1996). The multivariate skew-normal distribution. *Biometrika* **83**, 715–726.

AZZALINI, A. & MENARDI, G. (2014). Clustering via nonparametric density estimation: The R package pdfCluster. *Journal of Statistical Software* **57**, 1–26.

BANERJEE, A., DHILLON, I. S., GHOSH, J. & SRA, S. (2005). Clustering on the unit hypersphere using von Mises-Fisher distributions. *Journal of Machine Learning Research* **6**, 1345–1382.

BANFIELD, J. D. & RAFTERY, A. E. (1993). Model-based Gaussian and non-Gaussian clustering. *Biometrics* **49**, 803–821.

BARTOLUCCI, F. (2005). Clustering univariate observations via mixtures of unimodal normal mixtures. *Journal of Classification* **22**, 203–219.

BAUDRY, J.-P., RAFTERY, A., CELEUX, G., LO, K. & GOTTARDO, R. (2010). Combining mixture components for clustering. *Journal of Computational and Graphical Statistics* **2**, 332–353.

BIERNACKI, C., CELEUX, G. & GOVAERT, G. (2000). Assessing a mixture model for clustering with the integrated completed likelihood. *IEEE Transactions on Pattern Analysis and Machine Intelligence* **22**, 719–725.

BIERNACKI, C., CELEUX, G. & GOVAERT, G. (2003). Choosing starting values for the EM algorithm for getting the highest likelihood in multivariate Gaussian mixture models. *Computational Statistics & Data Analysis* **41**, 561–575.

BINDER, D. A. (1978). Bayesian cluster analysis. *Biometrika* **65**, 31–38.

BROCK, G., PIHUR, V., DATTA, S. & DATTA, S. (2008). clValid: An R package for cluster validation. *Journal of Statistical Software* **25**, 1–22.

BROWNE, R. P. & MCNICHOLAS, P. D. (2012). Model-based clustering, classification, and discriminant analysis of data with mixed type. *Journal of Statistical Planning and Inference* **142**, 2976–2984.

BRYANT, P. & WILLIAMSON, J. A. (1978). Asymptotic behaviour of classification maximum likelihood estimates. *Biometrika* **65**, 273–281.

CAI, J.-H., SONG, X.-Y., LAM, K.-H. & IP, E. H.-S. (2011). A mixture of generalized latent variable models for mixed mode and heterogeneous data. *Computational Statistics & Data Analysis* **55**, 2889–2907.

CELEUX, G. & GOVAERT, G. (1992). A classification EM algorithm for clustering and two stochastic versions. *Computational Statistics & Data Analysis* **14**, 315–332.

CELEUX, G. & GOVAERT, G. (1995). Gaussian parsimonious clustering models. *Pattern Recognition* **28**, 781–793.

CELEUX, G., MARTIN, O. & LAVERGNE, C. (2005). Mixture of linear mixed models for clustering gene expression profiles from repeated microarray experiments. *Statistical Modelling* **5**, 243–267.

CELEUX, G. & SOROMENHO, G. (1996). An entropy criterion for assessing the number of clusters in a mixture model. *Journal of Classification* **13**, 195–212.

CHAN, C., FENG, F., OTTINGER, J., FOSTER, D., WEST, M. & KEPLER, T. B. (2008). Statistical mixture modelling for cell subtype identification in flow cytometry. *Cytometry A* **73**, 693–701.

CHUNG, H., LOKEN, E. & SCHAFER, J. L. (2004). Difficulties in drawing inferences with finite-mixture models: A simple example with a simple solution. *American Statistician* **58**, 152–158.

COLLINS, L. M. & LANZA, S. T. (2010). *Latent Class and Latent Transition Analysis with Applications in the Social, Behavioral, and Health Sciences.* Hoboken, NJ: John Wiley.

DAVIES, D. L. & BOULDIN, D. W. (1979). A cluster separation measure. *IEEE Transactions on Pattern Analysis and Machine Intelligence* **1**, 224–227.

DAYTON, C. M. & MACREADY, G. B. (1988). Concomitant-variable latent-class models. *Journal of the American Statistical Association* **83**, 173–178.

DEAN, N. & RAFTERY, A. E. (2010). Latent class analysis variable selection. *Annals of the Institute of Statistical Mathematics* **62**, 11–35.

DI ZIO, M., GUARNERA, U. & ROCCI, R. (2007). A mixture of mixture models for a classification problem: The unity measure error. *Computational Statistics & Data Analysis* **51**, 2573–2585.

DIAS, J. G., VERMUNT, J. K. & RAMOS, S. (2015). Clustering financial time series: New insights from an extended hidden Markov model. *European Journal of Operational Research* **243**, 852–864.

DOLNICAR, S., GRÜN, B. & LEISCH, F. (2018). *Market Segmentation Analysis: Understanding It, Doing It, and Making It Useful.* Management for Professionals. Singapore: Springer.

DOLNICAR, S. & LEISCH, F. (2010). Evaluation of structure and reproducibility of cluster solutions using the bootstrap. *Marketing Letters* **21**, 83–101.

DOLNICAR, S. & LEISCH, F. (2014). Using graphical statistics to better understand market segmentation solutions. *International Journal of Market Research* **56**, 207–230.

DUNN, J. C. (1974). Well separated clusters and optimal fuzzy partitions. *Journal of Cybernetics* **4**, 95–104.

EVERITT, B. S., LANDAU, S., LEESE, M. & STAHL, D. (2011). *Cluster Analysis*. Wiley Series in Probability and Statistics. Chichester: John Wiley, 5th ed.

FOLLMANN, D. A. & LAMBERT, D. (1991). Identifiability of finite mixtures of logistic regression models. *Journal of Statistical Planning and Inference* **27**, 375–381.

FOWLKES, E. B. & MALLOWS, C. L. (1983). A method for comparing two hierarchical clusterings. *Journal of the American Statistical Association* **78**, 553–569.

FRAIMAN, R., JUSTEL, A. & SVARC, M. (2008). Selection of variables for cluster analysis and classification rules. *Journal of American Statistical Association* **103**, 1294–1303.

FRALEY, C. & RAFTERY, A. E. (2002). Model-based clustering, discriminant analysis and density estimation. *Journal of the American Statistical Association* **97**, 611–631.

FRANCZAK, B. C., BROWNE, R. P. & MCNICHOLAS, P. D. (2014). Mixtures of shifted asymmetric Laplace distributions. *IEEE Transactions in Pattern Analysis and Machine Intelligence* **36**, 1149–1157.

FRITSCH, A. & ICKSTADT, K. (2009). Improved criteria for clustering based on the posterior similarity matrix. *Bayesian Analysis* **4**, 367–391.

FRÜHWIRTH-SCHNATTER, S. (2001). Markov chain Monte Carlo estimation of classical and dynamic switching and mixture models. *Journal of the American Statistical Association* **96**, 194–209.

FRÜHWIRTH-SCHNATTER, S. (2006). *Finite Mixture and Markov Switching Models*. New York: Springer.

FRÜHWIRTH-SCHNATTER, S. (2011a). Dealing with label switching under model uncertainty. In *Mixtures: Estimation and Applications*, K. Mengersen, C. P. Robert & D. Titterington, eds., chap. 10. Chichester: John Wiley, pp. 193–218.

FRÜHWIRTH-SCHNATTER, S. (2011b). Panel data analysis: A survey on model-based clustering of time series. *Advances in Data Analysis and Classification* **5**, 251–280.

FRÜHWIRTH-SCHNATTER, S. & KAUFMANN, S. (2008). Model-based clustering of multiple time series. *Journal of Business & Economic Statistics* **26**, 78–89.

FRÜHWIRTH-SCHNATTER, S., PAMMINGER, C., WEBER, A. & WINTER-EBMER, R. (2012). Labor market entry and earnings dynamics: Bayesian inference using mixtures-of-experts Markov chain clustering. *Journal of Applied Econometrics* **27**, 1116–1137.

FRÜHWIRTH-SCHNATTER, S., PAMMINGER, C., WEBER, A. & WINTER-EBMER, R. (2016). Mothers' long-run career patterns after first birth. *Journal of the Royal Statistical Society Series A* **179**, 707–725.

FRÜHWIRTH-SCHNATTER, S. & PYNE, S. (2010). Bayesian inference for finite mixtures of univariate and multivariate skew-normal and skew-$t$ distributions. *Biostatistics* **11**, 317–336.

FRÜHWIRTH-SCHNATTER, S., TÜCHLER, R. & OTTER, T. (2004a). Bayesian analysis of the heterogeneity model. *Journal of Business & Economic Statistics* **22**, 2–15.

FRÜHWIRTH-SCHNATTER, S., TÜCHLER, R. & OTTER, T. (2004b). Capturing consumer heterogeneity in metric conjoint analysis using Bayesian mixture models. *International Journal of Research in Marketing* **21**, 285–297.

GNANADESIKAN, R., KETTENRING, J. R. & TSAO, S. L. (1995). Weighting and selection of variables for cluster analysis. *Journal of Classification* **12**, 113–136.

GOLLINI, I. & MURPHY, T. B. (2014). Mixture of latent trait analyzers for model-based clustering of categorical data. *Statistics and Computing* **24**, 569–588.

GOODMAN, L. A. (1974). Exploratory latent structure analysis using both identifiable and unidentifiable models. *Biometrika* **61**, 215–231.

GRÜN, B. & LEISCH, F. (2008). Identifiability of finite mixtures of multinomial logit models with varying and fixed effects. *Journal of Classification* **25**, 225–247.

GRÜN, B., SCHARL, T. & LEISCH, F. (2012). Modelling time course gene expression data with finite mixtures of linear additive models. *Bioinformatics* **28**, 222–228.

HALKIDI, M., BATISTAKIS, Y. & VAZIRGIANNIS, M. (2001). On clustering validation techniques. *Journal of Intelligent Information Systems* **17**, 107–145.

HANDCOCK, M. S., RAFTERY, A. E. & TANTRUM, J. M. (2007). Model-based clustering for social networks. *Journal of the Royal Statistical Society Series A* **170**, 301–322.

HENNIG, C. (2000). Identifiability of models for clusterwise linear regression. *Journal of Classification* **17**, 273–296.

HENNIG, C. (2004). Asymmetric linear dimension reduction for classification. *Journal of Computational and Graphical Statistics* **13**, 930–945.

HENNIG, C. (2007). Cluster-wise assessment of cluster stability. *Computational Statistics & Data Analysis* **52**, 258–271.

HENNIG, C. (2010). Methods for merging Gaussian mixture components. *Advances in Data Analysis and Classification* **4**, 3–34.

HENNIG, C. (2015). What are the true clusters? *Pattern Recognition Letters* **64**, 53–62.

HENNIG, C. & LIAO, T. F. (2013). How to find an appropriate clustering for mixed-type variables with application to socio-economic stratification. *Applied Statistics* **62**, 309–369.

HUBERT, L. & ARABIE, P. (1985). Comparing partitions. *Journal of Classification* **2**, 193–218.

HUNT, L. & JORGENSEN, M. (1999). Mixture model clustering using the MULTIMIX program. *Australian & New Zealand Journal of Statistics* **41**, 153–171.

JACCARD, P. (1912). The distribution of the flora in the alpine zone. *New Phytologist* **11**, 37–50.

JASRA, A., HOLMES, C. C. & STEPHENS, D. A. (2005). Markov chain Monte Carlo methods and the label switching problem in Bayesian mixture modelling. *Statistical Science* **20**, 50–67.

KAUFMAN, L. & ROUSSEEUW, P. J. (1990). *Finding Groups in Data.* New York: John Wiley.

KEBSCHULL, M., DEMMER, R. T., GRÜN, B., GUARNIERI, P., PAVLIDIS, P. & PAPAPANOU, P. N. (2014). Gingival tissue transcriptomes identify distinct periodontitis phenotypes. *Journal of Dental Research* **93**, 459–468.

KERIBIN, C. (2000). Consistent estimation of the order of mixture models. *Sankhyā: The Indian Journal of Statistics, Series A* **62**, 49–66.

KIM, K.-H., YUN, S.-T., PARK, S.-S., JOO, Y. & KIM, T.-S. (2014). Model-based clustering of hydrochemical data to demarcate natural versus human impacts on bedrock groundwater quality in rural areas, South Korea. *Journal of Hydrology* **519**, 626–636.

KIM, S., TADESSE, M. G. & VANNUCCI, M. (2006). Variable selection in clustering via Dirichlet process mixture models. *Biometrika* **93**, 877–893.

LAU, J. W. & GREEN, P. (2007). Bayesian model-based clustering procedures. *Journal of Computational and Graphical Statistics* **16**, 526–558.

LEE, S. & MCLACHLAN, G. J. (2013). Model-based clustering and classification with non-normal mixture distributions. *Statistical Methods and Applications* **22**, 427–454.

LEE, S. & MCLACHLAN, G. J. (2014). Finite mixtures of multivariate skew $t$-distributions: Some recent and new results. *Statistics and Computing* **24**, 181–202.

LEISCH, F. (2006). A toolbox for $k$-centroids cluster analysis. *Computational Statistics & Data Analysis* **51**, 526–544.

LEISCH, F. (2010). Neighborhood graphis, stripes, and shadow plots for cluster visualization. *Statistics and Computing* **20**, 457–469.

LI, J. (2005). Clustering based on a multilayer mixture model. *Journal of Computational and Graphical Statistics* **3**, 547–568.

LINZER, D. & LEWIS, J. (2011). poLCA: An R package for polytomous variable latent class analysis. *Journal of Statistical Software, Articles* **42**, 1–29.

LUAN, Y. & LI, H. (2003). Clustering of time-course gene expression data using a mixed-effects model with B-splines. *Bioinformatics* **19**, 474–482.

LUAN, Y. & LI, H. (2004). Model-based methods for identifying periodically expressed genes based on time course microarray gene expression data. *Bioinformatics* **20**, 332–339.

MALSINER-WALLI, G., FRÜHWIRTH-SCHNATTER, S. & GRÜN, B. (2016). Model-based clustering based on sparse finite Gaussian mixtures. *Statistics and Computing* **26**, 303–324.

MALSINER-WALLI, G., FRÜHWIRTH-SCHNATTER, S. & GRÜN, B. (2017). Identifying mixtures of mixtures using Bayesian estimation. *Journal of Computational and Graphical Statistics* **26**, 285–295.

MARIN, J.-M., MENGERSEN, K. & ROBERT, C. P. (2005). Bayesian modelling and inference on mixtures of distributions. In *Bayesian Thinking: Modeling and Computation*, D. Dey & C. Rao, eds., vol. 25 of *Handbook of Statistics*, chap. 16. Amsterdam: North-Holland, pp. 459–507.

MAUGIS, C., CELEUX, G. & MARTIN-MAGNIETTE, M.-L. (2009). Variable selection for clustering with Gaussian mixture models. *Biometrics* **65**, 701–709.

MCLACHLAN, G. J. (1982). The classification and mixture maximum likelihood approaches to cluster analysis. In *Handbook of Statistics: Classification Pattern Recognition and Reduction of Dimensionality*, P. R. Krishnaiah & L. N. Kanal, eds., vol. 2. Amsterdam: Elsevier, pp. 199–208.

MCLACHLAN, G. J., DO, K.-A. & AMBROISE, C. (2004). *Analyzing Microarray Gene Expression Data.* Wiley Series in Probability and Statistics. Chichester: John Wiley.

MCLACHLAN, G. J. & PEEL, D. (2000a). *Finite Mixture Models.* Chichester: John Wiley.

MCLACHLAN, G. J. & PEEL, D. (2000b). Mixtures of factor analyzers. In *Proceedings of the Seventeenth International Conference on Machine Learning.* San Francisco: Morgan Kaufmann.

MCLACHLAN, G. J., PEEL, D. & BEAN, R. W. (2003). Modelling high-dimensional data by mixtures of factor analyzers. *Computational Statistics & Data Analysis* **41**, 379–388.

MCNICHOLAS, P. D. (2016a). *Mixture Model-Based Classification.* Boca Raton, FL: CRC Press.

MCNICHOLAS, P. D. (2016b). Model-based clustering. *Journal of Classification* **33**, 331–373.

MCNICHOLAS, P. D. & MURPHY, T. B. (2008). Parsimonious Gaussian mixture models. *Statistics and Computing* **18**, 285–296.

MCNICHOLAS, P. D. & MURPHY, T. B. (2010). Model-based clustering of microarray expression data via latent Gaussian mixture models. *Bioinformatics* **26**, 2705–2712.

MELNYKOV, V. (2016). Merging mixture components for clustering through pairwise overlap. *Journal of Computational and Graphical Statistics* **25**, 66–90.

MOLITOR, J., PAPATHOMAS, M., JERRETT, M. & RICHARDSON, S. (2010). Bayesian profile regression with an application to the National Survey of Children's Health. *Biostatistics* **11**, 484–498.

MUTHÉN, B. & ASPAROUHOV, T. (2015). Growth mixture modeling with non-normal distributions. *Statistics in Medicine* **34**, 1041–1058.

MUTHÉN, B., BROWN, C. H., MASYN, K., JO, B., KHOO, S.-T. et al. (2002). General growth mixture modeling for randomized preventive interventions. *Biostatistics* **3**, 459–475.

NEWMAN, M. E. J. & LEICHT, E. A. (2007). Mixture models and exploratory analysis in networks. *Proceedings of the National Academy of Sciences of the USA* **104**, 9564–9569.

NG, S. K., MCLACHLAN, G. J., WANG, K., JONES, L. B.-T. & NG, S.-W. (2006). A mixture model with random-effects components for clustering correlated gene-expression profiles. *Bioinformatics* **22**, 1745–1752.

O'HAGAN, A., MURPHY, T. B., GORMLEY, I. C., MCNICHOLAS, P. D. & KARLIS, D. (2016). Clustering with the multivariate normal inverse Gaussian distribution. *Computational Statistics & Data Analysis* **93**, 18–30.

OHLSSEN, D. I., SHARPLES, L. D. & SPIEGELHALTER, D. J. (2006). Flexible random-effects models using Bayesian semi-parametric models: Applications to institutional comparisons. *Statistics in Medicine* **26**, 2088–2112.

PAN, W. & SHEN, X. (2007). Penalized model-based clustering with application to variable selection. *Journal of Machine Learning Research* **8**, 1145–1164.

PAPASTAMOULIS, P. (2016). label.switching: An R package for dealing with the label switching problem in MCMC outputs. *Journal of Statistical Software* **69**, 1–24.

PERERA, L. P. & MO, B. (2016). Data analysis on marine engine operating regions in relation to ship navigation. *Ocean Engineering* **128**, 163–172.

PYNE, S., HU, X., WANG, K., ROSSIN, E., LIN, T.-I. et al. (2009). Automated high-dimensional flow cytometric data analysis. *Proceedings of the National Academy of Sciences of the USA* **106**, 8519–8524.

RAFTERY, A. E. & DEAN, N. (2006). Variable selection for model-based clustering. *Journal of the American Statistical Association* **101**, 168–178.

RAND, W. M. (1971). Objective criteria for the evaluation of clustering methods. *Journal of the American Statistical Association* **66**, 846–850.

REDNER, R. A. & WALKER, H. F. (1984). Mixture densities, maximum likelihood and the EM algorithm. *SIAM Review* **26**, 195–239.

ROSEN, O. & TANNER, M. (1999). Mixtures of proportional hazards regression models. *Statistics in Medicine* **18**, 1119–1131.

ROUSSEEUW, P. J. (1987). Silhouettes: A graphical aid to the interpretation and validation of cluster analysis. *Computational and Applied Mathematics* **20**, 53–65.

SCHARL, T., GRÜN, B. & LEISCH, F. (2010). Mixtures of regression models for time-course gene expression data: Evaluation of initialization and random effects. *Bioinformatics* **26**, 370–377.

SCOTT, A. J. & SYMONS, M. J. (1971). Clustering methods based on likelihood ratio criteria. *Biometrics* **27**, 387–397.

SCRUCCA, L. (2010). Dimension reduction for model-based clustering. *Statistics and Computing* **20**, 471–484.

SCRUCCA, L. (2016). Identifying connected components in Gaussian finite mixture models for clustering. *Computational Statistics & Data Analysis* **93**, 5–17.

SKAKUN, S., FRANCH, B., VERMOTE, E., ROGER, J.-C., BECKER-RESHEF, I., JUSTICE, C. & KUSSUL, N. (2017). Early season large-area winter crop mapping using MODIS NDVI data, growing degree days information and a Gaussian mixture model. *Remote Sensing of Environment* **195**, 244–258.

SPÄTH, H. (1979). Algorithm 39 Clusterwise linear regression. *Computing* **22**, 367–373.

SPERRIN, M., JAKI, T. & WIT, E. (2010). Probabilistic relabelling strategies for the label switching problem in Bayesian mixture models. *Statistics and Computing* **20**, 357–366.

STAHL, D. & SALLIS, H. (2012). Model-based cluster analysis. *Wiley Interdisciplinary Reviews: Computational Statistics* **4**, 341–358.

STEINLEY, D. & BRUSCO, M. J. (2008a). A new variable weighting and selection procedure for *k*-means cluster analysis. *Psychometrika* **43**, 77–108.

STEINLEY, D. & BRUSCO, M. J. (2008b). Selection of variables in cluster analysis: An empirical comparison of eight procedures. *Psychometrika* **73**, 125–144.

STEPHENS, M. (2000). Dealing with label switching in mixture models. *Journal of the Royal Statistical Society, Series B* **62**, 795–809.

SYMONS, M. J. (1981). Clustering criteria and multivariate normal mixtures. *Biometrics* **37**, 35–43.

TADESSE, M. G., SHA, N. & VANUCCI, M. (2005). Bayesian variable selection in clustering high-dimensional data. *Journal of the American Statistical Association* **100**, 602–617.

TIBSHIRANI, R. (1996). Regression shrinkage and selection via the lasso. *Journal of the Royal Statistical Society, Series B* **58**, 267–288.

VRBIK, I. & MCNICHOLAS, P. D. (2014). Parsimonious skew mixture models for model-based clustering and classification. *Computational Statistics & Data Analysis* **71**, 196–210.

WADE, S. & GHARHAMANI, Z. (2018). Bayesian cluster analysis: Point estimation and credible balls (with discussion). *Bayesian Analysis* **13**, 559–626.

WEDEL, M. & DESARBO, W. S. (1995). A mixture likelihood approach for generalized linear models. *Journal of Classification* **12**, 21–55.

WEDEL, M. & DESARBO, W. S. (2002). Market segment derivation and profiling via a finite mixture model framework. *Marketing Letters* **13**, 17–25.

WEDEL, M. & KAMAKURA, W. A. (2001). *Market Segmentation – Conceptual and Methodological Foundations*. Boston: Kluwer Academic Publishers, 2nd ed.

WHITE, A., WYSE, J. & MURPHY, T. B. (2016). Bayesian variable selection for latent class analysis using a collapsed Gibbs sampler. *Statistics and Computing* **26**, 511–527.

YAO, F., FU, Y. & LEE, T. C. M. (2011). Functional mixture regression. *Biostatistics* **12**, 341–353.

YAU, C. & HOLMES, C. (2011). Hierarchical Bayesian nonparametric mixture models for clustering with variable relevance determination. *Bayesian Analysis* **6**, 329–352.

YEREBAKAN, H. Z., RAJWA, B. & DUNDAR, M. (2014). The infinite mixture of infinite Gaussian mixtures. In *Advances in Neural Information Processing Systems*, Z. Ghahramani, M. Welling, C. Cortes, N. D. Lawrence & K. Weinberger, eds., vol. 27 of *Proceedings from the Neural Information Processing Systems Conference*.

ZHAO, L., SHI, J., SHEARON, T. H. & LI, Y. (2015). A Dirichlet process mixture model for survival outcome data: Assessing nationwide kidney transplant centers. *Statistics in Medicine* **34**, 1404–1416.

ZHONG, S. & GHOSH, J. (2003). A unified framework for model-based clustering. *Journal of Machine Learning Research* **4**, 1001–1037.

# 9

# Mixture Modelling of Discrete Data

**Dimitris Karlis**

*Department of Statistics, Athens University of Economics and Business, Greece*

## CONTENTS

This chapter reviews the literature related to finite mixture models designed specifically for discrete data. By discrete data we mean a broad range of different data types, including counts, categorical data, rankings, and sequences. The chapter provides a broad overview over a wide range of models for these data types and discusses both univariate and multivariate models.

## 9.1   Introduction

The aim of the present chapter is to provide an overview of finite mixture models for discrete data. The central idea behind all models is to address explicitly the discreteness of the response variable under investigation. On the one hand, finite mixture models for discrete data have a lot in common with finite mixture models for continuous data, discussed in the other chapters, an example being the possibility of using EM type algorithms or Gibbs sampling for estimation. On the other hand, the very discreteness of the data creates specific difficulties and these difficulties become more prominent as the dimension of the data increases.

While the univariate Gaussian distribution is easily generalized to the multivariate Gaussian distribution, this is not the case even for the easiest of the discrete distributions, the Poisson distribution, as we will discuss later. This leads to a striking shortage of models for multivariate discrete data which can be employed as component-specific densities of a mixture model. This is perhaps the main reason why, compared to continuous data, not so many finite mixture models for multivariate discrete data have been suggested. The aim of this chapter is to present many of the existing models and to discuss the problems and challenges that arise in generalizing them

Under the general title "discrete data" we cover many distinct data types, namely (univariate) count data, categorical data, whether nominal or ordinal, multivariate count data, as well as rankings. The chapter is organized into several sections to provide a review of mixture models that have been suggested for each of these data types.

## 9.2   Mixtures of Univariate Count Data

### 9.2.1   Introduction

In this section we consider data sets that consist of counts, that is, they describe the number of occurrences of some event in some unit of time or space. Typical examples of such data are the number of accidents reported to an insurance company from each client, the number of earthquakes of a certain magnitude in a given area and time period, the number of purchases of some product from a supermarket, just to name a few. Such data occur in several disciplines, including epidemiology, marketing, geophysics and sport. Being counts, we can assume that these data sets take values in the non-negative integers, though extensions to relative integers (negative and positive) are straightforward.

The Poisson distribution is perhaps the simplest choice for modelling such data. However, since the mean of the Poisson distribution is equal to its variance, this choice may impose too restrictive an assumption in practice. A typical solution to deal with overdispersion (i.e. to allow the variance to be larger than the mean) is to consider mixtures of Poisson distributions; see Karlis & Xekalaki (2005) for a comprehensive review. In this section we consider finite mixtures of Poisson and related discrete distributions.

As in the other chapters in this book, the probability mass function (pmf) of a finite mixture model is defined as

$$p(y|\theta, \eta) = \sum_{g=1}^{G} \eta_g f_g(y|\theta_g), \quad y \in \mathbb{N}_0, \tag{9.1}$$

where $\mathbb{N}_0 = \{0, 1, \ldots, \}$ is the set of non-negative integers, $\theta = (\theta_1^\top, \ldots, \theta_G^\top)^\top \in (\Theta_1 \times \ldots \times \Theta_G)$ are the parameters related to the component distributions and $\eta = (\eta_1, \ldots, \eta_G)$, with $\eta_g \in (0, 1)$ and $\sum_{g=1}^G \eta_g = 1$, are the mixing proportions. Appropriate choices of $f_g(y|\theta_g)$ lead to various flexible models of moderate complexity.

## 9.2.2    Finite mixtures of Poisson and related distributions

Assuming that $f_g(\cdot|\theta)$ in (9.1) is a Poisson distribution gives rise to finite mixtures of Poisson distributions with pmf

$$p(y|\theta, \eta) = \sum_{g=1}^G \eta_g \frac{\exp(-\lambda_g)\lambda_g^y}{y!}, \quad y = 0, 1, \ldots, \tag{9.2}$$

where the $\lambda_g$ are the rates for each of the component-specific Poisson distributions $\mathcal{P}(\lambda_g)$. In order to make a finite mixture of Poisson distributions identifiable we need to impose some restriction on the ordering of the component-related parameters $\lambda_g$. Hence, we assume that $0 \leq \lambda_1 < \ldots < \lambda_G$; see Teicher (1963) for a proof of generic identifiability of Poisson mixtures.

As the Poisson distributions plays a similarly important role in discrete data modelling to that of the Gaussian distribution for continuous data, the model defined in (9.2) is perhaps the most commonly applied finite mixture model for count data. Compared to a single Poisson distribution, a Poisson mixture model offers several improvements. The most important one is their flexibility to model overdispersed data. Figure 9.1 depicts the pmf for several sets of parameters for $G = 2, 3, 5$, as described in the caption. An interesting feature is that in order to obtain, for example, a bimodal distribution we need to choose the $\lambda_g$ values far apart (see model (e)). Since, for each component, $\lambda_g$ corresponds both to the mean and the variance, increasing the mean also increases the variance and hence changes the shape of the mixture distribution.

Maximum likelihood estimation of a Poisson mixture model is straightforward using the EM algorithm, which involves only closed-form expressions. This makes the model easily accessible. A detailed treatment of this model in connection with choosing starting values for the EM algorithm can be found in Karlis & Xekalaki (2003). Bayesian estimation through Gibbs sampling is also straightforward (see, for example, Frühwirth-Schnatter, 2006, Section 9.2).

Interestingly, the sum of two (or more) random variables following (not necessarily identical) finite Poisson mixtures again follows a finite mixture of Poisson distributions, but with a larger number of components. The number of components depends on the number of distinct $\lambda_g$ values in the two mixtures. An obvious extension or modification of model (9.2) is to consider rates rather than counts.

Using a different pmf in (9.1), one can define finite mixtures of every other discrete distribution. An example is a finite mixture of binomial distributions defined as

$$p(y|\theta, \eta) = \sum_{g=1}^G \eta_g \binom{N}{y} \pi_g^y (1 - \pi_g)^{N-y}, \quad y = 0, \ldots, N. \tag{9.3}$$

Here, $\pi_g$ is the probability of success for the $g$th component-specific binomial distribution $\mathcal{B}(N, \pi_g)$, and the key idea is that the success probability varies among the components. Note that in this model $y$ can take only a finite number of distinct values, not infinitely many. This model was considered early on by Pearson (1915) who employed a mixture of two binomial distributions to model yeast cell count data. Discrete as well as continuous binomial

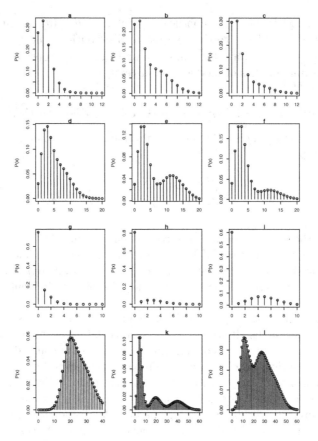

**FIGURE 9.1**
Finite mixtures of $G$ Poisson distributions with parameters $\lambda = (\lambda_1, \ldots, \lambda_G)$ and weights $\eta = (\eta_1, \ldots, \eta_G)$. The first two rows show mixtures with $G = 2$ and (a) $\lambda = (1, 2), \eta = (0.6, 0.4)$, (b) $\lambda = (1, 5), \eta = (0.6, 0.4)$, (c) $\lambda = (1, 5), \eta = (0.8, 0.2)$, (d) $\lambda = (3, 8), \eta = (0.6, 0.4)$, (e) $\lambda = (3, 12), \eta = (0.6, 0.4)$, and (f) $\lambda = (3, 12), \eta = (0.8, 0.2)$. The third row shows zero-inflated models with (g) $\lambda = (0, 1), \eta = (0.6, 0.4)$, (h) $\lambda = (0, 3), \eta = (0.8, 0.2)$, and (i) $\lambda = (0, 5), \eta = (0.6, 0.4)$. The final row shows mixtures with (j) $G = 2, \lambda = (20, 30), \eta = (0.6, 0.4)$, (k) $G = 3, \lambda = (5, 20, 40), \eta = (0.6, 0.2, 0.2)$, and (l) $G = 5, \lambda = (10, 15, 25, 30, 40), \eta = (0.2, 0.2, 0.2, 0.2, 0.2)$,

mixtures have been suggested as overdispersed alternatives to the binomial distribution. This binomial mixture model equally suffers from an issue of identifiability; see Teicher (1963) and Chapter 12, Section 12.5.1 for details. Bayesian inference for mixtures of binomial distributions is considered in Brooks (2001).

The model in (9.3) assumes that the probability parameter $\pi_g$ of the binomial distribution varies with $g$. If we further allow the size parameter $N$ to vary with $g$, implying a varying number of trials between the components, we end up with the so-called binomial $N$-mixture model which has found several applications and extensions in ecology and abundance models (see Royle, 2004).

### 9.2.3   Zero-inflated models

A very common extension of a finite mixture model for count data is the zero-inflated model. In fact, a zero-inflated model can be regarded as a finite mixture with one component degenerating at zero. The zero-inflated Poisson (ZIP) distribution, for instance, is defined as

$$p(y|\theta, \eta) = \begin{cases} \eta + (1 - \eta)\exp(-\lambda), & y = 0, \\ (1 - \eta)\dfrac{\exp(-\lambda)\lambda^y}{y!}, & y = 1, 2, \dots. \end{cases} \tag{9.4}$$

The model removes some probability mass, namely $\eta$, from the points $y = 1, 2, \dots$ and adds $\eta$ to the probability of observing a zero value. In this way, the model inflates the zero values by increasing their probability by $\eta$. Obviously, this is a two-component finite Poisson mixture with the mean of the first component being exactly zero, while the mean of the second components is equal to $\lambda$ (i.e. $\lambda_1 = 0$ and $\lambda_2 = \lambda$ in (9.2)), and all the probability mass in the first component is given to the zero point. By way of illustration, the third row of Figure 9.1 presents some zero-inflated Poisson models. It is apparent that the probability at zero is much higher since we have moved some probability to this point.

The ZIP model is relevant, for instance, in actuarial applications where the observed number of clients with no accidents is larger than would be predicted by a simple Poisson model. Zero inflation is caused by clients who did not have any accidents, for example, because they did not drive during the period of investigation or because they did not report any accidents that happened. In dental epidemiology statistics, ZIP distributions are used for modelling the number of decayed, missing and filled teeth (Böhning et al., 1999), known as the DMFT index.

It is obvious that one can consider zero inflation for any model, in particular a finite mixture model with $G$ components, by just allowing the first component to have zero values, only. Such a model can better model data sets with extra zeros, that is, data with more zeros than expected by a more simple model. Clearly, being a finite mixture, such a model can be used to model excess zeros and overdispersion at the same time.

Lambert (1992) introduced the ZIP regression model. Subsequently, a large number of applications of zero-inflated regression models based on several other distributions have been considered, including zero-inflated negative binomial regression models (Wang, 2003; Garay et al., 2011) and zero-inflated binomial regression models (Hall, 2000); see also Winkelmann (2008) for a review.

Models for inflation in the frequency of the ones (and not in the zeros) have been proposed in Godwin & Böhning (2017). In such models, we observe more ones than expected and we need to account for that. For example, in population size estimation, especially for capture–recapture type data, it is common that the frequency of one is higher than expected and we need to account for that inflation. The model is built in a similar mixture representation by migrating probability mass to one from the other values.

## 9.3   Extensions

### 9.3.1   Mixtures of time series count data

In many applications the observed data refer to counts observed over a time period. Examples of such data are the number of occurrences of a certain disease in an area over time, the number of daily purchases of some product and the number of earthquakes above a certain magnitude in a given area in a sequence of time points (say, months). Their common feature is that we observe time series where the variable of interest is a count. For such data, typical time series models for continuous data are not applicable since one must account for the discreteness of the data and the serial correlation at the same time. For a review of the literature on these models and methods for univariate count time series, see Fokianos (2012) and Weiss (2018).

Among the most common integer-valued time series models are the integer-valued autoregressive (INAR) models, introduced by McKenzie (1985). A sequence of random variables $\{X_t, t = 0, 1, \ldots\}$ is an INAR(1) process if it satisfies a difference equation of the form

$$X_t = \alpha \circ X_{t-1} + R_t, \quad t = 1, 2, \ldots,$$

where $\alpha \in [0, 1]$, $R_t$ is a sequence of uncorrelated non-negative integer-valued random variables, also called innovations, with mean $\mu$ and finite variance $\sigma^2$, and $X_0$ represents an initial value of the process. The operator "$\circ$" is defined by $\alpha \circ X = \sum_{i=1}^{X} Y_i$, where $Y_i \in \{0, 1\}$ are independently and identically distributed Bernoulli random variables with $P(Y_i = 1) = \alpha$. This operator, known as the binomial thinning operator, is due to Steutel & van Harn (1979) and mimics the scalar multiplication used for normal time series models while ensuring that only integer values will occur. A typical choice for the innovations $R_t$ is to assume a Poisson distribution, but the literature contains several other choices for $R_t$ to model specific characteristics, including different thinning operators and a variety of innovation distributions (see Weiss, 2018, Chapter 3).

Finite mixtures of such models have been considered in Böckenholt (1998), where mixing was used to create overdispersed INAR models. He proposed a model where in each subpopulation a separate INAR model with Poisson innovations was fitted, resulting in a model with a finite Poisson mixture as marginal distribution. In Pavlopoulos & Karlis (2008), the innovations are assumed to follow a finite mixture of Poisson distributions in order to create a model for overdispersed data. This model differs from the previous one in so far as mixing is introduced only in the innovations. A model with zero-inflated Poisson innovations was considered in Jazi et al. (2012).

Another model for count time series is the INGARCH model, which is a Poisson autoregressive model defined in Fokianos et al. (2009). The model has a feedback mechanism and is defined (in the simplest form of order 1) as

$$Y_t | \mathcal{F}_{t-1} \sim \mathcal{P}(\lambda_t), \quad \lambda_t = \delta + \alpha \lambda_{t-1} + \beta Y_{t-1},$$

for $t \geq 1$, where $(\delta, \alpha, \beta)$ are assumed to be positive model parameters and $Y_0$ and $\lambda_0$ are assume to be fixed. $\mathcal{F}_{t-1}$ is the information up to time $t-1$. Finite mixtures of such models were considered by Zhu et al. (2010) and Diop et al. (2016).

## 9.3.2 Hidden Markov models

Hidden Markov models (HMMs) are an important extension of finite mixture models to time series data and are discussed in detail in Chapter 13. HMMs are based on the assumption of existence of an unobservable variable that constitutes a Markov chain, and, depending on the state of this variable, we select the distribution attached to this state to produce the observable variables. In connection with mixture models, the unobservables are the components of the mixture. Selecting the state-specific distributions to be any discrete distribution, one can easily construct HMMs for discrete data. Leroux & Puterman (1992), for example, fitted an HMM based on the Poisson distribution to counts of epileptic seizures, while Spezia et al. (2014) modelled mussel counts in a river in Scotland using an HMM based on the negative binomial distribution.

In fact, an HMM with discrete distributions allows one to model time dependence, but also state dependence. For more details see the books by MacDonald & Zucchini (1997) and Frühwirth-Schnatter (2006, Chapter 10).

### 9.3.3 Mixture of regression models for discrete data

One possible extension of simple finite mixture models is to allow the component-specific parameters to be related to some covariates as in mixture of experts models. Typically, the effect of covariates is different for each component of the mixture. Detailed information for finite mixture regression models can be found in Chapter 12.

The case of finite mixtures of Poisson regressions is by far the best-known example of such a mixture and most broadly applied. It dates back to Wang et al. (1998) and assumes the following pmf for the $i$th individual:

$$p(y_i|\theta,\eta) = \sum_{g=1}^{G} \eta_g \frac{\exp(-\mu_{ig})\mu_{ig}^{y_i}}{y_i!}, \quad y_i = 0, 1, \ldots,$$

where $\theta = (\beta_1, \ldots, \beta_G)$ and $\mu_{ig} = \exp(x_i^\top \beta_g)$ is the mean of the $g$th component for individual $i$, given the covariate vector $x_i$. The model allows for overdispersion compared to the simple Poisson regression model; see Wang et al. (1998) and Grün & Leisch (2007) for more details. A further extension of this model is to assume covariates for the mixing proportions as well, by choosing, for example, a multinomial logistic regression for the mixing proportions; see also Chapter 12. However, the interpretation of such a model can be difficult especially if the same covariates are used for both parts, that is, both for the mixing proportion part and the components means.

A special case arises when the covariate effect is the same for each component. While we estimate a common set of regression coefficients, we allow the intercept and hence the mean for each component to be different. Such a model has a random effect interpretation: we assume a random effects regression model where the random effects have a finite mixture distribution; see, for example, Aitkin (1999).

Finite mixture regressions with other discrete distribution have been considered as well, in particular, finite mixtures of negative binomial regression models (Zou et al., 2013; Byung-Jung et al., 2014). Compared to finite mixtures of Poisson regressions, this model has an extra overdispersion parameter and hence allows for more flexible componentwise distributions. Finite mixtures of regression models have been widely used in several circumstances. A finite mixture of generalized Poisson regressions with two components has been described in Ma et al. (2008) for genetic data, while Sur et al. (2015) recently fitted a finite mixture of Conway Maxwell Poisson regression models to data describing the number of days spent in hospital for clients of an insurance company. Hennig (2000) discusses identifiability issues for finite mixture of regression models, which can be a problem especially when binary regressors are considered; see also Section 12.5.1 below.

As discussed already in Section 9.2.3, zero-inflated Poisson regression models can be regarded as a special case of what has been described above. Being zero, the mean of the inflated part need not to be estimated. The proportion of inflation can be related to some covariates. Dalrymple et al. (2003), for instance, use such a zero-inflated regression model for modelling the number of sudden infant death syndrome cases.

A mixture of Poisson regression models may be regarded as a mixture of generalized linear models (GLMs). Similarly, for binary or binomial data one may consider mixtures of logistic regression; see, for example, Follmann & Lambert (1989). However, such models need to be used with caution since identifiability problems may occur; see Follmann & Lambert (1991).

### 9.3.4   Other models

Another dependence structure may exhibit spatial dependence, where the data are dependent across a spatial domain. Examples include rates of a disease on a map, where the number of observed cases may be modelled using a Poisson distribution, the heterogeneity captured via a finite mixture, but the spatial relationship needs to be taken into account, as exemplified by the work of Forbes et al. (2013); see also Chapter 16 below. For example, risk mapping models focus on the estimated risk for each geographical unit. A risk classification, that is, a grouping of geographical units with similar risk, is then necessary to easily draw interpretable maps, with clear zones in which protection measures can be applied.

The underlying model is still a finite mixture of Poisson distributions. Denoting by $z_i$ the latent variable that indicates the component membership of the $i$th observation, the model further assumes a hidden Markov random field model such that the allocation $z_i$ of each observation relates to the allocations of its neighbouring points. The dependencies between neighbouring $z_i$ are then modelled by further assuming that the joint distribution of $\mathbf{z} = (z_1, \ldots, z_n)$ is a discrete Markov random field on the graph connecting contiguous locations (i.e. regions $i$ and $j$ are neighbours if they are spatially contiguous). This model implies spatial correlation between the observed counts. For a related model, see the work of Alfó et al. (2009).

A finite mixture of Poisson regressions for modelling censored data was described in Karlis et al. (2016). Censored discrete data may occur in questionnaires where answers of the type "10 or more" are allowed, circumstances when time is measured in discrete intervals, such as days in hospital, months of unemployment, etc.

Finally, in Bermúdez et al. (2017) finite mixtures of multiple distributions were introduced. The idea is to model excessive counts occurring at multiples of certain values (e.g. weeks or months). An example of this situation occurs with data on days of absence from work, where the duration of sick leave prescribed by physicians is often a multiple of weeks or months, inducing an increased frequency on certain numbers of days.

## 9.4   Mixtures of Multivariate Count Data

Multivariate count data appear in a wide range of fields where incidences of several related events are counted, as in epidemiology (e.g. different types of a disease), marketing (e.g. purchases of different products) and environmetrics (e.g. different kinds of plantation).

Moving from univariate to multivariate count data, we may consider bivariate (multivariate) extensions of univariate models and mixtures thereof. While in the continuous case the most commonly used Gaussian distribution easily generalizes to higher dimensions, this is unfortunately not the case for discrete data. For example, the Poisson distribution, the most common model for univariate count data, is not easily generalized to higher dimensions in a comfortable way. The aim of this section is to describe some models for multivariate count data, explain how finite mixtures can be considered and discuss the problems involved, which are the reason for their rather limited applicability so far.

### 9.4.1   Some models for multivariate counts

#### 9.4.1.1   Multivariate reduction approach

Consider, for example, the simplest extension of the Poisson distribution to the bivariate case. Following Kocherlakota & Kocherlakota (1992), the bivariate Poisson distribution has pmf given for $(y_1, y_2) \in \mathbb{N}_0 \times \mathbb{N}_0$ by

$$
\begin{aligned}
p(y_1, y_2|\lambda) &= \mathrm{P}(Y_1 = y_1, Y_2 = y_2|\lambda) \\
&= e^{-(\lambda_1+\lambda_2+\lambda_0)} \frac{\lambda_1^{y_1}}{y_1!} \frac{\lambda_2^{y_2}}{y_2!} \sum_{s=0}^{\min(y_1,y_2)} \binom{y_1}{s} \binom{y_2}{s} s! \left(\frac{\lambda_0}{\lambda_1\lambda_2}\right)^s, \quad (9.5)
\end{aligned}
$$

where $\lambda = (\lambda_0, \lambda_1, \lambda_2)$ are non-negative parameters. We will denote the pmf of this distribution by $\mathrm{BP}(y_1, y_2|\lambda)$. It is easy to verify that $\lambda_0$ is the covariance between $Y_1$ and $Y_2$, while the marginal means and variances are equal to $\lambda_1 + \lambda_0$ and $\lambda_2 + \lambda_0$, respectively. The marginal distributions are Poisson distributions. This bivariate Poisson distribution allows only positive correlation, and for $\lambda_0 = 0$ we obtain two independent Poisson distributions. One can easily recognize that this pmf involves a finite summation which can be computationally intensive for large counts. This model can be generalized to a certain extent by considering finite mixtures (Karlis & Meligkotsidou, 2007) or infinite mixtures (Chib & Winkelmann, 2001) of bivariate Poisson distributions.

Generalizations to higher dimensions become more and more demanding as the dimension increases. Let $Y = (Y_1, Y_2, \ldots, Y_d)^\top$ be a vector of $d$ discrete random variables. The definition of the multivariate Poisson distribution is based on the existence of a mapping $u : \mathbb{N}_0^q \to \mathbb{N}_0^d$, with $q \geq d$, such that $Y = u(S) = AS$, where $S = (S_1, \ldots, S_q)^\top$ with $S_r \sim \mathcal{P}(\lambda_r)$, $r = 1, \ldots, q$, being an independent sequence of univariate Poisson random variables with parameters $\lambda_r$, and $A$ is an $d \times q$ binary matrix with no duplicate columns. Then the vector $Y$ is said to follow a multivariate Poisson distribution with parameter $\lambda = (\lambda_1, \ldots, \lambda_q)^\top$. The mean and the variance–covariance matrix of $Y$ are given by

$$
\mathrm{E}(Y \mid \lambda) = A\lambda \quad \text{and} \quad \mathrm{Cov}(Y \mid \lambda) = A\Sigma A^\top,
$$

where $\Sigma = \mathrm{Diag}(\lambda_1, \lambda_2, \ldots, \lambda_q)$ is the covariance matrix of $S$, being diagonal because of the independence of the $S_r$. Each element $Y_i$ of $Y$ marginally follows a univariate Poisson distribution. For further details on this model, see Karlis & Meligkotsidou (2005).

Since the elements of $S$ follow independent univariate Poisson distributions, we obtain that the joint pmf is given by

$$
p(y \mid \lambda) = \sum_{\mathbf{s} \in u^{-1}(y)} \prod_{r=1}^{q} f_{\mathcal{P}}(s_r \mid \lambda_r), \quad (9.6)
$$

where $f_{\mathcal{P}}(s \mid \lambda)$ is the pmf of the $\mathcal{P}(\lambda)$ distribution, $\mathbf{s} = (s_1, \ldots, s_q)$, and $u^{-1}(y) \subseteq \mathbb{N}_0^q$ denotes the inverse image of $y$ under $u$.

The multivariate Poisson model for $d$ variables, derived by setting $A = [A_1 \mathbf{1}_d]$, where $A_1$ is the identity matrix of size $d \times d$ and $\mathbf{1}_d$ is the $d$-column vector of 1s, is frequently used. This model assumes that all the pairwise covariances are equal. However, this assumption is often not realistic.

A parsimonious model assumes that the matrix $A$ takes the form

$$
A = [A_1 \, A_2], \quad (9.7)
$$

where $A_1$ is the identity matrix of size $d \times d$ and $A_2$ is a $d \times \frac{d(d-1)}{2}$ binary matrix where each column of $A_2$ has exactly 2 ones and $d - 2$ zeros and no duplicate columns occur. The columns of $A_1$ and $A_2$ can be interpreted, respectively, as main effects and two-way covariance effects in an ANOVA type fashion. This model, which will be referred to as the multivariate Poisson model with two-way covariance structure, allows for different pairwise covariances. Therefore, it can be considered as a counterpart of the multivariate Gaussian distribution for multivariate count data. Henceforth, we assume that at least one of the elements of $\lambda$ is non-zero to avoid degenerate cases. In what follows, we will denote by $\mathrm{MP}_d(\lambda)$ and $\mathrm{MP}_d(\cdot \mid \lambda)$, respectively, the $d$-variate Poisson model with two-way covariance structure with parameter vector $\lambda$ and its joint probability function.

*Example*

Consider the case of the trivariate Poisson distribution. If the matrices $A_1$ and $A_2$ take the form

$$A_1 = \begin{bmatrix} 1 & 0 & 0 \\ 0 & 1 & 0 \\ 0 & 0 & 1 \end{bmatrix}, \quad A_2 = \begin{bmatrix} 1 & 1 & 0 \\ 1 & 0 & 1 \\ 0 & 1 & 1 \end{bmatrix},$$

then the model is

$$Y_1 = S_1 + S_4 + S_5,$$
$$Y_2 = S_2 + S_4 + S_6,$$
$$Y_3 = S_3 + S_5 + S_6,$$

where $S_i \sim \mathcal{P}(\lambda_i)$. One can easily verify that each $Y_j$, $j = 1, 2, 3$, follows a Poisson distribution and that the $Y_j$ are correlated due to the fact that they share some common $S$s. For example, since $S_4$ appears in the definition of both $Y_1$ and $Y_2$, it introduces correlation between $Y_1$ and $Y_2$, and the parameter related to this correlation is $\lambda_4$. Similar interpretations hold for $\lambda_5$ and $\lambda_6$. On the other hand, $S_i$, $i = 1, 2, 3$, appears only once for each $Y_i$, respectively, and hence the corresponding parameter $\lambda_i$ relates only to the mean of $Y_i$. This construction, known as multivariate reduction, leads to the following mean vector $\mu = \mathrm{E}(Y \mid \lambda)$ and closed-form matrix $\Sigma = \mathrm{Cov}(Y \mid \lambda)$ for $Y$:

$$\mu = \begin{pmatrix} \lambda_1 + \lambda_4 + \lambda_5 \\ \lambda_2 + \lambda_4 + \lambda_6 \\ \lambda_3 + \lambda_5 + \lambda_6 \end{pmatrix}, \quad \Sigma = \begin{pmatrix} \lambda_1 + \lambda_4 + \lambda_5 & \lambda_4 & \lambda_5 \\ \lambda_4 & \lambda_2 + \lambda_4 + \lambda_6 & \lambda_6 \\ \lambda_5 & \lambda_6 & \lambda_3 + \lambda_5 + \lambda_6 \end{pmatrix}.$$

### 9.4.1.2 Copulas approach

A different avenue for building multivariate models is to apply copulas. Copulas (see, for example, Nelsen, 2006) have found a remarkably large number of applications in several disciplines such as finance, hydrology and biostatistics, since they allow the derivation and application of flexible multivariate models with given marginal distributions. The key idea is that the marginal properties can be separated from the association properties, leading to a wealth of potential models. For the case of discrete data, copula-based modelling is less well developed; see Genest & Nešlehová (2007) for an excellent review. In recent years, copulas have been applied quite successfully to discrete data (Nikoloulopoulos & Karlis, 2009); see also the recent review by Nikoloulopoulos (2013).

It is important to keep in mind that some of the desirable properties of copulas are no longer valid when dealing with count data, an example being that the dependence properties depend on the marginal distributions. Also calculation of the pmf can be cumbersome in higher dimensions.

At their core, copulas are multivariate distributions with uniform marginals. Recall the inversion theorem, central to the simulation of standard distributions (see, for example, Robert & Casella, 2004, Chapter 2), where, starting from a uniform random variable and applying the inverse transform of a cumulative distribution function (cdf), we can generate whatever distribution we like. Copulas actually extend this idea in the sense that we start from two random variables that are correlated but marginally follow a uniform distribution. Transforming these two random variables by the inverse cdf produces correlated random variables with arbitrary marginals.

For instance, in the bivariate case, if $F_X(x)$ and $F_Y(y)$ are the cdfs of the univariate random variables $X$ and $Y$, then the copula transform $C(F_X(x), F_Y(y))$ is a bivariate distribution for $(X, Y)$ with marginal distributions $F_X$ and $F_Y$, respectively. Conversely, if $H$ is a bivariate cdf with univariate marginal cdfs $F_X, F_Y$, then, according to Sklar's (1959) theorem, there exists a bivariate copula $C$ such that, for all realizations of $(X, Y)$, $H(x, y) = C(F_X(x), F_Y(y))$. If $F_X, F_Y$ are continuous, then $C$ is unique, otherwise $C$ is uniquely determined on range$(F_X) \times$ range$(F_Y)$. This lack of uniqueness is not a problem in practical applications as it simply implies that there may exist two copulas with identical properties.

Actually, copulas provide a joint cumulative function. In order to derive the joint density (for continuous data) or the joint probability function (for discrete data), we need to take the derivatives or the finite differences of the copula, respectively. In the case of bivariate discrete data, the pmf is obtained by finite differences of the cdf through its copula representation (Genest & Nešlehová, 2007), namely

$$\begin{aligned} h(x, y; \alpha_1, \alpha_2, \delta) = &\, C(F_X(x; \alpha_1), F_Y(y; \alpha_2); \delta) - C(F_X(x-1; \alpha_1), F_Y(y; \alpha_2); \delta) \\ &- C(F_X(x; \alpha_1), F_Y(y-1; \alpha_2); \delta) + C(F_X(x-1; \alpha_1), F_Y(y-1; \alpha_2); \delta), \end{aligned}$$

where $F_X(\cdot)$ and $F_Y(\cdot)$ are the marginal cumulative functions, $\alpha_1$ and $\alpha_2$ the parameters associated with the marginal distributions and $\delta$ the parameter(s) of the copula.

If the selected copula is not available in closed form, further problems occur as the dimension increases. Recall once more that a copula is a cdf, and even for some well-known and often used copulas, such as the Gaussian copula, they require the repeated evaluation of multivariate integrals. In addition, adding and subtracting many small probabilities may lead to truncation errors.

For higher dimensions, this approach become much more demanding. When resorting to finite differences, one pdf evaluation requires us to assess the copula at eight locations in the trivariate case. For $d$ dimensions, one similarly needs to evaluate it $2^d$ times.

The pmf for dimension $d$ is obtained (see, for example, Panagiotelis et al., 2012) by applying finite differences of the distribution function as

$$p(y) = \sum_{\boldsymbol{v}} \operatorname{sgn}(\boldsymbol{v}) C(F_1(v_1), \ldots, F_d(v_d)), \qquad (9.8)$$

with $\boldsymbol{v} = (v_1, \ldots, v_d)$ vertices, where each $v_t$ is equal to either $y_t$ or $y_t - 1$ ($t = 1, \ldots, d$), and

$$\operatorname{sgn}(\boldsymbol{v}) = \begin{cases} 1, & \text{if } v_t = y_t - 1 \text{ for an even number of } ts, \\ -1, & \text{if } v_t = y_t - 1 \text{ for an odd number of } ts. \end{cases}$$

A particular issue relates to the failure of many copulas to easily allow for manageable correlation structures. For example, several popular multivariate copulas assume the same correlation for all pairs of variables, which is too restrictive in practice. Furthermore, if one needs to specify both positive and negative correlations, more restrictions apply. However, there are constructions of copulas with a relatively rich dependence structure and requiring moderate computing effort as discussed, for example, in Panagiotelis et al. (2012). Nonetheless, they all require numerical resolutions to achieve predetermined correlations.

### 9.4.1.3 Other approaches

In this section we have described some general directions for extending models to higher dimensions. While we focused on the Poisson distribution, the methods can be extended to any other discrete distribution. For example, the multivariate reduction approach can be used for other distributions but may lead to different marginal distributions. On the other hand, copulas can be used to derive multivariate discrete distributions with any collection of marginals.

There exist many more strategies for building flexible models for multivariate counts, such as models based on conditional distributions; see Berkhout & Plug (2004). Most of these approaches are far from simple and represent an increase in complexity when compared with continuous models, where the multivariate Gaussian distribution is a cornerstone allowing for great flexibility and feasible calculations. For a general approach to constructing bivariate discrete models, see Lai (2006).

### 9.4.2   Finite mixture for multivariate counts

#### 9.4.2.1   Conditional independence

As a starting point, let us consider a very simple model. As mentioned above, the main difficulty in constructing finite mixtures of multivariate distributions is the definition of appropriate multivariate distributions as component densities. A sensible and simplifying starting point is to assume that the variables are independent given component membership and hence within each component one can consider the product of simple distributions.

For example, let $y_i = (y_{i1}, \ldots, y_{id})^\top$ be a multivariate discrete observation for $d$ variables. We may assume that, conditional on the component indicator $z_i$, the $j$th variable follows a Poisson distribution independently of the other variables, that is, $Y_{ij}|z_i = g \sim \mathcal{P}(\lambda_{jg})$. Note that if we allow for covariates, then we need to add one more subscript to the $\lambda$s, since for each individual we obtain a different mean for each component and variable. We avoid this in order to keep the notation as simple as possible.

For the $g$th component and for each observation $y_i$ we obtain that

$$f_g(y_i) = \prod_{j=1}^{d} \frac{\exp(-\lambda_{jg})\lambda_{jg}^{y_{ij}}}{y_{ij}!},$$

and hence the mixture can be described as

$$p(y_i) = \sum_{g=1}^{G} \eta_g \prod_{j=1}^{d} \frac{\exp(-\lambda_{jg})\lambda_{jg}^{y_{ij}}}{y_{ij}!}.$$

The EM algorithm in this case is relatively straightforward, because the componentwise distributions are simple and, in particular for the Poisson case, we obtain closed-form expressions. As mentioned above, we can further use covariates for the component-specific means of the different variables to improve the structure. Models based on conditionally independent Poisson distributions have been described, for example, in Alfó et al. (2011). Rau et al. (2015) proposed alternative parameterizations to improve the applicability of such models in problems in genetics.

The main criticism of such models is that the assumption of conditional independence is not easily checked or dismissed for real data and it is rarely a reasonable assumption. Note, however, that, despite the simplicity of this model, the unconditional mixture distribution $p(y_i)$ produces correlation between the variables $(y_{i1}, \ldots, y_{id})$ and hence these mixtures can be used for modelling correlated data.

Hoben et al. (1993) consider a mixture of bivariate binomial distributions based on the conditional independence assumption for modelling the success of students in two separate tests. Mixture models with latent structures have been considered in Browne & McNicholas

(2012), but these models have limitations because of assumptions such as conditional independence.

### 9.4.2.2  Conditional dependence

One way to avoid defining high-dimensional structures is to assume that conditional on the component indicator only some of the variables are correlated. This is a compromise between considering the full structure and assuming conditional independence. This implies that there exist blocks of variables that, conditional on the component indicator, are dependent, while variables between blocks are independent. For instance, in Brijs et al. (2004) the number of purchases of some products from a supermarket were considered. Based on the nature of the product, it was assumed that within a cluster only some of the pairs were dependent, simplifying the underlying structure of the model considerably.

As a toy example, consider an application with four variables $(Y_1, Y_2, Y_3, Y_4)$. The model assumes that for each component $Y_1$ and $Y_2$ are dependent, following jointly a bivariate Poisson distribution, and similarly $Y_3$ and $Y_4$ are dependent. Therefore the conditional joint distribution of $(Y_1, Y_2, Y_3, Y_4)$ is the product of two bivariate Poisson distributions and hence has a relatively simple structure. A similar model was considered for a marketing application in Dippold & Hruschka (2013). In a recent paper, Marbac et al. (2015) considered blocks of conditionally dependent binary variables.

Such an approach can massively decrease the computational effort by reducing the number of parameters to be estimated. However, a challenging issue is to identify which of the correlations are needed and how the blocks of variables are constructed. In general, it is far from easy to check such assumptions, especially since it is based on a compromise with a view to avoiding complicated models.

### 9.4.2.3  Finite mixtures of multivariate Poisson distributions

It is possible to consider a model that assumes full structure conditionally on the component indicator; see Karlis & Meligkotsidou (2007). The pmf of a finite mixture of $G$ multivariate Poisson distributions is given by

$$p(y \mid \lambda, \eta) = \sum_{g=1}^{G} \eta_g \mathrm{MP}_d(y \mid \lambda_g), \quad y \in \mathbb{N}_0^d,$$

where $\mathrm{MP}_d(y \mid \lambda_g)$ is the pmf defined in (9.6), with $\lambda_g = (\lambda_{1g}, \ldots, \lambda_{qg})^\top$ being the $g$th component-specific parameter vector. The unknown parameters are $\lambda = (\lambda_1, \ldots, \lambda_G)$ and the mixing proportions $\eta = (\eta_1, \ldots, \eta_G)$. Without loss of generality, we assume that all of the mixture components are defined through the same matrix $A$ of the form (9.7). Under this mixture model, the unconditional expectation of $Y$ is given by

$$\mathrm{E}(Y \mid \lambda, \eta) = A \left( \sum_{g=1}^{G} \eta_g \lambda_g \right),$$

while its covariance matrix is given by

$$\mathrm{Cov}(Y \mid \lambda, \eta) = A \left[ \sum_{g=1}^{G} \eta_g \left( \Sigma_g + \lambda_g \lambda_g^\top \right) - \left( \sum_{g=1}^{G} \eta_g \lambda_g \right) \left( \sum_{g=1}^{G} \eta_g \lambda_g \right)^\top \right] A^\top,$$

where $\Sigma_g = \mathrm{Diag}(\lambda_{1g}, \ldots, \lambda_{qg})$. This model has interesting properties. One can verify that the finite mixture of multivariate Poisson distributions leads to overdispersion in all marginals. It can also be seen that such a model is overdispersed when compared to a simple multivariate Poisson model, in the sense that the generalized variance is larger than that expected from a single multivariate Poisson model. An interesting feature is that the marginal distributions are themselves finite mixture of Poisson distributions. This property can be used as an *a posteriori* model checking device, that is, for selecting the number of components in the mixture by considering which model gives a structure closest to the observed covariance.

As mentioned above, an EM type algorithm is available in closed form. However, for large dimensions a multivariate Poisson model with full two-way covariance structure involves many parameters and time-consuming summations in the joint pmf. An application concerning different types of crimes is given in Karlis & Meligkotsidou (2007). A Bayesian approach including selecting the number of components with a transdimensional MCMC method is given in Meligkotsidou (2007).

### 9.4.3    Zero-inflated multivariate models

For the univariate case, we described zero-inflated models in detail in Section 9.2.3. The idea can be generalized in several directions for the multivariate case. Figure 9.2 depicts some inflated situations for the bivariate case. As a start, one may inflate the $(0,0)$ cell only (case (a)); see, for example, Gurmu & Elder (2008). Alternatively, one may observe more zeros than expected for one of the variables and hence inflate not only the $(0,0)$ cell but also all the cells with zero for one variable (cases (b) and (c)) or for both (case (d)); see, for example, Arab et al. (2012). Other models assume diagonal inflation (case (e)), that is, observe larger frequencies on the diagonal. Examples occur when modelling sports outcomes, such as football scores; see Karlis & Ntzoufras (2003). Finally, modelling inflation in certain areas as in case (f) proved to be useful; see Dixon & Coles (1997). Clearly, in all cases, the marginal distributions differ.

Multivariate zero inflation for the multivariate case is analysed in Li et al. (1999). Recently, in the actuarial literature and for ratemaking purposes, Bermúdez (2009) and Bermúdez & Karlis (2011) dealt with bivariate and multivariate versions of the zero-inflated Poisson regression models, respectively. As for other distributions, Yau et al. (2003) consider, for example, zero-inflated bivariate negative binomial distributions. Extensions to any other bivariate or multivariate discrete distribution are obvious.

### 9.4.4    Copula-based models

Finite mixture models using copulas have been exploited in a recent paper by Kosmidis & Karlis (2016). The copula-based mixture model is defined as in (9.1) but the parameter $\theta_g$ is partitioned as $\theta_g = (\gamma_g^\top, \psi_g^\top)^\top$, where $\gamma_g$ contains the parameter vectors for all marginals and $\psi_g$ contains the parameter of the copula used for the $g$th component. As before, $f_g(y|\theta_g)$ is the pmf of the $g$th component and is given as in (9.8) in the case of copulas.

The key idea is that the multivariate distributions in each of the components are defined through copulas, which provides a series of advantages such as flexibility in the choice of the marginal distributions and a choice of a wide range of potential dependence structures by choosing an appropriate copula.

The underlying idea is that one can construct a multivariate distribution with, say, Poisson marginals and consider this as the distribution for each component. An appropriate choice of copulas will lead to a dependence structure that describes the data sufficiently well. For example, some copulas such as the Gaussian allow for a full structure, with a different dependence parameter for each pair for variables, while other copulas such as the

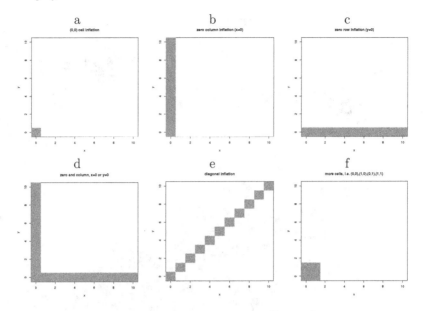

**FIGURE 9.2**
Zero-inflated and related models for the bivariate case.

Archimedean allow only for one parameter for all pairs. There are also copulas between the two cases.

*Example*

Suppose we have only two mixture components ($G = 2$). For each component, we assume a bivariate distribution created by considering Poisson marginals and a Frank copula. Each component actually has different marginal properties (different marginal Poisson distributions) and different dependence structures. Figure 9.3 presents such an example. The two component distributions can be seen at the top: the marginals are Poisson with means equal to 2 for both marginals for the first component and equal to 6 for the second component. The Frank copula parameter equals 3 and −3 for the two components, implying relatively large positive and negative correlation, respectively. One can verify from the image plots at the top what kind of correlation pattern is implied. The darker the colour in the image plot, the larger the probability. The example is interesting, as it demonstrates that we can get negative correlation for some components, which is not the case for finite mixtures of multivariate Poisson distributions.

Such a mixture can be fitted using an EM algorithm (Kosmidis & Karlis, 2016). The M step is not given in closed form and one needs to solve instead a maximization problem numerically at each iteration, which is equivalent to fitting a weighted likelihood approach to the data for each component. As implied by (9.8), one drawback of the model might be the complicated form of the joint probability function, which for large dimensions may imply the calculation either of a multivariate integral or an extensive summation.

In Kosmidis & Karlis (2016), trivariate data were considered related to cognitive diagnosis modelling. The data set referred to the responses of 536 middle school students on 20 items of a fraction subtraction test. Each item can belong to more than one attribute that one wants to measure. Hence, attribute scores for each student can be obtained by counting

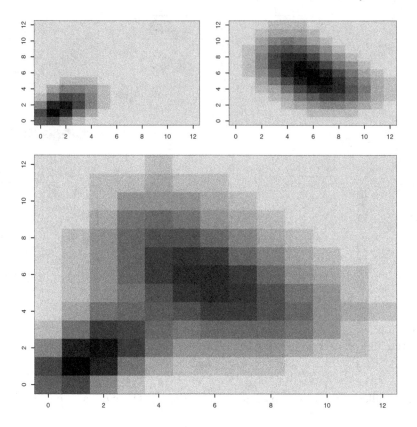

**FIGURE 9.3**
Example of a finite mixture model using copulas. At the top one can see the two components, which correspond to Poisson marginals with mean equal to 2 and a Frank copula with parameter 3 (left), and Poisson marginals with mean equal to 6 and a Frank copula with parameter $-3$ (right). The resulting copula with mixing proportions 0.3 and 0.7 can be seen below. The darker the colour, the higher the probability.

the number of successful items out of the total items that belong to each attribute. The model assumed a trivariate distribution with binomial marginals, coupled using different copulas. A finite mixture model was used to cluster the students based on their behaviour.

### 9.4.5 Finite mixture of bivariate Poisson regression models

As in the univariate case, we may consider extensions allowing for covariates. A finite mixture of bivariate regression models can be described as follows. We define as a $G$-finite mixture of bivariate Poisson distributions the distribution with joint probability function

$$p(y_{i1}, y_{i2}) = \sum_{g=1}^{G} \eta_g \mathrm{BP}(y_{i1}, y_{i2} | \lambda_{ig}),$$

where $\eta_g > 0$, $g = 1, \ldots, G$, are the mixing proportions with $\sum_{g=1}^{G} \eta_g = 1$, and $\lambda_{ig} = (\lambda_{i0g}, \lambda_{i1g}, \lambda_{i2g})$ is the component-specific parameters for the $i$th individual. The joint pmf of the bivariate Poisson distribution $\mathrm{BP}(y_1, y_2 | \lambda)$ is given in (9.5). We further assume that for each component the parameters $\lambda_{ijg}$ depend on some covariate information by assuming that

$$\log \lambda_{ijg} = x_{ij}^\top \beta_{jg}, \quad i = 1, \dots, n, \quad j = 0, 1, 2, \quad g = 1, \dots, G,$$

where $\beta_{jg}$ is a vector of coefficients for the $j$th variable and the $g$th component, while $x_{ij}$ is a vector of covariates for the $i$th individual and the $j$th variable. In fact, we assume that we can have different covariates for different variables. The model is described in detail in Bermúdez & Karlis (2012) with an application to accident data of different types.

## 9.5 Other Mixtures for Discrete Data

### 9.5.1 Latent class models

Suppose that the data refer to a binary multivariate random variable $Y = (Y_1, \dots, Y_d)^\top$, that is, the $Y_i$ take values 0 and 1. In a latent class model (also known as a latent structure model) the correlation between the elements $Y_1, \dots, Y_d$ of $Y$ is assumed to be caused by a discrete latent variables $z_i$, also called the latent class. It is then assumed that the variables $Y_1, \dots, Y_d$, which are also called manifest variables, are stochastically independent conditional on the latent variable $z_i$. Latent structure analysis is closely related to multivariate mixture modelling, as marginally the distribution of $Y$ is a multivariate discrete mixture given by

$$p(y_i|\theta) = \sum_{g=1}^{G} \eta_g \prod_{m=1}^{d} \theta_{g,m}^{y_{im}} (1 - \theta_{g,m})^{1-y_{im}},$$

where $\theta_{g,m}$ is the probability related to the $m$th variable for the $g$th group, and $y_i = (y_{i1}, \dots, y_{id})^\top$ is a realization of $Y$.

The model relates to factor analysis for continuous data. It also relates to the analysis of contingency tables since the co-occurrence of the many binary variables can be put in the form of a contingency table. Hence, the model can be seen as a mixture of many independent contingency tables, given the latent class variables. More details can be found in Vermunt & Magidson (2002).

In a closely related manner, Rudas et al. (1994) proposed a measure of fit for contingency tables in the sense that one can write any contingency table as a mixture of an independent contingency table plus some other. If the mixing proportion allocated to the second table is zero this means that the independent table is sufficient and hence the underlying assumption corresponds to independence. For finite mixtures in a contingency table, see the work of Govaert & Nadif (2007) and the references therein.

The reader is referred to Section 11.5 below for a detailed presentation of the latent class model and to Section 11.8.2 for a complete illustration of the model.

### 9.5.2 Mixtures for ranking data

Ranking data arise when experts are asked to rank some or all of a group of objects. Examples can be found in many areas, including electoral systems, college admission systems, and stated preferences. Candidates for a job might be asked to rank a list of various criteria for choosing a job in order of importance.

In many examples, the population of experts is more likely to be heterogeneous rather than homogeneous. In such cases, the heterogeneity can, as usual, be modelled by considering finite mixtures of simple models for ranking data.

Following Murphy & Martin (2003), distance-based models for ranking data have two parameters, a central ranking $R$ and a precision parameter $\xi$. The probability of a ranking $r$, say, is

$$f(r|R,\xi) = c(\xi)\exp\left(-\xi D(r,R)\right),$$

where $D(r,R)$ is a distance between the ranking $r$ and the central ranking $R$, while $c(\xi)$ is a normalizing constant. Different distances can be used. A mixture model assumes that

$$p(r) = \sum_{g=1}^{G} \eta_g f(r|R_g,\xi_g).$$

Murphy & Martin (2003) describe an EM type algorithm for this model and examine different distances. Furthermore, an application to election voting is given. In a similar application, Gormley & Murphy (2008) used a mixture model in which the model parameters are functions of covariates; see also Section 12.4.2 below. The Benter model for rank data is employed as the family of component densities within the mixture model. Busse et al. (2007) adapt these models for tied and partial rankings. Lee & Philip (2012) consider a weighted version of this mixture model family with applications in political studies. While mixtures of multistage models lead to interesting adequacy power, mixtures of distance-based models have more meaningful parameters and are easier to implement.

Mollica & Tardella (2014) used another model for the probability of a ranking, the so-called Plackett–Luce model. This expression moves from the decomposition of the ranking process in independent stages, one for each rank that has to be assigned, combined with the underlying assumption of a standard forward procedure on the ranking elicitation. In fact, a ranking can be elicited through a series of sequential comparisons in which a single item is preferred to all the remaining ones and, after being selected, is removed from the next comparisons. For a Bayesian approach for this model, see Mollica & Tardella (2017).

D'Elia & Piccolo (2005) proposed considering $r$ as the realization of a shifted binomial random variable and considered mixtures of such shifted binomial distributions. Jacques & Biernacki (2014) proposed a different probability model for rankings called the insertion sorting rank model suitable for multivariate and partial rankings.

### 9.5.3 Mixtures of multinomial distributions

Suppose one has to choose between $d$ distinct options, with probability $\pi_j$ of choosing $j$, $j = 1,\ldots,d$, and $\sum_{j=1}^d \pi_j = 1$. If we count the number of times that each of the distinct choices was actually made, the distribution that describes this vector is the multinomial distribution $\mathcal{M}(n,\pi)$, with $n = \sum_{j=1}^d y_j$ and $\pi = (\pi_1,\ldots,\pi_d)$, generalizing the widely used binomial distribution (which occurs when only two choices are available).

A multinomial distribution has pmf given by

$$p(y_1,\ldots,y_d|\pi) = \frac{n!}{y_1!\ldots y_d!} \prod_{j=1}^d \pi_j^{y_j},$$

which will be denoted by $\mathcal{M}(y|n,\pi)$. Such a model can describe the number of different categories of sites visited during a web session, the number of products from different categories purchased during a shopping visit to a supermarket or the number of persons of a particular age observed in a census, to name but a few examples.

The idea of finite mixtures of multinomial distributions implies that the population is not homogeneous and that there are groups of individuals who can be characterized by a different pattern of probabilities $\pi_g = (\pi_{g1}, \ldots, \pi_{gd})$. Hence, such a mixture takes the form

$$p(y) = \sum_{g=1}^{G} \eta_g \mathcal{M}(y|n, \pi_g).$$

As in every finite mixture, EM type algorithms in the classical approach and Gibbs samplers for the Bayesian approach (Frühwirth-Schnatter, 2006, Section 9.3) can be easily constructed to fit such a model. For applications, the reader is referred to an internet traffic clustering application (Jorgensen, 2004) as well as an application in demography (Jorgensen, 2013). In Grün & Leisch (2008), a model with covariates is discussed together with identifiability issues.

Since the multinomial model relates to contingency tables, mixtures of multinomials relate to latent class mixtures; see, for example, Vermunt & Magidson (2002). Bayesian analysis of finite mixtures of multinomial and negative multinomial distributions can be found in Rufo et al. (2007). Morel & Nagaraj (1993) used such a model for capturing multinomial extra variation. Kamakura & Russell (1989) applied a multinomial logit mixture regression model in marketing research to model consumers' choices among a set of brands and identified segments of consumers that differ in price sensitivity. Finally, to model a panel of categorical time series, Pamminger & Frühwirth-Schnatter (2010) considered finite mixtures of products of independent multinomial distrubutions within the framework of Dirichlet multinomial clustering.

Recall that in Section 11.5 a detailed presentation of the latent class model is provided for both binomial and multinomial distributions and that a complete illustration of this model is presented in Section 11.8.2.

### 9.5.4 Mixtures of Markov chains

Clustering sequences of categorical data is a challenging problem that recently attracted the attention of several researchers. One particular area of application in which this framework occurs is grouping web users by their navigation patterns. Such data are known as click stream data (Cadez et al., 2003).

The idea behind such models implies the existence of different first-order Markov chains with different transition probabilities. The problem can be stated as follows: the data refer to different kinds of some attribute (e.g. different categories of sites). What we need to do is to model how the user navigates between sites of different types. The behaviour can be described via a transition probability matrix; however, due to inhomogeneity it makes sense to consider mixtures of transition matrices which correspond to different subpopulations.

Cadez et al. (2003) described such a mixture model, providing an EM algorithm for estimation, and discussed practical aspects of its implementation using click stream data. One problem for such data is that the transition matrices can be very large, creating practical problems. Melnykov (2016) described a biclustering approach to overcome some computational problems.

A model for clustering categorical time series has been proposed in Pamminger & Frühwirth-Schnatter (2010) using mixtures of Markov chain models within a Bayesian framework. This model has been applied in various papers to analyse earnings profiles in panels of labour market data; see, for example, Frühwirth-Schnatter et al. (2012, 2016).

## 9.6    Concluding Remarks

In this chapter we have tried to provide an overview of finite mixture models for discrete data, including counts, categorical data and ranking data in one or more dimensions. We have also outlined in detail the restrictions and obstacles encountered when working with such models. While attempting to cover many models, this review is by no means complete and we have most certainly overlooked interesting models and important applications.

Finally, note that related issues occur for mixed mode data, that is, data of different types involving, for example, both continuous and discrete variables. When trying to work with such data, many of the problems mentioned in this chapter for discrete data are relevant. Recent works towards solving the problem often rely on assuming a latent model, as in McParland & Gormley (2016). The corresponding model implies that for the discrete data there is a latent continuous variable which is then modelled jointly with the observed continuous variable; see also Chapter 11 belowfor further details.

# Bibliography

AITKIN, M. (1999). A general maximum likelihood analysis of variance components in generalized linear models. *Biometrics* **55**, 117–128.

ALFÓ, M., MARUOTTI, A. & TROVATO, G. (2011). A finite mixture model for multivariate counts under endogenous selectivity. *Statistics and Computing* **21**, 185–202.

ALFÓ, M., NIEDDU, L. & VICARI, D. (2009). Finite mixture models for mapping spatially dependent disease counts. *Biometrical Journal* **51**, 84–97.

ARAB, A., HOLAN, S. H., WIKLE, C. & WILDHABER, M. L. (2012). Semiparametric bivariate zero-inflated Poisson models with application to studies of abundance for multiple species. *Environmetrics* **23**, 183–196.

BERKHOUT, P. & PLUG, E. (2004). A bivariate Poisson count data model using conditional probabilities. *Statistica Neerlandica* **58**, 349–364.

BERMÚDEZ, L. (2009). A priori ratemaking using bivariate Poisson regression models. *Insurance: Mathematics and Economics* **44**, 135–141.

BERMÚDEZ, L. & KARLIS, D. (2011). Bayesian multivariate Poisson models for insurance ratemaking. *Insurance: Mathematics and Economics* **48**, 226–236.

BERMÚDEZ, L. & KARLIS, D. (2012). A finite mixture of bivariate Poisson regression models with an application to insurance ratemaking. *Computational Statistics & Data Analysis* **56**, 3988–3999.

BERMÚDEZ, L., KARLIS, D. & SANTOLINO, M. (2017). A finite mixture of multiple discrete distributions for modelling heaped count data. *Computational Statistics & Data Analysis* **112**, 14–23.

BÖCKENHOLT, U. (1998). Mixed INAR(1) Poisson regression models: Analyzing heterogeneity and serial dependencies in longitudinal count data. *Journal of Econometrics* **89**, 317–338.

BÖHNING, D., DIETZ, E., SCHLATTMANN, P., MENDONCA, L. & KIRCHNER, U. (1999). The zero-inflated Poisson model and the decayed, missing and filled teeth index in dental epidemiology. *Journal of the Royal Statistical Society Series A* **162**, 195–209.

BRIJS, T., KARLIS, D., SWINNEN, G., VANHOOF, K., WETS, G. & MANCHANDA, P. (2004). A multivariate Poisson mixture model for marketing applications. *Statistica Neerlandica* **58**, 322–348.

BROOKS, S. (2001). On Bayesian analyses and finite mixtures for proportions. *Statistics and Computing* **11**, 179–190.

BROWNE, R. & MCNICHOLAS, P. (2012). Model-based clustering, classification, and discriminant analysis of data with mixed type. *Journal of Statistical Planning and Inference* **142**, 2976–2984.

BUSSE, L. M., ORBANZ, P. & BUHMANN, J. M. (2007). Cluster analysis of heterogeneous rank data. In *Proceedings of the 24th International Conference on Machine Learning*. New York: ACM Press.

BYUNG-JUNG, P., LORD, D. & LEE, C. (2014). Finite mixture modeling for vehicle crash data with application to hotspot identification. *Accident Analysis & Prevention* **71**, 319–326.

CADEZ, I., HECKERMAN, D., MEEK, C., SMYTH, P. & WHITE, S. (2003). Model-based clustering and visualization of navigation patterns on a web site. *Data Mining and Knowledge Discovery* **7**, 399–424.

CHIB, S. & WINKELMANN, R. (2001). Markov chain Monte Carlo analysis of correlated count data. *Journal of Business & Economic Statistics* **19**, 428–435.

DALRYMPLE, M. L., HUDSON, I. & FORD, R. P. K. (2003). Finite mixture, zero-inflated Poisson and hurdle models with application to SIDS. *Computational Statistics & Data Analysis* **41**, 491–504.

D'ELIA, A. & PICCOLO, D. (2005). A mixture model for preferences data analysis. *Computational Statistics & Data Analysis* **49**, 917–934.

DIOP, M. L., DIOP, A. & DIONGUE, A. K. (2016). A mixture integer-valued GARCH model. *REVSTAT – Statistical Journal* **14**, 245–271.

DIPPOLD, K. & HRUSCHKA, H. (2013). A parsimonious multivariate Poisson model for market basket analysis. *Review of Managerial Science* **7**, 393–415.

DIXON, M. J. & COLES, S. G. (1997). Modelling association football scores and inefficiencies in the football betting market. *Applied Statistics* **46**, 265–280.

FOKIANOS, K. (2012). Count time series models. In *Time Series Analysis: Methods and Applications*, T. Subba Rao, S. Subba Rao & C. Radhakrishna Rao, eds., vol. 30 of *Handbook of Statistics*. Amsterdam: North-Holland, pp. 315–347.

FOKIANOS, K., RAHBEK, A. & TJØSTHEIM, D. (2009). Poisson autoregression. *Journal of the American Statistical Association* **104**, 1430–1439.

FOLLMANN, D. A. & LAMBERT, D. (1989). Generalizing logistic regression by nonparametric mixing. *Journal of the American Statistical Association* **84**, 295–300.

FOLLMANN, D. A. & LAMBERT, D. (1991). Identifiability of finite mixtures of logistic regression models. *Journal of Statistical Planning and Inference* **27**, 375–381.

FORBES, F., CHARRAS-GARRIDO, M., AZIZI, L., DOYLE, S. & ABRIAL, D. (2013). Spatial risk mapping for rare disease with hidden Markov fields and variational EM. *Annals of Applied Statistics* **7**, 1192–1216.

FRÜHWIRTH-SCHNATTER, S. (2006). *Finite Mixture and Markov Switching Models*. New York: Springer-Verlag.

FRÜHWIRTH-SCHNATTER, S., PAMMINGER, C., WEBER, A. & WINTER-EBMER, R. (2012). Labor market entry and earnings dynamics: Bayesian inference using mixtures-of-experts Markov chain clustering. *Journal of Applied Econometrics* **27**, 1116–1137.

FRÜHWIRTH-SCHNATTER, S., PAMMINGER, C., WEBER, A. & WINTER-EBMER, R. (2016). Mothers' long-run career patterns after first birth. *Journal of the Royal Statistical Society Series A* **179**, 707–725.

GARAY, A., HASHIMOTO, E., ORTEGA, E. & LACHOS, V. (2011). On estimation and influence diagnostics for zero-inflated negative binomial regression models. *Computational Statistics & Data Analysis* **55**, 1304–1318.

GENEST, C. & NEŠLEHOVÁ, J. (2007). A primer on copulas for count data. *ASTIN Bulletin* **37**, 475–515.

GODWIN, R. T. & BÖHNING, D. (2017). Estimation of the population size by using the one-inflated positive Poisson model. *Applied Statistics* **66**, 425–448.

GORMLEY, I. C. & MURPHY, T. B. (2008). A mixture of experts model for rank data with applications in election studies. *Annals of Applied Statistics* **2**, 1452–1477.

GOVAERT, G. & NADIF, M. (2007). Clustering of contingency table and mixture model. *European Journal of Operational Research* **183**, 1055–1066.

GRÜN, B. & LEISCH, F. (2007). Fitting finite mixtures of generalized linear regressions in R. *Computational Statistics & Data Analysis* **51**, 5247–5252.

GRÜN, B. & LEISCH, F. (2008). Identifiability of finite mixtures of multinomial logit models with varying and fixed effects. *Journal of Classification* **25**, 225–247.

GURMU, S. & ELDER, J. (2008). A bivariate zero-inflated count data regression model with unrestricted correlation. *Economics Letters* **100**, 245–248.

HALL, D. B. (2000). Zero-inflated Poisson and binomial regression with random effects: A case study. *Biometrics* **56**, 1030–1039.

HENNIG, C. (2000). Identifiablity of models for clusterwise linear regression. *Journal of Classification* **17**, 273–296.

HOBEN, T., ARNOLD, L. & BRAINERD, C. J. (1993). *Modeling Growth and Individual Differences in Spatial Tasks*, vol. 237 of *Monographs of the Society for Research in Child Development*. Chicago: University of Chicago Press.

JACQUES, J. & BIERNACKI, C. (2014). Model-based clustering for multivariate partial ranking data. *Journal of Statistical Planning and Inference* **149**, 201–217.

JAZI, M., JONES, G. & LAI, C. (2012). First-order integer valued AR processes with zero inflated Poisson innovations. *Journal of Time Series Analysis* **33**, 954–963.

JORGENSEN, M. (2004). Using multinomial mixture models to cluster internet traffic. *Australian & New Zealand Journal of Statistics* **46**, 205–218.

JORGENSEN, M. (2013). Forming clusters from census areas with similar tabular statistics. *Communications in Statistics – Theory and Methods* **42**, 2136–2151.

KAMAKURA, W. A. & RUSSELL, G. (1989). A probabilistic choice model for market segmentation and elasticity structure. *Journal of Marketing Research* **26**, 379–390.

KARLIS, D. & MELIGKOTSIDOU, L. (2005). Multivariate Poisson regression with covariance structure. *Statistics and Computing* **15**, 255–265.

KARLIS, D. & MELIGKOTSIDOU, L. (2007). Finite multivariate Poisson mixtures with applications. *Journal of Statistical Planning and Inference* **137**, 1942–1960.

KARLIS, D. & NTZOUFRAS, L. (2003). Analysis of sports data by using bivariate Poisson models. *The Statistician* **52**, 381–393.

KARLIS, D., PAPATLA, P. & ROY, S. (2016). Finite mixtures of censored Poisson regression models. *Statistica Neerlandica* **70**, 100–122.

KARLIS, D. & XEKALAKI, E. (2003). Choosing initial values for the EM algorithm for finite mixtures. *Computational Statistics & Data Analysis* **41**, 577–590.

KARLIS, D. & XEKALAKI, E. (2005). Mixed Poisson distributions. *International Statistical Review* **73**, 35–58.

KOCHERLAKOTA, S. & KOCHERLAKOTA, K. (1992). *Bivariate Discrete Distributions*, vol. 132 of *Statistics: Textbooks and Monographs*. New York: Markel Dekker.

KOSMIDIS, I. & KARLIS, D. (2016). Model-based clustering using copulas with applications. *Statistics and Computing* **26**, 1079–1099.

LAI, C.-D. (2006). Constructions of discrete bivariate distributions. In *Advances in Distribution Theory, Order Statistics, and Inference*, B. C. Arnold, N. Balakrishnan, E. Castillo & J.-M. Sarabia, eds. Boston: Birkhäuser, pp. 29–58.

LAMBERT, D. (1992). Zero-inflated Poisson regression, with an application to defects in manufacturing. *Technometrics* **34**, 1–14.

LEE, P. H. & PHILIP, L. (2012). Mixtures of weighted distance-based models for ranking data with applications in political studies. *Computational Statistics & Data Analysis* **56**, 2486–2500.

LEROUX, B. G. & PUTERMAN, M. L. (1992). Maximum-penalized-likelihood estimation for independent and Markov-dependent mixture models. *Biometrics* **48**, 545–558.

LI, C.-S., LU, J.-C., PARK, J., KIM, K., BRINKLEY, P. & PETERSON, J. (1999). Multivariate zero-inflated Poisson models and their applications. *Technometrics* **41**, 29–38.

MA, C. X., QIBIN, Y., BERG, A., DROST, D., NOVAES, E. et al. (2008). A statistical model for testing the pleiotropic control of phenotypic plasticity for a count trait. *Genetics* **179**, 627–636.

MACDONALD, I. L. & ZUCCHINI, W. (1997). *Hidden Markov and Other Models for Discrete-valued Time Series*. London: Chapman & Hall.

MARBAC, M., BIERNACKI, C. & VANDEWALLE, V. (2015). Model-based clustering for conditionally correlated categorical data. *Journal of Classification* **32**, 145–175.

MCKENZIE, E. (1985). Some simple models for discrete variate time series. *Water Resources Bulletin* **21**, 645–650.

MCPARLAND, D. & GORMLEY, I. C. (2016). Model based clustering for mixed data: clustmd. *Advances in Data Analysis and Classification* **10**, 155–169.

MELIGKOTSIDOU, L. (2007). Bayesian multivariate Poisson mixtures with an unknown number of components. *Statistics and Computing* **17**, 93–107.

MELNYKOV, V. (2016). Model-based biclustering of clickstream data. *Computational Statistics & Data Analysis* **93**, 31–45.

MOLLICA, C. & TARDELLA, L. (2014). Epitope profiling via mixture modeling of ranked data. *Statistics in Medicine* **33**, 3738–3758.

MOLLICA, C. & TARDELLA, L. (2017). Bayesian Plackett–Luce mixture models for partially ranked data. *Psychometrika* **82**, 442–458.

MOREL, J. G. & NAGARAJ, N. K. (1993). A finite mixture distribution for modelling multinomial extra variation. *Biometrika* **80**, 363–371.

MURPHY, T. B. & MARTIN, D. (2003). Mixtures of distance-based models for ranking data. *Computational Statistics & Data Analysis* **41**, 645–655.

NELSEN, R. (2006). *An Introduction to Copulas.* New York: Springer–Verlag, 2nd ed.

NIKOLOULOPOULOS, A. K. (2013). Copula-based models for multivariate discrete response data. In *Copulae in Mathematical and Quantitative Finance.* Berlin: Springer, pp. 231–249.

NIKOLOULOPOULOS, A. K. & KARLIS, D. (2009). Finite normal mixture copulas for multivariate discrete data modeling. *Journal of Statistical Planning and Inference* **139**, 3878–3890.

PAMMINGER, C. & FRÜHWIRTH-SCHNATTER, S. (2010). Model-based clustering of categorical time series. *Bayesian Analysis* **5**, 345–368.

PANAGIOTELIS, A., CZADO, C. & JOE, M. (2012). Pair copula constructions for multivariate discrete data. *Journal of the American Statistical Association* **107**, 1063–1072.

PAVLOPOULOS, H. & KARLIS, D. (2008). INAR (1) modeling of overdispersed count series with an environmental application. *Environmetrics* **19**, 369–393.

PEARSON, K. (1915). On certain types of compound frequency distributions in which the components can be individually described by binomial series. *Biometrika* **11**, 139–144.

RAU, A., MAUGIS-RABUSSEAU, C., MARTIN-MAGNIETTE, M.-L. & CELEUX, G. (2015). Co-expression analysis of high-throughput transcriptome sequencing data with Poisson mixture models. *Bioinformatics* **31**, 1420–1427.

ROBERT, C. P. & CASELLA, G. (2004). *Monte Carlo Statistical Methods.* New York: Springer-Verlag, 2nd ed.

ROYLE, J. A. (2004). N-mixture models for estimating population size from spatially replicated counts. *Biometrics* **60**, 108–115.

RUDAS, T., CLOGG, C. C. & LINDSAY, B. G. (1994). A new index of fit based on mixture methods for the analysis of contingency tables. *Journal of the Royal Statistical Society, Series B* , 623–639.

RUFO, M., PÉREZ, C. & MARTÍN, J. (2007). Bayesian analysis of finite mixtures of multinomial and negative-multinomial distributions. *Computational Statistics & Data Analysis* **51**, 5452–5466.

SKLAR, M. (1959). Fonctions de répartition à *n* dimensions et leurs marges. *Publication de l'Institut de Statistique de l'Université de Paris* **8**, 229–231.

SPEZIA, L., COOKSLEY, S., BREWER, M., DONNELLY, D. & TREE, A. (2014). Modelling species abundance in a river by negative binomial hidden Markov models. *Computational Statistics & Data Analysis* **71**, 599–614.

STEUTEL, F. & VAN HARN, K. (1979). Discrete analogues of self-decomposability and stability. *Annals of Probability* **7**, 893–899.

SUR, P., SHMUELI, G., BOSE, S. & DUBEY, P. (2015). Modeling bimodal discrete data using Conway-Maxwell-Poisson mixture models. *Journal of Business & Economic Statistics* **33**, 352–365.

TEICHER, H. (1963). Identifiability of finite mixtures. *Annals of Mathematical Statistics* **34**, 1265–1269.

VERMUNT, J. K. & MAGIDSON, J. (2002). Latent class cluster analysis. *Applied Latent Class Analysis* **11**, 89–106.

WANG, P. (2003). A bivariate zero-inflated negative binomial regression model for count data with excess zeros. *Economics Letters* **78**, 373–378.

WANG, P., COCKBURN, I. & PUTERMAN, M. (1998). Analysis of patent data: A mixed Poisson regression model approach. *Journal of Business & Economic Statistics* **16**, 27–36.

WEISS, C. (2018). *An Introduction to Discrete-Valued Time Series*. Hoboken, NJ: John Wiley.

WINKELMANN, R. (2008). *Econometric Analysis of Count Data*. New York: Springer, 4th ed.

YAU, K., WANG, K. & LEE, A. (2003). Zero-inflated negative binomial mixed regression modeling of over-dispersed count data with extra zeros. *Biometrical Journal* **45**, 437–452.

ZHU, F., LI, Q. & WANG, D. (2010). A mixture integer-valued ARCH model. *Journal of Statistical Planning and Inference* **140**, 2025–2036.

ZOU, Y., ZHANG, Y. & LORD, D. (2013). Application of finite mixture of negative binomial regression models with varying weight parameters for vehicle crash data analysis. *Accident Analysis & Prevention* **50**, 1042 – 1051.

# 10

## Continuous Mixtures with Skewness and Heavy Tails

**David Rossell and Mark F. J. Steel**

*University of Warwick, UK; Universitat Pompeu Fabra, Spain, and University of Warwick, UK*

## CONTENTS

The multivariate normal distribution remains a central choice for modelling continuous multivariate data. This is partially due to its providing a useful representation for the underlying data-generating mechanism in many applications, but also to convenience in terms of interpretation and ease of computation. An alternative to add flexibility is to use parametric distributions with skewness and thick tails that robustify inference while keeping interpretation and computations manageable. We review some of the available options and, by way of illustration, we use a novel formulation based on two-piece families that contains the normal and Student $t$ distributions as particular cases, adds a reduced set of easily interpretable parameters, leads to simple model fitting and is unimodal, a convenient property for clustering. As also described by others before us, our examples illustrate that in the presence of asymmetries or heavy tails normal mixtures may lead to biased parameter estimates, poor clustering, or suggest the addition of spurious components, whereas more flexible mixtures tend to better capture the underlying subpopulations. The methodology is implemented in the R package twopiece, freely available at https://r-forge.r-project.org/projects/twopiece.

## 10.1    Introduction

Let $y_i \in \mathbb{R}^d$, $i = 1, \dots, n$, be independent realizations from a mixture with $G$ components and probability density function (pdf)

$$p(y_i) = \sum_{g=1}^{G} \eta_g f_g(y_i \mid \theta_g),$$

where $f_g(\cdot)$ is a pdf indexed by $\theta_g \in \Theta$, and $\eta_g \in [0,1]$ with $\sum_{g=1}^{G} \eta_g = 1$ and $G \in \mathbb{N}$. For now we assume $G$ to be fixed (possibly to a large value), but in our examples we illustrate strategies to set $G$ and point to open questions for future research. We denote by $\eta = (\eta_1, \ldots, \eta_G)$ the vector of weights and by $\theta = (\theta_1, \ldots, \theta_G)$ the other parameters in the model. When constructing a mixture model one is often interested in unimodal densities $f_g(\cdot)$ as this helps interpret each component as a distinct subpopulation underlying the observed data.

A popular strategy is to assume that $f_g(y_i \mid \theta_g) = \mathcal{N}(y_i|\mu_g, \Sigma_g)$ is a normal density with mean $\mu_g \in \mathbb{R}^d$ and positive definite covariance $\Sigma_g \in \mathbb{R}^{d \times d}$. This choice is unimodal and practical to implement; however, the data may exhibit departures from normality that render this choice suboptimal. For instance, the presence of strong asymmetries may bias parameter estimates or cause some observations to have an unduly large influence on inference, whereas thick tails may suggest the need to add extra components (i.e., larger $G$), which adversely affects parsimony and interpretability.

To address these issues, numerous authors have suggested more flexible choices for $f_g(\cdot)$. Here we focus on parametric families $f_g(\cdot)$, since nonparametric alternatives are discussed in Chapters 6 and 14. Most parametric strategies are based on extensions of the normal distribution that incorporate skewness and thicker-than-normal tails, for example, skew-normal, skew-Laplace, skew-$t$ or normal-inverse-Gaussian (NIG) distributions, which we now briefly review. In the $d = 1$ case a skew-normal random variable can be expressed as

$$y = \mu_g + \Sigma_g^{1/2} x, \quad \text{where } x = u + v \text{ and } u \sim \mathcal{N}(0,1), \ v \sim \mathcal{N}(0, \alpha_g) \mathbb{I}(v \geq 0)$$

are independent draws from a standard and truncated normal distributions respectively (Azzalini & Dalla Valle, 1996). The symmetric case is obtained by setting $\alpha_g = 0$. Multivariate versions can be similarly built from location–scale transforms, for example, letting $u$ and $v$ in the formulation above be vectors with independent components rather than scalars (Arnold & Beaver, 2002). Back to the $d = 1$ case, the skew-Laplace can be obtained by letting $u \sim \mathcal{N}(0, w)$, $v \sim \mathcal{N}(0, w\alpha_g) \mathbb{I}(v \geq 0)$, where $w$ is drawn from a gamma distribution (Franczak et al., 2014), and the skew-$t$ similarly by drawing $w$ from an inverse gamma (Sahu et al., 2003). The NIG provides yet another option by letting $x \sim \mathcal{N}(\alpha_g w, w)$, where $\alpha_g = 0$ corresponds to the symmetric case (Karlis & Santourian, 2009).

One caveat with the families described so far is that, while adding flexibility and often leading to efficient computation, communication and prior elicitations may be hampered by parameters not being straightforward to interpret. For instance, the mean and mode of the skew-normal depend on $\mu_g$, $\Sigma_g$ and $\alpha_g$ and the asymmetry is a non-trivial function of $\alpha_g$. As long as $f_g(\cdot)$ allows for asymmetry and heavy tails, most of these parametric choices often lead to similar inference; however, for our illustrations we develop an alternative based on two-piece distributions, a family that leads to simple parameter interpretation. Before doing so, we review some of the existing proposals in the context of mixture models.

Frühwirth-Schnatter & Pyne (2010) proposed mixtures of skew-normal and skew-$t$s (Azzalini & Dalla Valle, 1996) and derived a convenient two- or three-block Gibbs sampler (respectively) to fit the model. Cabral et al. (2008) proposed mixtures of skew-$t$-normal distributions, an extension of the skew-normal with an increased skewness range, derived a Markov chain Monte Carlo (MCMC) algorithm to fit the model and computed marginal likelihoods via bridge-sampling (Frühwirth-Schnatter, 2004) to help choose $G$. Lin (2009) and Lin et al. (2014) proposed mixtures of slightly more general skew-normal and skew-$t$-normal distributions and derived algorithms to obtain maximum likelihood estimates and asymptotic covariances, although the theoretical validity of the latter is restricted to the case where $G$ is known (Redner & Walker, 1984). For an interesting review of skew-$t$ distributions and model-fitting strategies, see Lee & McLachlan (2014). As an alternative, Kosmidis & Karlis

(2016) proposed multivariate distributions where dependence is introduced via copulas; see also Chapter 9 above.

To conclude our review we note that one may define non-normal clusters by combining multiple normal components into a single cluster. Such combination can be obtained in a post-processing step (e.g. Baudry et al., 2010) or in a hierarchical framework where a carefully formulated prior distribution encourages the formation of groups (clusters) such that within-group components are tightly knit and between-groups components are well separated (Malsiner-Walli et al., 2017), a strategy referred to as *mixture of mixtures*. Some interesting advantages are that nonparametric choices are needed for $f_g(\cdot)$, that available algorithms for normal mixtures can be easily extended and that the formulation alleviates label-switching issues, whereas considerations for further research include that $f_g(\cdot)$ may no longer be unimodal or that computations for the required larger $G$ may be costly.

Here, we consider densities $f_g(\cdot)$ which are unimodal but can capture asymmetry and thick tails via multivariate two-piece skew-$t$ distributions (Ferreira & Steel, 2007a,b), which we review in Section 10.2. Section 10.3 discusses the prior formulation and Section 10.4 outlines an MCMC algorithm to fit the model. Section 10.5 illustrates the methodology with some examples and Section 10.6 offers some concluding remarks.

## 10.2   Skew-$t$ Mixtures

Following Ferreira & Steel (2007b), we model asymmetries via two-piece distributions (see Arellano-Valle et al., 2005; Rubio & Steel, 2014). These have a long history and basically introduce skewness into a symmetric unimodal continuous distribution with support on $\mathbb{R}$ by using different scale parameters on either side of the mode. Let $\phi(x)$ denote the univariate standard normal density evaluated at $x \in \mathbb{R}$. The two-piece skew-normal with zero location parameter, variance $\omega$ and asymmetry parameter $\alpha_g \in [-1, 1]$ is given by the pdf

$$s_{\mathcal{N}}(x \mid \alpha_g, \omega) = \frac{1}{\sqrt{\omega}} \left[ \phi \left( \frac{x}{(1 - \alpha_g)\sqrt{\omega}} \right) \mathbb{I}(x \geq 0) + \phi \left( \frac{x}{(1 + \alpha_g)\sqrt{\omega}} \right) \mathbb{I}(x < 0) \right]. \quad (10.1)$$

As desired, (10.1) has a single mode at $x = 0$ and allows asymmetry to be captured by combining two normal pdfs with variance $\omega(1 - \alpha_g)^2$ for $x > 0$ and $\omega(1 + \alpha_g)^2$ for $x < 0$ (note that (10.1) is continuous at $x = 0$). The particular case $\omega = 1$ and $\alpha_g = 0$ simplifies to $s_{\mathcal{N}}(x \mid 0, 1) = \phi(x)$, whereas positive skewness corresponds to $\alpha_g < 0$ and negative skewness to $\alpha_g > 0$. The Arnold–Groeneveld (AG) measure of skewness is an easily interpretable quantity, defined as $\text{AG} = 1 - 2F(m) \in [-1, 1]$ for a univariate random variable with mode $m$ and cdf $F(\cdot)$. In our parameterization it takes the form $\text{AG} = -\alpha_g$, immediately giving $\alpha_g$ a clear interpretation as a skewness parameter.

The specific parameterization (called $\epsilon$-skew parameterization) of the differing scales in (10.1) is due to Mudholkar & Hutson (2000), but other possibilities are available, such as the inverse scale factor parameterization suggested by Fernández & Steel (2008). A further advantage of the $\epsilon$-skew parameterization is that it leads to orthogonality between $\alpha_g$ and $\sigma_g$, where $\sigma_g$ is a scale parameter defined after (10.2); for example, the maximum likelihood estimator of $\alpha_g$ does not depend on $\sigma_g$ (see Jones & Anaya-Izquierdo, 2011).

Thick tails may be obtained by mixing over $\omega$. This immediately defines a family of skewed scale mixtures of normals. In particular, if we let $\omega \sim \mathcal{IG}(v_g/2, v_g/2)$ be an inverse

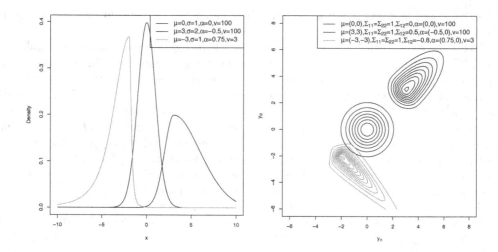

**FIGURE 10.1**

Density for three mixture components with various asymmetry/tail parameters in the univariate (left) and bivariate (right) cases. The solid black lines in both panels correspond to a symmetric standard $t$ distribution with $v_g = 100$. The dashed black lines correspond to positive skewness in the first axis ($\alpha_g = -0.5$), $v_g = 100$ and correlation equals 0.5 in the bivariate case, whereas the grey line corresponds to negative skewness ($\alpha_g = 0.75$), negative correlation equals $-0.8$ and thick tails ($v_g = 3$).

gamma random variable where $v_g \in \mathbb{R}^+$ we obtain the marginal pdf

$$s_t(x \mid \alpha_g, v_g) = \left[ t_{v_g}\left( \frac{x}{1 - \alpha_g} \right) \mathbb{I}(x \geq 0) + t_{v_g}\left( \frac{x}{1 + \alpha_g} \right) \mathbb{I}(x < 0) \right], \qquad (10.2)$$

where $t_v(\cdot)$ is the pdf of the univariate $t$ distribution with $v$ degrees of freedom. If thick tails are not desired, one simply sets $\omega = 1$ in (10.1). The univariate skew-normal and skew-$t$ with general location $\mu_g \in \mathbb{R}$ and scale $\sigma_g \in \mathbb{R}^+$ are defined with the generative model $y_i = \sqrt{\sigma_g} x_i + \mu_g$, where $x_i$ is drawn from (10.1) or (10.2) (respectively). In a slight abuse of notation we distinguish the skew-normal pdf,

$$f_g(y_i \mid \mu_g, \sigma_g, \alpha_g) = \sigma_g^{-1/2} s_N\left( (y_i - \mu_g)\sigma_g^{-1/2} \mid \alpha_g, 1 \right),$$

from the skew-$t$,

$$f_g(y_i \mid \mu_g, \sigma_g, \alpha_g, v_g) = \sigma_g^{-1/2} s_t\left( (y_i - \mu_g)\sigma_g^{-1/2} \mid \alpha_g, v_g \right),$$

by $v_g$ appearing as an argument or not. An interesting feature of two-piece distributions is leading to easily interpretable parameters. Beyond the discussed connection between $\alpha_g$ and the AG asymmetry coefficient, $\mu_g$ is the mode, both mean and median are linear in $\mu_g$ (e.g. the mean for the two-piece normal is $\mu_g - \alpha_g \sqrt{8\sigma_g/\pi}$) and the variance is proportional to $\sigma_g$. Moments and quantiles for an ample range of two-piece families are provided in Arellano-Valle et al. (2005).

For multivariate $y_i \in \mathbb{R}^d$ let $\Sigma_g \in \mathbb{R}^{d \times d}$ be a symmetric positive definite matrix and $\Sigma_g = A_g^\top D_g A_g$ its eigendecomposition, where $D_g = \text{Diag}(\delta_{g1}, \dots, \delta_{gd})$, $\delta_{gj} > 0$ is the $j$th eigenvalue and $A_g \in \mathbb{R}^{d \times d}$ the non-singular eigenvector matrix. Following two possible definitions for the symmetric multivariate $t$ distribution, we consider the *iskew-t* built upon independent realizations from (10.2) and the *dskew-t* built upon uncorrelated but dependent realizations.

We first discuss the iskew-*t*. Let $y_i = A_g^\top D_g^{1/2} x_i + \mu_g$, where $\mu_g \in \mathbb{R}^d$, $x_i = (x_{i1}, \ldots, x_{id})^\top$ and $x_{ij}$ are independent draws from $s_t(x_{ij} \mid \alpha_{gj}, v_g)$. Then it is straightforward to see that

$$f_g(y_i \mid \mu_g, \Sigma_g, \alpha_g, v_g) = |\Sigma_g|^{-1/2} \prod_{j=1}^{d} s_t \left( \delta_{gj}^{-1/2} a_{gj}^\top (y_i - \mu_g) \mid \alpha_{gj}, v_g \right), \qquad (10.3)$$

where $\alpha_g = (\alpha_{g1}, \ldots, \alpha_{gd})$, $v_g \in \mathbb{R}^+$ and $a_{gj} \in \mathbb{R}^d$ is the $j$th row of $A_g$. Note that (10.3) implies that $x_i = D_g^{-1/2} A_g (y_i - \mu_g)$ has independent components. As in Ferreira & Steel (2007a,b), it is possible to generalize the specification to include different degrees of freedom parameters for each dimension, that is, use $v_{gj}$ in (10.3), which would allow for different tail behaviour in each dimension. This is easily implemented and could be a useful extension in some applications; however, for simplicity here we focus on the common $v_g$ case.

As an alternative, the dskew-*t* uses the multivariate version of the two-piece normal in (10.1) to define a scale mixture with pdf

$$f_g(y_i \mid \mu_g, \Sigma_g, \alpha_g, v_g) = \int f_g(y_i \mid \mu_g, \Sigma_g, \alpha_g, \omega_i) \mathcal{IG} \left( \omega_i \middle| \frac{v_g}{2}, \frac{v_g}{2} \right) d\omega_i$$

$$= |\Sigma_g|^{-1/2} \int \left[ \prod_{j=1}^{d} \frac{\exp\left( -\frac{\xi_{gij} x_{ij}^2}{2\omega_i} \right)}{\omega_i^{1/2} \sqrt{2\pi}} \right] \mathcal{IG} \left( \omega_i \middle| \frac{v_g}{2}, \frac{v_g}{2} \right) d\omega_i$$

$$= \frac{|\Sigma_g|^{-1/2} \Gamma\left( \frac{d+v_g}{2} \right)}{\Gamma\left( \frac{v_g}{2} \right) \pi^{d/2} v_g^{d/2}} \left( 1 + \frac{1}{v_g} x_i^\top \Xi_{gi} x_i \right)^{-(d+v_g)/2}, \qquad (10.4)$$

where $\xi_{gij} = \frac{1}{(1-\alpha_{gj})^2} \mathbb{I}(x_{ij} \geq 0) + \frac{1}{(1+\alpha_{gj})^2} \mathbb{I}(x_{ij} < 0)$ can be interpreted as weights, $\Xi_{gi} = \text{Diag}(\xi_{gi1}, \ldots, \xi_{gid})$ and $\Gamma(\cdot)$ is the gamma function. In (10.4) there is a shared latent $\omega_i \in \mathbb{R}$ inducing dependence across $j = 1, \ldots, d$, whereas (10.3) was implicitly built upon a latent vector $\omega_i = (\omega_{i1}, \ldots, \omega_{id}) \in \mathbb{R}^d$ in (10.1) and (10.2) with a different scale parameter for each dimension. For both the iskew-*t* and dskew-*t* specifications we denote by $\Omega_i$ the $d \times d$ diagonal matrix containing these latent parameters, that is, $\Omega_i = \omega_i I$ for the dskew-*t* and $\Omega_i = \text{Diag}(\omega_{i1}, \ldots, \omega_{id})$ for the iskew-*t*.

In what follows, we impose two identifiability constraints for the iskew-*t* and dskew-*t* in (10.3) and (10.4). First, changing the sign of an eigenvector (row in $A_g$) would alter the interpretation of the corresponding element in $\alpha_g$, hence we restrict all elements in the first column of $A_g$ to be positive. Clearly this is without loss of generality, as changes in eigenvector signs do not affect $\Sigma_g$. The second constraint is $\delta_{g1} > \ldots > \delta_{gd}$. This is also without loss of generality since the $j$th element in $y_i$ is

$$y_{ij} = \sum_{l=1}^{d} \delta_{gl}^{1/2} a_{glj} x_{il} + \mu_j,$$

where $a_{glj} \in \mathbb{R}$ is the $j$th element in $a_{gl}$. Given that $x_{i1}, \ldots, x_{id}$ are exchangeable, any reordering of the eigenvectors and eigenvalues in $A_g$ and $D_g$ gives rise to the same probability distribution for $y_i$, hence we may choose any arbitrary order. Note also that we shall assume a continuous prior distribution for $\Sigma_g$ in Section 10.3, hence there is zero probability of having 0 entries in $A_g$ and also of having $\delta_{gj} = \delta_{gj'}$ for $j \neq j'$, that is, there exists a strict ordering $\delta_{g1} > \ldots > \delta_{gd}$ with probability one. An advantage of the identifiability constraint is that $\Sigma_g$ becomes a one-to-one function of $(A_g, D_g)$, hence we may assume a probability model for $\Sigma_g$ and bypass technical difficulties in dealing with $A_g$ and $D_g$ separately (Ferreira & Steel, 2007b).

The skew-$t$ densities (10.3) and (10.4) allow for thick tails and different degrees of asymmetry $\alpha_{g1}, \ldots, \alpha_{gd}$ in the directions defined by the eigenvectors. We note that Ferreira & Steel (2007b) considered a more general family where the tail behaviour and asymmetry can be introduced along different axes. This extra flexibility requires a separate treatment of $A_g$ and $D_g$ that can be cumbersome for prior specification and model fitting. We view (10.3) and (10.4) as convenient models that are flexible enough in most practical settings; see also the discussion in Ferreira & Steel (2007a). As an example, Figure 10.1 shows various univariate (left) and bivariate (right) pdfs.

The mixture density is thus

$$p(y_i) = \sum_{g=1}^{G} \eta_g f_g(y_i \mid \theta_g),$$

where $\theta_g = (\mu_g, \Sigma_g, \alpha_g, v_g)$ and $f_g(\cdot)$ is given by (10.3) or (10.4). Note that we allow each component to have different asymmetry and tail thickness $(\alpha_g, v_g)$. We denote by $\theta = (\theta_1, \ldots, \theta_G)$ the whole set of component-specific parameters and by $\eta = (\eta_1, \ldots, \eta_G)$ the weights. We now complete the Bayesian probability model by discussing the prior formulation.

---

## 10.3    Prior Formulation

We consider $p(\theta, \eta) = p(\eta) \prod_{g=1}^{G} p(\mu_g, \Sigma_g) p(\alpha_g) p(v_g)$ with the usual symmetric Dirichlet prior $p(\eta) = \mathcal{D}_G(\eta | e_0)$, $e_0 \in \mathbb{R}^+$, and the normal-inverse Wishart prior $p(\mu_g, \Sigma_g) = \mathcal{N}(\mu_g | 0, t\Sigma_g) \mathcal{W}^{-1}(\Sigma_g | q, Q)$, where $q > d - 1$ is required to ensure that the prior is proper.

By default we recommend $e_0 = 1/G$ to encourage the posterior distribution of the weights of unnecessary components to shrink to zero (Chambaz & Rousseau, 2008), which is important when setting $G$ to a fixed value. Alternatively, one could conduct formal Bayesian model selection on $G$, in which case it is critical to adopt a prior formulation that adequately induces sparsity; see Fuquene et al. (2016) for a recent review and a proposal based on non-local priors. For simplicity, here we focus on fixed $G$ and leave model choice for robust mixtures as a topic for future research. The MCMC algorithm described in Section 10.4 facilitates inferring the number of components by counting the number of non-empty clusters in each iteration, as suggested in Malsiner-Walli et al. (2016). A comprehensive review of choosing $G$ in a mixture analysis framework is also given in Chapter 7 above.

Also by default we set the relatively weak informative value $t = 1$ (the prior information is equivalent to that provided by one observation that we knew to arise from component $g$) and $q = d + 1$ as this leads to thick tails (undefined expectation, whereas for $q > d + 1$ it is $E(\Sigma_g) = Q/(q - d - 1)$) and $Q = I$ to reflect that in the absence of prior information it seems reasonable not to encourage any particular departure from the canonical basis for the axes in $\Sigma_g$, as these drive the asymmetry in the data.

Regarding $p(v_g)$ numerous proposals are available, each with its own set of appealing properties (see Villa & Walker, 2014; Rubio & Steel, 2015, for a review). Following the objective prior recently proposed by Villa & Walker (2014), we recommend that $p(v_g)$ should be decreasing in $v_g$ based on the fact that the Kullback–Leibler divergence between $t_{v_g}(\cdot)$ and $t_{v_g+1}(\cdot)$ decreases with $v_g$, that is, one should not assign high prior probability to multiple large $v_g$ as these define similar sampling models. We also follow their recommendation of

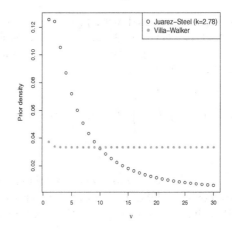

**FIGURE 10.2**
Juárez–Steel and Villa-Walker prior probabilities $p(v_g)$ for $v_g \in \{1,\ldots,30\}$, $d=1$, $\mu_g = 0$, $\Sigma_g = 1$.

discretizing the support to $v_g \in \{1,\ldots,v_{\max}\}$ since small changes in $v_g$ have a small effect on $t_{v_g}(\cdot)$. By default we set $v_{\max} = 30$, as for larger values $t_{v_g}(\cdot)$ is practically indistinguishable from a normal. However, the Villa–Walker prior depends on $\mu_g, \Sigma_g$ and requires numerical integration, creating practical difficulties, and further it is fairly flat (Figure 10.2 shows $p(v_g)$ for $d=1$, $\mu_g = 0$, $\Sigma_g = 1$), leading to potential sensitivity of posterior inference to $v_{\max}$ (Fonseca et al., 2008).

Instead, we propose discretizing the gamma-gamma prior of Juárez & Steel (2010), arising from $v_g \sim \mathcal{G}(2,l)$ and $l \sim \mathcal{E}(k)$, which gives

$$p(v_g) \propto v_g/(v_g + k)^3.$$

The prior is easy to evaluate and decreasing for $v_g \geq 1$ when $k \leq 2.84$. By default we set $k = 2.78$ (see Figure 10.2), which gives $\mathrm{P}(v_g \leq 2) = 0.25$ (infinite second moments) and prior median equal to 5. Decreasing $k$ further assigns a higher prior probability to infinite second moments, which does not seem sensible by default, whereas larger values of $k$ would soon violate $k \leq 2.84$. Although we recommend a careful prior elicitation, in our examples in Section 10.5 posterior inference was fairly robust to $p(v_g)$ for moderate sample sizes.

Finally, for the asymmetry parameters we set independent priors $p(\alpha_g) = \prod_{j=1}^{d} p(\alpha_{gj})$, relying on the Arnold–Groeneveld measure of skewness (1995). Recall that $\mathrm{AG} = -\alpha_{gj}$ in our parameterization. As a simple yet flexible choice we consider $\frac{1}{2}(\alpha_{gj}+1) \sim \mathcal{B}e(a,b)$, where by default we set $a = b = 2$ so that the prior is centred on the symmetric case $\mathrm{AG} = 0$ and places zero density on extreme asymmetries for which $\mathrm{AG} \in \{-1,1\}$.

## 10.4   Model Fitting

We propose a Metropolis–Hastings (MH) within Gibbs algorithm to draw from the posterior distribution of $(\theta, v)$ given an independent sample $\mathbf{y} = (y_1, \ldots, y_n)^\top$ from the mixture with pdf $p(y_i) = \sum_{g=1}^{G} \eta_g f_g(y_i \mid \theta_g)$. As already detailed in Chapter 1, $p(y_i)$ can be equivalently expressed as the marginal pdf associated with $p(y_i, z_i) = f_g(y_i \mid \theta_g)P(z_i = g)$, where $P(z_i = g) = \eta_g$ and $z_i \in \{1, \ldots, G\}$, $i = 1, \ldots, n$, are latent cluster indicators. Letting $\mathbf{z} = (z_1, \ldots, z_n)$ and $\boldsymbol{\omega} = (\omega_1, \ldots, \omega_n)$, the augmented model allows for sampling from the joint posterior pdf $p(\mathbf{z}, \boldsymbol{\omega}, \theta, \eta \mid \mathbf{y})$ proportional to

$$\prod_{g=1}^{G} \left( \prod_{i:z_i=g} f_g(y_i \mid \mu_g, \Sigma_g, \alpha_g, \omega_i)p(\omega_i) \right) \eta_g^{n_g} \mathcal{N}(\mu_g|0, \Sigma_g) \mathcal{W}^{-1}(\Sigma_g|q, Q)p(\alpha_g)\frac{v_g}{(v_g + k)^3}$$

$$(10.5)$$

where $p(\omega_i) = \prod_{j=1}^{d} \mathcal{IG}\left(\omega_{ij}|\frac{v_g}{2}, \frac{v_g}{2}\right)$ for (10.3), $p(\omega_i) = \mathcal{IG}\left(\omega_i|\frac{v_g}{2}, \frac{v_g}{2}\right)$ for (10.4) and $n_g = \sum_{i=1}^{n} \mathbb{I}(z_i = g)$ is the number of individuals allocated to cluster $g$. As usual for mixture models, sampling from (10.5) has the advantage of being conceptually simpler than drawing from $p(\theta, \eta \mid \mathbf{y})$ directly.

The Gibbs steps to sample $\mathbf{z}$ and $\eta$ are straightforward and outlined below. For the remaining parameters previous skew-$t$ algorithms combined Gibbs draws for $(v, \boldsymbol{\omega})$ with MH steps for $\mu$, $\Sigma$ and $\alpha$ (Ferreira & Steel, 2007a,b).

To avoid practical difficulties in setting the MH proposal, which can be particularly challenging for $\Sigma$, we propose a novel algorithm with MH steps for $\mu$, $\Sigma$ and $\alpha$ automatically calibrated to attain high acceptance rates. Algorithm 10.1 obtains $L$ draws from our proposed MH-within-Gibbs algorithm, starting from an arbitrary initial value $(\mu^{(0)}, \Sigma^{(0)}, \alpha^{(0)}, v^{(0)}, \eta^{(0)})$. In our examples, the initialization was based on the $k$-medians algorithm (Section 10.5). For simplicity, $A_g$ and $D_g$ denote the eigenvectors and eigenvalues of $\Sigma_g$ evaluated at its current value, and likewise $x_i = D_g^{-1/2}A_g(y_i - \mu_g)$, $\xi_{gij} = \frac{\mathbb{I}(x_{ij} \geq 0)}{(1-\alpha_{gj})^2} + \frac{\mathbb{I}(x_{ij} < 0)}{(1+\alpha_{gj})^2}$ and the $\Omega_i$, defined after (10.4), are evaluated at the current $(\mu_g, \Sigma_g, \alpha_g, \boldsymbol{\omega})$.

Several remarks are in order. To improve the mixing of the Markov chain, steps 1 and 2 of Algorithm 10.1 sample from the conditionals of $\mathbf{z}$ and $\eta$ after marginalizing $\boldsymbol{\omega}$, and step 6 samples from the joint of $(\omega, v)$ given all other parameters and $\mathbf{y}$. Steps 3 and 4 reduce to Gibbs updates in the particular case $\alpha_g = (0, \ldots, 0)$. Specifically, in step 3 the proposal parameters are

$$V_g^{-1} = \frac{1}{t}(\Sigma_g^{(l-1)})^{-1} + \sum_{i:z_i=g} V_{gi}^{-1},$$

$$V_{gi}^{-1} = A_g^{\top} D_g^{-1/2} \Omega_i^{-1} \Xi_{gi} D_g^{-1/2} A_g,$$

$$m_g = V_g \sum_{i:z_i=g} V_{gi}^{-1} y_i. \tag{10.6}$$

Note that one may write

$$V_g^{-1} = A_g^{\top} D_g^{-1/2} \left( \frac{1}{t}I + \sum_{i:z_i=g} \Omega_i^{-1} \Xi_{gi} \right) D_g^{-1/2} A_g,$$

which only requires recomputing the inner summation, and $m_g = V_g A_g^{\top} D_g^{-1/2} (\Phi_g^{\top} \circ D_g^{-1/2} A_g \mathbf{y}) \mathbf{1}_n$ recomputing $\Phi_{gij} = \xi_{gij}/\omega_{ij}$, where $\circ$ is the entrywise product and $\mathbf{1}_n$ the

---

**Algorithm 10.1** MH-within-Gibbs simulation of the skew-$t$ mixture parameters

---

Iterate the following steps for $l = 1, \ldots, L$:

1 Sample from $p(\mathbf{z} \mid \mathbf{y}, \theta, \eta)$. For $i = 1, \ldots, n$, set $z_i^{(l)} = g$ with probability proportional to

$$\eta_g^{(l-1)} f_g\left(y_i \mid \mu_g^{(l-1)}, \Sigma_g^{(l-1)}, \alpha_g^{(l-1)}, v_g^{(l-1)}\right).$$

2 Sample from $\eta^{(l)}$ from $p(\eta \mid \mathbf{y}, \mathbf{z}, \theta) \sim \mathcal{D}(e_0 + n_1, \ldots, e_0 + n_G)$ where $n_g = \sum_{i=1}^{n} \mathbb{I}(z_i^{(l)} = g)$.

3 Sample from $p(\mu \mid \mathbf{y}, \mathbf{z}, \Sigma, \alpha, \boldsymbol{\omega}, v)$. For $g = 1, \ldots, G$, propose $\mu_g^* \sim \mathcal{N}(m_g, V_g)$, where $V_g$, $V_{gi}$ and $m_g$ are defined in (10.6). Recall that $\Xi_{gi}$ depends on $\mu_g^{(l-1)}$, let $\Xi_{gi}^*$ be its counterpart for $\mu_g^*$ and define $V_{gi}^*$, $V_g^*$ and $m_g^*$ as in (10.7). Set $\mu_g^{(l)} = \mu_g^{(l-1)}$ with probability $1 - \min\{1, u\}$ and else set $\mu_g^{(l)} = \mu_g^*$ where $u$ is given by

$$\frac{\phi(\mu_g^{(l-1)} \mid m_g^*, V_g^*) \exp\left\{-\frac{1}{2}\left(\frac{1}{t}(\mu_g^*)^\top \Sigma_g^{-1} \mu_g^* + \sum_{i:z_i=g}(y_i - \mu_g^*)^\top (V_{gi}^*)^{-1}(y_i - \mu_g^*)\right)\right\}}{\phi(\mu_g^* \mid m_g, V_g) \exp\left\{-\frac{1}{2}\left(\frac{1}{t}(\mu_g^{(l-1)})^\top \Sigma_g^{-1} \mu_g^{(l-1)} + \sum_{i:z_i=g}(y_i - \mu_g^{(l-1)})^\top V_{gi}^{-1}(y_i - \mu_g^{(l-1)})\right)\right\}}.$$

4 Sample from $p(\Sigma \mid \mathbf{y}, \mathbf{z}, \mu, \alpha, \boldsymbol{\omega}, v)$. For $g = 1, \ldots, G$, let $\bar{y}_g = \frac{1}{n_g}\sum_{i:z_i=g} y_i$, consider the eigendecomposition of $\sum_{i:z_i=g}(y_i - \bar{y}_g)(y_i - \bar{y}_g)^\top = \widetilde{A}^\top D_y \widetilde{A}$ and let $\widetilde{x}_i = \widetilde{\Xi}_i^{1/2} \Omega_i^{-1/2} \widetilde{A}(y_i - \mu_g)$, where $\widetilde{\Xi}_i$ is defined according to the sign of the elements in $\widetilde{A}(y_i - \mu_g)$. Define $\widetilde{D} = \mathrm{Diag}\left(\sum_{i:z_i=g} \widetilde{x}_i \widetilde{x}_i^\top\right)$, let $S = Q + \mu_g^{(l)}(\mu_g^{(l)})^\top + (\widetilde{A})^\top \widetilde{D}\widetilde{A}$, propose $\Sigma_g^* \sim \mathcal{W}^{-1}(q + d + n_g, S)$ and let $\Sigma_g^* = (A_g^*)^\top D_g^* A_g^*$ be its eigendecomposition. Set $\Sigma_g^{(l)} = \Sigma_g^{(l-1)}$ with probability $1 - \min\{1, u\}$ and else set $\Sigma_g^{(l)} = \Sigma_g^*$, where $u$ is given by

$$\frac{\left|\Sigma_g^{(l-1)}\right|^{\frac{n_g+q+2d+1}{2}} e^{\frac{1}{2}\left(l(A_g, D_g, \mu_g^{(l)}) + \mathrm{tr}(Q(\Sigma_g^{(l)})^{-1})\right)}}{\left|\Sigma_g^*\right|^{\frac{n_g+q+2d+1}{2}} e^{\frac{1}{2}\left(l(A_g^*, D_g^*, \mu_g^{(l)}) + \mathrm{tr}(Q(\Sigma_g^*)^{-1})\right)}} \frac{\mathcal{W}^{-1}\left(\Sigma_g^{(l-1)} \mid q + n_g + d, S\right)}{\mathcal{W}^{-1}\left(\Sigma_g^* \mid q + n_g + d, S\right)},$$

where $l(A, D, \mu) = \frac{1}{t}\mu^\top A^\top D^{-1} A \mu + \sum_{i:z_i=g}(y_i - \mu)^\top A^\top D^{-1/2} \Omega_i^{-1} \Xi_i D^{-1/2} A(y_i - \mu)$.

5 Sample from $p(\alpha \mid \mathbf{y}, \mathbf{z}, \mu, \Sigma, \boldsymbol{\omega}, v)$. For $g = 1, \ldots, G$, $j = 1, \ldots, d$, propose $\alpha_{gj}^* \sim t_{\tilde{n}}(\tilde{m}, \tilde{v})\mathbb{I}(-1 < \alpha_{gj}^* < 1)$ where $\tilde{n} = \max\{n_g, 3\}$, and $(\tilde{m}, \tilde{v})$ are given in (10.9) and (10.10). Set $\alpha_{gj}^{(l)} = \alpha_{gj}^{(l-1)}$ with probability $1 - \min\{1, u\}$ and else set $\alpha_{gj}^{(l)} = \alpha_{gj}^*$, where

$$u = \frac{h(\alpha_{gj}^*; s_1, s_2)(1 + \alpha_{gj}^*)^{a-1}(1 - \alpha_{gj}^*)^{b-1} t_{\tilde{n}}(\alpha_{gj}^{(l-1)}; \tilde{m}, \tilde{v})}{h(\alpha_{gj}^{(l-1)}; s_1, s_2)(1 + \alpha_{gj}^{(l-1)})^{a-1}(1 - \alpha_{gj}^{(l-1)})^{b-1} t_{\tilde{n}}(\alpha_{gj}^*; \tilde{m}, \tilde{v})},$$

and $h(\alpha_{gj}; s_1, s_2)$ is defined in (10.8).

6 Sample from $p(v, \boldsymbol{\omega} \mid \mathbf{y}, \mathbf{z}, \mu, \Sigma, \alpha)$. Set $v_g^{(l)} = u$ with probability proportional to

$$\frac{u}{(u+k)^3} \prod_{i:z_i=g} f_g(y_i \mid \mu_g, \Sigma_g, \alpha_g, v_g = u),$$

where $f_g(\cdot)$ is either the iskew-$t$ (10.3) or the dskew-$t$ (10.4). Then for (10.3) draw $\omega_{ij}^{(l)} \mid z_i^{(l)} = g \sim \mathcal{IG}\left(\frac{v_g^{(l)}+1}{2}, \frac{v_g^{(l)}+x_{ij}^2 \xi_{gij}}{2}\right)$; for (10.4) draw $\omega_i^{(l)} \mid z_i^{(l)} = g \sim \mathcal{IG}\left(\frac{v_g^{(l)}+d}{2}, \frac{v_g^{(l)}+\sum_{j=1}^{d} x_{ij}^2 \xi_{gij}}{2}\right)$.

---

$n \times 1$ unit vector. Analogously the reverse proposal parameters are

$$(V_g^*)^{-1} = \frac{1}{t}(\Sigma_g^{(l-1)})^{-1} + \sum_{i:z_i=g} (V_{gi}^*)^{-1},$$

$$(V_{gi}^*)^{-1} = A_g^\top D_g^{-1/2} \Omega_i^{-1} \Xi_{gi}^* D_g^{-1/2} A_g,$$

$$m_g^* = V_g^* \sum_{i:z_i=g} (V_{gi}^*)^{-1} y_i. \tag{10.7}$$

In fact, if $V_{gi}$ did not depend on $\mu_g$ via $\Xi_{gi}$ we would also obtain a Gibbs step. Because $\xi_{gij}$ depends on $\mu_g$ only through the sign of $x_{ij}$, moderate changes in $\mu_g$ often result in $V_{gi} = V_{gi}^*$, which in turns leads to a cancellation of terms in the acceptance probability $u$ in step 3 and thus a high acceptance rate. In step 4 the intuition is that $\widetilde{A}$ captures the correlation in $y_i$, which allows approximating $x_i$ by $\tilde{x}_i$ and estimating $D_g$ by $\widetilde{D}$ (conditional on the current $(\mathbf{z}, \mu_g, \alpha_g, \boldsymbol{\omega})$).

Regarding Step 5, the Student $t_{\tilde{n}}(\tilde{m}, \tilde{v})$ proposal is based on a second-order Taylor approximation to the likelihood of $\alpha$ given all other parameters (which factors across $g$ and $j$). This proposal has thicker tails than the target's exponential tails, which we empirically observed to lead to good mixing of the Markov chain. Specifically, define

$$h(\alpha; s_1, s_2) = \exp\left\{ -\frac{1}{2}\left( \frac{s_1}{(1-\alpha)^2} + \frac{s_2}{(1+\alpha)^2} \right) \right\}, \tag{10.8}$$

where

$$s_1 = \sum_{i:z_i=g} \frac{x_{ij}^2}{\omega_{ij}^{(l-1)}} \mathbb{I}(x_{ij} \geq 0), \quad s_2 = \sum_{i:z_i=g} \frac{x_{ij}^2}{\omega_{ij}^{(l-1)}} \mathbb{I}(x_{ij} < 0)$$

for the iskew-$t$ and replacing $\omega_{ij} = \omega_i$ for the dskew-$t$. Then the proposal mean $\tilde{m} = \text{argmax}_\alpha h(\alpha; s_1, s_2)$ is a real root of

$$(s_1 + s_2)\alpha^3 + 3(s_1 - s_2)\alpha^2 + 3(s_1 + s_2)\alpha + s_1 - s_2 = 0 \tag{10.9}$$

and

$$\tilde{v}^{-1} = \frac{3s_2}{(1+\tilde{m})^4} + \frac{3s_1}{(1-\tilde{m})^4} + \frac{a-1}{(1+\tilde{m})^2} + \frac{b-1}{(1-\tilde{m})^2}. \tag{10.10}$$

Finally, as usual with mixture models (Chapter 1), the posterior $p(\theta, \eta \mid \mathbf{y})$ is invariant to permutations of the component indices. This non-identifiability issue can complicate the interpretation of the output of Algorithm 10.1; for instance, under the presence of label switching the index $g$ associated with an underlying component may change from one iteration to the next. The two main strategies to address this issue are imposing artificial identifiability constraints or relabelling the component indices, either during or after running the MCMC (Frühwirth-Schnatter, 2001; Jasra et al., 2005).

The issue with identifiability constraints is that they need not improve interpretability; in our experience they often reduce interpretability when the mixture components are not well separated in the direction defined by the constraint. Instead, we used the equivalence class representation (ECR; Papastamoulis & Iliopoulos, 2010) implemented in the R package label.switching (Papastamoulis, 2015). Briefly, after running Algorithm 10.1, the ECR permutes the group labels to make the latent cluster allocations $\mathbf{z}^{(l)}$ as similar as possible to a pivot value (taken to be $\mathbf{z}^{(L)}$). As an alternative, our R package twopiece also implements identifiability constraints based on the projection of cluster means on the

**TABLE 10.1**

Proportion of correct classifications (CC) and posterior medians for various prior settings in the $\mu_2^\top = (1, -2)$ simulation. VW: Villa–Walker $p(v_g)$

|  | CC | $\mu_1$ | $\mu_2$ | $\alpha_1$ | $\alpha_2$ | $v_1$ | $v_2$ |
|---|---|---|---|---|---|---|---|
| Default | 0.93 | 0.19, −0.02 | 0.84, −2.05 | 0.12, 0.18 | −0.52, −0.26 | 19 | 7 |
| $t = 0.1$ | 0.87 | 0.23, −0.04 | 0.55, −1.49 | 0.13, 0.37 | −0.48, −0.43 | 19 | 11 |
| $t = 0.5$ | 0.93 | 0.19, −0.04 | 0.82, −2.02 | 0.11, 0.17 | −0.52, −0.31 | 19 | 8 |
| $t = 2$ | 0.93 | 0.17, 0.01 | 0.84, −2.12 | 0.09, 0.13 | −0.54, −0.31 | 19 | 7 |
| $t = 10$ | 0.93 | 0.19, −0.04 | 0.85, −2.10 | 0.11, 0.17 | −0.53, −0.27 | 19 | 7 |
| $k = 1$ | 0.94 | 0.19, −0.04 | 0.84, −2.06 | 0.11, 0.16 | −0.52, −0.30 | 19 | 7 |
| $k = 5$ | 0.93 | 0.21, −0.04 | 0.83, −2.06 | 0.12, 0.19 | −0.52, −0.27 | 20 | 8 |
| VW | 0.93 | 0.19, −0.04 | 0.83, −2.05 | 0.11, 0.18 | −0.52, −0.29 | 23 | 8 |
| $a = b = 1$ | 0.94 | 0.11, 0.03 | 0.80, −2.10 | 0.01, 0.11 | −0.56, −0.42 | 17 | 8 |
| $a = b = 2$ | 0.93 | 0.19, −0.03 | 0.85, −2.06 | 0.11, 0.17 | −0.51, −0.27 | 19 | 7 |

first principal component and a recent proposal based on defining a loss function that seeks a relabelling that minimizes the changes in cluster means across iterations (Rodriguez & Walker (2014), also implemented in `label.switching`). We found the ECR to exhibit the most satisfactory performance among these three methods in our examples and to be computationally fast.

## 10.5 Examples

For all examples below we used our R package twopiece implementing Algorithm 10.1. To initialize $(\mathbf{z}^{(0)}, \mu^{(0)}, \Sigma^{(0)})$ we used the $k$-medians (Kaufman & Rousseeuw, 1987) cluster allocations and parameter estimates (R package `flexclust`, function cclust; Leisch, 2006), set $\eta^{(0)}$ to the proportion of observations in each cluster and $\alpha_{gj}^{(0)} = 1$, $\omega_{ij}^{(0)} = 1$, $\nu_g = 30$ for all $g, i, j$. Trace plots and multiple independent runs (not shown) suggested that $L = 10{,}000$ iterations with a 1000 burn-in were sufficient to obtain stable estimates of parameters and clustering probabilities.

### 10.5.1 Simulation study

Our basic rationale for considering skew-$t$ mixtures is that in the presence of asymmetries or heavy tails the performance of normal mixtures may suffer. We therefore aim to gain intuition as to when this indeed becomes a practically relevant issue.

We first focus on the ideal case where the number of components $G$ is known. In our experience, parameter estimates based on normal mixtures are biased (particularly for $\mu$), but as long as the clusters are moderately well separated, the fitted model usually remains effective in terms of clustering. However, as the overlap between clusters increases, normal-based mixtures may fail to identify the underlying structure. As an example, we drew $n = 1000$ bivariate observations from a well-separated mixture with two iskew-$t$ components and $\eta = (2/3, 1/3)$, $\mu_1^\top = (0, 0)$, $\mu_2^\top = (1, -2)$, $\alpha_1^\top = (0, 0)$, $\alpha_2^\top = (-0.5, -0.5)$, $\nu_1 = 100$,

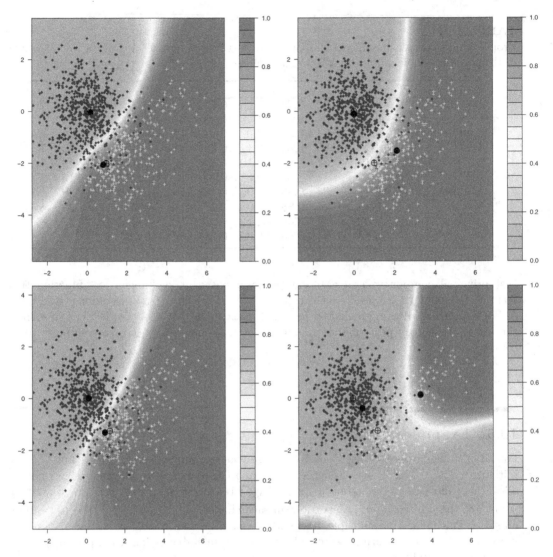

**FIGURE 10.3**
Posterior probability of cluster 2 in simulation with $\mu_2^\top = (1, -2)$ (top) and $\mu_2^\top = (1.25, -1.25)$ (bottom) based on skew-$t$ (left) and normal (right) mixtures. Solid circles: estimated cluster means. Cross-filled circles: true cluster means. Small dots (crosses) show observations truly from cluster 1 (cluster 2).

$\nu_2 = 10$ and

$$\Sigma_1 = \begin{pmatrix} 1 & 0 \\ 0 & 1 \end{pmatrix}, \quad \Sigma_2 = \begin{pmatrix} 1 & 0.5 \\ 0.5 & 1 \end{pmatrix}.$$

Figure 10.3 shows the simulated data and the estimated posterior probability $P(z_i = 2 \mid y_i)$ of belonging to cluster 2 under an iskew-$t$ (upper left) fitted with Algorithm 10.1 and a normal mixture (upper right) fitted by maximum likelihood estimation (Fraley & Raftery, 2003). Although there were minor differences in the equal probability $P(z_i = 2 \mid y_i) = 0.5$ boundary and the maximum likelihood estimate $\hat{\mu}_2$ is slightly shifted, when assigning each observation to the most probable cluster we obtained 93.3% correct classifications both for skew-$t$ and normal mixtures. The proportion of accepted Metropolis–Hastings moves was as high as 0.978 and 0.999 for $\mu$ and $\alpha$, respectively. The automatically calibrated proposals performed satisfactorily also for $\Sigma$ with an acceptance rate of 0.343.

These results were obtained with the default prior settings in Section 10.3, which, as discussed, were set to rather uninformative values that aim to help regularize inference while not dominating the information contained in the data. While these defaults performed well in our examples, we nevertheless conducted an informal sensitivity analysis by comparing the results with those under different prior parameters. Specifically, we started by repeating the analyses under $t = 0.1, 0.5, 2, 10$ in $p(\mu_g \mid \Sigma_g)$ while keeping the remaining parameters constant. Table 10.1 shows the proportion of correctly classified observations (CCs) and posterior means for $(\mu, \alpha, v)$. As $t$ decreased, $p(\mu)$ and consequently $p(\mu \mid \mathbf{y})$ became more concentrated around 0, resulting in somewhat reduced CCs for really small $t$. The shrinking effect is more noticeable for cluster 2, indicating that a small $t$ value may unduly bias estimates in small clusters, (e.g. $t = 0.1$ corresponds to $p(\mu_g)$ being as informative as 10 observations from component $g$). Next, we assessed sensitivity to $p(v_g)$ by changing the default $k = 2.78$ to $k = 1, 5$ and also by setting $p(v_g)$ to the Villa–Walker proposal (Figure 10.2). Although these are moderately large changes in $p(v_g)$ (for $k > 2.84$, $p(v_g)$ is no longer monotone decreasing), posterior inference was fairly robust (Table 10.1). Finally, the results also remained robust to $p(\alpha_g)$ when changing the default $a = b = 2$ to $a = b = 1$ (uniform) and $a = b = 4$.

The bottom panels in Figure 10.3 show the analogous results when generating observations from a less well-separated mixture ($\mu_2^\top = (1.25, -1.25)$ and remaining parameters as in the earlier simulation). The skew-$t$ mixture (bottom left) identified the underlying clusters whereas the normal mixture (bottom right) returned a shifted estimate $\hat{\mu}_2^\top = (3.42, 0.13)$ and inflated estimates of Diag($\Sigma_1$), Diag($\Sigma_2$) and $\hat{\eta}_1 = 0.893$. CC was 90.6% for the skew-$t$ and 72.3% for the normal mixture. Trace plots (not shown) again exhibited a a similar behaviour of Algorithm 10.1 as before, with acceptance rates of 0.977, 0.374 and 0.999 for $\mu$, $\Sigma$ and $\alpha$, respectively.

Next, we suppose that in most practical situations the number of components $G$ is not known and needs to be inferred from the data. In this situation, specifying flexible $f_g(\cdot)$ gains importance, else spurious components are likely to be added. A likelihood- or posterior-based fitted mixture inherently aims to approximate the underlying data-generating pdf rather than the number of clusters; for example, a single non-normal cluster may be captured by introducing several normal components. As a result one may need to merge overlapping normal components in a post-processing step to obtain interpretable clusters (Baudry et al., 2010). As another token when data arise from a two-component skew-normal model, Miller & Dunson (2015) found that the highest posterior probability value of $G$ increases with $n$. To illustrate this issue in our simulated example with $\mu_2^\top = (1, -2)$ we computed the Bayesian information criterion (BIC) for normal mixture models with $G = 1, \ldots, 5$ components, which favoured the three-component model (BIC was 6940.2, 6913.3 and 6937.0 for $G = 2, 3, 4$, respectively). The fitted $G = 3$ normal model (Figure 10.4, middle right) shows that cluster 1 truly centred at $\mu_1^\top = (0, 0)$ was correctly identified, while cluster 2 was split into two components, one of which captured cluster 2 (albeit underestimating its weight) and the other being spread across the two clusters. The fact that the latter has a large estimated weight ($\hat{\eta}_3 = 0.225$) and is clearly distinct from the other two components is problematic, as it hints at an incorrect underlying clustering structure. Similar issues are observed when fitting $G = 4$ normal components (Figure 10.4 (bottom right)). In contrast, the clustering structure inferred from skew-$t$ mixtures was more robust to the chosen $G$ (Figure 10.4, left). For instance, for $G = 3$, component 3 receives a low weight (E($\eta_3 \mid \mathbf{y}$) = 0.041) and strongly overlaps with component 2, jointly capturing the underlying cluster 1. Similar

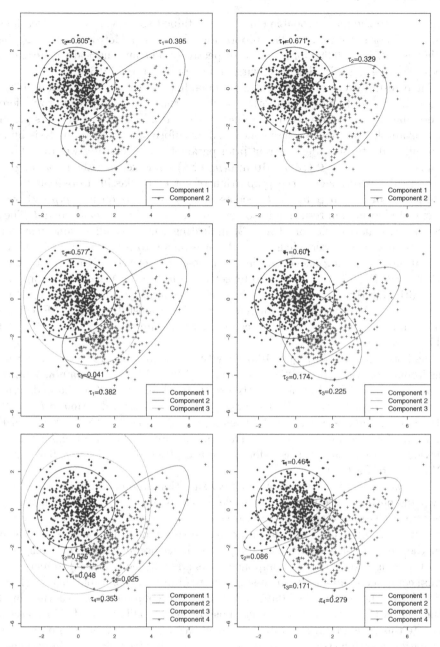

**FIGURE 10.4**
Estimated 90% contours under skew-$t$ (left) and normal (right) when setting $G = 2, 3, 4$ (top, middle, bottom) in $\mu_2^\top = (1, -2)$ simulated data.

results were observed for $G = 4$, where cluster 1 was recovered by components 1–3 (note that $\mathrm{E}(\eta_1 \mid \mathbf{y}) = 0.048$, $\mathrm{E}(\eta_2 \mid \mathbf{y}) = 0.575$, $\mathrm{E}(\eta_3 \mid \mathbf{y}) = 0.025$) and cluster 2 by component 4.

We note that it is possible to extend our framework either by considering repulsive mixtures (Petralia et al., 2012), that is, introducing a prior penalty for poorly separated compo-

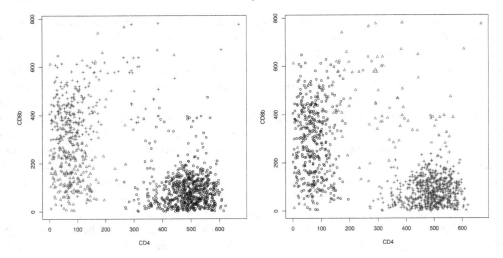

**FIGURE 10.5**
(CD4,CD8b) measurements for $n = 1126$ CD3+ cells. Symbols indicate the most likely
fitted skew-$t$ (left) and normal (right) component.

nents to enhance interpretability and help discard spurious components, or by conducting
formal Bayesian model selection on $G$. For the latter our view is that the usage of non-local
prior distributions (Johnson & Rossell, 2010), which in this setting are strongly connected
to repulsive mixtures, should help obtain sparse and interpretable solutions as we recently
showed for normal mixtures (Fuquene et al., 2016). While exciting, these extensions require a
careful treatment that is beyond the scope of this chapter and is currently the topic of future
research. (See Chapter 7 for an entry to the general issue of inferring about $G$.) As a practi-
cal alternative we defined the criterion $B_G = -2\mathrm{E}\left(\log(p(\mathbf{y} \mid \theta, \eta, v)) \mid \mathbf{y}\right) + \nu_G \log(n)$ where
the expectation is under a model with $G$ components and $\nu_G = 2dG + 0.5d(d+1)G + 2G - 1$
is the number of parameters. As usual, the first term in $B_G$ measures goodness of fit and the
second penalizes model complexity. On our simulated data we found that $B_G$ was 7006.3,
6094.2, 6966.4 and 7027.7 for $G = 1, 2, 3, 4$ respectively, correctly suggesting that $G = 2$
skew-$t$ components suffice.

### 10.5.2 Experimental data

We apply skew-$t$ mixtures to the graft-versus-host flow cytometry data of Brinkman et al.
(2007). Briefly, a flow cytometer is similar to a microscope but allows screening thousands
of cells simultaneously to detect the presence of certain biomarkers in each cell. This data
set has $d = 4$ continuous variables measuring four biomarkers named CD3, CD4, CD8b and
CD8, and was analysed by Baudry et al. (2010) and provided as part of their R package
`Rmixmodcombi`. The goal of the study was to identify cell subpopulations which were positive
for CD3, CD4 and CD8b (CD3+/CD4+/CD8b+), that is, presented high levels for the first
three variables. To this end, the authors created a control sample specifically designed not to
contain any CD3+/CD4+/CD8b+ cells. The idea was to use this control sample to identify
subpopulations other than CD3+/CD4+/CD8b+ characterizing the background signal in
the experiment.

We followed the analysis of Baudry et al. (2010) and selected the $n = 1126$ cells in
their four CD3+ subpopulations (obtained by thresholding the CD3 marker above 280 – for

discussion, see Brinkman et al., 2007; Baudry et al., 2010). Following the above reasoning, these should not contain a subpopulation of CD4+/CD8b+ cells. Both BIC and $B_G$ (defined in Section 10.5.1) selected $G = 3$ components for the normal and iskew-$t$ mixture, respectively. Analogously to (Baudry et al., 2010, Figure 13), Figure 10.5 shows the (CD4,CD8b) values for the 1126 cells, where the plotting symbols indicate the most probable component for each cell. For instance, a cell located in the bottom left corner presents low CD4 and CD8b, indicating it was likely CD4−/CD8b−. Interestingly, the subvectors with estimated (CD4,CD8b) location parameters under a skew-$t$ fit were (518.1, 40.5), (32.8, 434.3) and (46.0, 91.1) for components $1, 2, 3$, respectively. The interpretation is that these components are CD4+/CD8b−, CD4−/CD8b+ and CD4−/CD8b− subpopulations, that is, none of them is CD4+/CD8b+ (Figure 10.5, left). In contrast the location parameters under a normal fit were (490.0,80.5), (75.6,295.6) and (295.6,366.3). Here the first component corresponds to CD4+/CD8b− as before, the second to a mixture of CD4−/CD8b− and CD4−/CD8b+ and the third to CD4+/CD8b+, which is not supposed to be present (Figure 10.5, right). The plot highlights an interesting difference between skew-$t$ and normal-based clustering. The former accommodates outlying observations via asymmetric or heavy-tailed components, whereas the latter tends to do so by introducing new components. Naturally, which strategy ends up being preferable will depend on the application at hand, but in this example the skew-$t$ interpretation seems to align considerably better with the biological experiment.

## 10.6   Concluding Remarks

Model-based clustering can often be adjusted to account for deviations from normality in a manner that is conceptually straightforward and retains both computational tractability and ease of interpretation. For instance, our illustrations are based upon a simple extension of the $d$-dimensional multivariate normal and Student $t$ distribution. The generalization introduces $d$ new parameters directly interpretable in terms of the Arnold–Groeneveld measure of skewness, and allows adapting existing computational strategies. Although not explored here, the same construction can be used to generalize other multivariate distributions. For instance, a family of skewed scale mixture of normals is immediately generated by varying the mixing distribution and, following Ferreira & Steel (2007a,b), we can choose other skewed univariate distributions in (10.3).

In our examples we focused on situations where poor separation between components, skewness or heavy tails can create particular difficulties for normal-based mixtures, but we emphasize that in many situations normality remains a sensible, useful choice. As might be expected, our results suggest that in these situations skew-$t$ mixtures possess better robustness properties. One interesting topic for future research is the selection of $G$, where, particularly for clustering, one is really interested in minimally well-separated components. In this sense it will be interesting to investigate the addition of repulsive forces between components, to encourage solutions that are simultaneously robust and parsimonious.

# Bibliography

ARELLANO-VALLE, R., GÓMEZ, H. & QUINTANA, F. (2005). Statistical inference for a general class of asymmetric distributions. *Journal of Statistical Planning and Inference* **128**, 427–443.

ARNOLD, B. & BEAVER, R. (2002). Skewed multivariate models related to hidden truncation and/or selective reporting (with discussion). *Test* **11**, 7–54.

ARNOLD, B. C. & GROENEVELD, R. A. (1995). Measuring skewness with respect to the mode. *American Statistician* **49**, 34–38.

AZZALINI, A. & DALLA VALLE, A. (1996). The multivariate skew-normal distribution. *Biometrika* **8**, 715–726.

BAUDRY, J., RAFTERY, A., CELEUX, G., LO, K. & GOTTARDO, R. (2010). Combining mixture components for clustering. *Journal of Computational and Graphical Statistics* **19**, 332–353.

BRINKMAN, R. R., GASPARETTO, M., LEE, S.-J., RIBICKAS, A., PERKINS, J. et al. (2007). High-content flow cytometry and temporal data analysis for defining a cellular signature of graft-versus-host disease. *Biology of Blood and Marrow Transplantation* **13**, 691–700.

CABRAL, C., BOLFARINE, H. & PEREIRA, J. (2008). Bayesian density estimation using skew student-t-normal mixtures. *Computational Statistics & Data Analysis* **52**, 5075–5090.

CHAMBAZ, A. & ROUSSEAU, J. (2008). Bounds for Bayesian order identification with application to mixtures. *Annals of Statistics* **36**, 928–962.

FERNÁNDEZ, C. & STEEL, M. (2008). On Bayesian modeling of fat tails and skewness. *Journal of the American Statistical Association* **93**, 359–371.

FERREIRA, J. & STEEL, M. (2007a). Model comparison of coordinate-free multivariate skewed distributions with an application to stochastic frontiers. *Journal of Econometrics* **137**, 641–673.

FERREIRA, J. & STEEL, M. (2007b). A new class of skewed multivariate distributions with applications to regression analysis. *Statistica Sinica* **17**, 505–529.

FONSECA, T., FERREIRA, M. & MIGON, H. (2008). Objective Bayesian analysis for the Student-t regression model. *Biometrika* **95**, 325–333.

FRALEY, C. & RAFTERY, A. (2003). Enhanced software for model-based clustering, density estimation, and discriminant analysis: MCLUST. *Journal of Classification* **20**, 263–286.

FRANCZAK, B., BROWNE, R. & MCNICHOLAS, P. (2014). Mixtures of shifted asymmetric Laplace distributions. *IEEE Transactions in Pattern Analysis and Machine Intelligence* **36**, 1149–1157.

FRÜHWIRTH-SCHNATTER, S. (2001). Markov chain Monte Carlo estimation of classical and dynamic switching and mixture models. *Journal of the American Statistical Association* **96**, 194–209.

FRÜHWIRTH-SCHNATTER, S. (2004). Estimating marginal likelihoods for mixture and Markov switching models using bridge sampling techniques. *Econometrics Journal* **7**, 143–167.

FRÜHWIRTH-SCHNATTER, S. & PYNE, S. (2010). Bayesian inference for finite mixtures of univariate and multivariate skew-normal and skew-t distributions. *Biostatistics* **11**, 317–336.

FUQUENE, J. A., STEEL, M. F. J. & ROSSELL, D. (2016). On choosing mixture components via non-local priors. Preprint, arXiv:1604.00314.

JASRA, A., HOLMES, C. C. & STEPHENS, D. A. (2005). Markov chain Monte Carlo methods and the label switching problem in Bayesian mixture modeling. *Statistical Science* **20**, 50–67.

JOHNSON, V. & ROSSELL, D. (2010). On the use of non-local prior densities for default Bayesian hypothesis tests. *Journal of the Royal Statistical Society, Series B* **72**, 143–170.

JONES, M. & ANAYA-IZQUIERDO, K. (2011). On parameter orthogonality in symmetric and skew models. *Journal of Statistical Planning and Inference* **141**, 758–770.

JUÁREZ, M. & STEEL, M. (2010). Model-based clustering of non-Gaussian panel data based on skew-t distributions. *Journal of Business and Economic Statistics* **28**, 52–66.

KARLIS, D. & SANTOURIAN, A. (2009). Model-based clustering with non-elliptically contoured distributions. *Statistics and Computing* **19**, 73–83.

KAUFMAN, L. & ROUSSEEUW, P. J. (1987). Clustering by means of medoids. In *Statistical Data Analysis Based on the $L_1$-Norm and Related Methods*, Y. Dodge, ed. Amsterdam: North-Holland, pp. 405–416.

KOSMIDIS, I. & KARLIS, D. (2016). Model-based clustering using copulas with applications. *Statistics and Computing* **26**, 1079–1099.

LEE, S. & MCLACHLAN, G. (2014). Finite mixtures of multivariate skew t-distributions: Some recent and new results. *Statistics and Computing* **24**, 181–202.

LEISCH, F. (2006). A toolbox for k-centroids cluster analysis. *Computational Statistics & Data Analysis* **51**, 526–544.

LIN, T. I. (2009). Maximum likelihood estimation for multivariate skew normal mixture models. *Journal of Multivariate Analysis* **100**, 257–265.

LIN, T. I., HO, H. J. & LEE, C. R. (2014). Flexible mixture modelling using the multivariate skew-t-normal distribution. *Statistics and Computing* **24**, 531–546.

MALSINER-WALLI, G., FRÜHWIRTH-SCHNATTER, S. & GRÜN, B. (2016). Model-based clustering based on sparse finite Gaussian mixtures. *Statistics and Computing* **26**, 303–324.

MALSINER-WALLI, G., FRÜHWIRTH-SCHNATTER, S. & GRÜN, B. (2017). Identifying mixtures of mixtures using Bayesian estimation. *Journal of Computational and Graphical Statistics* **26**, 285–295.

MILLER, J. & DUNSON, D. (2015). Robust Bayesian inference via coarsening. Preprint, arXiv:1506.06101.

MUDHOLKAR, G. & HUTSON, A. (2000). The epsilon-skew-normal distribution for analyzing near-normal data. *Journal of Statistical Planning and Inference* **83**, 291–309.

PAPASTAMOULIS, P. (2015). *label.switching: Relabelling MCMC Outputs of Mixture Models*. R package version 1.4.

PAPASTAMOULIS, P. & ILIOPOULOS, G. (2010). An artificial allocations based solution to the label switching problem in Bayesian analysis of mixtures of distributions. *Journal of Computational and Graphical Statistics* **19**, 313–331.

PETRALIA, F., RAO, V. & DUNSON, D. B. (2012). Repulsive mixtures. In *Advances in Neural Information Processing Systems 25*, F. C. N. Pereira, C. J. C. Burges, L. Bottou & K. Q. Weinberger, eds. Red Hook, NY: Curran Associates, Inc., pp. 1889–1897.

REDNER, R. & WALKER, H. (1984). Mixture densities, maximum likelihood and the EM algorithm. *SIAM Review* **24**, 195–239.

RODRIGUEZ, C. & WALKER, S. (2014). Label switching in Bayesian mixture models: Deterministic relabeling strategies. *Journal of Computational and Graphical Statistics* **23**, 25–45.

RUBIO, F. J. & STEEL, M. F. J. (2014). Inference in two-piece location-scale models with Jeffreys priors (with discussion). *Bayesian Analysis* **9**, 1–22.

RUBIO, F. J. & STEEL, M. F. J. (2015). Bayesian modelling of skewnewess and kurtosis with two-piece scale and shape distributions. *Electronic Journal of Statistics* **9**, 1884–1912.

SAHU, S., DEY, D. & BRANCO, M. (2003). A new class of multivariate skew distributions with applications to Bayesian regression models. *Canadian Journal of Statistics* **31**, 129–150.

VILLA, C. & WALKER, S. (2014). Objective prior for the number of degrees of freedom of a t distribution. *Bayesian Analysis* **9**, 197–200.

# 11

## Mixture Modelling of High-Dimensional Data

**Damien McParland and Thomas Brendan Murphy**

*University College Dublin, Ireland*

## CONTENTS

High-dimensional data arise in a wide range of application areas, including medicine, biology, genetics and food science. The advent of inexpensive sensors and measurement instruments that can simultaneously measure many features for a given observation has led to a proliferation of high-dimensional data.

    This chapter focuses on the finite mixture models that have been used in high-dimensional settings and on the approaches that have been used for fitting and inference for these models. A number of mixture models that have been proposed for high-dimensional data are reviewed, with particular attention paid to mixture models for high-dimensional continuous, categorical and mixed data. Software for fitting mixture models in a high-dimensional setting is also discussed, with particular attention given to R (R Core Team, 2017) packages.

## 11.1  Introduction

In recent years there has been a growth in the availability of high-dimensional data sets in a wide range of subject areas, including biomedicine, food science and social sciences. Modeling such data offers many challenges due to complex dependencies between the observed variables, low sample sizes and heterogeneous data sources.

Mixture models are an important tool in the modeling of high-dimensional data because they can account for heterogeneous data sources and uncover subgroups. Mixture models are a core tool in the development of model-based clustering and classification methods for high-dimensional data.

In Section 11.2 motivating data examples are introduced as examples of high-dimensional continuous, categorical and mixed data. Some issues involved in modeling high-dimensional data are discussed in Section 11.3, including some properties of high-dimensional spaces and the number of parameters in high-dimensional data models. Models for high-dimensional continuous data (Section 11.4), categorical data (Section 11.5) and mixed data (Section 11.6) are reviewed. Some variable selection approaches for mixture modeling are discussed in Section 11.7. The motivating examples are investigated using mixture models in Section 11.8, and it is shown that mixture models are able to appropriately model substructures in these data. The chapter concludes, in Section 11.9, with a summary.

Two important high-dimensional mixture modeling areas which are not covered in this chapter are non-normal mixtures and mixtures for high-dimensional count data. Recently, much work has been done developing non-normal mixture models for high-dimensional continuous data. McNicholas (2016) and Lee & McLachlan (2013) provide detailed reviews of these advances; see also Chapter 10 of this volume for a detailed discussion. High-dimensional count data are becoming increasingly common in bioinformatics and event monitoring applications. Mixture modeling for high-dimensional count data has recently received attention, including in Rau et al. (2015) and White & Murphy (2016); see Chapter 8 of this volume for a comprehensive review.

## 11.2  High-Dimensional Data

The mixture models reviewed in this chapter are demonstrated using a number of data examples. The Italian wine data (Forina et al., 1986) are introduced in Section 11.2.1 and used as an example of high-dimensional continuous data. A multivariate categorical data set of clinical indicators of back pain (Smart et al., 2011) is introduced in Section 11.2.2 as an example of high-dimensional categorical data. A multivariate data set of mixed type from prostate cancer patients (Byar & Green, 1980) is introduced in Section 11.2.3 and used as an example of multivariate mixed data.

**TABLE 11.1**

A cross-tabulation of region versus year for the wine samples in the Forina et al. (1986) wine data

|  | 70 | 71 | 72 | 73 | 74 | 75 | 76 | 78 | 79 |
|---|---|---|---|---|---|---|---|---|---|
| Barbera |  |  |  |  | 9 |  | 5 | 29 | 5 |
| Grignolino | 9 | 9 | 7 | 9 | 16 | 9 | 12 |  |  |
| Barolo |  | 19 |  | 20 | 20 |  |  |  |  |

**TABLE 11.2**

The 27 variables included in the wine data

| Variables | | |
|---|---|---|
| Alcohol | Potassium | Color intensity |
| Sugar-free extract | Calcium | Hue |
| Fixed acidity | Magnesium | OD280/OD315 of diluted wines |
| Tartaric acid | Phosphate | OD280/OD315 of flavanoids |
| Malic acid | Chloride | Glycerol |
| Uronic acids | Total phenols | 2,3-butanediol |
| pH | Flavanoids | Total nitrogen |
| Ash | Non-flavanoid phenols | Proline |
| Alcalinity of ash | Proanthocyanins | Methanol |

### 11.2.1 Continuous data: Italian wine

Forina et al. (1986) collected a number of physical and chemical measurements from wine samples from three regions in Italy (Barbera, Grignolino and Barolo). The wine samples were taken from nine different years and a cross-tabulation of region versus year is given in Table 11.1. The data consists of $n = 178$ samples where $d = 27$ continuous measurements are recorded in each observation; the names of the variables recorded are shown in Table 11.2. These data are available in the pgmm R package (McNicholas et al., 2015).

Fitting a mixture model to these data allows for investigating if wines from the same region cluster together, whether the year of the wine also leads to clustering structure, or whether the wines form groups of another form.

### 11.2.2 Categorical data: lower back pain

Smart et al. (2011) present the results of a study into the diagnosis of lower back pain. The aim of the original study was to identify which clinical indicators distinguish between three established types of back pain (nociceptive, peripheral neuropathic and central neuropathic). The data from the study consist of $n = 425$ observations and $d = 36$ binary clinical indicators ($1 = $ present, $0 = $ absent); a description of the 36 variables is available in Smart et al. (2011, Table 1).

While the established types of back pain are considered reasonably distinct, there may exist subgroups within the pain types and some patients may have more than one type of pain. Thus, modeling the data using a mixture model will help establish if the clinical indicators suggest the existence of subgroups within the population, whether the subgroups correspond to the established pain types, and whether the established pain types have subgroups within them.

**TABLE 11.3**
The 12 variables in the Byar & Green (1980) prostate cancer data. For variable type, C
denotes continuous, O denotes ordinal and N denotes nominal

| Variable | Type | Levels |
|---|---|---|
| Age | C | |
| Weight | C | |
| Systolic blood pressure | C | |
| Diastolic blood pressure | C | |
| Serum hemoglobin | C | |
| Size of primary tumor | C | |
| Index of tumor stage and histologic grade | C | |
| Serum prostatic acid phosphatase | C | |
| Performance rating | O | 4 |
| Cardiovascular disease history | O | 2 |
| Bone metastasis | O | 2 |
| Electrocardiogram code | N | 3 |

### 11.2.3    Mixed data: prostate cancer

Byar & Green (1980) present $d = 12$ mixed type measurements for $n = 475$ prostate cancer
patients who were diagnosed as having either stage 3 or 4 prostate cancer; eight of the
variables are continuous, three are ordinal, and one is nominal (Table 11.3). These data are
available in the clustMD R package (McParland & Gormley, 2017).

Fitting a mixture model to these data will help establish if the stage 3 and stage 4
prostate cancer patients are distinct subgroups and whether there are other subgroups
within the population.

## 11.3    Mixtures for High-Dimensional Data

Let $\mathbf{y} = (y_1, y_2, \ldots, y_n)$ be an independently and identically distributed sample of size $n$
from some population that can be modeled using $p(y_i)$. The finite mixture model can be
expressed as

$$p(y_i) = \sum_{g=1}^{G} \eta_g f_g(y_i | \theta_g),$$

where $\eta_g$ is the mixing proportion for group $g$ and $f_g(y_i | \theta_g)$ is the component density for
group $g$. We typically assume that each component density is from the same parametric
family of distributions, so we can write

$$p(y_i) = \sum_{g=1}^{G} \eta_g f(y_i | \theta_g).$$

The most commonly used mixture model for high-dimensional continuous data is the normal or Gaussian mixture model (Section 11.4); other mixture models are used for high-
dimensional discrete data (Section 11.5) and high-dimensional mixed data (Section 11.6).

Yeung et al. (2001) review a number of approaches for fitting mixture models to high-
dimensional gene expression data, with particular attention being paid to the use of the

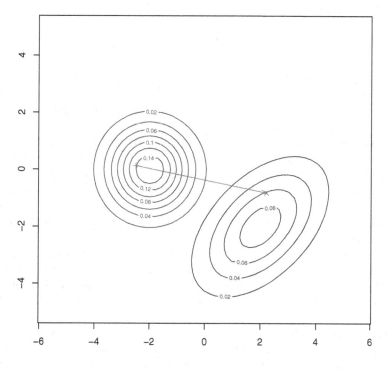

**FIGURE 11.1**

An illustration of the distance between two points from two normal distributions in two dimensions. The contours of each normal distribution and a randomly sampled point from each normal are shown.

mixture models for clustering. Further, comprehensive reviews of mixture models with clustering applications has been provided by McLachlan & Basford (1988), McLachlan & Peel (2000), Bouveyron & Brunet-Saumard (2014) and McNicholas (2016); see also Chapter 8 above for a comprehensive review.

## 11.3.1 Curse of dimensionality/modeling issues

An important feature of high-dimensional data is that the data points become sparsely distributed as the data dimensionality increases. To illustrate this point, consider two independent random vectors in $\mathbb{R}^d$, where $X \sim \mathcal{N}(\mu_x, \Sigma_x)$ and $Y \sim \mathcal{N}(\mu_y, \Sigma_y)$ (see Figure 11.1). The expected squared Euclidean distance between $X$ and $Y$ is

$$\mathrm{E}(||X - Y||^2) = ||\mu_x - \mu_y||^2 + \mathrm{tr}(\Sigma_x + \Sigma_y),$$

and this grows linearly with the dimensionality $d$. Thus, we expect data points to be sparsely distributed in high-dimensional spaces.

Further properties of high-dimensional spaces are reviewed in Zimek et al. (2012), with particular reference to the problem of outlier detection in high-dimensional settings. A result of Beyer et al. (1999) quoted therein, which further exemplifies the sparsity of data points in high-dimensional settings, is that

$$\lim_{d \to \infty} \frac{D_{\max} - D_{\min}}{D_{\min}} = 0,$$

where $D_{\max}$ is the function that gives the distance from a particular point to the furthest data point and $D_{\min}$ is the function that gives the distance of a particular point to the closest data point. Thus, the difference in the nearest and furthest point distance functions diminishes in high-dimensional settings.

An important feature that arises when modeling high-dimensional data is that the number of model parameters typically grows with the data dimension ($d$), and often at a faster than linear rate. For example, consider the multivariate normal distribution with mean $\mu$ and covariance $\Sigma$. The mean parameter, $\mu$, is a vector of length $d$ and the covariance is a symmetric matrix which has $d(d+1)/2$ parameters. Thus, the number of parameters in the model grows quadratically in the data dimension. This is an example of the so-called *curse of dimensionality* (Bellman, 1957).

A consequence of the large number of parameters required to model high-dimensional data is that the number of observations required to fit a high-dimensional model can be very large. However, many high-dimensional data settings are characterized by having small sample size $n$ and high dimensionality $d$; this is the so-called *large d, small n* problem (West, 2003).

A number of solutions have been proposed to resolve the large $d$, small $n$ problem, and these fall into two categories. The first approach is to use dimension reduction to reduce the data dimensionality before or during the modeling process and thus effectively reduce the value of $d$. A second approach is to use parsimonious models which reduce the burden of the number of parameters required for data modeling. Both of these approaches are considered in the subsequent sections of this chapter.

### 11.3.2    Data reduction

One approach to modeling high-dimensional data using mixture models is to incorporate data reduction to reduce the data dimension. The main approaches for data reduction either use variable selection (or screening) or a projection of the data into a lower-dimensional subspace.

Kohavi & John (1997) provide a detailed comparison of two approaches to data reduction in the context of supervised classification. Data reduction can be implemented as a pre-processing step before model fitting, which is an approach called *filtering*. In contrast, the data reduction can be implemented as part of the model fitting algorithm, and this is called the *wrapper* approach. While filtering approaches are computationally efficient and easy to implement, there is much evidence to suggest that wrapper approaches achieve more accurate results. Building upon the work of Kohavi & John (1997), Dy & Brodley (2004) considered feature selection in the context of unsupervised clustering and similarly concluded that wrapper approaches are preferred. More recently, Steinley & Brusco (2008a,b) compared and contrasted a number of approaches for variable selection in a clustering context, with particular reference to psychometric applications.

Chang (1983) demonstrated the difficulty with the filtering approach to data reduction using principal component analysis. Chang provided a theoretical argument why the leading principal components may not contain clustering information and provided a simulation example to demonstrate this. In particular, he simulated 300 observations from a mixture of two normal distributions in $\mathbb{R}^{15}$ with $\eta = (0.8, 0.2)$, where the parameters $\mu_1$ and $\mu_2$ are given by

$$\mu_1 = -\mu_2 = \tag{11.1}$$
$$(0.45, 0.425, 0.40, 0.375, 0.35, 0.325, 0.30, 0.275, 0.25, 0.225, 0.20, 0.175, 0.15, 0.125, 0.10)^\top,$$

and the covariances $\Sigma_1$ and $\Sigma_2$ are selected as:

$$\Sigma_1 = \Sigma_2 =$$

$$\begin{pmatrix}
1.00 & -0.11 & -0.11 & -0.11 & -0.11 & -0.11 & -0.11 & -0.11 & 0.06 & 0.06 & 0.06 & 0.06 & 0.06 & 0.06 & 0.06 \\
-0.11 & 1.00 & -0.11 & -0.11 & -0.11 & -0.11 & -0.11 & -0.11 & 0.06 & 0.06 & 0.06 & 0.06 & 0.06 & 0.06 & 0.06 \\
-0.11 & -0.11 & 1.00 & -0.11 & -0.11 & -0.11 & -0.11 & -0.11 & 0.06 & 0.06 & 0.06 & 0.06 & 0.06 & 0.06 & 0.06 \\
-0.11 & -0.11 & -0.11 & 1.00 & -0.11 & -0.11 & -0.11 & -0.11 & 0.06 & 0.06 & 0.06 & 0.06 & 0.06 & 0.06 & 0.06 \\
-0.11 & -0.11 & -0.11 & -0.11 & 1.00 & -0.11 & -0.11 & -0.11 & 0.06 & 0.06 & 0.06 & 0.06 & 0.06 & 0.06 & 0.06 \\
-0.11 & -0.11 & -0.11 & -0.11 & -0.11 & 1.00 & -0.11 & -0.11 & 0.06 & 0.06 & 0.06 & 0.06 & 0.06 & 0.06 & 0.06 \\
-0.11 & -0.11 & -0.11 & -0.11 & -0.11 & -0.11 & 1.00 & -0.11 & 0.06 & 0.06 & 0.06 & 0.06 & 0.06 & 0.06 & 0.06 \\
-0.11 & -0.11 & -0.11 & -0.11 & -0.11 & -0.11 & -0.11 & 1.00 & 0.06 & 0.06 & 0.06 & 0.06 & 0.06 & 0.06 & 0.06 \\
0.06 & 0.06 & 0.06 & 0.06 & 0.06 & 0.06 & 0.06 & 0.06 & 1.00 & -0.03 & -0.03 & -0.03 & -0.03 & -0.03 & -0.03 \\
0.06 & 0.06 & 0.06 & 0.06 & 0.06 & 0.06 & 0.06 & 0.06 & -0.03 & 1.00 & -0.03 & -0.03 & -0.03 & -0.03 & -0.03 \\
0.06 & 0.06 & 0.06 & 0.06 & 0.06 & 0.06 & 0.06 & 0.06 & -0.03 & -0.03 & 1.00 & -0.03 & -0.03 & -0.03 & -0.03 \\
0.06 & 0.06 & 0.06 & 0.06 & 0.06 & 0.06 & 0.06 & 0.06 & -0.03 & -0.03 & -0.03 & 1.00 & -0.03 & -0.03 & -0.03 \\
0.06 & 0.06 & 0.06 & 0.06 & 0.06 & 0.06 & 0.06 & 0.06 & -0.03 & -0.03 & -0.03 & -0.03 & 1.00 & -0.03 & -0.03 \\
0.06 & 0.06 & 0.06 & 0.06 & 0.06 & 0.06 & 0.06 & 0.06 & -0.03 & -0.03 & -0.03 & -0.03 & -0.03 & 1.00 & -0.03 \\
0.06 & 0.06 & 0.06 & 0.06 & 0.06 & 0.06 & 0.06 & 0.06 & -0.03 & -0.03 & -0.03 & -0.03 & -0.03 & -0.03 & 1.00
\end{pmatrix}.$$

In the resulting simulation, the best separation between the data from the two mixture components was achieved by examining the first and 15th principal component scores; Figure 11.2 shows scatter plots of the first versus second principal component scores and the first versus 15th principal component scores from a reproduction of his simulation. It is clear that the two observations from the two mixture components exhibit greater separation when the first and 15th principal component scores are plotted.

In Section 11.7 we review a number of wrapper-based approaches for feature selection which avoid the problems with filtering approaches to data reduction in the context of mixture modeling.

## 11.4 Mixtures for Continuous Data

The most common approach for mixture modeling of continuous data is to use the normal mixture model. The normal mixture model assumes that component $g$ has a normal density function with mean $\mu_g$ and covariance $\Sigma_g$, that is,

$$p(y_i) = \sum_{g=1}^{G} \eta_g \phi(y_i | \mu_g, \Sigma_g).$$

Estimating the normal mixture model with maximum likelihood, using the EM algorithm (Dempster et al., 1977), is straightforward, and Bayesian estimation using Markov chain Monte Carlo (MCMC) is also possible (Diebolt & Robert, 1994; Frühwirth-Schnatter, 2006).

Ferguson (1983) showed that any density on $\mathbb{R}^d$ can be approximated to arbitrary accuracy using a normal mixture model, which illustrates its flexibility as a model for density estimation. The normal mixture model is the most commonly used mixture for model-based data clustering.

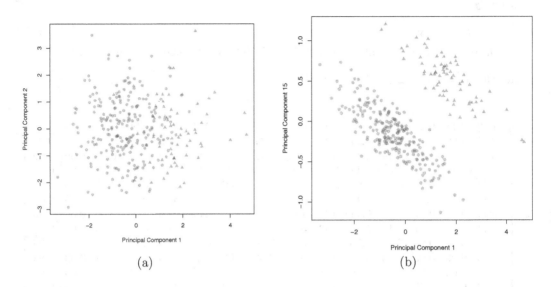

**FIGURE 11.2**
A scatter plot of the principal component scores from a reproduction of the Chang (1983) simulation: (a) scatter plot of the second versus first principal components; (b) scatter plot of the 15th versus first principal components. The points from the two different components are distinguished using different plot symbols.

The normal mixture model has $(G-1)+Gd+Gd(d+1)/2$ parameters when no constraints are put on the model parameters. Thus, the model has a large number of parameters when $d$ is large. In particular, the number of covariance parameters is $Gd(d+1)/2$, which grows quadratically in $d$. Further, evaluating the normal component densities in a normal mixture model involves inverting a $d \times d$ covariance matrix for each of the component densities; when $d$ is large this calculation is computationally intensive and requires that the matrix is of full rank. These issues limit the applicability of the general normal mixture model in high-dimensional settings, but a number of special cases of the normal mixture with parsimonious covariance structure have been proposed to reduce the number of covariance parameters and thus facilitate applying normal mixtures in high-dimensional settings. A number of these parsimonious models are reviewed in this section.

### 11.4.1   Diagonal covariance

A special case of the normal mixture model assumes that the covariance matrix $\Sigma_g$ is diagonal, That is,

$$\Sigma_g = \mathrm{Diag}(\sigma_{g1}^2, \sigma_{g2}^2, \ldots, \sigma_{gd}^2).$$

This assumption implies that the variables within each normal mixture component are independent and the number of covariance parameters reduces to $Gd$ instead of $Gd(d+1)/2$; being diagonal, the covariance matrices are straightforward to invert. Further reduction in the number of parameters can be achieved by further constraining the diagonal covariance (see Hartigan, 1975, Chapter 5);[1] examples of these constrained diagonal covariances are given in Table 11.4, where $I_d$ is the $d \times d$ identity matrix. Note that evaluating the component

---

[1] In fact, Hartigan (1975, Chapter 5) also considers letting $\Sigma_g = \Sigma$ for $g = 1, 2, \ldots, G$ as a further approach to reduce the number of parameters in a normal mixture model.

**TABLE 11.4**

Diagonal covariance structures for the normal mixture model component densities for high-dimensional data

| Covariance | Description | Parameters |
|---|---|---|
| $\Sigma_g = \mathrm{Diag}(\sigma_{g1}^2, \sigma_{g2}^2, \ldots, \sigma_{gd}^2)$ | Diagonal | $Gd$ |
| $\Sigma_g = \mathrm{Diag}(\sigma_1^2, \sigma_2^2, \ldots, \sigma_d^2)$ | Diagonal and equal | $d$ |
| $\Sigma_g = \sigma_g^2 \times I_d$ | Isotropic | $G$ |
| $\Sigma_g = \sigma^2 \times I_d$ | Isotropic and equal | $1$ |

densities for a normal mixture with diagonal covariance is simplified, because inverting a diagonal covariance matrix is straightforward.

An important connection can be made between the isotropic and equal covariance model (Table 11.4) and the $k$-means clustering method (Hartigan & Wong, 1978). Suppose we consider a special case of the normal mixture model where each group has the same probability, and an isotropic and equal covariance is assumed, that is,

$$p(y_i) = \sum_{g=1}^{G} \frac{1}{G} \phi(y_i | \mu_g, \sigma^2 I_d).$$

If this model is fitted using the CEM algorithm (Celeux & Govaert, 1992), then the resulting algorithm is exactly the same as the $k$-means algorithm. Thus, the $k$-means algorithm implicitly assumes a model structure of this type. This connection gives an important insight into when the $k$-means algorithm works well and when it works poorly for clustering high-dimensional data.

## 11.4.2 Eigendecomposed covariance

Banfield & Raftery (1993) and Celeux & Govaert (1995) proposed an alternative strategy for reducing the number of covariance parameters in the normal mixture model. Their approach utilizes a modified eigendecomposition of the component distribution covariance of the form

$$\Sigma_g = \lambda_g D_g A_g D_g^\top,$$

where $\lambda_g > 0$ is a scalar, $D_g$ is an orthogonal $d \times d$ matrix, and $A_g$ is a $d$-dimensional diagonal matrix with positive diagonal entries and $\det(A_g) = 1$. The components of this decomposition can be interpreted as $\lambda_g$ controlling the volume of the contours of the component density, $D_g$ controlling the orientation of the contours and $A_g$ controlling the shape of the contours.

The eigendecomposition can be used to reduce the number of parameters in the model by forming a family of models where the components of the decomposition can be constrained to be equal or variable across components. Further, there is the possibility of allowing $D_g = I_d$ or $A_g = I_d$ because these conform to the structure of the decomposition. Considering all possible constraints on the eigenvalue decomposition gives a family of 14 models which are outlined in Table 11.5. These models are easily fitted using the mclust (Scrucca et al., 2016) and Rmixmod (Lebret et al., 2015) R packages.

It is worth noting that the VVV model is the normal mixture model with unconstrained covariance and that the VVI, EEI, VII, EII models are equivalent to those given in Table 11.4. The six models whose mclust name finishes with an "I" have diagonal covariance and thus the number of covariance parameters is linear in the data dimension; however, the number of covariance parameters in the remaining eight models is still quadratic in the

**TABLE 11.5**
The family of models with parsimonious eigendecomposed covariance as implemented in mclust and Rmixmod

| mclust | Rmixmod | Volume | Shape | Orientation | Covariance parameters |
|--------|---------|--------|-------|-------------|----------------------|
| EII | L_I | Equal | Identity | Identity | $1$ |
| VII | Lk_I | Variable | Identity | Identity | $G$ |
| EEI | L_b | Equal | Equal | Identity | $d$ |
| VEI | Lk_B | Variable | Equal | Identity | $G + (d-1)$ |
| EVI | L_Bk | Equal | Variable | Identity | $1 + G(d-1)$ |
| VVI | Lk_Bk | Variable | Variable | Identity | $Gd$ |
| EEE | L_C | Equal | Equal | Equal | $d(d+1)/2$ |
| EVE | L_D_Ak_D | Equal | Variable | Equal | $1 + d(d-1)/2 + G(d-1)$ |
| VEE | Lk_C | Variable | Equal | Equal | $G + d(d-1)/2 + (d-1)$ |
| VVE | Lk_D_Ak_D | Variable | Variable | Equal | $G + d(d-1)/2 + G(d-1)$ |
| EEV | L_Dk_A_Dk | Equal | Equal | Variable | $1 + Gd(d-1)/2 + (d-1)$ |
| VEV | Lk_Dk_A_Dk | Variable | Equal | Variable | $G + Gd(d-1)/2 + (d-1)$ |
| EVV | L_Ck | Equal | Variable | Variable | $1 + Gd(d-1)/2 + G(d-1)$ |
| VVV | Lk_Ck | Variable | Variable | Variable | $Gd(d+1)/2$ |

data dimension. The Rmixmod framework also allows for possibly constraining the mixing proportions to be identical (i.e. $\eta_g = 1/G$, for $g = 1, 2, \ldots, G$), yielding a model class with 28 models; for these models, the nomenclature shown in Table 11.5 can be preceded with a p_k or p depending on whether the mixing proportions are assumed to be unequal or equal, respectively.

In summary, the parsimonious eigendecomposed covariance structures reduce the number of parameters in the normal mixture model, which increases the applicability of the models in higher-dimensional settings. However, when the data dimensionality gets very large, only the six diagonal models are directly applicable unless the sample size is very large. Using a Bayesian approach with carefully designed priors, the normal mixture model can be applied in high dimensions, as exemplified in Chapter 5, Section 5.5.

### 11.4.3 Mixtures of factor analyzers and probabilistic principal components analyzers

The mixture of factor analyzers (MFA) model (Ghahramani & Hinton, 1997; McLachlan et al., 2003) assumes that data within each component can be modeled using a factor analysis model with group-specific loading and uniqueness parameters. Thus, the model assumes that each component has a covariance matrix of the form

$$\Sigma_g = \Lambda_g \Lambda_g^\top + \Psi_g, \tag{11.2}$$

where $\Lambda_g$ is a $d \times q$ loading matrix and $\Psi_g$ is a $d \times d$ diagonal uniqueness matrix. Thus, each component covariance matrix has $dq + d$ parameters. However, the loadings matrix needs to be further constrained to ensure identifiability, which reduces the number of covariance parameters to $dq - q(q-1)/2 + d$ parameters. This identifiability issue arises because if $\Lambda_g$ is replaced by $\Lambda_g R_g$, where $R_g$ is an orthogonal matrix, then the model is unchanged. Thus, the MFA for each $q$ has a total of $(G-1) + Gd + G(dq - q(q-1)/2 + d)$ parameters, and this is linear in the data dimension.

The Woodbury (1950) identity (11.3) can be used to find the inverse of the component covariance in an efficient manner because it only involves inverting $q \times q$ matrices and $d \times d$ diagonal matrices. In the context of the factor analysis covariance structure, the Woodbury identity is of the form

$$\Sigma_g^{-1} = (\Lambda_g \Lambda_g^\top + \Psi_g)^{-1} = \Psi_g^{-1} + \Psi_g^{-1} \Lambda_g (I_q + \Lambda_g^\top \Psi_g^{-1} \Lambda_g)^{-1} \Lambda_g^\top \Psi_g^{-1}, \qquad (11.3)$$

and the determinant of the component covariance is of the form

$$|\Sigma_g| = |\Lambda_g \Lambda_g^\top + \Psi_g| = |\Psi_g|/|I_q - \Lambda_g^\top (\Lambda_g \Lambda_g^\top + \Psi_g)^{-1} \Lambda_g|.$$

Tipping & Bishop (1999) developed the *mixture of probabilistic principal components analyzers* model which corresponds to the MFA model with $\Psi_g = \psi_g I_d$ which is an isotropic covariance structure.

McNicholas & Murphy (2008) and McNicholas et al. (2010) developed a family of models with parsimonious Gaussian mixture models (PGMMs) by constraining components of the covariance matrix (11.2) to be equal or unequal across groups. Further, they allowed the $\Psi_g$ matrix to be constrained to isotropic or not in their model family. The list of models in the PGMM family is detailed in Table 11.6. A further four models were added to the family in McNicholas & Murphy (2010) by constraining the uniqueness matrix further in a manner equivalent to mclust. The pgmm R package (McNicholas et al., 2015) provides routines fitting models in this family of normal mixture models.

Nyamundanda et al. (2010a) developed extensions of the MFA models to incorporate covariates, using the mixture of experts modeling framework; an R package called Metabol-Analyse is available for fitting this model family (Nyamundanda et al., 2010b).

McLachlan et al. (2011) provide a recent overview of models that utilize the MFA framework.

### 11.4.4 High-dimensional models

Bouveyron et al. (2007) proposed an alternative method for reducing the number of parameters in the component covariance in order to model high-dimensional data using normal mixtures. In this approach, the spectral decomposition of the covariance is constrained to take the form

$$\Sigma_g = D_g^\top \Delta_g D_g$$

, where $D_g$ is a matrix of eigenvectors and $\Delta_g$ is assumed to have the form

$$\Delta_g = \begin{pmatrix} a_{g1} & 0 & \cdots & 0 & & & & & \\ 0 & a_{g2} & \cdots & 0 & & & & & \\ \vdots & \vdots & \ddots & \vdots & & & & & \\ 0 & 0 & \cdots & a_{gd_g} & & & & & \\ & & & & b_g & 0 & 0 & \cdots & 0 \\ & & & & 0 & b_g & 0 & \cdots & 0 \\ & & & & 0 & 0 & b_g & \cdots & 0 \\ & & & & \vdots & \vdots & \vdots & \ddots & \vdots \\ & & & & 0 & 0 & 0 & \cdots & b_g \end{pmatrix},$$

in which $a_{gj} > b_g$, for all $j = 1, 2, \ldots, d_g$, and the off-diagonal blocks of entries are all zero. The number of parameters used for all of the covariance is

$$\sum_{g=1}^{G} d_g \left( d - \frac{d_g + 1}{2} \right) + 2G + \sum_{g=1}^{G} d_g,$$

and this is a linear function in the data dimension. Similarly to the mclust and Rmixmod

**TABLE 11.6**
The models in the PGMM family of normal mixture models

| pgmm | Loading | Uniqueness | Isotropic | Covariance parameters |
|------|---------|------------|-----------|----------------------|
| CCC | Constrained | Constrained | Constrained | $(dq - q(q-1)/2) + 1$ |
| CCU | Constrained | Constrained | Unconstrained | $(dq - q(q-1)/2) + d$ |
| CUC | Constrained | Unconstrained | Constrained | $(dq - q(q-1)/2) + G$ |
| CUU | Constrained | Unconstrained | Unconstrained | $(dq - q(q-1)/2) + Gd$ |
| UCC | Unconstrained | Constrained | Constrained | $G(dq - q(q-1)/2) + 1$ |
| UCU | Unconstrained | Constrained | Unconstrained | $G(dq - q(q-1)/2) + d$ |
| UUC | Unconstrained | Unconstrained | Constrained | $G(dq - q(q-1)/2) + G$ |
| UUU | Unconstrained | Unconstrained | Unconstrained | $G(dq - q(q-1)/2) + Gd$ |

models (Section 11.4.2) and the `pgmm` models (Section 11.4.3), further constraints can be put on components of the covariance decomposition to be equal or unequal across groups. This leads to a family of 14 models that are available in the `HDclassif` R package (Bergé et al., 2012) which can be used to fit this family models to high-dimensional data; the nomenclature used to summarize these models is given in Bouveyron et al. (2007, Table 1).

### 11.4.5  Sparse models

An alternative approach to fitting normal mixture models to high-dimensional data is to find a sparse model fit where the number of parameters is reduced from $(G-1)+Gd+Gd(d+1)/2$ to a smaller number.

Pan & Shen (2007) developed a framework for fitting normal mixtures with common diagonal covariance (Section 11.4.1) but where penalized maximum likelihood estimation is used and the penalty on the means is of the form

$$\lambda_1 \sum_{g=1}^{G} \sum_{j=1}^{d} |\mu_{gj}|,$$

where $\lambda_1 > 0$. This penalty shrinks the mean parameter estimates to the overall data mean if the overall data has been standardized to have mean zero and variance one. Thus, the mean parameters of variables that take similar values across groups are shrunken together to a common value and the number of parameters is reduced in the model. Bhattacharya & McNicholas (2014) developed an approach using an alternative penalty that accounts for the component weights, where the penalty is of the form

$$\lambda_1 \sum_{g=1}^{G} \eta_g \sum_{j=1}^{d} |\mu_{gj}|.$$

The method was applied to the PGMM family of models (Table 11.6), and properties of the resulting solutions were investigated.

Xie et al. (2008a) extended this approach to models with unequal diagonal component covariance matrices by also penalizing the variance parameters using penalties of the form

$$\lambda_1 \sum_{g=1}^{G} \sum_{j=1}^{d} |\mu_{gj}| + \lambda_2 \sum_{g=1}^{G} \sum_{j=1}^{d} |\sigma_{gj}^2 - 1|$$

and

$$\lambda_1 \sum_{g=1}^{G} \sum_{j=1}^{d} |\mu_{gj}| + \lambda_2 \sum_{g=1}^{G} \sum_{j=1}^{d} |\log \sigma_{gj}^2|,$$

where $\lambda_1 > 0$ and $\lambda_2 > 0$. Thus, the number of variance parameters is also reduced because some are shrunken to the value 1; Xie et al. (2008b) also considered alternative penalties on the mean and variance parameters.

Zhou et al. (2009) developed an important extension of these penalized approaches to account for a common unconstrained component covariance structure. The approach taken in this penalized approach is to consider a penalty of the form

$$\lambda_1 \sum_{g=1}^{G} \sum_{j=1}^{d} |\mu_{gj}| + \lambda_2 \sum_{j=1}^{d} \sum_{j'=1}^{d} |\Delta_{jj'}|,$$

where $\Delta = \Sigma^{-1}$ and $\Sigma = \Sigma_g$ for all $g = 1, \ldots, G$; thus $\Delta$ is the common component precision matrix. Thus, this model has the same penalty on the common component precision matrix as the graphical lasso (Friedman et al., 2008).

Xie et al. (2010) applied a similar penalization approach to the MFA model (Section 11.4.3) where they penalized the mean parameters and the terms in the loading matrix. In their approach the penalty has the form

$$\lambda_1 \sum_{g=1}^{G} \sum_{j=1}^{d} |\mu_{gj}| + \lambda_2 \sum_{g=1}^{G} \sum_{j=1}^{d} \sqrt{\sum_{j'=1}^{d} \Lambda_{gjj'}^2},$$

where $\Lambda_{gjj'}$ is the $(j, j')$th element of the $g$th loading matrix in the MFA model.

Galimberti et al. (2009) also considered penalties on the loading matrix in a special case of the MFA model with common loading matrix; their model also includes a lower-dimensional representation of the component means. Thus, this approach achieves a reduction of the data dimension and variable selection simultaneously. In this work, an approximation of the absolute value terms in the penalty (Fan & Li, 2001; Hunter & Li, 2005) is considered to improve computational efficiency of this approach.

In a Bayesian context, Yau & Holmes (2011) used a double exponential prior and Malsiner-Walli et al. (2016) used a normal-gamma prior to induce sparsity in the parameters of mixture models.

Sparse versions of $k$-means clustering have been explored in Witten & Tibshirani (2010) and Sun et al. (2012). These sparse $k$-means algorithms are similarly connected with fitting mixture models with common isotropic diagonal covariance as discussed in Section 11.4.1.

## 11.5 Mixtures for Categorical Data

Let $y_i = (y_{i1}, y_{i2}, \ldots, y_{id})$ be a vector of values where $y_{ij}$ takes a categorical value from the set $\{1, 2, \ldots, C_j\}$, $j = 1, 2, \ldots, d$. If the values are unordered, then the variables are nominal variables, whereas if the values are ordered, then the variables are ordinal variables. Bartholomew et al. (2011) provides an excellent review of latent variable models for categorical data. The latent class analysis model for categorical data is summarized in Section 11.5.1, and other models for high-dimensional categorical data are summarized in Section 11.5.2.

### 11.5.1 Local independence models and latent class analysis

The latent class analysis (LCA) model (Lazarsfeld & Henry, 1968; Goodman, 1974; Clogg, 1995) is the most commonly used mixture model for high-dimensional categorical data.

The model is developed around the concept of *local independence* between the variables within each component density, that is, the variables within the vector $y_i$ are modeled as independent conditional on the component membership being known. In LCA, the components are termed classes and the latent variable indicating component membership is called the *latent class* variable. Thus, in the LCA model we have a component density of the form

$$f(y_i|\theta_g) = \prod_{j=1}^{d}\prod_{c=1}^{C_j} \theta_{gjc}^{\mathbb{I}(y_{ij}=c)}.$$

Thus, the LCA model takes the form

$$p(y_i) = \sum_{g=1}^{G}\eta_g f(y_i|\theta_g) = \sum_{g=1}^{G}\eta_g \prod_{j=1}^{d}\prod_{c=1}^{C_j} \theta_{gjc}^{\mathbb{I}(y_{ij}=c)}.$$

It has a total of $(G-1) + G\sum_j(C_j-1)$ parameters and thus the number of parameters grows linearly in the data dimension. A necessary, but not sufficient, condition for the LCA model to be identifiable is

$$\prod_{j=1}^{d} C_j > G\left(\sum_{j=1}^{d} C_j - d + 1\right).$$

The model can be fitted in a straightforward manner using the EM algorithm (Bartholomew et al., 2011, Section 6.5) or using a Bayesian approach (White & Murphy, 2014; Pandolfi et al., 2014). A number of R packages are available for fitting LCA models, including poLCA (Linzer & Lewis, 2011) and BayesLCA (White & Murphy, 2014). The use of priors in the BayesLCA R package allows for regularization of the algorithm to avoid situations where probabilities are estimated to be zero or one.

### 11.5.2 Other models

Celeux & Govaert (1991) developed a parsimonious version of the LCA model by constraining the item parameters in a manner analogous to the parsimonious normal mixture models, discussed in Sections 11.4.2–11.4.4. Routines for fitting this family of models are provided in the Rmixmod R package (Lebret et al., 2015).

Many mixture models for multivariate categorical data have been proposed, with particular focus being put on relaxing the local independence assumption of the LCA model. The primary way to relax the local independence assumption is to use extra latent variables to induce dependence between variables. Rost (1990) developed a mixture of Rasch models for modeling multivariate binary data. Skrondal & Rabe-Hesketh (2004) and Vermunt (2007) developed general latent variable modeling frameworks that include mixture models for multivariate categorical data, where the mixture components use a latent Gaussian random variable to induce dependence between variables; these models are implemented in the GLLAMM (Rabe-Hesketh et al., 2004) and LatentGold (Vermunt & Magidson, 2013) software packages.

More recently, Cagnone & Viroli (2012) developed a latent variable model for multivariate binary data, where a continuous latent variable induces dependence between the binary variables and a normal mixture model is assumed for the continuous latent variable. Gollini & Murphy (2014) developed a mixture model for high-dimensional binary data where dependence between variables was accounted for using a latent trait analysis model structure; the model is called a mixture of latent trait analyzers model (MLTA) and it can be seen as a categorical analogue of the MFA model. They also developed a variational EM algorithm for efficient inference for this model; Tang et al. (2015) developed further extensions of this model.

McParland & Gormley (2013) developed a mixture of item response models for modeling multivariate ordinal categorical data. In this model, the ordinal outcomes are modeled using latent normal random variables, where the observed ordinal outcome depends on which of the latent normal random variables is in a particular interval.

## 11.6   Mixtures for Mixed Data

Suppose $y_i = (y_{i1}, y_{i2}, \ldots, y_{id_0}, y_{id_0+1}, \ldots, y_{id})$, where $y_{ij}$ takes a categorical value from the set $\{1, 2, \ldots, C_j\}$ for $j = 1, 2, \ldots, d_0$, and takes a continuous value for $j = d_0+1, d_0+2, \ldots, d$. In this case, the data are mixed, in that there are $d_0$ categorical variables and $d - d_0$ continuous variables.

Everitt (1984) proposed a latent class model for mixed data that assumes that the $d$ variables are conditionally independent given the group membership. In particular, each of the $d_0$ categorical variables has a multinomial distribution with group-specific parameters, and each of the continuous variables has a normal distribution with group-specific mean and variance, that is,

$$p(y_i) = \sum_{g=1}^{G} \eta_g \left[ \prod_{j=1}^{d_0} \prod_{c=1}^{C_j} \theta_{gjc}^{\mathbb{I}(y_{ij}=c)} \prod_{j=d_0+1}^{d} \phi\left(y_{ij} | \mu_{gj}, \sigma_{gj}^2\right) \right].$$

Hunt & Jorgensen (1999, 2003) developed a framework for mixture modeling of mixed data using the location model; Lawrence & Krzanowski (1996) developed a similar mixture model for mixed data. In this model, the $d_0$ categorical variables $(y_{i1}, y_{i2}, \ldots, y_{id_0})$ are collapsed into a single categorical variable $y_i^c$ that takes $S = \prod_{j=1}^{d_0} C_j$ levels which is modeled as a multinomial random variable; thus the model can account for arbitrary dependence between these variables. Further, the model assumes that the continuous variables have distribution $(y_{id_0+1}, y_{id_0+2}, \ldots, y_{id}) | y_i^c = s \sim \mathcal{N}(\mu_s, \Sigma)$.

Specifically, the location model assumes that $p(y_i) = \sum_{g=1}^{G} \eta_g f(y_i | \theta_g, \mu_g, \Sigma_g)$, where the mixture component is of the form

$$
\begin{aligned}
f(y_i | \theta_g, \mu_g, \Sigma_g) &= f(y_{i1}, y_{i2}, \ldots, y_{id_0}, y_{id_0+1}, \ldots, y_{id} | \theta_g, \mu_g, \Sigma_g) \\
&= f(y_{i1}, y_{i2}, \ldots, y_{id_0} | \theta_g) f(y_{id_0+1}, \ldots, y_{id} | y_{i1}, y_{i2}, \ldots, y_{id_0}, \mu_g, \Sigma_g) \\
&= f(y_i^c | \theta_g) f(y_{id_0+1}, \ldots, y_{id} | y_i^c, \mu_g, \Sigma_g) \\
&= \prod_{s=1}^{S} \theta_{gs}^{\mathbb{I}(y_{ij}^c=s)} \times \prod_{s=1}^{S} \phi\left(y_{id_0+1}, \ldots, y_{id} | \mu_{gs}, \Sigma_g\right)^{\mathbb{I}(y_{ij}^c=s)} \\
&= \prod_{s=1}^{S} \left[ \theta_{gs} \phi\left(y_{id_0+1}, \ldots, y_{id} | \mu_{gs}, \Sigma_g\right) \right]^{\mathbb{I}(y_{ij}^c=s)},
\end{aligned}
\tag{11.4}
$$

with $\theta_g = (\theta_{g1}, \ldots, \theta_{gS})$, $\mu_g$ and $\Sigma_g$ being component-specific parameters of the location model.

The structure of the location model makes it suitable for mixture modeling of mixed data when the number of categorical variables is relatively low because the number of model parameters depends heavily on the value of $S$; the location model is implemented in the MULTIMIX software (Hunt & Jorgensen, 1999).

More recently, a number of mixture models have been developed for mixed data. Browne & McNicholas (2012) developed a mixture model for mixed nominal and continuous random vectors. The model uses latent normal random variables within each component to induce dependence between variables in a manner analogous to the MLTA model (Section 11.5.2).

McParland et al. (2014, 2017) developed an alternative model for mixed nominal, ordinal, and continuous random vectors. They assumes that each observed ordinal categorical variable is modeled using an item response theory model with one latent normal random variable, each observed nominal categorical variable is a function of $C_j - 1$ latent normal random variables, and each observed continuous variable is a rescaled version of a latent normal random variable. Thus, in total the $d$-dimensional data vector is modeled as a manifestation of a $q = \sum_{j=1}^{d_0}(C_j - 1) + (d - d_0)$-dimensional latent random vector that follows a normal mixture model. In a manner analogous to pgmm (Section 11.4.3), a parsimonious factor-analytic covariance structure is assumed for the latent normal mixture model. Inference for the model is facilitated using a Metropolis-within-Gibbs MCMC algorithm.

McParland & Gormley (2016) develop an alternative model for mixed data that also assumes a latent normal structure analogous to the above, but where the latent variable mixture model has an eigendecomposed covariance structure similar to mclust (Section 11.4.2). The model is fitted using a Monte Carlo EM algorithm, and functions for fitting the model are available in the clustMD R package (McParland & Gormley, 2017).

## 11.7 Variable Selection

In Section 11.3.2 we discussed *filtering* methods that find combinations of variables that potentially contain information about subgroups in the data and thus are candidate methods for variable selection (or data reduction) prior to data modeling. The example from Chang (1983) which is described in (11.1) is one where filtering methods can fail (as illustrated in Figure 11.2). In this section, we review the *wrapper* approach which implements variable selection during the mixture modeling process rather than as a pre-processing step. In particular, we review wrapper-based variable selection methods for continuous data in Section 11.7.2 and for categorical data in Section 11.7.3.

### 11.7.1 Wrapper-based methods

Liu et al. (2003) proposed a normal mixture model that included a wrapper-based feature selection and variable transformation, in the context of analyzing high-dimensional data. Their approach involved assuming that the vector $y_i = (y_{i1}, y_{i2}, \dots, y_{id})$ can be partitioned into two parts where $d_0$ variables are modeled using a normal mixture model and the remaining $d - d_0$ variables are modeled using a single normal distribution. Suppose, without loss of generality, that the first $d_0$ variables are modeled using a normal mixture model and the remaining by the single normal distribution. This yields

$$p(y_i) = \sum_{g=1}^{G} \eta_g \phi\left(y_{i1}, \dots, y_{id_0} | \mu_g, \Sigma_g\right) \phi\left(y_{id_0+1}, \dots, y_{id} | \mu_0, \Sigma_0\right)$$

for the data model, and thus they assume independence between the clustering and non-clustering variables. Further, they assume that the identity of the clustering variables is

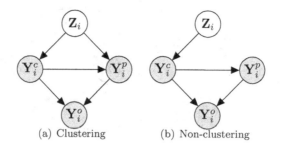

(a) Clustering      (b) Non-clustering

**FIGURE 11.3**
A graphical model representation of the two models being compared in the Raftery & Dean (2006) variable selection method. The observed values are shaded and the latent variable is unshaded.

unknown and can be estimated as part of a Bayesian model fitting procedure using MCMC; that is, the variable selection for clustering is implemented using a wrapper approach. Liu et al. (2003) apply their modeling approach to high-dimensional data, from image processing and gene expression studies, but where some data reduction has been implemented using principal component analysis prior to mixture modeling.

Tadesse et al. (2005) proposed a similar model for wrapper-based variable selection for mixture-based clustering; however, they marginalized the posterior distribution to find the posterior for the unknown number of components, component membership vector, and the indicator of which variables are modeled using a mixture or single normal distribution. An MCMC procedure was developed to sample from this posterior so that posterior estimates of these quantities could be found.

Law et al. (2004) pursued a similar approach but assumed conditional independence between all variables in the model, giving

$$p(y_i) = \sum_{g=1}^{G} \eta_g \prod_{j=1}^{d_0} \phi\left(y_{ij}|\mu_{gj}, \sigma_{gj}^2\right) \prod_{j=d_0+1}^{d} \phi\left(y_{ij}|\mu_{0j}, \sigma_{0j}^2\right).$$

An EM algorithm was used to find the posterior mode of the model parameters and a minimum message length procedure was used for selecting the number of clusters and clustering variables.

White et al. (2016) developed an analogous approach to that of Tadesse et al. (2005) for the LCA model and developed an efficient collapsed MCMC algorithm for sampling from the marginalized posterior. In addition, they developed a post-hoc method of finding approximate values for the posterior mean and variance for any model parameters from the resulting collapsed MCMC output.

### 11.7.2 Stepwise approaches for continuous data

Raftery & Dean (2006) developed a wrapper-based method for variable selection for normal mixture models, with a particular emphasis on model-based clustering. The approach taken utilized a forward stepwise variable selection procedure, where the selection of variable was based on a model comparison procedure.

In particular, the approach is based on partitioning the vector $y_i$ into three components: the $y_i^c$ are the variables currently included in the mixture model, $y_i^p$ is the variable proposed for inclusion in the mixture model, and $y_i^o$ are the other variables – thus, $y_i$ can be written

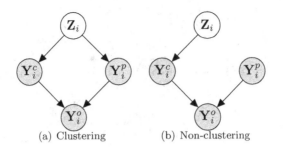

(a) Clustering                    (b) Non-clustering

**FIGURE 11.4**
A graphical model representation of the two models being compared in the Dean & Raftery (2010) variable selection method for LCA. The observed values are shaded and the latent variable is unshaded.

as $y_i = (y_i^c, y_i^p, y_i^o)$. If $z_i$ is the latent indicator variable for which component (or cluster) observation $i$ comes from, the procedure is based on the following model comparison:

Clustering:  $f(y_i|z_i) = f(y_i^c, y_i^p, y_i^o|z_i) = f(y_i^c|z_i)f(y_i^p|y_i^c, z_i)f(y_i^o|y_i^c, y_i^p),$

Non-clustering:  $f(y_i|z_i) = f(y_i^c, y_i^p, y_i^o|z_i) = f(y_i^c|z_i)f(y_i^p|y_i^c)f(y_i^o|y_i^c, y_i^p).$

Further, the form of $f(y_i^o|y_i^c, y_i^p)$ is assumed to be the same for both models, and thus does not need to be specified. The model can be represented using a graphical model (Figure 11.3) and the difference between the clustering and non-clustering models is indicated by the presence/absence of a direct edge between $Z_i$ and $Y_i^p$. More specifically, the models are of the form:

Clustering:  $$(y_i^c, y_i^p) \sim \sum_{g=1}^{G} \eta_g \mathcal{N}(\mu_g^{c,p}, \Sigma_g^{c,p}),$$

Non-clustering:  $$y_i^c \sim \sum_{g=1}^{G} \eta_g \mathcal{N}(\mu_g^c, \Sigma_g^c), \quad y_i^p|y_i^c \sim \mathcal{N}(\alpha + \beta^\top y_i^c, \sigma^2).$$

The models are fitted using an EM algorithm to find maximum likelihood estimates of the parameters, and model comparison is made using the Bayesian information criterion (BIC; Schwarz, 1978). The stepwise algorithm proceeds by searching over all non-clustering variables to find the variable (if any) with the greatest evidence of clustering. This variable is added to the clustering set and then a search is implemented to see if any variable should be removed from the clustering set (based on a similar procedure to that outlined). The algorithm continues until no more variables are added to or removed from the clustering set.

Maugis et al. (2009a,b) developed a similar variable selection procedure and showed that the Raftery & Dean (2006) approach can be improved by including a variable selection step in the regression part of the non-clustering model. Maugis et al. (2011) developed a variable selection method for model-based classification using the same framework. Murphy et al. (2010) adapted the variable selection procedure for semi-supervised model-based classification.

Celeux et al. (2011) provided a comparison of sparse $k$-means clustering (Witten & Tibshirani, 2010) to these stepwise variable selection procedures and demonstrated improved performance in terms of clustering. Celeux et al. (2014) provided a recent review of variable selection for clustering, and the clustvarsel (Scrucca & Raftery, 2018) and SelvarMix (Sedki et al., 2015) R packages implement these stepwise variable selection approaches.

**TABLE 11.7**

The Bayesian information criterion for different normal mixture models when fitted to the Italian wine data

| Family | Model | $G$ | BIC |
|---|---|---|---|
| mclust | VVE | 3 | $-11{,}945.08$ |
| Rmixmod | p_Lk_D_Ak_D | 3 | $-11{,}941.74$ |
| pgmm | CCUU ($q = 5$) | 4 | $-11{,}548.55$ |
| HDclassif | ABQD | 7 | $-12{,}369.70$ |

**TABLE 11.8**

A cross-tabulation of type with the component membership for the Italian wine data

| Type | Component 1 | 2 | 3 | 4 |
|---|---|---|---|---|
| Barbera | | | 1 | 47 |
| Grignolino | | 48 | 23 | |
| Barolo | 58 | | 1 | |

### 11.7.3 Stepwise approaches for categorical data

Dean & Raftery (2010) developed a stepwise approach to variable selection for the LCA model in an analogous manner to the methods outlined in Section 11.7.2. The LCA model has a local independence assumption (Section 11.5.1), so the procedure is set up to maintain this structure for the variables selected. A forward stepwise variable selection procedure is developed based on comparing clustering and non-clustering models of the form:

$$\text{Clustering:} \quad f(y_i|z_i) = f(y_i^c, y_i^p, y_i^o|z_i) = f(y_i^c|z_i)f(y_i^p|z_i)f(y_i^o|y_i^c, y_i^p),$$
$$\text{Non-clustering:} \quad f(y_i|z_i) = f(y_i^c, y_i^p, y_i^o|z_i) = f(y_i^c|z_i)f(y_i^p)f(y_i^o|y_i^c, y_i^p).$$

The models $f(y_i^c|z_i)$, $f(y_i^c, y_i^p|z_i)$ assume that all of the variables in these parts of the partition of variables are conditionally independent multinomial random variables given $z_i$. The model can be represented using a graphical model (Figure 11.4) where the difference between the models is due to the presence/absence of an edge between $Z_i$ and $Y_i^p$. The models are fitted using an EM algorithm to find maximum likelihood parameter estimates, and model comparison is implemented using BIC; the implementation mirrors that outlined in Section 11.7.2. Bartolucci et al. (2018) developed a similar variable selection procedure and extended it to account for missing data in Bartolucci et al. (2016).

Recently, Fop et al. (2017) extended the modeling approach of Dean & Raftery (2010) to include the concept of redundant variables in LCA. This approach tends to select a smaller set of clustering variables than the Dean & Raftery (2010) approach. An R package called LCAvarsel (Fop & Murphy, 2017) has been developed to implement the Fop et al. (2017) approach and the Dean & Raftery (2010) approach for variable selection.

## 11.8 Examples

A number of the mixture models described in this chapter are applied to the three example data sets from Section 11.2.

**TABLE 11.9**
A cross-tabulation of type and year with the component membership for the Italian wine data

| Type | Year | Component |  |  |  |
|------|------|---|---|---|---|
|      |      | 1 | 2 | 3 | 4 |
| Barbera | 74 |  |  | 1 | 8 |
| Barbera | 76 |  |  |  | 5 |
| Barbera | 78 |  |  |  | 29 |
| Barbera | 79 |  |  |  | 5 |
| Grignolino | 70 |  | 3 | 6 |  |
| Grignolino | 71 |  | 3 | 6 |  |
| Grignolino | 72 |  | 5 | 2 |  |
| Grignolino | 73 |  | 9 |  |  |
| Grignolino | 74 |  | 12 | 4 |  |
| Grignolino | 75 |  | 5 | 4 |  |
| Grignolino | 76 |  | 11 | 1 |  |
| Barolo | 71 | 19 |  |  |  |
| Barolo | 73 | 19 |  | 1 |  |
| Barolo | 74 | 20 |  |  |  |

**TABLE 11.10**
The selected variables included in the `clustervarsel` mixture model for the Italian wine data

| Selected Variables | | | |
|------|------|------|------|
| Chloride | Malic acid | Proline | Fixed acidity |
| Flavanoids | Color intensity | Uronic acids | |

### 11.8.1 Continuous data: Italian wine

A number of different normal mixture models were fitted to the Italian wine data from Section 11.2.1; the variables in the data were standardized prior to analysis. In particular, the normal mixture models implemented in the mclust (Scrucca et al., 2016), Rmixmod (Lebret et al., 2015; Langrognet et al., 2016), pgmm (McNicholas et al., 2015), and HDclassif (Bergé et al., 2012, 2016) R packages were fitted. The resulting BIC values are shown in Table 11.7; the model types fitted within these packages are outlined in Sections 11.4.2–11.4.4.

The model with the highest BIC is the $G = 4$ component pgmm model with CCUU covariance structure and with $q = 5$; this model is one of the extended PGMM models proposed by McNicholas & Murphy (2010, Table 2). This model has covariance structure of the form $\Sigma_g = \Lambda\Lambda^\top + \omega_g\Delta$, which has $(G+d-1)+(dq-q(q-1)/2) = 155$ parameters and gives a different non-diagonal covariance for each mixture component with a relatively low number of parameters. In contrast, a $G = 5$ component mixture model with unconstrained covariance matrices has 1512 covariance parameters.

A cross-tabulation of the wine type versus maximum *a posteriori* component membership is given in Table 11.8, and a more detailed cross-tabulation versus the combination of type and year is given in Table 11.9. The results show that the component membership has a close correspondence with wine type and the type and year combination; the adjusted Rand indices (Hubert & Arabie, 1985) are 0.639 and 0.296, respectively. The Barolo wines are mainly assigned to component 1, the Barbera wines are mainly assigned to component 4, and the Grignolino wines are divided across components 2 and 3. It is worth noting that

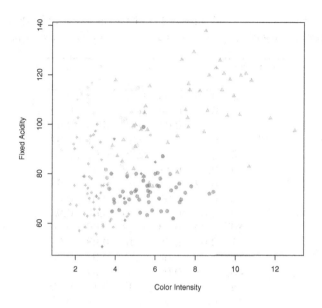

**FIGURE 11.5**
A scatter plot of the color intensity and fixed acidity of the Italian wines with the wine types indicated by different plot symbols and the component memberships indicated using color.

**TABLE 11.11**
A cross-tabulation of the wine type versus component membership for the clustvarsel mixture model

| Type | Component | | | | |
|------|---|---|---|---|---|
| | 1 | 2 | 3 | 4 | 5 |
| Barbera | | 1 | | | 47 |
| Grignolino | | 49 | 6 | 15 | 1 |
| Barolo | 43 | 2 | 6 | 8 | |

the Grignolino wines from earlier years are more likely to belong to component 3 than to component 2.

Further, the variable selection method as implemented in **clustervarsel** (Section 11.7.2) was also applied to these data. The resulting model has $G = 5$ components with a VVI covariance structure based on seven variables (Table 11.10). Similarly, the component memberships are closely related to wine type (Table 11.11).

A scatter plot of the color intensity versus fixed acidity is shown in Figure 11.5, where the wine types are shown using different symbols and the component memberships are shown using different colors; the relationship between the components and wine types can be seen in the plot.

Thus, modeling the wine data using normal mixture models provides a methodology for capturing the heterogeneous nature of these data and where the mixture components closely correspond to the wine types. The fitted mixture models indicated that the Grignolino wine type is quite heterogeneous and requires more than one normal component

**TABLE 11.12**
The parameter estimates of the LCA model fitted to the back pain data. The parameter estimates are the probability of "present" for each variable. The mixing proportions are also shown.

| Variable | Class | | | | |
|---|---|---|---|---|---|
| | 1 | 2 | 3 | 4 | 5 |
| 1 | 0.00 | 0.03 | 0.06 | 0.34 | 0.17 |
| 2 | 0.00 | 0.86 | 0.06 | 0.99 | 0.95 |
| 3 | 0.08 | 0.04 | 0.29 | 0.14 | 0.98 |
| 4 | 0.96 | 0.10 | 0.84 | 0.01 | 0.01 |
| 5 | 0.04 | 0.89 | 0.40 | 0.97 | 0.90 |
| 6 | 0.98 | 0.11 | 0.62 | 0.00 | 0.12 |
| 7 | 0.52 | 0.15 | 0.58 | 0.21 | 0.63 |
| 8 | 0.10 | 0.95 | 0.52 | 0.97 | 0.49 |
| 9 | 0.00 | 0.13 | 0.25 | 0.06 | 0.98 |
| 10 | 0.90 | 0.03 | 0.42 | 0.00 | 0.00 |
| 11 | 0.02 | 0.98 | 0.10 | 0.98 | 0.94 |
| 12 | 0.00 | 0.04 | 0.16 | 0.06 | 0.95 |
| 13 | 0.98 | 0.00 | 0.82 | 0.01 | 0.00 |
| 14 | 0.45 | 0.17 | 0.47 | 0.11 | 0.39 |
| 15 | 0.33 | 0.08 | 0.37 | 0.04 | 0.48 |
| 16 | 0.96 | 0.06 | 0.48 | 0.01 | 0.20 |
| 17 | 0.18 | 0.12 | 0.36 | 0.13 | 0.81 |
| 18 | 0.96 | 0.46 | 0.75 | 0.27 | 0.68 |
| 19 | 0.19 | 0.75 | 0.48 | 0.82 | 0.69 |
| 20 | 0.00 | 0.01 | 0.00 | 0.76 | 0.48 |
| 21 | 1.00 | 0.87 | 0.92 | 0.01 | 0.33 |
| 22 | 0.94 | 0.35 | 0.60 | 0.05 | 0.14 |
| 23 | 0.92 | 0.24 | 0.77 | 0.03 | 0.06 |
| 24 | 0.92 | 0.13 | 0.65 | 0.05 | 0.34 |
| 25 | 0.73 | 0.16 | 0.28 | 0.14 | 0.46 |
| 26 | 0.00 | 0.93 | 0.19 | 0.98 | 0.96 |
| 27 | 0.11 | 0.14 | 0.31 | 0.10 | 0.90 |
| 28 | 1.00 | 0.00 | 0.65 | 0.01 | 0.00 |
| 29 | 0.06 | 0.05 | 0.14 | 0.05 | 0.72 |
| 30 | 0.00 | 0.89 | 0.59 | 0.79 | 0.75 |
| 31 | 1.00 | 0.02 | 0.45 | 0.04 | 0.00 |
| 32 | 0.46 | 0.02 | 0.17 | 0.01 | 0.04 |
| 33 | 0.71 | 0.20 | 0.50 | 0.19 | 0.27 |
| 34 | 0.25 | 0.22 | 0.34 | 0.13 | 0.26 |
| 35 | 0.21 | 0.09 | 0.19 | 0.06 | 0.60 |
| 36 | 1.00 | 0.27 | 0.73 | 0.04 | 0.08 |
| $\eta_g$ | 0.24 | 0.11 | 0.24 | 0.11 | 0.29 |

for modeling. Further, the variable selection methods demonstrate that the data can be clustered effectively using a small subset of the 27 variables.

## 11.8.2 Categorical data: lower back pain

An LCA model was fitted to the back pain data from Section 11.2.2 for $G = 1, 2, \ldots, 6$ using the poLCA R package (Linzer & Lewis, 2011, 2014). Maximum likelihood estimates of the model parameters were found using an EM algorithm and the model with the highest BIC

**TABLE 11.13**

A cross-tabulation of the maximum *a posteriori* class estimates for the LCA model and the known pain groups for the back pain data

| Type | Class | | | | |
|---|---|---|---|---|---|
| | 1 | 2 | 3 | 4 | 5 |
| Nociceptive | 10 | | 97 | 3 | 125 |
| Peripheral neuropathic | 89 | | 1 | 4 | 1 |
| Central neuropathic | 1 | 48 | 5 | 41 | |

**TABLE 11.14**

A cross-tabulation of the maximum *a posteriori* component estimates for the MLTA model and the known pain groups for the back pain data

| Type | Component | | |
|---|---|---|---|
| | 1 | 2 | 3 |
| Nociceptive | 5 | 197 | 33 |
| Peripheral neuropathic | | | 95 |
| Central neuropathic | 81 | 7 | 7 |

value was selected. The resulting model had $G = 5$ latent classes and the model parameters are shown in Table 11.12 and Figure 11.6. The maximum *a posteriori* class estimates were produced for each observation and a cross-tabulation of the resulting class estimates against the known pain classes shows strong agreement (Table 11.13); the adjusted Rand index is 0.502. A close inspection of the cross-tabulation shows that the nociceptive patients are mainly in classes 2 and 5, the peripheral neuropathic patients are mainly in class 1, and the central neuropathic patients are mainly in classes 3 and 4. Thus, the LCA suggests that some of the pain types may have subtypes.

The variable selection method of Dean & Raftery (2010) was applied to this data and the optimal model, selected using BIC, also has $G = 5$ classes. Further, it was found that variable 34 was removed and assigned as a non-clustering variable. This variable exhibits very little variation in the probability of "present" in Table 11.12, so its removal using this procedure is unsurprising.

Both analyses assume local independence within mixture components, and this may lead to an overestimation of the number of clusters in the data. Thus, MLTA models (Gollini & Murphy, 2014) were fitted to the data with $G = 1, 2, 3, 4$ and $q = 1, 2, 3, 4$ (where $q$ is the dimensionality of the latent trait that accounts for dependence within groups). Maximum likelihood estimation of the model parameters was achieved using a variational EM algorithm, with BIC used for model selection.

The selected model had $G = 3$ components and a $q = 2$ dimensional latent trait. The maximum *a posteriori* component membership values were estimated for each observation and a cross-tabulation of the resulting estimates against the pain classes shows very strong agreement (Table 11.14); the adjusted Rand index is 0.658, which is higher than for the LCA model.

The MLTA components have strong correspondence with the pain groups; however, some nociceptive patients are being put in component 3 which has predominantly peripheral neuropathic patients. The pain type assigned to each of the patients is the primary pain type recorded by the physiotherapist. Thus, the nociceptive patients that are assigned to class 3 could have both nociceptive and peripheral neuropathic pain, and thus could have symptoms more like peripheral neuropathic patients.

Thus, the mixture models were able to effectively model the heterogeneous nature of the

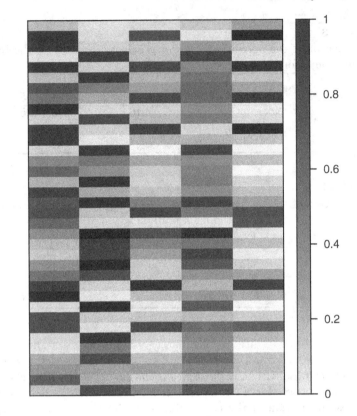

**FIGURE 11.6**
A heatmap representation of parameter estimates of the LCA model fitted to the back pain data. The darker the shading, the higher the probability of recording "present" for that variable.

back pain data. The mixture components had a strong correspondence with the pain types. The LCA model tended to overestimate the number of groups because of the assumed local independence structure, but the MLTA model which accounts for dependence using a latent trait was able to cluster the patients more accurately.

### 11.8.3   Mixed data: prostate cancer

The Byar & Green (1980) prostate cancer data, introduced in Section 11.2.3, were analyzed using the clustMD family of models introduced in Section 11.6. Prior to mixture modeling, the size of the primary tumor variable was transformed using the square root and the serum prostatic acid phosphatase variable was transformed using the logarithm, and all of the continuous variables were scaled to have mean zero and variance one.

The six models available in the clustMD R package were fitted to the data with $G = 1, 2, \ldots, 5$ and the BIC for each model was computed. The model with the highest approximate BIC ($\widehat{\mathrm{BIC}} = -12{,}870.84$) is a mixture model with $G = 3$ and an EVI covariance structure for the latent variable; the exact BIC is not available because the likelihood is not maximized in the model fitting procedure.

**TABLE 11.15**

A cross-tabulation of the prostate cancer stage versus maximum *a posteriori* component membership

| Stage | Component | | |
|---|---|---|---|
| | 1 | 2 | 3 |
| 3 | 209 | 43 | 21 |
| 4 | 16 | 12 | 174 |

**TABLE 11.16**

A cross-tabulation of, respectively, the cardiovascular disease score and the electrocardiogram score versus maximum *a posteriori* component membership for the prostate cancer data

| Cardiovascular disease | Component | | |
|---|---|---|---|
| | 1 | 2 | 3 |
| 1 | 126 | 12 | 130 |
| 2 | 99 | 43 | 65 |
| Electrocardiogram score | Component | | |
| | 1 | 2 | 3 |
| 1 | 84 | 7 | 70 |
| 2 | 43 | 9 | 46 |
| 3 | 98 | 39 | 79 |

The maximum *a posteriori* component membership for each patient has strong correspondence with the stage of the cancer for the patient (Table 11.15); the adjusted Rand index is 0.525. Most of the stage 3 patients are assigned to component 1, whereas most of the stage 4 patients are assigned to component 3. Component 2 contains both stage 3 and stage 4 patients, but with stage 3 patients being in the majority.

Inspecting the observations assigned to components 1 and 2 indicates that those assigned to component 2 are more likely to have a history of cardiovascular disease and their electrocardiogram score is more likely to indicate a serious anomaly than those assigned to component 1; see Table 11.16.

Thus, the mixture model is modeling a combination of prostate cancer stage and cardiovascular issues of the patients in the study. Interestingly, by the end of the clinical trial more of the patients assigned to component 2 had died of cardiovascular issues than prostate cancer or any other causes.

## 11.9   Concluding Remarks

In this chapter, we have reviewed a number of mixture models that can be used for high-dimensional data. We focused on normal mixture models for continuous data, latent class models (and extensions) for categorical data, and latent variable mixture models for mixed data.

Mixture models were applied to example data of each type and the resulting mixture revealed interesting grouping structure in each case. The grouping has a close relationship to known group structure but also revealed further group structure. The mixture component distributions provided a characterization of group structure that was found.

# Bibliography

BANFIELD, J. D. & RAFTERY, A. E. (1993). Model-based Gaussian and non-Gaussian clustering. *Biometrics* **49**, 803–821.

BARTHOLOMEW, D. J., KNOTT, M. & MOUSTAKI, I. (2011). *Latent Variable Models and Factor Analysis: A Unified Approach.* New York: John Wiley, 3rd ed.

BARTOLUCCI, F., MONTANARI, G. E. & PANDOLFI, S. (2016). Item selection by latent class-based methods: An application to nursing home evaluation. *Advances in Data Analysis and Classification* **10**, 245–262.

BARTOLUCCI, F., MONTANARI, G. E. & PANDOLFI, S. (2018). Latent ignorability and item selection for nursing home case-mix evaluation. *Journal of Classification* **35**, 172–193.

BELLMAN, R. E. (1957). *Dynamic Programming.* Princeton, NJ: Princeton University Press.

BERGÉ, L., BOUVEYRON, C. & GIRARD, S. (2012). HDclassif: An R package for model-based clustering and discriminant analysis of high-dimensional data. *Journal of Statistical Software* **46**, 1–29.

BERGÉ, L., BOUVEYRON, C. & GIRARD, S. (2016). *HDclassif: High Dimensional Supervised Classification and Clustering.* R package version 2.0.2.

BEYER, K., GOLDSTEIN, J., RAMAKRISHNAN, R. & SHAFT, U. (1999). When is "nearest neighbor" meaningful? In *Database Theory – ICDT '99: 7th International Conference.* Berlin: Springer-Verlag.

BHATTACHARYA, S. & MCNICHOLAS, P. D. (2014). A LASSO-penalized BIC for mixture model selection. *Advances in Data Analysis and Classification* **8**, 45–61.

BOUVEYRON, C. & BRUNET-SAUMARD, C. (2014). Model-based clustering of high-dimensional data: A review. *Computational Statistics and Data Analysis* **71**, 52–78.

BOUVEYRON, C., GIRARD, S. & SCHMID, C. (2007). High-dimensional data clustering. *Computational Statistics and Data Analysis* **52**, 502–519.

BROWNE, R. P. & MCNICHOLAS, P. D. (2012). Model-based clustering, classification, and discriminant analysis of data with mixed type. *Journal of Statistical Planning and Inference* **142**, 2976–2984.

BYAR, D. & GREEN, S. (1980). The choice of treatment for cancer patients based on covariate information: Applications to prostate cancer. *Bulletin du Cancer* **67**, 477–490.

CAGNONE, S. & VIROLI, C. (2012). A factor mixture analysis model for multivariate binary data. *Statistical Modelling* **12**, 257–277.

CELEUX, G. & GOVAERT, G. (1991). Clustering criteria for discrete data and latent class model. *Journal of Classification* **8**, 157–176.

CELEUX, G. & GOVAERT, G. (1992). A classification EM algorithm for clustering and two stochastic versions. *Computational Statistics and Data Analysis* **14**, 315–332.

CELEUX, G. & GOVAERT, G. (1995). Gaussian parsimonious clustering models. *Pattern Recognition* **28**, 781–793.

CELEUX, G., MARTIN-MAGNIETTE, M.-L., MAUGIS, C. & RAFTERY, A. E. (2011). Letter to the editor: "A framework for feature selection in clustering". *Journal of the American Statistical Association* **106**, 383.

CELEUX, G., MARTIN-MAGNIETTE, M.-L., MAUGIS-RABUSSEAU, C. & RAFTERY, A. E. (2014). Comparing model selection and regularization approaches to variable selection in model-based clustering. *Journal de la Société Française de Statistique* **155**, 57–71.

CHANG, W.-C. (1983). On using principal components before separating a mixture of two multivariate normal distributions. *Applied Statistics* **32**, 267–275.

CLOGG, C. C. (1995). Latent class models. In *Handbook of Statistical Modeling for the Social and Behavioral Sciences*, G. Arminger, C. C. Clogg & M. E. Sobel, eds., chap. 6. New York: Plenum Press, pp. 311–360.

DEAN, N. & RAFTERY, A. E. (2010). Latent class analysis variable selection. *Annals of the Institute of Statistical Mathematics* **62**, 11–35.

DEMPSTER, A. P., LAIRD, N. M. & RUBIN, D. B. (1977). Maximum likelihood from incomplete data via the EM algorithm (with discussion). *Journal of the Royal Statistical Society, Series B* **39**, 1–38.

DIEBOLT, J. & ROBERT, C. P. (1994). Estimation of finite mixture distributions through Bayesian sampling. *Journal of the Royal Statistical Society, Series B* **56**, 363–375.

DY, J. G. & BRODLEY, C. E. (2004). Feature selection for unsupervised learning. *Journal of Machine Learning Research* **5**, 845–889.

EVERITT, B. (1984). *An Introduction to Latent Variable Models*. London: Chapman & Hall.

FAN, J. & LI, R. (2001). Variable selection via nonconcave penalized likelihood and its oracle properties. *Journal of the American Statistical Association* **96**, 1348–1360.

FERGUSON, T. S. (1983). Bayesian density estimation by mixtures of normal distributions. In *Recent Advances in Statistics*, M. H. Rizvi, J. S. Rustagi & D. Siegmund, eds. New York: Academic Press, pp. 287–302.

FOP, M. & MURPHY, T. (2017). *LCAvarsel: Variable selection for latent class analysis*. R package version 1.0.

FOP, M., SMART, K. & MURPHY, T. B. (2017). Variable selection for latent class analysis with application to low back pain diagnosis. *Annals of Applied Statistics* **11**, 2080–2110.

FORINA, M., ARMANINO, C., CASTINO, M. & UBIGLI, M. (1986). Multivariate data analysis as a discriminating method of the origin of wines. *Vitis* **25**, 189–214.

FRIEDMAN, J., HASTIE, T. & TIBSHIRANI, R. (2008). Sparse inverse covariance estimation with the graphical lasso. *Biostatistics* **9**, 432–441.

FRÜHWIRTH-SCHNATTER, S. (2006). *Finite Mixture and Markov Switching Models.* New York: Springer.

GALIMBERTI, G., MONTANARI, A. & VIROLI, C. (2009). Penalized factor mixture analysis for variable selection in clustered data. *Computational Statistics and Data Analysis* **53**, 4301–4310.

GHAHRAMANI, Z. & HINTON, G. E. (1997). The EM algorithm for factor analyzers. Tech. Rep. CRG-TR-96-1, University of Toronto, Toronto.

GOLLINI, I. & MURPHY, T. B. (2014). Mixture of latent trait analyzers for model-based clustering of categorical data. *Statistics and Computing* **24**, 569–588.

GOODMAN, L. A. (1974). Exploratory latent structure analysis using both identifiable and unidentifiable models. *Biometrika* **61**, 215–231.

HARTIGAN, J. A. (1975). *Clustering Algorithms.* Wiley Series in Probability and Mathematical Statistics. New York: John Wiley.

HARTIGAN, J. A. & WONG, M. A. (1978). Algorithm AS 136: A $k$-means clustering algorithm. *Applied Statistics* **28**, 100–108.

HUBERT, L. & ARABIE, P. (1985). Comparing partitions. *Journal of Classification* **2**, 193–218.

HUNT, L. & JORGENSEN, M. (1999). Mixture model clustering using the MULTIMIX program. *Australian and New Zealand Journal of Statistics* **41**, 154–171.

HUNT, L. & JORGENSEN, M. (2003). Mixture model clustering for mixed data with missing information. *Computational Statistics and Data Analysis* **41**, 429–440.

HUNTER, D. & LI, R. (2005). Variable selection using MM algorithms. *Annals of Statistics* **33**, 1617–1642.

KOHAVI, R. & JOHN, G. H. (1997). Relevance wrappers for feature subset selection. *Artificial Intelligence* **97**, 273–324.

LANGROGNET, F., LEBRET, R., POLI, C. & IOVLEFF, S. (2016). *Rmixmod: Supervised, unsupervised, semi-supervised classification with MIXture MODelling (interface of MIXMOD software).* R package version 2.1.1.

LAW, M., FIGUEIREDO, M. & JAIN, A. (2004). Simultaneous feature selection and clustering using mixture models. *IEEE Transactions on Pattern Analysis and Machine Intelligence* **26**, 1154–1166.

LAWRENCE, C. J. & KRZANOWSKI, W. J. (1996). Mixture separation for mixed-mode data. *Statistics and Computing* **6**, 85–92.

LAZARSFELD, P. & HENRY, N. (1968). *Latent Structure Analysis.* Boston: Houghton Mifflin.

LEBRET, R., IOVLEFF, S., LANGROGNET, F., BIERNACKI, C., CELEUX, G. & GOVAERT, G. (2015). Rmixmod: The R package of the model-based unsupervised, supervised, and semi-supervised classification Mixmod library. *Journal of Statistical Software* **67**, 1–29.

LEE, S. X. & MCLACHLAN, G. J. (2013). Model-based clustering and classification with non-normal mixture distributions (with discussion). *Statistical Methods and Applications* **22**, 427–479.

LINZER, D. A. & LEWIS, J. B. (2011). poLCA: An R package for polytomous variable latent class analysis. *Journal of Statistical Software* **42**, 1–29.

LINZER, D. A. & LEWIS, J. B. (2014). *poLCA: Polytomous Variable Latent Class Analysis.* R package version 1.4.1.

LIU, J. S., LIANG, J. L., PALUMBO, M. J. & LAWRENCE, C. E. (2003). Bayesian clustering with variable and transformation selections. In *Bayesian Statistics 7*, J. M. Bernardo, M. J. Bayarri, J. O. Berger, A. P. Dawid, D. Heckerman, A. F. M. Smith & M. West, eds. Oxford: Oxford University Press, pp. 249–275.

MALSINER-WALLI, G., FRÜHWIRTH-SCHNATTER, S. & GRÜN, B. (2016). Model-based clustering based on sparse finite Gaussian mixtures. *Statistics and Computing* **26**, 303–324.

MAUGIS, C., CELEUX, G. & MARTIN-MAGNIETTE, M.-L. (2009a). Variable selection for clustering with Gaussian mixture models. *Biometrics* **65**, 701–709.

MAUGIS, C., CELEUX, G. & MARTIN-MAGNIETTE, M.-L. (2009b). Variable selection in model-based clustering: A general variable role modeling. *Computational Statistics and Data Analysis* **53**, 3872–3882.

MAUGIS, C., CELEUX, G. & MARTIN-MAGNIETTE, M.-L. (2011). Variable selection in model-based discriminant analysis. *Journal of Multivariate Analysis* **102**, 1374–1387.

MCLACHLAN, G. J., BAEK, J. & RATHNAYAKE, S. I. (2011). Mixtures of factor analysers for the analysis of high-dimensional data. In *Mixtures: Estimation and Applications*, K. Mengersen, C. Robert & M. Titterington, eds., Wiley Series in Probability and Statistics. Chichester: John Wiley, pp. 189–212.

MCLACHLAN, G. J. & BASFORD, K. E. (1988). *Mixture Models: Inference and Applications to Clustering.* New York: Marcel Dekker.

MCLACHLAN, G. J. & PEEL, D. (2000). *Finite Mixture Models.* New York: John Wiley.

MCLACHLAN, G. J., PEEL, D. & BEAN, R. W. (2003). Modelling high-dimensional data by mixtures of factor analyzers. *Computational Statistics and Data Analysis* **41**, 379–388.

MCNICHOLAS, P. D. (2016). *Mixture Model-Based Classification.* Boca Raton, FL: Chapman & Hall/CRC Press.

MCNICHOLAS, P. D., ELSHERBINY, A., MCDAID, A. F. & MURPHY, T. B. (2015). *pgmm: Parsimonious Gaussian mixture models.* R package version 1.2.

MCNICHOLAS, P. D. & MURPHY, T. B. (2008). Parsimonious Gaussian mixture models. *Statistics and Computing* **18**, 285–296.

MCNICHOLAS, P. D. & MURPHY, T. B. (2010). Model-based clustering of microarray expression data via latent Gaussian mixture models. *Bioinformatics* **26**, 2705–2712.

MCNICHOLAS, P. D., MURPHY, T. B., MCDAID, A. F. & FROST, D. (2010). Serial and parallel implementations of model-based clustering via parsimonious Gaussian mixture models. *Computational Statistics and Data Analysis* **54**, 711–723.

MCPARLAND, D. & GORMLEY, I. C. (2013). Clustering ordinal data via latent variable models. In *Algorithms from and for Nature and Life*, B. Lausen, D. V. den Poel & A. Ultsch, eds. Cham: Springer, pp. 127–135.

MCPARLAND, D. & GORMLEY, I. C. (2016). Model based clustering for mixed data: clustMD. *Advances in Data Analysis and Classification* **10**, 155–169.

MCPARLAND, D. & GORMLEY, I. C. (2017). *clustMD: Model based clustering for mixed data*. R package version 1.2.1.

MCPARLAND, D., GORMLEY, I. C., MCCORMICK, T. H., CLARK, S. J., KABUDULA, C. W. & COLLINSON, M. A. (2014). Clustering South African households based on their asset status using latent variable models. *Annals of Applied Statistics* **8**, 747–776.

MCPARLAND, D., PHILLIPS, C. M., BRENNAN, L., ROCHE, H. M. & GORMLEY, I. C. (2017). Clustering high dimensional mixed data to uncover sub-phenotypes: Joint analysis of phenotypic and genotypic data. *Statistics in Medicine* **36**, 4548–4569.

MURPHY, T. B., DEAN, N. & RAFTERY, A. E. (2010). Variable selection and updating in model-based discriminant analysis for high dimensional data with food authenticity applications. *Annals of Applied Statistics* **4**, 396–421.

NYAMUNDANDA, G., BRENNAN, L. & GORMLEY, I. C. (2010a). Probabilistic principal component analysis for metabolomic data. *BMC Bioinformatics* **11**, 1–11.

NYAMUNDANDA, G., GORMLEY, I. C. & BRENNAN, L. (2010b). *MetabolAnalyze: Probabilistic principal components analysis for metabolomic data*. R package version 1.3.

PAN, W. & SHEN, X. (2007). Penalized model-based clustering with application to variable selection. *Journal of Machine Learning Research* **8**, 1145–1164.

PANDOLFI, S., BARTOLUCCI, F. & FRIEL, N. (2014). A generalized multiple-try version of the reversible jump algorithm. *Computational Statistics and Data Analysis* **72**, 298–314.

R CORE TEAM (2017). *R: A Language and Environment for Statistical Computing*. Vienna: R Foundation for Statistical Computing.

RABE-HESKETH, S., SKRONDAL, A. & PICKLES, A. (2004). GLLAMM manual. UC Berkeley Division of Biostatistics Working Paper Series. Working Paper 160.

RAFTERY, A. E. & DEAN, N. (2006). Variable selection for model-based clustering. *Journal of the American Statistical Association* **101**, 168–178.

RAU, A., MAUGIS-RABUSSEAU, C., MARTIN-MAGNIETTE, M.-L. & CELEUX, G. (2015). Co-expression analysis of high-throughput transcriptome sequencing data with Poisson mixture models. *Bioinformatics* **31**, 1420—1427.

ROST, J. (1990). Rasch models in latent classes: An integration of two approaches to item analysis. *Applied Psychological Measurement* **14**, 271–282.

SCHWARZ, G. (1978). Estimating the dimension of a model. *Annals of Statistics* **6**, 461–464.

SCRUCCA, L., FOP, M., MURPHY, T. B. & RAFTERY, A. E. (2016). mclust 5: Clustering, classification and density estimation using Gaussian finite mixture models. *R Journal* **8**, 289–317.

SCRUCCA, L. & RAFTERY, A. E. (2018). clustvarsel: A package implementing variable selection for model-based clustering in R. *Journal of Statistical Software* **84**, 1–28.

SEDKI, M., CELEUX, G. & MAUGIS-RABUSSEAU, C. (2015). *SelvarMix: Regularization for variable selection in model-based clustering and discriminant analysis.* R package version 1.1.

SKRONDAL, A. & RABE-HESKETH, S. (2004). *Generalized Latent Variable Modeling: Multilevel, Longitudinal and Structural Equation Models.* Boca Raton, FL: Chapman & Hall/CRC.

SMART, K. M., BLAKE, C., STAINES, A. & DOODY, C. (2011). The discriminative validity of "nociceptive", "peripheral neuropathic", and "central sensitization" as mechanisms-based classifications of musculoskeletal pain. *Clinical Journal of Pain* **27**, 655–663.

STEINLEY, D. & BRUSCO, M. J. (2008a). A new variable weighting and selection procedure for $k$-means cluster analysis. *Multivariate Behavioral Research* **43**, 77–108.

STEINLEY, D. & BRUSCO, M. J. (2008b). Selection of variables in cluster analysis: An empirical comparison of eight procedures. *Psychometrika* **73**, 125–144.

SUN, W., WANG, J., FANG, Y. et al. (2012). Regularized k-means clustering of high-dimensional data and its asymptotic consistency. *Electronic Journal of Statistics* **6**, 148–167.

TADESSE, M. G., SHA, N. & VANNUCCI, M. (2005). Bayesian variable selection in clustering high-dimensional data. *Journal of the American Statistical Association* **100**, 602–617.

TANG, Y., BROWNE, R. P. & MCNICHOLAS, P. D. (2015). Model based clustering of high-dimensional binary data. *Computational Statistics and Data Analysis* **87**, 84–101.

TIPPING, M. E. & BISHOP, C. M. (1999). Mixtures of probabilistic principal component analysers. *Neural Computation* **11**, 443–482.

VERMUNT, J. (2007). Multilevel mixture item response theory models: An application in education testing. In *Proceedings of the 56th Session of the International Statistical Institute, Lisbon, Portugal.* Voorburg, Netherlands: International Statistical Institute.

VERMUNT, J. K. & MAGIDSON, J. (2013). *LG-syntax user's guide: Manual for Latent GOLD 5.0 syntax module.* Belmont, MA: Statistical Innovations Inc.

WEST, M. (2003). Bayesian factor regression models in the "large $p$, small $n$" paradigm. In *Bayesian Statistics 7*, J. M. Bernardo, M. J. Bayarri, J. O. Berger, A. P. Dawid, D. Heckerman, A. F. M. Smith & M. West, eds. Oxford: Oxford University Press, pp. 733–742.

WHITE, A. & MURPHY, T. B. (2014). BayesLCA: An R package for Bayesian latent class analysis. *Journal of Statistical Software* **61**.

WHITE, A. & MURPHY, T. B. (2016). Exponential family mixed membership models for soft clustering of multivariate data. *Advances in Data Analysis and Classification* **10**, 521–540.

WHITE, A., WYSE, J. & MURPHY, T. B. (2016). Bayesian variable selection for latent class analysis using a collapsed Gibbs sampler. *Statistics and Computing* **26**, 511–527.

WITTEN, D. M. & TIBSHIRANI, R. (2010). A framework for feature selection in clustering. *Journal of the American Statistical Association* **105**, 713–726.

WOODBURY, M. A. (1950). Inverting modified matrices. Tech. rep., Statistical Research Group, Princeton University.

XIE, B., PAN, W. & SHEN, X. (2008a). Penalized model-based clustering with cluster-specific diagonal covariance matrices and grouped variables. *Electronic Journal of Statistics* **2**, 168.

XIE, B., PAN, W. & SHEN, X. (2008b). Variable selection in penalized model-based clustering via regularization on grouped parameters. *Biometrics* **64**, 921–930.

XIE, B., PAN, W. & SHEN, X. (2010). Penalized mixtures of factor analyzers with application to clustering high-dimensional microarray data. *Bioinformatics* **26**, 501–508.

YAU, C. & HOLMES, C. (2011). Hierarchical Bayesian nonparametric mixture models for clustering with variable relevance determination. *Bayesian Analysis* **6**, 329–352.

YEUNG, K. Y., FRALEY, C., MURUA, A., RAFTERY, A. E. & RUZZO, W. L. (2001). Model-based clustering and data transformations for gene expression data. *Bioinformatics* **17**, 977–987.

ZHOU, H., PAN, W. & SHEN, X. (2009). Penalized model-based clustering with unconstrained covariance matrices. *Electronic Journal of Statistics* **3**, 1473.

ZIMEK, A., SCHUBERT, E. & KRIEGEL, H.-P. (2012). A survey on unsupervised outlier detection in high-dimensional numerical data. *Statistical Analysis and Data Mining* **5**, 363–387.

# 12

# Mixture of Experts Models

**Isobel Claire Gormley and Sylvia Frühwirth-Schnatter**

*University College Dublin, Ireland; Vienna University of Economics and Business, Austria*

## CONTENTS

Mixture of experts models provide a framework in which covariates may be included in mixture models. This is achieved by modelling the parameters of the mixture model as functions of the concomitant covariates. Given their mixture model foundation, mixture of experts models possess a diverse range of analytic uses, from clustering observations to capturing parameter heterogeneity in cross-sectional data. This chapter focuses on delineating the mixture of experts modelling framework and demonstrates the utility and flexibility of mixture of experts models as an analytic tool.

## 12.1    Introduction

The term *mixture of experts models* encapsulates a broad class of mixture models in which the model parameters are modelled as functions of concomitant covariates. While the response variable $y$ is modelled via a mixture model, model parameters are modelled as functions of other, related, covariates $x$ from the context under study.

The mixture of experts (ME) nomenclature has its origins in the machine learning literature (Jacobs et al., 1991), but ME models appear in many different guises, including switching regression models (Quandt, 1972), concomitant variable latent class models (Dayton & Macready, 1988), latent class regression models (DeSarbo & Cron, 1988), and mixed models (Wang et al., 1996). Li et al. (2011) discuss finite smooth mixtures, a special case of ME modelling. McLachlan & Peel (2000) and Frühwirth-Schnatter (2006) provide background to a range of ME models; Masoudnia & Ebrahimpour (2014) survey the ME literature from a machine learning perspective.

The ME framework facilitates flexible modelling, allowing a wide range of applications. ME models for rank data (Gormley & Murphy, 2008b, 2010a), for network data (Gormley & Murphy, 2010b), for time series data (Waterhouse et al., 1996; Huerta et al., 2003; Frühwirth-Schnatter et al., 2012), for non-normal data (Villani et al., 2009, 2012; Chamroukhi, 2015) and for longitudinal data (Tang & Qu, 2015), among others, have been developed. Peng et al. (1996) employed a hierarchical ME model in a speech recognition context. The general ME framework has also been incorporated into the mixed membership model setting, giving rise to a mixed membership of experts model (White & Murphy, 2016), and into the infinite mixture model setting (Rasmussen & Ghahramani, 2002). Cluster weighted models (Ingrassia et al., 2015; Subedi et al., 2013; Gershenfeld, 1997) are also closely related to ME models.

This chapter introduces the generic ME framework in Section 12.2, and describes approaches to inference for ME models in Section 12.3. A broad range of illustrative data analyses are given in Section 12.4, and an overview of existing software packages which fit ME models is provided. Section 12.5 discusses identifiability issues for ME models. The chapter concludes with some discussion of the benefits and issues of the ME framework, and of some areas ripe for future development.

## 12.2    The Mixture of Experts Framework

Any mixture model which incorporates covariates or concomitant variables falls within the ME framework.

### 12.2.1    A mixture of experts model

Let $y_1, \ldots, y_n$ be an independent and identically distributed (i.i.d.) sample of outcome variables from a population modelled by a $G$-component finite mixture model. Depending on the application context, the outcome variable can be univariate or multivariate, discrete or continuous, or of a more general structure such as time series or network data. Each component $g$ (for $g = 1, \ldots, G$) is modelled by the probability density function $f_g(\cdot|\theta_g)$ with parameters denoted by $\theta_g$, and has weight $\eta_g$ where $\sum_{g=1}^{G} \eta_g = 1$. Observation $y_i$ ($i = 1, \ldots, n$) has $q$ associated covariates, which are denoted $x_i$. The ME model extends the standard finite mixture model introduced in Chapter 1 of this volume by allowing model parameters to be functions of the concomitant variables $x_i$:

$$p(y_i|x_i) = \sum_{g=1}^{G} \eta_g(x_i) f_g(y_i|\theta_g(x_i)). \tag{12.1}$$

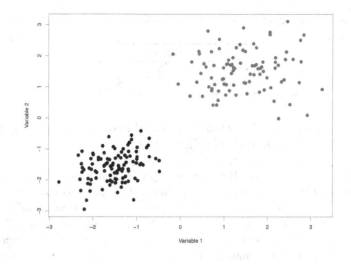

**FIGURE 12.1**
A two-dimensional simulated data set, from a $G = 2$ ME model. Black observations belong to cluster 1, and grey to cluster 2, based on the maximum *a posteriori* clustering from fitting a $G = 2$ mixture of bivariate Gaussian distributions.

ME models can be considered as a member of the class of conditional mixture models (Bishop, 2006); for a given set of covariates $x_i$, the distribution of $y_i$ is a finite mixture model. Jacobs et al. (1991) consider the component densities $f_g(y_i|\theta_g(x_i))$ as the *experts*, which model different parts of the input space, and the component weights $\eta_g(x_i)$ as the *gating networks*, hence the *mixture of experts* terminology.

The models for $\eta_g(x_i)$ and for $\theta_g(x_i)$ in (12.1) vary and are typically application specific. For example, Jacobs et al. (1991) model the component weights using a multinomial logit (MNL) regression model, and the component densities using generalized linear models. Young & Hunter (2010) provide further flexibility by allowing the mixing proportions to be modelled nonparametrically, as a function of the covariates.

## 12.2.2 An illustration

A simple simulated data set is employed here to introduce the ME framework. Figure 12.1 shows $n = 200$ two-dimensional continuous-valued observations, $y_1, \ldots, y_n$, simulated from an ME model with $G = 2$ components. A single ($q = 1$) categorical covariate $x_i$ is associated with each observation representing, for example, gender where level 0 denotes female. Interest lies in clustering the observations and exploring any relations between the resulting clusters and the associated covariate.

It is common that a clustering method is implemented on the outcome variables of interest, $y_1, \ldots, y_n$, without reference to the covariate information. Once a clustering has been produced, the user typically probes the clusters to investigate their structure. Interpretations of the clusters are produced with reference to values of the model parameters within each cluster and with reference to the covariates that were not used in the construction of the clusters. Therefore, a natural approach to modelling the data in Figure 12.1 is to cluster them by fitting a two-component mixture of bivariate Gaussian distributions to $y_1, \ldots, y_n$.

**TABLE 12.1**

Cross-tabulation of MAP cluster memberships and the gender covariate for the simulated data of Figure 12.1

|           | Female | Male |
|-----------|--------|------|
| Cluster 1 | 75     | 17   |
| Cluster 2 | 23     | 85   |

The maximum *a posteriori* (MAP) cluster membership of each observation resulting from fitting such a model is also illustrated in Figure 12.1.

A cross-tabulation of the MAP cluster memberships and the gender covariate is given in Table 12.1. It is clear that females have a strong presence in cluster 1, and males in cluster 2. However, the mixture of Gaussians model fitted does not incorporate or quantify this relationship or its associated uncertainty. It is in such a setting that an ME model is useful.

The model from which the data in Figure 12.1 are simulated is an ME model where $f_g(y_i|\theta_g) = \phi(y_i|\mu_g, \Sigma_g)$ is the density of a bivariate normal distribution and in which the component weights arise from a multinomial logit model with $G$ categories with gender as covariate $x_i$, that is,

$$\log \left[ \frac{\eta_g(x_i)}{\eta_1(x_i)} \right] = \gamma_{g0} + \gamma_{g1} x_i, \tag{12.2}$$

where cluster 1 is the baseline cluster with $\gamma_1 = (\gamma_{10}, \gamma_{11})^\top = (0,0)^\top$, and $g = 2, \ldots, G$. In our example, where $G = 2$, model (12.2) reduces to a binary logit model. The parameter $\gamma_{g1}$ (and its associated uncertainty) quantifies the relationship between the gender covariate and membership of cluster $g$, with $\gamma_{g1} = 0$ corresponding to independence between cluster membership and the gender covariate. Note that such a model easily extends to $q > 1$ covariates $x_i = (x_{i1}, \ldots, x_{iq})$ with associated parameter $\gamma_g = (\gamma_{g0}, \ldots, \gamma_{gq})^\top$ for cluster $g$.

Fitting such an ME model to the simulated data results in a MAP clustering unchanged from that reported in Table 12.1 and gives the maximum likelihood estimate (MLE) $\hat{\gamma}_{21} = 2.79$, with standard error 0.36. (Details of the maximum likelihood estimation process and standard error derivation follow in Section 12.3.1.) Thus, the odds of a male belonging to cluster 2 are $\exp(2.79) \approx 16$ times greater than the odds of a female belonging to cluster 2. Thus the ME model has clustering capabilities and provides insight into the type of observation which characterizes each cluster.

## 12.2.3  The suite of mixture of experts models

The ME model outlined in Section 12.2.2 involves modelling the component weights as a function of covariates. This is one model type (termed a *simple mixture of experts model*) from the ME framework. Figure 12.2 shows a graphical model representation of the suite of four models in the ME framework, based on a latent variable representation of the mixture model (12.1), involving the latent cluster membership of each observation, denoted $z_i$, where $z_i = g$ if observation $y_i$ belongs to cluster $g$. The indicator variable $z_i$ therefore has a multinomial distribution with a single trial and success probabilities equal to $\eta_g(x_i)$ for $g = 1, \ldots, G$ and the latent variable representation reads

$$y_i|x_i, z_i = g \sim f_g(y_i|\theta_g(x_i)), \quad \mathrm{P}(z_i = g|x_i) = \eta_g(x_i). \tag{12.3}$$

This suite of models ranges from a standard ME regression model (in which all model parameters are functions of covariates) to the special cases where some of the model parameters do not depend on covariates. The four models in the ME framework have the following interpretations (see also Figure 12.2):

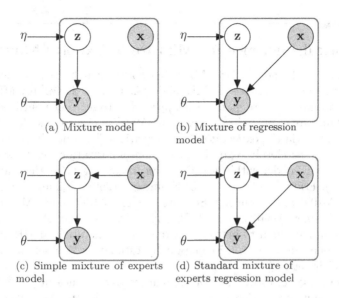

(a) Mixture model

(b) Mixture of regression model

(c) Simple mixture of experts model

(d) Standard mixture of experts regression model

**FIGURE 12.2**
The graphical model representation of mixture of experts models. The differences between the four special cases are due to the presence or absence of edges between the covariates $x$ and the latent variable $z$ and response variable $y$. For model (a) $p(y, z|x) = p(y|z)p(z)$; for model (b) $p(y, z|x) = p(y|x, z)p(z)$; for model (c) $p(y, z|x) = p(y|z)p(z|x)$; and for model (d) $p(y, z|x) = p(y|x, z)p(z|x)$.

(a) Mixture models, where the outcome variable distribution depends on the latent cluster membership, denoted $z$. The model is independent of the covariates $x$; that is, $p(y_i, z_i|x_i) = f_{z_i}(y_i|\theta_{z_i})\eta_{z_i}$.

(b) Mixture of regression models, where the outcome variable distribution depends on both the covariates $x$ and the latent cluster membership variable $z$. The distribution of the latent variable is independent of the covariates; that is, $p(y_i, z_i|x_i) = f_{z_i}(y_i|\theta_{z_i}(x_i))\eta_{z_i}$.

(c) Simple mixture of experts models, where the outcome variable distribution depends on the latent cluster membership variable $z$ and the distribution of the latent variable $z$ depends on the covariates $x$; that is, $p(y_i, z_i|x_i) = f_{z_i}(y_i|\theta_{z_i})\eta_{z_i}(x_i)$.

(d) Standard mixture of experts regression models, where the outcome variable distribution depends on both the covariates $x$ and on the latent cluster membership variable $z$. Additionally, the distribution of the latent variable $z$ depends on the covariates $x$; that is, $p(y_i, z_i|x_i) = f_{z_i}(y_i|\theta_{z_i}(x_i))\eta_{z_i}(x_i)$.

The manner in which the different models within the ME framework depend on the covariates is typically application-specific. The component weights are usually modelled using an MNL model, but this need not be the case; Geweke & Keane (2007) employ a model similar to an ME model, where the component weights have a multinomial probit structure. The form of the distribution $f_g(y_i|\theta_g(x_i))$ depends on the type of outcome data under study. The applications of the ME framework outlined in Section 12.4 include cases where the outcome data range from a categorical time series to rank data and network data.

## 12.3   Statistical Inference for Mixture of Experts Models

Before illustrating the breadth of the ME framework through illustrative applications in Section 12.4, the issue of inference for ME models is addressed. For any ME model that is underpinned by a finite mixture model, the approaches to inference outlined in Chapters 2 and 5 in this volume are applicable. Jacobs et al. (1991) and Jordan & Jacobs (1994) derive MLEs for ME models via the expectation-maximization (EM) algorithm; Gormley & Murphy (2008a) employ the closely related expectation-minorization-maximization algorithm. Estimation of the ME model within the Bayesian framework is detailed, among others, in Peng et al. (1996), Frühwirth-Schnatter & Kaufmann (2008), Villani et al. (2009), Gormley & Murphy (2010a) and Frühwirth-Schnatter et al. (2012) where Markov chain Monte Carlo methods (Tanner, 1996) are used; Bishop & Svensen (2003) use variational methods in the Bayesian paradigm to perform inference for a hierarchical ME model. Hunter & Young (2012) present an algorithm for parameter estimation in a semiparametric mixture of regression models setting.

In this section, a general overview of approaches to inference in the ME framework is provided. Throughout the section, $\mathbf{y} = (y_1, \ldots, y_n)$ will denote the collection of outcome variables and $\mathbf{x} = (x_1, \ldots, x_n)$ the associated covariates. The latent cluster membership indicators introduced in (12.3) are denoted by $\mathbf{z} = (z_1, \ldots, z_n)$, whereas $\theta = \{\theta_1, \ldots, \theta_G\}$ refers to the collection of the $G$ component parameters and $\gamma = \{\gamma_2, \ldots, \gamma_G\}$ to the unknown parameters in the $G$ component weights.

The exact manner in which an ME model is estimated again depends on the nature of the ME model and the outcome variable. The simple simulated data example of Section 12.2.2 is used here to delineate approaches to inference; more detailed application-specific estimation approaches are outlined in Section 12.4.

### 12.3.1   Maximum likelihood estimation

The EM algorithm (Dempster et al., 1977) provides an efficient approach to deriving MLEs in ME models. The EM algorithm is most commonly known as a technique to produce MLEs in settings where the data under study are incomplete or when optimization of the likelihood would be simplified if an additional set of variables were known. The iterative EM algorithm consists of an expectation (E) step followed by a maximization (M) step. Generally, during the E step the conditional expectation of the complete (i.e. observed and unobserved) data log likelihood is computed, given the data and current parameter values. In the M step the expected log likelihood is maximized with respect to the model parameters. The imputation of latent variables often makes maximization of the expected log likelihood more feasible. The parameter estimates produced in the M step are then used in a new E step and the cycle continues until convergence. The parameter estimates produced on convergence are estimates that achieve a stationary point of the likelihood function of the data, which is at least a local maximum but may be a saddle point.

The component weights of the simple ME model outlined in Section 12.2.2 are given by

$$\eta_g(x_i|\gamma) = \frac{\exp\left(\tilde{x}_i \gamma_g\right)}{\sum_{g'=1}^{G} \exp\left(\tilde{x}_i \gamma_{g'}\right)}, \tag{12.4}$$

where $\tilde{x}_i = (1, x_i)$ and $\gamma_g = (\gamma_{g0}, \gamma_{g1})^\top$. Note that this is a special case of the multinomial logit model. For the normal distribution, $\theta_g = \{\mu_g, \Sigma_g\}$ and the observed-data likelihood

**Algorithm 12.1** EM algorithm for a simple Gaussian mixture of experts model

Let $s = 0$. Choose initial estimates for the component weight parameters $\gamma^{(0)} = (0, \gamma_2^{(0)}, \dots, \gamma_G^{(0)})$ and, for each $g = 1, \dots, G$, for the component parameters $\mu_g^{(0)}$ and $\Sigma_g^{(0)}$.

1 **E step.** For $i = 1, \dots, n$ and $g = 1, \dots, G$, compute the estimates

$$z_{ig}^{(s+1)} = \frac{\eta_g^{(s)}(x_i|\gamma^{(s)})\phi(y_i|\mu_g^{(s)}, \Sigma_g^{(s)})}{\sum_{g'=1}^{G} \eta_{g'}^{(s)}(x_i|\gamma^{(s)})\phi(y_i|\mu_{g'}^{(s)}, \Sigma_{g'}^{(s)})}.$$

2 **M step.** Substituting the $z_{ig}^{(s+1)}$ values obtained in the E step into the log of the complete-data likelihood (12.5) forms the so-called "Q function"

$$Q = \sum_{i=1}^{n} \sum_{g=1}^{G} z_{ig}^{(s+1)} \left[ \tilde{x}_i \gamma_g - \log\left\{ \sum_{g'=1}^{G} \exp(\tilde{x}_i \gamma_{g'}) \right\} \right.$$
$$\left. -\frac{d}{2}\log(2\pi) - \frac{1}{2}\log|\Sigma_g| - \frac{1}{2}(y_i - \mu_g)^\top \Sigma_g^{-1}(y_i - \mu_g) \right],$$

with $d = \dim(y_i)$, which is maximized with respect to the model parameters.

(a) The updates of the $g = 1, \dots, G$ component means and covariances are, respectively,

$$\mu_g^{(s+1)} = \frac{\sum_{i=1}^{n} z_{ig}^{(s+1)} y_i}{\sum_{i=1}^{n} z_{ig}^{(s+1)}},$$

$$\Sigma_g^{(s+1)} = \frac{\sum_{i=1}^{n} z_{ig}^{(s+1)} (y_i - \mu_g^{(s+1)})(y_i - \mu_g^{(s+1)})^\top}{\sum_{i=1}^{n} z_{ig}^{(s+1)}}.$$

(b) The update for the component weight parameters is obtained via a numerical optimization (e.g. Newton–Raphson) step, where, for $g = 2, \dots, G$,

$$\gamma_g^{(s+1)} = \gamma_g^{(s)} - (H(\gamma_g^{(s)}))^{-1} Q'(\gamma_g^{(s)})$$

and $Q'$ and $H$ denote the first and second derivatives of $Q$ with respect to $\gamma_g$, respectively. Note that this M step is equivalent to fitting a generalized linear model with weights provided by the E step.

3 If converged, stop. Otherwise, increment $s$ and return to step 1.

function of the simple ME model is

$$L_o(\gamma, \theta; G) = p(\mathbf{y}|\mathbf{x}, \gamma, \theta) = \prod_{i=1}^{n} \sum_{g=1}^{G} \eta_g(x_i|\gamma)\phi(y_i|\mu_g, \Sigma_g),$$

where $\phi(y_i|\mu_g, \Sigma_g)$ is the pdf of the $d$-variate normal distribution and $d = \dim(y_i)$. It is difficult to directly obtain MLEs from this likelihood. To alleviate this, the data are augmented by imputing for each observation $y_i$, $i = 1, \dots, n$, the latent group membership indicator $z_i$. For the EM algorithm, this latent variable is represented through $G$ binary variables $(z_{i1}, \ldots, z_{iG})$ where $z_{ig} = \mathbb{I}(z_i = g)$ takes the value 1 if observation $y_i$ is a member of component $g$ and the value 0 otherwise. This provides the complete-data likelihood

$$L_c(\gamma, \theta, \mathbf{z}; G) = p(\mathbf{y}, \mathbf{z} | \mathbf{x}, \gamma, \theta) = \prod_{i=1}^{n} \prod_{g=1}^{G} \{\eta_g(x_i | \gamma) \phi(y_i | \mu_g, \Sigma_g)\}^{z_{ig}}, \qquad (12.5)$$

the expectation of (the log of) which is obtained in the E step of the EM algorithm. As the complete-data log likelihood is linear in the latent variable, the E step simply involves replacing, for each $i = 1, \ldots, n$, the missing data $z_i$ with their expected values $\hat{z}_i$. In the M step the complete-data log likelihood, computed with the estimates $\hat{\mathbf{z}} = (\hat{z}_1, \ldots, \hat{z}_n)$, is maximized to provide estimates of the component weight parameters $\hat{\gamma}$ and the component parameters $\hat{\theta}$.

The EM algorithm for fitting ME models is straightforward in principle, but the M step is often difficult in practice. This is usually due to a complex component density and/or component weights model, or a large parameter set. A modified version of the EM algorithm, the expectation and conditional maximization (ECM) algorithm (Meng & Rubin, 1993) is therefore often employed. In the ECM algorithm, the M step consists of a series of conditional maximization steps. In the context of the simple ME example considered here, these maximizations are not straightforward with regard to the $\gamma$ parameters; as in any MNL model, no closed-form expression for the parameter MLEs is available. Thus, while the conditional M steps for $\mu_g$ and $\Sigma_g$ for all $g = 1, \ldots, G$ are available in closed form, the conditional M step for $\gamma$ requires the use of a numerical optimization technique, or, as in Gormley & Murphy (2008b), the MM algorithm (Hunter & Lange, 2004) in which a minorizing function is iteratively maximized and updated. In summary, to fit the simple ME example outlined in Section 12.2.2 the EM algorithm proceeds as described in Algorithm 12.1. In the simulated data example, $d = 2$.

McLachlan & Peel (2000) outline a number of approaches to assessing convergence in step 3; typically it is assessed by tracking the change in the log likelihood as the algorithm proceeds. Standard errors of the resulting parameter estimates are not automatically produced by the EM algorithm, but they can be approximately computed after convergence, for example, by computing and inverting the observed information matrix (McLachlan & Peel, 2000). For a detailed discussion of EM algorithms in a mixture context, see Chapters 2 and 3 of this volume.

## 12.3.2   Bayesian estimation

Estimation of ME models can be achieved within the Bayesian paradigm, either using a Markov chain Monte Carlo (MCMC) algorithm or via a variational approach. The reader is directed to Bishop & Svensen (2003) for details on the variational approach; this section focuses on inference using MCMC methods. Both the Gibbs sampler (Geman & Geman, 1984) and the Metropolis–Hastings algorithm (Chib & Greenberg, 1995; Metropolis et al., 1953) are typically required. Again, the specific MCMC algorithm, and the form of the prior distributions, depend on the nature of the ME model under study and on the type of the response data. As is standard in Bayesian estimation of mixture models (Diebolt & Robert, 1994) and of mixture of regression models (Hurn et al., 2003), fitting ME models is greatly simplified by augmenting the observed data with the latent group indicator variable $z_i$ for each observation $y_i$.

Performing inference on the illustrative simple ME model of Section 12.2.2 is again straightforward in principle, but can be difficult in practice. To begin, priors for the model parameters $\mu_g$, $\Sigma_g$ ($g = 1, \ldots, G$) and $\gamma_g$ ($g = 2, \ldots, G$) require specification. Positing a conditional $d$-variate normal prior $\mathcal{N}(\mu_0, \Lambda_0)$ on the group means $\mu_g$ and an inverse Wishart

---

**Algorithm 12.2** Metropolis–Hastings-within-Gibbs MCMC inference for a simple Gaussian mixture of experts model

---

Iterate the following steps for $m = 1, \ldots, M$:

1. For $g = 1, \ldots, G$, draw $\mu_g$ from the $d$-variate normal posterior $\mathcal{N}(\mu_{ng}, \Lambda_{ng})$ where $\Lambda_{ng} = (\Lambda_0^{-1} + n_g \Sigma_g^{-1})^{-1}$ and $\mu_{ng} = \Lambda_{ng}(\Lambda_0^{-1}\mu_0 + \Sigma_g^{-1} n_g \bar{y}_g)$, in which $n_g = \sum_{i=1}^{n} \mathbb{I}(z_i = g)$ and $n_g \bar{y}_g = \sum_{i=1}^{n} y_i \mathbb{I}(z_i = g)$.

2. For $g = 1, \ldots, G$, draw $\Sigma_g$ from $\mathcal{W}^{-1}(\nu_{ng}, S_{ng})$ where $\nu_{ng} = \nu_0 + n_g$ and $S_{ng} = S_0 + \sum_{i=1}^{n} \mathbb{I}(z_i = g)(y_i - \mu_g)(y_i - \mu_g)^{\top}$.

3. For $i = 1, \ldots, n$, draw $z_i$ from a multinomial distribution $\mathcal{M}(1, p_{i1}, \ldots, p_{iG})$ with success probabilities $(p_{i1}, \ldots, p_{iG})$ where

$$p_{ig} = \frac{\eta_g(x_i|\gamma)\phi(y_i|\mu_g, \Sigma_g)}{\sum_{g'=1}^{G} \eta_{g'}(x_i|\gamma)\phi(y_i|\mu_{g'}, \Sigma_{g'})}.$$

4. For $g = 2, \ldots, G$, the component weight parameters $\gamma_g$ are updated via a Metropolis–Hastings step, while holding the remaining component weight parameters $\gamma_{-g}$ fixed. Typically, a multivariate normal proposal distribution $q(\gamma_g^*|\gamma_g, \gamma_{-g})$ is employed:

  (a) Propose $\gamma_g^* \sim \mathcal{N}(\tilde{\mu}_\gamma, \tilde{\Lambda}_\gamma)$ from a $(q+1)$-variate normal distribution where $\tilde{\mu}_\gamma$ and $\tilde{\Lambda}_\gamma$ are specified by the user and might depend on the current value of $\gamma$.

  (b) If $U \sim U[0, 1]$ is such that

$$U \leq \min\left\{\frac{p(\mathbf{z}|\gamma_g^*, \gamma_{-g}, \mathbf{x})p(\gamma_g^*)q(\gamma_g \mid \gamma_g^*, \gamma_{-g})}{p(\mathbf{z}|\gamma_g, \gamma_{-g}, \mathbf{x})p(\gamma_g)q(\gamma_g^* \mid \gamma_g, \gamma_{-g})}, 1\right\},$$

  then set $\gamma_g = \gamma_g^*$; otherwise leave $\gamma_g$ unchanged.

---

prior $\mathcal{W}^{-1}(\nu_0, S_0)$ on the group covariances $\Sigma_g$ provides conjugacy for these parameters (Hoff, 2009). The full conditional distributions for these parameters are therefore available in closed form, and thus Gibbs sampling can be used to draw samples.

A $(q+1)$-variate normal $\mathcal{N}(\mu_\gamma, \Lambda_\gamma)$ is an intuitive prior for the component weight parameters $\gamma_g$, but it is non-conjugate. Hence the full conditional distribution is not available in closed form and a Metropolis–Hastings (MH) step can be applied to sample the component weight parameters. One sweep of such a Metropolis-within-Gibbs sampler required to fit the simple ME model of Section 12.2.2 in a Bayesian framework is outlined in Algorithm 12.2. Note that the full conditional distribution of the latent indicator variable $z_i$, for $i = 1, \ldots, n$, is also available in closed form and thus Gibbs sampling is possible in step 3 of Algorithm 12.2.

Sampling the component weight parameters in step 4 through an MH algorithm brings issues such as choosing suitable proposal distributions $q(\gamma_g^*|\gamma_g, \gamma_{-g})$ and tuning parameters, which may make fitting ME models troublesome. Gormley & Murphy (2010b) detail an approach to deriving proposal distributions with attractive properties, within the context of an ME model for network data. Villani et al. (2009, 2012) introduce a highly efficient MCMC scheme based on a Metropolis-within-Gibbs sampler that exploits a few Newton iterations to construct suitable proposal distributions.

Alternatively, Frühwirth-Schnatter et al. (2012) exploit data augmentation of the MNL model (12.4) based on the differenced random utility model representation in the context of ME models to implement step 4. As shown by Frühwirth-Schnatter & Frühwirth (2010), for each $g = 1, \ldots, G$ the MNL model has the following representation as a binary logit model conditional on knowing $\lambda_{hi} = \exp(\tilde{x}_i \gamma_h)$ for all $h \neq g$:

$$u_{gi} = \tilde{x}_i \gamma_g - \log\left(\sum_{h \neq g} \lambda_{hi}\right) + \varepsilon_{gi}, \tag{12.6}$$

$$D_i^g = \mathbb{I}(u_{gi} \geq 0).$$

In (12.6), $u_{gi}$ is a latent variable, $\varepsilon_{gi}$ are i.i.d. errors following a logistic distribution, while $D_i^g = \mathbb{I}(z_i = g)$ is a binary outcome variable indicating whether the group indicator $z_i$ is equal to $g$. Note that $\gamma_1 = 0$ for the baseline, hence $\lambda_{1i} = 1$. In a data-augmented implementation of step 4, the latent variables $(u_{2i}, \ldots, u_{Gi})$ are introduced, for each $i = 1, \ldots, n$, as unknowns. Given $\lambda_{2i}, \ldots, \lambda_{Gi}$ and $z_i$, $(u_{2i}, \ldots, u_{Gi})$ can be sampled in closed form from exponentially distributed random variables. Following Scott (2011), natural proposal distributions are available to implement an MH step to sample $\gamma_g | \gamma_{-g}, \mathbf{z}, \mathbf{u}_g$, for all $g = 2, \ldots, G$, conditional on $\mathbf{u}_g = \{u_{g1}, \ldots, u_{gn}\}$ from the linear, non-Gaussian regression model (12.6).

To avoid any MH-step, Frühwirth-Schnatter et al. (2012) apply auxiliary mixture sampling as introduced by Frühwirth-Schnatter & Frühwirth (2010) to (12.6) and approximate for each $\varepsilon_{gi}$ the logistic distribution by a 10-component scale mixture of normal distributions with zero means and parameters $(s_r^2, w_r)$, $r = 1, \ldots, 10$. In a second step of data augmentation, the component indicator $r_{gi}$ is introduced as yet another latent variable. Conditional on the latent variables $\mathbf{u}_g$ and the indicators $\mathbf{r}_g = \{r_{g1}, \ldots, r_{gn}\}$ the binary logit model (12.6) reduces to a linear Gaussian regression model. Hence, the posterior $\gamma_g | \gamma_{-g}, \mathbf{z}, \mathbf{u}_g, \mathbf{r}_g$ is Gaussian and a Gibbs step is available to sample $\gamma_g$ for all $g = 2, \ldots, G$ conditional on $\mathbf{u}_g$ and $\mathbf{r}_g$. Finally, each component indicator $r_{gi}$ is sampled from a discrete distribution conditional on $u_{gi}$ and $\gamma$.

Chapter 13 in this volume details Bayesian estimation of informative regime switching models which can be regarded as an extension of ME models to hidden Markov models in time series analysis.

As in any mixture model setting, the label switching problem already mentioned in earlier chapters (see also Stephens, 2000; Frühwirth-Schnatter, 2011a) must be considered when employing such Gibbs based algorithms; see Chapter 5 in this volume. This identifiability issue, along with others, is discussed in Section 12.5.

### 12.3.3   Model selection

Within the suite of ME models outlined in Section 12.2.3 the question of which, how and where covariates are used naturally arises. This is a challenging problem as the space of ME models is potentially very large, once variable selection for the covariates entering the component weights and the mixture components is considered. Thus in practice only models where covariates enter all mixture components and/or all component weights as main effects are typically considered in order to restrict the size of the model search space. In fact, even for this reduced model space, there are a maximum of $G \times 2^q \times 2^q$ possible models to consider. As shown by Villani et al. (2012), in ME models involving generalized linear models of covariates, variable selection approaches can be used to prevent overfitting and find the optimal model. Practical approaches to this issue are detailed in the illustrative applications in Section 12.4. Note that the manner in which covariates enter the ME model may also be guided by the question of interest in the application under study.

If the number of components $G$ is unknown, the model search space increases again. Approaches such as marginal likelihood evaluation, or information criteria, are useful for choosing the optimal $G$ in ME models; the reader is referred to Chapter 7 in this volume which addresses model selection and selecting the number of components in a mixture model in great detail.

## Marginal likelihood computation for mixture of experts models

As discussed in Chapter 7, Section 7.2.3.2, highly accurate sampling-based approximations to the marginal likelihood are available, if $G$ is not too large. For instance, Frühwirth-Schnatter & Kaufmann (2008) apply bridge sampling (Frühwirth-Schnatter, 2004) to compute marginal likelihoods for an ME model with a single covariate ($q = 1$) with up to four components. Frühwirth-Schnatter (2011b) combines auxiliary mixture sampling (Frühwirth-Schnatter & Wagner, 2008) with importance sampling to compute marginal likelihoods for ME models. A detailed summary of this approach is provided below.

Permutation sampling is applied to ensure that all equivalent modes of the posterior distribution are visited. Consider a permutation $\mathfrak{s} \in \mathfrak{S}(G)$, where $\mathfrak{S}(G)$ denotes the set of the $G!$ permutations of $\{1, \ldots, G\}$. To relabel all parameters in an ME model according to the permutation $\mathfrak{s}$, define $\theta_g^\star = \theta_{\mathfrak{s}(g)}$ and $\eta_g^\star(\tilde{x}_i) = \eta_{\mathfrak{s}(g)}(\tilde{x}_i)$ for $g = 1, \ldots, G$. Special attention has to be given to the correct relabelling of the coefficients $\gamma_g$ in the MNL model when applying the permutation $\mathfrak{s}$. The coefficients $(\gamma_1, \ldots, \gamma_G)$ and $(\gamma_1^\star, \ldots, \gamma_G^\star)$ defining, respectively, the MNL models $\eta_g(\tilde{x}_i)$ and $\eta_g^\star(\tilde{x}_i)$ are related through

$$
\begin{aligned}
\tilde{x}_i \gamma_g^\star &= \log\left[\frac{\eta_g^\star(\tilde{x}_i)}{\eta_{g_0}^\star(\tilde{x}_i)}\right] = \log\left[\frac{\eta_{\mathfrak{s}(g)}(\tilde{x}_i)}{\eta_{\mathfrak{s}(g_0)}(\tilde{x}_i)}\right] = \log\left[\frac{\eta_{\mathfrak{s}(g)}(\tilde{x}_i)}{\eta_{g_0}(\tilde{x}_i)}\right] - \log\left[\frac{\eta_{\mathfrak{s}(g_0)}(\tilde{x}_i)}{\eta_{g_0}(\tilde{x}_i)}\right] \\
&= \tilde{x}_i(\gamma_{\mathfrak{s}(g)} - \gamma_{\mathfrak{s}(g_0)}).
\end{aligned}
$$

To ensure that the baseline $g_0$ (assumed to be equal to $g_0 = 1$ throughout this chapter) remains the same, despite relabelling, the coefficients are permuted in the following way:

$$
\gamma_g^\star = \gamma_{\mathfrak{s}(g)} - \gamma_{\mathfrak{s}(g_0)}, \quad g = 1, \ldots, G,
$$

which indeed implies that $\gamma_{g_0}^\star = 0$. For $G = 2$, the sign of all coefficients of $\gamma_2$ is simply flipped if $\mathfrak{s} = (2, 1)$, and remains unchanged otherwise.

Frühwirth-Schnatter & Wagner (2008) discuss various importance sampling estimators of the marginal likelihood for non-Gaussian models such as logistic models. Using auxiliary mixture sampling, one of their approaches constructs the importance density from the Gaussian full conditional densities appearing in the augmented Gibbs sampler. This approach is easily extended to ME models. As discussed in Section 12.3.2, auxiliary mixture sampling yields Gaussian posteriors $p(\gamma_g|\gamma_{-g}, \mathbf{z}, \mathbf{u}_g, \mathbf{r}_g)$ for the MNL coefficients $\gamma_g$ in an ME model, conditional on the latent utilities $\mathbf{u}_g$ and the latent indicators $\mathbf{r}_g$. This allows for the construction of an importance density $q_G(\theta)$ as in Section 7.2.3.2 above, but it is essential that $q_G(\theta)$ covers all symmetric modes of the mixture posterior. A successful strategy is to apply random permutation sampling, where each sampling step is concluded by relabelling as described above, using a randomly selected permutation $\mathfrak{s} \in \mathfrak{S}(G)$. The corresponding importance density reads

$$
q_G(\theta) = \frac{1}{S} \sum_{s=1}^{S} \prod_{g=2}^{G} p(\gamma_g|\gamma_{-g}^{(s)}, \mathbf{u}_g^{(s)}, \mathbf{r}_g^{(s)}, \mathbf{z}^{(s)}) \prod_{g=1}^{G} p(\theta_g|\mathbf{z}^{(s)}, \mathbf{y}), \tag{12.7}
$$

where $\{\gamma^{(s)}, \mathbf{u}_2^{(s)}, \ldots, \mathbf{u}_G^{(s)}, \mathbf{r}_2^{(s)}, \ldots, \mathbf{r}_G^{(s)}, \mathbf{z}^{(s)}\}$, $s = 1, \ldots, S$, is a subsequence of posterior draws. Only if $S$ is large compared to $G!$ are all symmetric modes covered by random permutation sampling, with the number of visits per mode being on average $S/G!$. The construction of this importance density is fully automatic and it is sufficient to store the moments of the various conditional densities (rather than the allocations $\mathbf{z}$ and the latent utilities $\mathbf{u}_g$ and indicators $\mathbf{r}_g$ themselves) during MCMC sampling for later evaluation. This importance density is used to compute importance sampling estimators of the marginal likelihood; see the illustrative application in Section 12.4.1.

## 12.4  Illustrative Applications

The utility of mixture of experts models is illustrated in this section through the use of several applications. ME Markov chain models for categorical time series, ME models for ranked preference data, and ME models for network data, all of which are members of the ME model framework, are applied.

### 12.4.1  Analysing marijuana use through mixture of experts Markov chain models

Lang et al. (1999) studied data on the marijuana use of 237 teenagers taken from five annual waves (1976–80) of the US National Youth Survey. The respondents were 13 years old in 1976 and for five consecutive years reported their marijuana use in the past year as a categorical variable with the three categories "never", "not more than once a month" and "more than once a month". Hence, for $i = 1, \ldots, 237$, the outcome variable is a categorical time series $y_i = (y_{i0}, y_{i1}, \ldots, y_{i4})$ with three states, labelled 1 for non-users, 2 for light and 3 for heavy users.

To identify groups of teenagers with similar marijuana use behaviour, Frühwirth-Schnatter (2011b) applied an ME approach based on Markov chain models (Frühwirth-Schnatter et al., 2012) and considered each time series $y_i$ as a single entity belonging to one of $G$ underlying classes. Various types of ME Markov chain models were applied to capture dependence in marijuana use over time and to investigate if the gender of the teenagers can be associated with a certain type of marijuana use.

Given the times series nature of the categorical outcome variable $y_i$, the component density $f_g(\cdot)$ in the ME model (12.1) must have an appropriate form, and various models are considered. Model $\mathcal{M}_1$ is a standard finite mixture of time-homogeneous Markov chain models of order one (Pamminger & Frühwirth-Schnatter, 2010) where each component-specific density $f_g(\cdot)$ in (12.1) is characterized by a transition matrix $\xi_g$ with $J = 3$ rows and the weight distribution $\eta_1, \ldots, \eta_G$ is independent of any covariates. Each row $\xi_{g,j\cdot} = (\xi_{g,j1}, \ldots, \xi_{g,j3})$, $j = 1, \ldots, J$, of the matrix $\xi_g$ represents a probability distribution over the three categories of marijuana use with

$$\xi_{g,jk} = \mathrm{P}(y_{it} = k | y_{i,t-1} = j, z_i = g), \quad k = 1, \ldots, 3.$$

This model is extended in various ways to include covariate information into the transition behaviour. First, an inhomogeneous model (labelled model $\mathcal{M}_2$) is considered, where the transition matrix in each group depends on the gender $x_i$ of the teenager. If all $J = 6$ possible combinations $\mathcal{H}_{it} = (y_{i,t-1}, x_i)$ of the immediate past $y_{i,t-1}$ at time $t$ and the gender $x_i$ are indexed by $j = 1, \ldots, J$, then the component-specific density $f_g(y_i | \xi_g)$ in (12.1) can be described by a generalized transition matrix $\xi_g$ with six rows, with the $j$th row $\xi_{g,j\cdot} = (\xi_{g,j1}, \ldots, \xi_{g,j3})$ describing again the conditional distribution of $y_{it}$, given that the state of the history $\mathcal{H}_{it}$ equals $j$:

$$\xi_{g,jk} = \mathrm{P}(y_{it} = k | \mathcal{H}_{it} = j, z_i = g), \quad k = 1, \ldots, 3.$$

**TABLE 12.2**
Marijuana data; marginal likelihood $\log p(\mathbf{y}|\mathcal{M}_k)$ for various finite mixtures of homogeneous $(\mathcal{M}_1)$ and inhomogeneous $(\mathcal{M}_2, \mathcal{M}_3, \mathcal{M}_4)$ Markov chain models with an increasing number $G$ of classes (best values for each model in bold type)

| | | | | $G$ | |
| Model | Covariates | $J$ | 1 | 2 | 3 |
|---|---|---|---|---|---|
| $\mathcal{M}_1$ | – | 3 | −605.5 | **−600.0** | −600.3 |
| $\mathcal{M}_2$ | $x_i$ | 6 | −610.0 | **−601.3** | −603.6 |
| $\mathcal{M}_3$ | $t$ | 12 | −613.7 | **−596.5** | −599.4 |
| $\mathcal{M}_4$ | $t, x_i$ | 24 | −619.8 | −602.7 | **−601.1** |

Evidently, the component-specific distribution reads

$$f_g(y_i|\xi_g) = \prod_{j=1}^{J} \prod_{k=1}^{3} \xi_{g,jk}^{n_{i,jk}}, \tag{12.8}$$

where, for each time series $i$, $n_{i,jk} = \sum_{t=1}^{4} \mathbb{I}(y_{it} = k, \mathcal{H}_{it} = j)$ is the number of transitions into state $k$ given a history of type $j$. Note that (12.8) is formulated conditional on the first observation $y_{i0}$.

Alternative component-specific distributions can be constructed by defining the history $\mathcal{H}_{it}$ through different combinations of past values and covariates. Choosing $\mathcal{H}_{it} = (y_{i,t-1}, t)$, for instance, defines a time-inhomogeneous Markov chain model, labelled model $\mathcal{M}_3$, with $J = 12$ different covariate combinations. This model is able to capture the effect that the transition behaviour between the states might change as the teenagers grow older.

The most complex model, labelled model $\mathcal{M}_4$, extends model $\mathcal{M}_3$ by assuming additional dependence on gender, $\mathcal{H}_{it} = (y_{i,t-1}, t, x_i)$, with $J = 24$ different covariate combinations. Both models $\mathcal{M}_3$ and $\mathcal{M}_4$ are characterized by component-specific generalized transition matrices $\xi_g$ with, respectively, 12 and 24 rows. For each of the models $\mathcal{M}_2, \mathcal{M}_3, \mathcal{M}_4$, it is assumed that the weight distribution $\eta_1, \ldots, \eta_G$ is independent of any covariate, leading to various finite mixtures of inhomogeneous Markov chain models.

Bayesian inference is carried out for all models $\mathcal{M}_1, \ldots, \mathcal{M}_4$ for an increasing number $G = 1, 2, 3$ of classes. MCMC estimation as described in Section 12.3.2 is easily applied, as the $J$ rows $\xi_{g,j\cdot}$ of $\xi_g$ are conditionally independent under the conditionally conjugate Dirichlet prior $\xi_{g,j\cdot} \sim \mathcal{D}(d_{0,j1}, \ldots, d_{0,j3})$. Given $\mathbf{z}$ and $\mathbf{y}$, the generalized transition matrix $\xi_g$ is sampled row-by-row from a total of $JG$ Dirichlet distributions:

$$\xi_{g,j\cdot}|\mathbf{z}, \mathbf{y} \sim \mathcal{D}(d_{0,j1} + n_{j1}^g, \ldots, d_{0,j3} + n_{j3}^g), \quad j = 1, \ldots, J, \ g = 1, \ldots, G, \tag{12.9}$$

where $n_{jk}^g = \sum_{i:z_i=g} n_{i,jk}$ is the total number of transitions into state $k$ observed in class $g$, given a history of type $j$.

For model comparison, the marginal likelihood is computed explicitly for $G = 1$, while importance sampling as described in Section 12.3.3 is applied for $G = 2, 3$, using the importance density

$$q_G(\theta) = \frac{1}{S} \sum_{s=1}^{S} p(\eta|\mathbf{z}^{(s)}) \prod_{g=1}^{G} \prod_{j=1}^{J} p(\xi_{g,j\cdot}|\mathbf{z}^{(s)}, \mathbf{y}),$$

where $p(\xi_{g,j\cdot}|\mathbf{z}, \mathbf{y})$ is equal to the full conditional Dirichlet posterior of $\xi_{g,j\cdot}$ given in (12.9). Random permutation sampling is applied to ensure that all $G!$ symmetric modes are visited and $S = 10,000$. The marginal likelihoods reported in Table 12.2 select $G = 2$ for all models

**TABLE 12.3**
Marijuana data; finite mixture of time-inhomogeneous Markov chain models (model $\mathcal{M}_3$) with $G = 2$ classes. the estimated posterior mean $E(\xi_{g,.}|\mathbf{y})$ is arranged for each $t = 1, \ldots, 4$ as a $3 \times 3$ matrix; the estimated class sizes $\hat{\eta}_g$ are equal to the posterior mean $E(\eta_g|\mathbf{y})$

| | $t = 1$ | | | $t = 2$ | | | $t = 3$ | | | $t = 4$ | | |
|---|---|---|---|---|---|---|---|---|---|---|---|---|
| Group 1 | 0.93 | 0.04 | 0.03 | 0.89 | 0.09 | 0.02 | 0.90 | 0.04 | 0.06 | 0.93 | 0.04 | 0.03 |
| ($\hat{\eta}_1 = 0.56$) | 0.50 | 0.17 | 0.34 | 0.10 | 0.33 | 0.57 | 0.17 | 0.65 | 0.18 | 0.20 | 0.64 | 0.16 |
| | 0.22 | 0.18 | 0.60 | 0.10 | 0.17 | 0.73 | 0.04 | 0.27 | 0.69 | 0.15 | 0.12 | 0.74 |
| Group 2 | 0.76 | 0.21 | 0.03 | 0.70 | 0.24 | 0.06 | 0.75 | 0.18 | 0.07 | 0.45 | 0.43 | 0.12 |
| ($\hat{\eta}_2 = 0.44$) | 0.34 | 0.15 | 0.51 | 0.23 | 0.39 | 0.38 | 0.31 | 0.41 | 0.28 | 0.46 | 0.43 | 0.11 |
| | 0.18 | 0.23 | 0.59 | 0.10 | 0.22 | 0.68 | 0.13 | 0.15 | 0.72 | 0.05 | 0.10 | 0.85 |

**TABLE 12.4**
Marijuana data; ME model with $\tilde{x}_i = (1, x_i, D_{i0})$ (model $\mathcal{M}_5$), extending model $\mathcal{M}_3$ with $G = 2$ classes. Posterior expectation and 95% HPD region of the component weight parameters $\gamma_{2j}$ in the ME model (12.4)

| Covariate $\tilde{x}_{ij}$ | $E(\gamma_{2j}|\mathbf{y})$ | 95% HPD region of $\gamma_{2j}$ |
|---|---|---|
| constant | $-0.69$ | $(-1.75, 0.35)$ |
| male (baseline: female) | 0.28 | $(-0.71, 1.22)$ |
| marijuana use in 1976 (baseline: no) | $-0.07$ | $(-1.70, 1.43)$ |
| $\log p(\mathbf{y}|\mathcal{M}_5)$ | -598.5 | |

except for $\mathcal{M}_4$, where $G - 3$ is selected. Among all models, the marginal likelihood is the highest for model $\mathcal{M}_3$ with $G = 2$ classes.

Hence, a time-inhomogeneous Markov chain model which does not depend on gender best describes the transition behaviour in each class. Table 12.3 reports the corresponding posterior means $E(\xi_{g,.}|\mathbf{y})$ and $E(\eta_g|\mathbf{y})$ for each of the two groups. Label switching was resolved by applying $k$-means clustering to a vector constructed from all persistence probabilities at all time points. Both groups are roughly of equal size, with the first group being slightly larger. A characteristic difference is evident for the two groups of teenagers. In group 1, non-users have a high probability $\xi_{t,11}$ of remaining non-users throughout the whole observation period, whereas this probability is much smaller for the second group right from the beginning and drops to only 45% in the last year.

To investigate if gender is associated with group membership, model $\mathcal{M}_3$ with $G = 2$ classes is combined with the ME model (12.4), by including gender as subject-specific covariate $x_i$ as in the example in Section 12.2.2. This model is labelled model $\mathcal{M}_5$. Additionally, a dummy variable $D_{i0}$ is included, indicating if the teenager was a light or heavy marijuana user in the first year. As $G = 2$, the ME model (12.4) reduces to a binary logit model with regression coefficients $\gamma_2 = (\gamma_{20}, \gamma_{21}, \gamma_{22})$, each assumed to follow a standard normal prior distribution.

From posterior inference in Table 12.4, we find that male teenagers have a slightly higher probability of belonging to the second group, because $E(\gamma_{21}|\mathbf{y}) > 0$; however, the coefficient $\gamma_{21}$ is not significantly different from 0. Similarly, the initial state from which a teenager started in 1976 does not have a significant influence on the probability of belonging to the second group. This suggests that the ME time-inhomogeneous Markov chain model actually reduces to a standard mixture of time-inhomogeneous Markov chain models, which is confirmed by comparing the log marginal likelihood of both models, being equal to $-596.5$ for a standard mixture model with $G = 2$ groups (see Table 12.2) and $-598.5$ for an ME model with $G = 2$ groups (see Table 12.4).

The marginal likelihood estimator for the ME model is based on importance sampling using the importance density (12.7) derived from auxiliary mixture sampling:

$$q_G(\theta) = \frac{1}{S} \sum_{s=1}^{S} p(\gamma_2|\mathbf{u}_2^{(s)}, \mathbf{r}_2^{(s)}, \mathbf{z}^{(s)}) \prod_{g=1}^{2} \prod_{j=1}^{J} p(\xi_{g,j}.|\mathbf{z}^{(s)}, \mathbf{y}),$$

where $p(\gamma_2|\mathbf{u}_2, \mathbf{r}_2, \mathbf{z})$ is conditionally Gaussian and $p(\xi_{g,j}.|\mathbf{z}, \mathbf{y})$ is equal to the full conditional Dirichlet posterior of $\xi_{g,j}.$ given in (12.9). Again, random permutation sampling is applied to ensure that the two equivalent modes are visited.

To sum up, this investigation shows that teenagers may indeed be clustered into two groups with different behaviour with respect to marijuana use, the first being a non-user group, while the second group has a much higher risk of becoming a user. Preference for a standard mixture of Markov chain models over an ME Markov chain model based on gender shows that the two types of marijuana use cannot be associated with the gender of the teenager. Both male and female teenagers have about the same risk of belonging to the second group. Unobserved factors, rather than gender, are relevant for the membership of a teenager in one group or the other.

## 12.4.2 A mixture of experts model for ranked preference data

Mary McAleese served as the eighth president of Ireland from 1997 to 2011 and was elected under the single transferable vote electoral system. Under this system voters rank, in order of their preference, some or all of the electoral candidates. The vote counting system which results in the elimination of candidates and the subsequent election of the president is an intricate process involving the transfer of votes between candidates as specified by the voters' ballots. Details of the electoral system, the counting process and the 1997 Irish presidential election are given in Coakley & Gallagher (2004), Sinnott (1995, 1999) and Marsh (1999).

The 1997 presidential election race involved five candidates: Mary Banotti, Mary McAleese, Derek Nally, Adi Roche and Rosemary Scallon. Derek Nally and Rosemary Scallon were independent candidates, while Mary Banotti and Adi Roche were endorsed by the then current opposition parties Fine Gael and Labour, respectively. Mary McAleese was endorsed by the Fianna Fáil party who were in power at that time. In terms of candidate type, McAleese and Scallon were deemed to be conservative candidates, with the other candidates regarded as liberal. Gormley & Murphy (2008a,b, 2010a,b) provide further details on the 1997 presidential election and on the candidates.

One month prior to election day a survey was conducted by Irish Marketing Surveys with 1083 respondents. Respondents were asked to list some or all of the candidates in order of preference, as if they were voting on the day of the poll. In addition, pollsters gathered data on attributes of the respondents as detailed in Table 12.5.

Interest lies in determining if groups of voters with similar preferences (i.e. voting blocs) exist within the electorate. If such voting blocs do exist, the influence the recorded socio-economic variables may have on the clustering structure and/or on the preferences which characterize a voting bloc is also of interest. Jointly modelling the rank preference votes and the covariates through an ME model for rank preference data when clustering the electorate provides this insight.

Given the rank nature of the outcome variables or votes $y_i$ ($i = 1, \ldots, n = 1083$) the component density $f_g(\cdot)$ in the ME model (12.1) must have an appropriate form. The Plackett–Luce or exploded logit model (Plackett, 1975; Gormley & Murphy, 2006) for rank data provides a suitable model; Benter's model (Benter, 1994) provides another alternative. Let $y_i = [c(i,1), \ldots, c(i, m_i)]$ denote the ranked ballot of voter $i$, where $c(i, j)$ denotes the candidate ranked in $j$th position by voter $i$ and $m_i$ is the number of candidates ranked by

**TABLE 12.5**
Covariates recorded for each respondent in the Irish Marketing Surveys poll

| Age | Area | Gender | Government satisfaction | Marital status | Social class |
|---|---|---|---|---|---|
| – | City | Housewife | No opinion | Married | AB |
| | Rural | Male | Not satisfied | Single | C1 |
| | Town | Non-housewife | Satisfied | Widowed | C2 |
| | | | | | DE |
| | | | | | F50+ |
| | | | | | F50− |

**TABLE 12.6**
The model with smallest BIC within each type of ME model for ranked preference data applied to the 1997 Irish presidential election data

| | BIC | G | Covariates |
|---|---|---|---|
| Simple mixture of experts model | 8491 | 4 | $\eta_g$: government satisfaction, age. |
| Standard mixture of experts Regression model | 8512 | 3 | $\eta_g$: government satisfaction, age. $p_g$: Age |
| Mixture model | 8513 | 3 | – |
| Mixture of regression models | 8528 | 1 | $p_g$: government satisfaction |

voter $i$. Under the Plackett–Luce model, given that voter $i$ is a member of voting bloc $g$ and given the "support parameter" $p_g = (p_{g1}, \ldots, p_{gM})$, the probability of voter $i$'s ballot is

$$p(y_i|p_g) = \frac{p_{g,c(i,1)}}{\sum_{s=1}^{M} p_{g,c(i,s)}} \cdot \frac{p_{g,c(i,2)}}{\sum_{s=2}^{M} p_{g,c(i,s)}} \cdots \frac{p_{g,c(i,m_i)}}{\sum_{s=m_i}^{M} p_{g,c(i,s)}},$$

where $M = 5$ denotes the number of candidates in the electoral race. The support parameter $p_{gj}$ (typically restricted such that $\sum_{j=1}^{M} p_{gj} = 1$) can be interpreted as the probability of ranking candidate $j$ first, out of the currently available choice set. Hence, the Plackett–Luce model models the ranking of candidates by a voter as a set of independent choices by the voter, conditional on the cardinality of the choice set being reduced by 1 after each choice is made.

In the standard ME regression model, the parameters of the component densities are modelled as a function of covariates. Here the support parameters are modelled as a logistic function of the covariates

$$\log\left[\frac{p_{gj}(x_i)}{p_{g1}(x_i)}\right] = \beta_{gj0} + \beta_{gj1}x_{i1} + \cdots + \beta_{gjq}x_{iq},$$

where $x_i = (x_{i1}, \ldots, x_{iq})$ is the set of $q$ covariates associated with voter $i$ and $\beta_{gj} = (\beta_{gj0}, \ldots, \beta_{gjq})^\top$ are unknown parameters for $j = 2, \ldots, M$. Note that for identifiability reasons candidate 1 is used as the baseline choice and $\beta_{g1} = (0, \ldots, 0)$ for all $g = 1, \ldots, G$. The component weights are also modelled as a function of covariates, in a similar vein to the example used in Section 12.2.2, that is,

$$\log\left[\frac{\eta_g(x_i)}{\eta_1(x_i)}\right] = \gamma_{g0} + \gamma_{g1}x_{i1} + \cdots + \gamma_{gq}x_{iq},$$

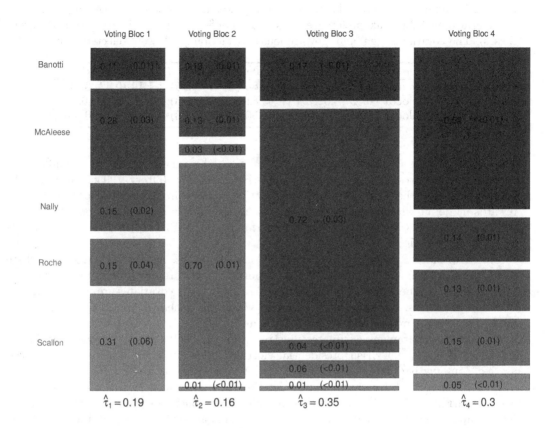

**FIGURE 12.3**
A mosaic plot representation of the parameters of the component densities of the simple mixture of experts model for rank preference data. The width of each block is proportional to the marginal probability of component membership $(\hat{\eta}_g = \sum_{i=1}^{n} \eta_g(x_i|\hat{\gamma})/n)$. The blocks are divided in proportion to the Plackett–Luce support parameters which are detailed therein. Standard errors are provided in parentheses.

where voting bloc 1 is used as the baseline voting bloc.

The suite of four ME models in the ME framework (Figure 12.2) arise from modelling the component parameters and/or the component weights as functions of covariates, or as constant with respect to covariates. In this application, each model is fitted in a maximum likelihood framework using the EM algorithm; approximate standard errors for the model parameters are derived from the empirical information matrix (McLachlan & Peel, 2000) after the EM algorithm has converged. Model fitting details for each model are outlined in Gormley & Murphy (2008a,b, 2010a,b).

Each of the four ME models for rank preference data were fitted to the data from the electorate in the Irish presidential election poll. A range of models with $G = 1, \ldots, 5$

**TABLE 12.7**
Odds ratios $(\exp(\gamma_g)/\exp(\gamma_1))$ for the component weight parameters in the simple mixture of experts model for rank preference data (95% confidence intervals are given in parentheses). The covariates 'age' and 'government satisfaction level' were selected as influential

|                | Age              | Not satisfied      | Satisfied          |
| -------------- | ---------------- | ------------------ | ------------------ |
| Voting bloc 2  | 0.01 (0.00, 0.05) | 2.80 (0.77, 10.15) | 1.14 (0.42, 3.11)  |
| Voting bloc 3  | 0.95 (0.32, 2.81) | 3.81 (0.90, 16.13) | 3.12 (0.94, 10.31) |
| Voting bloc 4  | 1.56 (0.35, 6.91) | 3.50 (1.07, 11.43) | 0.35 (0.12, 0.98)  |

was considered and a forward stepwise selection method was employed to choose influential covariates. The Bayesian information criterion (BIC) (Kass & Raftery, 1995; Schwarz, 1978) was used to select the optimal model; this criterion is a penalized likelihood criterion which rewards model fit while penalizing non-parsimonious models; see also Section 7.2.2 of this volume. Small BIC values indicate a preferable model. Table 12.6 details the optimal models for each type of ME model fitted.

Based on the BIC values, the optimal model is a simple ME model with four groups where "age" and "government satisfaction" are important covariates for determining group or "voting bloc" membership. Under this simple ME model, the covariates are not informative within voting blocs, but only in determining voting bloc membership. The MLEs of the model parameters are reported in Figure 12.3 and Table 12.7.

The support parameter estimates illustrated in Figure 12.3 have an interpretation in the context of the 1997 Irish presidential election. Voting bloc 1 could be characterized as the "conservative voting bloc" due to its large support parameters for McAleese and Scallon. Voting bloc 2 has large support for the liberal candidate Adi Roche. Voting bloc 3 is the largest voting bloc in terms of marginal component weights and intuitively has larger support parameters for the high-profile candidates McAleese and Banotti. These candidates were endorsed by the two largest political parties in the country at that time. Voters belonging to voting bloc 4 favor Banotti and have more uniform levels of support for the other candidates. A detailed discussion of this optimal model is also given in Gormley & Murphy (2008b).

Table 12.7 details the odds ratios computed from the component weight parameters $\gamma = \{\gamma_2, \gamma_3, \gamma_4\}$. In the model, voting bloc 1 (the conservative voting bloc) is the baseline voting bloc and $\gamma_1 = (0, \ldots, 0)^\top$. Two covariates were selected as influential: age and government satisfaction levels. In the "government satisfaction" covariate, the baseline was chosen to be "no opinion".

Interpreting the odds ratios provides insight into the type of voter which characterizes each voting bloc. For example, older (and generally more conservative) voters are much less likely to belong to the liberal voting bloc 2 than to the conservative voting bloc 1 $(\exp(\gamma_{21}) = 0.01)$. Also, voters with some interest in government are more likely to belong to voting bloc 3 $(\exp(\gamma_{32}) = 3.81$ and $\exp(\gamma_{33}) = 3.12)$, the bloc favouring candidates backed by large government parties, than to voting bloc 1. Voting bloc 1 had high levels of support for the independent candidate Scallon. The component weight parameter estimates further indicate that voters dissatisfied with the current government are more likely to belong to voting bloc 4 than to voting bloc 1 $(\exp(\gamma_{42}) = 3.50)$. This is again intuitive as voting bloc 4 favours Mary Banotti who was backed by the main government opposition party, while voting bloc 1 favours the government-backed Mary McAleese. Further interpretation of the component weight parameters is given in Gormley & Murphy (2008b).

**TABLE 12.8**

Covariates associated with the 71 lawyers in the US corporate law firm. The last category in each categorical covariate is treated as the baseline category in all analyses.

| Covariate | Levels |
| --- | --- |
| Age | – |
| Gender | 1 = male |
| | 2 = female |
| Law school | 1 = Harvard or Yale |
| | 2 = University of Connecticut |
| | 3 = other |
| Office | 1 = Boston |
| | 2 = Hartford |
| | 3 = Providence |
| Practice | 1 = litigation |
| | 2 = corporate |
| Seniority | 1 = partner |
| | 2 = associate |
| Years with the firm | – |

### 12.4.3 A mixture of experts latent position cluster model

The latent position cluster model (Handcock et al., 2007) develops the idea of the latent social space (Hoff et al., 2002) by extending it to accommodate clusters of actors in the latent space. Under the latent position cluster model, the latent location of each actor is assumed to be drawn from a finite normal mixture model, each component of which represents a cluster of actors. In contrast, the model outlined in Hoff et al. (2002) assumes that the latent positions are normally distributed. Thus, the latent position cluster model offers a more flexible version of the latent space model for modelling heterogeneous social networks.

The latent position cluster model provides a framework in which actor covariates may be explicitly included in the model – the probability of a link between two actors may be modelled as a function of both their separation in the latent space and of their relative covariates. However, the covariates may contribute more to the structure of the network than solely through the link probabilities – the covariates may influence both the cluster membership of an actor and their link probabilities. A latent position cluster model in which the cluster membership of an actor is modelled as a function of their covariates lies within the ME framework.

Specifically, social network data take the form of a set of relations $\{y_{i,j}\}$ between a group of $i, j = 1, \ldots, n$ actors, represented by an $n \times n$ sociomatrix $\mathbf{y}$. Here it is assumed that the relation $y_{i,j}$ between actors $i$ and $j$ is a binary relation, indicating the presence or absence of a link between the two actors; the ME latent position cluster model is easily extended to other forms of relation (such as count data). Covariate data $x_i = (x_{i1}, \ldots, x_{iq})$ associated with actor $i$ are assumed to be available, where $q$ denotes the number of observed covariates.

Each actor $i$ is assumed to have a location $w_i = (w_{i1}, \ldots, w_{iD})$ in the $D$-dimensional latent social space. The probability of a link between any two actors is assumed to be independent of all other links in the network, given the latent locations of the actors. Let $x_{i,j} = (x_{ij1}, \ldots, x_{ijq})$ denote a $q$ vector of dyadic specific covariates where $x_{ijk} = d(x_{ik}, x_{jk})$ is a measure of the similarity in the value of the $k$th covariate for actors $i$ and $j$. Given the link probabilities parameter vector $\beta$, the likelihood function is then

$$p(\mathbf{y}|\mathbf{w}, \mathbf{x}, \beta) = \prod_{i=1}^{n} \prod_{j \neq i} p(y_{i,j}|w_i, w_j, x_{i,j}, \beta),$$

where $\mathbf{w}$ is the $n \times D$ matrix of latent locations and $\mathbf{x}$ is the matrix of dyadic specific covariates. The probability of a link between actors $i$ and $j$ is then modelled using a logistic regression model where both dyadic specific covariates and Euclidean distance in the latent space are covariates:

$$\log\left[\frac{\mathrm{P}(y_{i,j} = 1)}{\mathrm{P}(y_{i,j} = 0)}\right] = \beta_0 + \beta_1 x_{ij1} + \cdots + \beta_q x_{ijq} - \|w_i - w_j\|.$$

To account for clustering of actor locations in the latent space, it is assumed that the latent locations $w_i$ are drawn from a finite mixture model. Moreover, in the ME latent position cluster model, the latent locations are assumed drawn from a finite mixture model in which actor covariates may influence the mixing proportions:

$$w_i \sim \sum_{g=1}^{G} \eta_g(x_i|\gamma)\phi(w_i|\mu_g, \sigma_g^2 I),$$

where

$$\eta_g(x_i|\gamma) = \frac{\exp(\gamma_{g0} + \gamma_{g1}x_{i1} + \cdots + \gamma_{gq}x_{iq})}{\sum_{g'=1}^{G} \exp(\gamma_{g'0} + \gamma_{g'1}x_{i1} + \cdots + \gamma_{g'q}x_{iq})}$$

and $\gamma_1 = (0, \ldots, 0)^\top$. This model has an intuitive motivation: the covariates of an actor may influence their cluster membership, their cluster membership influences their latent location, and in turn their latent location determines their link probabilities.

The ME latent position cluster model can be fitted within the Bayesian paradigm; as outlined in Section 12.3.2 a Metropolis-within-Gibbs sampler can be employed to draw samples from the posterior distribution of interest. Model issues such as likelihood invariance to distance-preserving transformations of the latent space and label switching must be considered during the model fitting process – an approach to dealing with such model identifiability and full model fitting details are available in Gormley & Murphy (2010b). In this application, model choice concerns not only the number $G$ of clusters, but also the dimension $D$ of the latent space.

An example of the ME latent position cluster model methodology is provided here through the analysis of a network data set detailing interactions between a set of 71 lawyers in a corporate law firm in the USA (Lazega, 2001). The data include measurements of the co-worker network, an advice network and a friendship network. Covariates associated with each lawyer in the firm are also included and are detailed in Table 12.8. Interest lies in identifying social processes within the firm such as knowledge sharing and organizational structures, and examining the potential influence of covariates on such processes.

Under the ME model framework outlined in Section 12.2.3, a suite of four ME latent position cluster models is available. This suite of models was fitted to the advice network; data in this network detail links between lawyers who sought basic professional advice from each other over the previous 12 months. Gormley & Murphy (2010b) explore the co-workers network data set and the friendship network data set using similar methodology. Figure 12.4 illustrates the resulting latent space locations of the lawyers under each fitted model with $(G, D) = (2, 2)$. These values were selected using the BIC after fitting a range of latent position cluster models (with no covariates) to the network data only (Handcock et al., 2007). Table 12.9 details the resulting regression parameter estimates and their associated uncertainty for the four fitted models.

The models are compared through the AICM, the posterior simulation-based analogue of Akaike's information criterion (AIC) (Akaike, 1973; Raftery et al., 2007). In this implementation the optimal model is that with the highest AICM and is the model with covariates

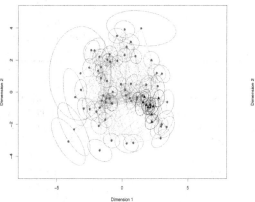

(a) Latent position cluster model

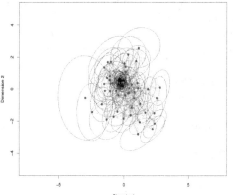

(b) Mixture of regressions latent position cluster model

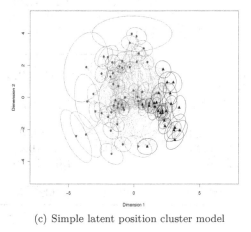

(c) Simple latent position cluster model

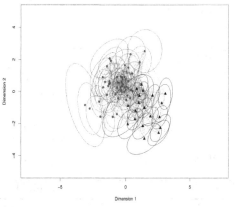

(d) Standard mixture of experts latent position cluster model

**FIGURE 12.4**

Estimates of clusters and latent positions of the lawyers from the advice network data. The ellipses are 50% posterior sets illustrating the uncertainty in the latent locations. Lawyers who are members of the same cluster are illustrated using the same shade and symbol. Observed links between lawyers are also illustrated.

in the link probabilities and in the component weights. The results of the analysis show some interesting patterns. The coefficients of the covariates in the link probabilities are very similar in models (b) and (d) in Table 12.9. These coefficients indicate that a number of factors have a positive or negative effect on whether a lawyer asks another for advice. In summary, lawyers who are similar in seniority, gender, office location and practice type are more likely to ask each other for advice. The effects of years and age seem to have a negative effect, but these variables are correlated with seniority and with each other, so their marginal effects are more difficult to interpret.

Importantly, the latent positions are very similar in models (a) and (c) which do not have covariates in the link probabilities, and models (b) and (d) which do have covariates in the link probabilities. This can be explained because of the different role that the latent

**TABLE 12.9**
Posterior mean parameter estimates for the four mixture of experts models fitted to the lawyer advice data as detailed in Figure 12.4. Standard deviations are given in parentheses. Note that cluster 1 was used as the baseline cluster in the case of the cluster membership parameters. Baseline categories for the covariates are detailed in Table 12.8

|  | Model (a) | Model (b) | Model (c) | Model (d) |
|---|---|---|---|---|
| **Link Probabilities** | | | | |
| Intercept | 1.26 (0.10) | −2.87 (0.17) | 1.23 (0.10) | −2.65 (0.17) |
| Age | | −0.02 (0.004) | | −0.02 (0.004) |
| Gender | | 0.60 (0.09) | | 0.62 (0.09) |
| Office | | 2.02 (0.10) | | 1.97 (0.10) |
| Practice | | 1.63 (0.10) | | 1.57 (0.10) |
| Seniority | | 0.89 (0.11) | | 0.81 (0.11) |
| Years | | −0.04 (0.005) | | −0.04 (0.005) |
| | | | | |
| **Cluster Memberships** | | | | |
| Intercept | −1.05 (1.75) | 0.94 (0.79) | −0.62 (1.23) | 1.27 (1.29) |
| Age | | | −0.09 (0.04) | −0.14 (0.06) |
| Office (=1) | | | 1.94 (1.02) | 2.40 (1.14) |
| Office (=2) | | | −2.08 (1.09) | −0.97 (1.19) |
| Practice | | | 3.18 (0.85) | 2.14 (1.08) |
| | | | | |
| **Latent Space Model** | | | | |
| Cluster 1 mean | −0.50 (0.52) | 0.09 (0.19) | −1.09 (0.31) | −0.54 (0.21) |
| | 0.21 (0.58) | −0.09 (0.26) | 0.40 (0.28) | 0.40 (0.20) |
| Cluster 1 variance | 3.35 (1.29) | 2.12 (0.77) | 3.19 (0.58) | 1.25 (0.34) |
| | | | | |
| Cluster 2 mean | 1.66 (0.92) | −0.24 (0.20) | 2.10 (0.30) | 1.32 (0.51) |
| | −0.67 (0.58) | 0.35 (0.23) | −0.77 (0.30) | −0.98 (0.47) |
| Cluster 2 variance | 1.29 (1.58) | 0.27 (0.68) | 1.16 (0.40) | 1.63 (0.69) |
| | | | | |
| AICM | −3644.24 | −3346.87 | −3682.71 | −3325.95 |

space plays in the models with covariates in the link probabilities and those that do not have such covariates. When the covariates are in the link probabilities, the latent space is modelling the network structure that could not be explained by the link covariates, whereas in the other case the latent space is modelling much of the network structure.

Interestingly, in the model with the highest AICM value, there are covariates in the cluster membership probabilities as well as in the link probabilities. This means that the structure in the latent space, which is modelling what could not be explained directly in the link probabilities, has structure that can be further explained using the covariates. The office location, practice and age of the lawyers retain explanatory power in explaining the clustering found in the latent social space.

The difference in the cluster membership coefficients in models (c) and (d) is due to the different interpretation of the latent space in these models. However, it is interesting to note that in this application the signs of the coefficients are identical because the cluster memberships shown for these models in Figure 12.4(c) and Figure 12.4(d) are similar; this phenomenon does not hold generally (see Gormley & Murphy, 2010b, Section 5.3).

The results of this analysis offer a cautionary message in automatically selecting the type of ME latent position cluster model for analysing the lawyer advice network. The role of the latent space in the model is very different depending on how the covariates enter the model. So, if the latent space is to be interpreted as a social space that explains network structure, then the covariates should not directly enter the link probabilities. However, if the latent space is being used to find interesting or anomalous structure in the network that cannot be explained by the covariates, then one should consider allowing the covariates enter the cluster membership probabilities.

### 12.4.4 Software

As demonstrated in this section, the approach to fitting an ME model depends on the application setting and on the form of the ME model itself. Therefore, a single software capable of fitting any ME model is not currently available.

In R (R Core Team, 2018), the MEclustnet package (Gormley & Murphy, 2018) fits the ME latent position cluster model detailed in Section 12.4.3. The flexmix package (Grün & Leisch, 2008b) has model fitting capabilities for a range of mixture of regression models, which include covariates (or concomitant variables), as does the mixreg package (Turner, 2014). In addition, mixtools (Benaglia et al., 2009) facilitates fitting of a $G = 2$ mixture of regressions model in which the component weights are modelled as an inverse logit function of the covariates. The cluster weighted models which are closely related to ME models can be fitted using the flexCWM package (Mazza et al., 2017). All packages are freely available through the Comprehensive R Archive Network (CRAN) at https://cran.r-project.org.

In MATLAB, the bayesf package (Frühwirth-Schnatter, 2018) allows a broad range of mixture models to be estimated using either finite mixtures, ME or Markov switching models as a model for the hidden group indicators **z**.

As to other packages, the FMM procedure in SAS also facilitates ME model fitting, and stand-alone packages such as Latent GOLD (Vermunt & Magidson, 2005) and Mplus (Muthén & Muthén, 2011) fit closely related latent class models.

## 12.5 Identifiability of Mixture of Experts Models

For a finite mixture distribution one has to distinguish three types of non-identifiability (Frühwirth-Schnatter, 2006, Section 1.3): invariance to relabelling the components of the mixture distribution (the label switching problem); non-identifiability due to potential over-fitting; and generic non-identifiability which occurs only for certain classes of mixture distributions.

Consider a standard mixture distribution with $G$ components with non-zero weights $\eta_1, \ldots, \eta_G$ generated by distinct parameters $\theta_1, \ldots, \theta_G$. Assume that for all possible realizations $y$ from this mixture distribution the identity

$$\sum_{g=1}^{G} \eta_g f_g(y|\theta_g) = \sum_{g=1}^{G^\star} \eta_g^\star f_g(y|\theta_g^\star)$$

holds, where the right-hand side is a mixture distribution from the same family with $G^\star$ components with non-zero weights $\eta_1^\star, \ldots, \eta_{G^\star}^\star$ generated by distinct parameters $\theta_1^\star, \ldots, \theta_{G^\star}^\star$. Then generic identifiability implies that $G^\star = G$ and the two mixtures' parameters $\theta = (\eta_1, \ldots, \eta_G, \theta_1, \ldots, \theta_G)$ and $\theta^\star = (\eta_1^\star, \ldots, \eta_G^\star, \theta_1^\star, \ldots, \theta_G^\star)$ are identical up to relabelling the component indices. Common finite mixture distributions such as Gaussian and Poisson mixtures are generically identified, see Teicher (1963), Yakowitz & Spragins (1968), and Chandra (1977) for a detailed discussion.

Discrete mixtures often suffer from generic non-identifiability for certain parameter configurations, well-known examples being mixtures of binomial distributions (see Section 12.5.1) and mixtures of multinomial distributions (Grün & Leisch, 2008c). Somewhat unexpectedly, mixture of regression models suffer from generic non-identifiability (Hennig, 2000; Grün & Leisch, 2008a), as will be discussed in more detail in Section 12.5.2. Little is known about generic identifiability of ME models, and some results are presented in Section 12.5.3. However, ensuring generic identifiability for general ME models remains a challenging issue.

Identifiability problems for mixtures with nonparametric components are discussed in Chapter 14 of this volume.

### 12.5.1  Identifiability of binomial mixtures

For binomial mixtures the component densities arise from $\mathcal{B}(N, \pi)$ distributions, where $N$ is commonly assumed to be known, whereas $\pi$ is heterogeneous across the components:

$$Y \sim \eta_1 \mathcal{B}(N, \pi_1) + \cdots + \eta_G \mathcal{B}(N, \pi_G). \tag{12.10}$$

The probability mass function (pmf) of this mixture takes on $N+1$ different support points:

$$p(y|\theta) = \mathrm{P}(Y = y|\theta) = \sum_{g=1}^{G} \eta_g \binom{N}{y} \pi_g^y (1 - \pi_g)^{N-y}, \quad y = 0, 1, \ldots, N, \tag{12.11}$$

with $2G - 1$ independent parameters $\theta = (\pi_1, \ldots, \pi_G, \eta_1, \ldots, \eta_G)$, with $\eta_G = 1 - \sum_{g=1}^{G-1} \eta_g$.

Given data $\mathbf{y} = (y_1, \ldots, y_n)$ from mixture (12.10), the only information available to estimate $\theta$ are $N$ (among the $N+1$ observed) relative frequencies $h_n(Y = y)$ $(y = 0, 1, \ldots, N)$. As $n \to \infty$ (while $N$ is fixed), $h_n(Y = y)$ converges to $\mathrm{P}(Y = y|\theta)$ by the law of large numbers, but the number of support points remains fixed. Hence, the data provide only $N$ statistics, given by the relative frequencies, to estimate $2G - 1$ parameters. Simple counting yields the following necessary condition for identifiability for a binomial mixture, which has been shown by Teicher (1961) to be also sufficient:

$$2G - 1 \leq N \quad \Leftrightarrow \quad G \leq (N + 1)/2. \tag{12.12}$$

Consider, for illustration, a mixture of two binomial distributions,

$$Y \sim \eta \times \mathcal{B}(N, \pi_1) + (1 - \eta) \times \mathcal{B}(N, \pi_2), \tag{12.13}$$

with three unknown parameters $\theta = (\eta, \pi_1, \pi_2)$, and assume that the population indeed contains two different groups, that is, $\pi_1 \neq \pi_2$ and $\eta > 0$. Assuming $N = 2$ obviously violates condition (12.12). Lack of identification can be verified directly from the pmf which is different from zero only for the three outcomes $y \in \{0, 1, 2\}$:

$$\begin{aligned} \mathrm{P}(Y = 0|\theta) &= \eta(1 - \pi_1)^2 + (1 - \eta)(1 - \pi_2)^2, \\ \mathrm{P}(Y = 1|\theta) &= 2\eta\pi_1(1 - \pi_1) + 2(1 - \eta)\pi_2(1 - \pi_2), \\ \mathrm{P}(Y = 2|\theta) &= \eta\pi_1^2 + (1 - \eta)\pi_2^2. \end{aligned} \tag{12.14}$$

Since $\sum_y \mathrm{P}(Y = y|\theta) = 1$, only two linearly independent equations remain to identify the three parameters $(\eta, \pi_1, \pi_2)$. Hence parameters $\theta = (\pi_1, \pi_2, \eta) \neq \theta^\star = (\pi_1^\star, \pi_2^\star, \eta^\star)$ fulfilling equations (12.14) exist which imply the same distribution for $Y$, that is, $\mathrm{P}(Y = y|\theta) = \mathrm{P}(Y = y|\theta^\star)$, for $y = 0, 1, 2$, but are not related to each other by simple relabelling of the component indices.

Such generic non-identifiability severely impacts statistical estimation of the mixture parameters $\theta$ from observations $\mathbf{y} = (y_1, \ldots, y_n)$, even if $G$ is known, and goes far beyond label switching. Assume, for illustration, that $\mathbf{y}$ is the realization of a random sample $(Y_1, \ldots, Y_n)$ from the two-component binomial mixture (12.13) with $N = 2$ and true parameter $\theta^{\text{true}} = (\pi_1^{\text{true}}, \pi_2^{\text{true}}, \eta^{\text{true}})$, and consider the corresponding observed-data likelihood $p(\mathbf{y}|\theta) = \prod_{i=1}^{n} \mathrm{P}(Y_i = y_i|\theta)$. Generic non-identifiability of the underlying mixture distribution implies that the observed-data likelihood is the same for any pair $\theta \neq \theta^\star$ of distinct parameters satisfying (12.14), for any possible sample $\mathbf{y}$ in the sampling space $\mathcal{Y} = \{0, 1, 2\}^n$, that is, $p(\mathbf{y}|\theta) = p(\mathbf{y}|\theta^\star)$, for all $\mathbf{y} \in \mathcal{Y}$. Since this holds for arbitrary sample size $n = 1, 2, \ldots$, the true parameter $\theta^{\text{true}}$ cannot be recovered, even if $n \to \infty$, and both maximum likelihood estimation and Bayesian inference suffer from non-identifiability problems for such a mixture.

This example motivates the following more formal definition of generic non-identifiability. For a given $\theta$, any subset $U(\theta)$ of the parameter space $\Theta$ of a mixture model, defined as $U(\theta) = \{\theta^\star \in \Theta : p(\mathbf{y}|\theta^\star) = p(\mathbf{y}|\theta), \forall \mathbf{y} \in \mathcal{Y}\}$, is called a non-identifiability set if it contains at least one point $\theta^\star$ which is not related to $\theta$ by simple relabelling of the component indices. Let $\theta^{\text{true}}$ be the true parameter value of a mixture model with $G$ distinct parameters (i.e. $\theta_g \neq \theta_{g'}$, for $g \neq g'$). If $U(\theta^{\text{true}})$ is a non-identifiability set in the sense defined above, then $\theta^{\text{true}}$ cannot be recovered from data, even as $n$ goes to infinity.

Such generic non-identifiability has important implications for practical mixture analysis. For finite $n$, the observed-data likelihood function $p(\mathbf{y}|\theta)$ has a ridge close to $U(\theta^{\text{true}})$ instead of $G!$ isolated modes and no unique maximum, leading to inconsistent estimates of $\theta^{\text{true}}$. In a Bayesian framework, this leads to a posterior distribution that does not concentrate around $G!$ isolated, equivalent modes as $n$ increases, as for identifiable models (see Section 4.3). Rather, the posterior concentrates over the entire non-identifiability set $U(\theta^{\text{true}})$ which has a complex geometry and can be represented as the union of $G!$ symmetric subspaces; see, for example, the left of Figure 12.5 for a binomial mixture with $G = 2$ and $N = 2$. The prior $p(\theta)$ provides information beyond the data and might influence how the posterior concentrates on each of these $G!$ subspaces $U(\theta^{\text{true}})$, in particular, if the prior $p(\theta)$ is not constant over $U(\theta)$.

While generic non-identifiability has important practical implications for mixture analysis, it is rarely as easily diagnosed as for mixtures of binomial distributions and can easily go unnoticed for more complex mixture models, in particular for maximum likelihood estimation, whereas MCMC-based Bayesian inference often provides indications of potential identifiability problems, as the following example demonstrates.

## MCMC inference for an example: binomial mixtures

By way of further illustration, we perform MCMC inference (based on 10,000 draws after a burn-in of 5000 iterations) for two data sets simulated from a mixture of two binomial distributions with logit $\pi_1 = -1$ and logit $\pi_2 = 1.5$ using random permutation sampling as explained in Section 5.2. We assume that $G = 2$ and $N$ is known, whereas all other parameters in mixture (12.13) are unknown. Bayesian inference is based on the priors $\pi_g \sim \mathcal{U}(0, 1)$ and $(\eta_1, \eta_2) \sim \mathcal{D}(1, 1)$. The two data sets were generated with, respectively, $N = 2, n = 250$ and $N = 5, n = 100$, implying the same total number $n \times N = 500$ of experiments.

MCMC inference is summarized in Figure 12.5, showing scatter plots of $\pi_2$ against $\pi_1$ for both values of $N$. For $N = 5$, the mixture is generically identified and the posterior draws concentrate around two symmetric modes, centred at the true values $(0.269, 0.818)$ and $(0.818, 0.269)$. Non-identifiability due to label switching is resolved by applying $k$-means clustering to the posterior draws; see the lower part of Figure 12.5 showing identified posterior draws of the group-specific success probabilities $\pi_1$ and $\pi_2$.

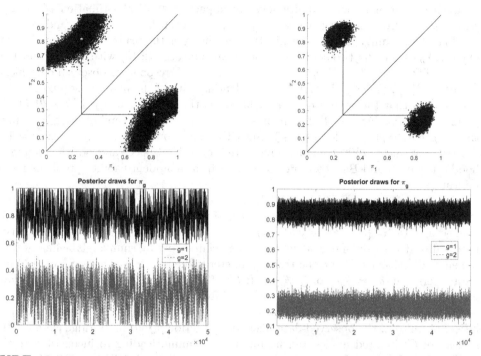

**FIGURE 12.5**
MCMC inference for data simulated from a mixture of two binomial distributions with $N = 2$ (left) and $N - 5$ (right). Top: scatter plot of $\pi_2$ against $\pi_1$ (true values indicated by a circle). Bottom: posterior draws of the group-specific probabilities $\pi_1$ and $\pi_2$ after resolving label switching in the scatter plot of $\pi_2$ against $\pi_1$ through $k$-means clustering.

For $N = 2$, a similar scatter plot of $\pi_2$ against $\pi_1$ clearly indicates severe identifiability issues, showing that the posterior draws arise from two symmetric unidentifiability sets, rather than concentrating around two symmetric modes centered at the true values. When we apply $k$-means clustering to resolve label switching, we obtain the posterior draws of the success probabilities $\pi_1$ and $\pi_2$ shown in the lower part of Figure 12.5, also indicating problems with identifying $\pi_1$ and $\pi_2$ from the data for $N = 2$.

### 12.5.2    Identifiability for mixtures of regression models

Consider a mixture of $G$ regression models for $i = 1, \ldots, n$ outcomes $y_i$, arising from $G$ different groups,

$$y_i | \tilde{x}_i \sim \sum_{g=1}^{G} \eta_g \phi(y | \mu_{i,g}(\tilde{x}_i), \sigma_g^2), \tag{12.15}$$

where, for each $g = 1, \ldots, G$, the group-specific mean $\mu_{i,g}(\tilde{x}_i) = \tilde{x}_i \beta_g$ depends on a group-specific regression parameter $\beta_g$ and on the $(1 \times (q+1))$-dimensional row vector $\tilde{x}_i$ containing the $q$ covariates $x_i$ and a constant. For a fixed design point $x = \tilde{x}_i$, (12.15) is a standard finite Gaussian mixture distribution and as such generically identified. Hence, if the identity

$$\sum_{g=1}^{G} \eta_g \phi(y | \mu_{i,g}(x), \sigma_g^2) = \sum_{g=1}^{G} \eta_g^{\star} \phi(y | \mu_{i,g}^{\star}(x), \sigma_g^{2,\star}) \tag{12.16}$$

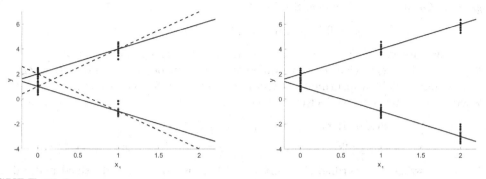

**FIGURE 12.6**
Data simulated from a mixture of two regression lines under design 1 (left) and design 2 (right). The solid lines indicate the true underlying model used to generate 100 data points (black dots). For the unidentified design 1, a second solution exists which is indicated by the dashed lines.

holds, then the two mixtures are related to each other by relabelling, that is, $\mu_{i,g}^\star(x) = \mu_{i,\mathfrak{s}_x(g)}(x) = x\beta_{\mathfrak{s}_x(g)}$, $\sigma_g^{2,\star} = \sigma_{\mathfrak{s}_x(g)}^2$, and $\eta_g^\star = \eta_{\mathfrak{s}_x(g)}$, for $g = 1, \ldots, G$, for some permutation $\mathfrak{s}_x \in \mathfrak{S}(G)$, where $\mathfrak{S}(G)$ denotes the set of the $G!$ permutations of $\{1, \ldots, G\}$. Note that $\mathfrak{s}_x$ depends on the covariate $x$ and that there is no guarantee that $\mathfrak{s}_x$ is identical across different values of $x$ which can cause *intra-component label switching*. One such example is displayed on the left in Figure 12.6.

Nevertheless, assume for the moment that $\mathfrak{s}_x \equiv \mathfrak{s}_\star$ is the same for all possible covariates $x$. Then (12.16) implies $\tilde{x}_i\beta_g^\star = \tilde{x}_i\beta_{\mathfrak{s}_\star(g)}$, for all $i = 1, \ldots, n$, and $X\beta_g^\star = X\beta_{\mathfrak{s}_\star(g)}$, where the rows of the matrix $X$ are equal to $\tilde{x}_1, \ldots, \tilde{x}_n$. If the usual condition in regression modelling, that $X^\top X$ has full rank is satisfied, then it follows immediately that the regression coefficients are determined up to relabelling: $\beta_g^\star = \beta_{\mathfrak{s}_\star(g)}$.

Hence, generic identifiability for a mixture of regression models can be verified through sufficient conditions guaranteeing that $\mathfrak{s}_x$ is indeed identical across all values of $x$. Mathematically, one such condition is the assumption that either the error variances $\sigma_1^2, \ldots, \sigma_G^2$ or the weights $\eta_1, \ldots, \eta_G$ satisfy a strict order constraint. However, in practice such constraints are rarely fulfilled, and forcing an order constraint on one coefficient does not necessarily prevent label switching for the other coefficients in Bayesian posterior sampling; see, for example, Frühwirth-Schnatter (2006, Section 2.4).

Hence, several papers have focused on conditions for generic identifiability through the regression part of the model (Hennig, 2000; Grün & Leisch, 2008a,c). Assume that the covariates $\tilde{x}_i$ take $p$ different values in a design space $\{x_1, \ldots, x_p\}$ for the observed outcome $y_i$, for $i = 1, \ldots, n$. Identifiability through the regression part requires enough variability in the design space and is guaranteed under so-called *coverage conditions*. These conditions require that the number of clusters $G$ is exceeded by the minimum number of distinct $q$-dimensional hyperplanes needed to cover the covariates (excluding the constant). For $q = 1$, for instance, the coverage condition is satisfied if the number of design points $p$ (i.e. the number of distinct values of the univariate covariate) is larger than the number of clusters $G$. These identifiability conditions go far beyond the usual condition that $X^\top X$ has full rank and are often violated for regression models with too few design points, a common example being regression models with 0/1 dummy variables as covariates which are identifiable for $G = 1$, but not for $G > 1$, as the following examples with $q = 1$ demonstrate.

Consider the following special case of the mixture of regression models (12.15), investigated in Grün & Leisch (2008a, Section 3.1):

$$y_i \sim 0.5\phi(y|\mu_{i,1}(\tilde{x}_i), 0.1) + 0.5\phi(y|\mu_{i,2}(\tilde{x}_i), 0.1), \tag{12.17}$$

with covariate vector $\tilde{x}_i = (1 \; d_i)$ and group-specific regression parameters $\beta_1 = (2 \; 2)^{\top}$ and $\beta_2 = (1 \; -2)^{\top}$. Consider two different regression designs – design 1, where $d_i$ is a 0/1 dummy variable capturing the effect of gender (with female as baseline); and design 2, where $d_i$ captures a time effect over three periods (with $t = 0$ serving as baseline):

$$\text{design 1: } x_1 = (\; 1 \quad 0 \;), \quad x_2 = (\; 1 \quad 1 \;),$$
$$\text{design 2: } x_1 = (\; 1 \quad 0 \;), \quad x_2 = (\; 1 \quad 1 \;), \quad x_3 = (\; 1 \quad 2 \;).$$

In the following, it is verified that mixture (12.17) is generically identified under design 2 (which contains three design points), but generically unidentified under design 1 (which contains only two design points).

We first consider design 1. According to (12.16), $\mu_{j,1}$ and $\mu_{j,2}$ are identified for $j = 1$ and $j = 2$ up to label switching arising from two permutations $\mathfrak{s}_1$ and $\mathfrak{s}_2$, where we may assume without loss of generality that $\mathfrak{s}_1$ is equal to the identity:

$$x_1\beta_1 = \mu_{1,1}, \qquad\qquad\qquad x_1\beta_2 = \mu_{1,2}, \tag{12.18}$$
$$x_2\beta_1 = \mu_{2,\mathfrak{s}_2(1)}, \qquad\qquad\qquad x_2\beta_2 = \mu_{2,\mathfrak{s}_2(2)}.$$

If $\mathfrak{s}_2$ is identical to $\mathfrak{s}_1$, then the original values $\beta_1$ and $\beta_2$ are recovered through

$$\beta_1 = X_{1,2}^{-1}\begin{pmatrix} \mu_{1,1} \\ \mu_{2,1} \end{pmatrix} = \begin{pmatrix} 2 \\ 2 \end{pmatrix}, \quad \beta_2 = X_{1,2}^{-1}\begin{pmatrix} \mu_{1,2} \\ \mu_{2,2} \end{pmatrix} = \begin{pmatrix} 1 \\ -2 \end{pmatrix}, \tag{12.19}$$

since the design matrix

$$X_{1,2} = \begin{pmatrix} x_1 \\ x_2 \end{pmatrix} = \begin{pmatrix} 1 & 0 \\ 1 & 1 \end{pmatrix}$$

is invertible. However, as mentioned above, $\mathfrak{s}_2$ need not be identical to $\mathfrak{s}_1$, in which case $\mathfrak{s}_2(1) = 2, \mathfrak{s}_2(2) = 1$, and a second solution emerges:

$$\beta_1^{\star} = X_{1,2}^{-1}\begin{pmatrix} \mu_{1,1} \\ \mu_{2,2} \end{pmatrix} = \begin{pmatrix} 2 \\ -3 \end{pmatrix}, \quad \beta_2^{\star} = X_{1,2}^{-1}\begin{pmatrix} \mu_{1,2} \\ \mu_{2,1} \end{pmatrix} = \begin{pmatrix} 1 \\ 3 \end{pmatrix}.$$

Evidently, the group-specific slopes of this second solution are different from the original ones and the unidentifiability set $U(\theta^{\text{true}})$ contains two points. The two possible solutions are depicted on the left in Figure 12.6, which also shows a balanced sample of $n = 100$ observations simulated from mixture (12.17) under design 1.

For design 2, the first two design points are as before and a third point is added, with $\mu_{3,1}$ and $\mu_{3,2}$ being identified up to label switching according to a permutation $\mathfrak{s}_3$:

$$x_3\beta_1 = \mu_{3,\mathfrak{s}_3(1)}, \qquad\qquad\qquad x_3\beta_2 = \mu_{3,\mathfrak{s}_3(2)}. \tag{12.20}$$

As only two different permutations exist for $G = 2$, at least two of the three permutations $\mathfrak{s}_1$, $\mathfrak{s}_2$ and $\mathfrak{s}_3$ in (12.18) and (12.20) have to be identical (assuming again without loss of generality that $\mathfrak{s}_1$ is equal to the identity). Assume, for example, that $\mathfrak{s}_1 = \mathfrak{s}_2$. Then the true parameters $\beta_1$ and $\beta_2$ are recovered from $(\mu_{j,1}, \mu_{j,2})$, $j = 1, 2$, as in (12.19) and can be used to uniquely predict $\mu_{3,1} = x_3\beta_1$ and $\mu_{3,2} = x_3\beta_2$ in both groups. Comparing these predictions with (12.20), it is clear that $\mathfrak{s}_3(1) = 1$ and $\mathfrak{s}_3(2) = 2$, hence $\mathfrak{s}_1 = \mathfrak{s}_2 = \mathfrak{s}_3$. A similar proof holds for any pair of identical permutations $\mathfrak{s}_j = \mathfrak{s}_l$, $j \neq l$, as long as the matrix $X_{j,l}^{\top} = (x_j^{\top} \; x_l^{\top})$ is invertible and generic identifiability of design 2 follows.

The only possible solution under design 2 is depicted on the right in Figure 12.6, which also shows a balanced sample of $n = 100$ observations simulated from mixture (12.17) under this design.

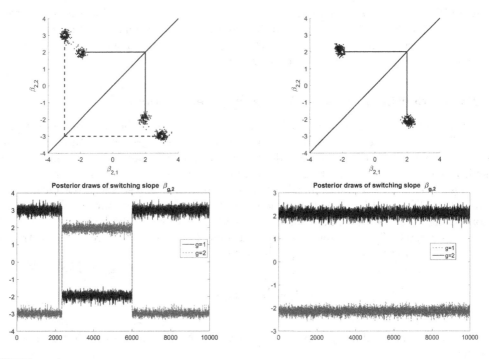

**FIGURE 12.7**
MCMC inference for data simulated from a mixture of two regression models under the generically unidentified design 1 (left) and under the generically identified design 2 (right). Top: scatter plot of the group-specific slopes $\beta_{2,2}$ against $\beta_{1,2}$. Bottom: posterior draws of the group-specific slopes $\beta_{1,2}$ and $\beta_{2,2}$ after resolving label switching through $k$-means clustering in the posterior draws.

*MCMC inference for an example: a mixture of regression models*

By way of further illustration, we perform MCMC inference (based on 10,000 draws after a burn-in of 5000 iterations) for both data sets shown in Figure 12.6 using random permutation sampling. We assume that $G = 2$ is known, whereas all other parameters in mixture (12.15) are unknown. Bayesian inference is based on the priors $(\eta_1, \eta_2) \sim \mathcal{D}(4, 4)$ and $\beta_g \sim \mathcal{N}(0, 100 \times I)$, $\sigma_g^2 \sim \mathcal{IG}(2.5, 1.25 s_y^2)$, for $g = 1, 2$, where $s_y^2$ is the data variance of $(y_1, \ldots, y_n)$. The upper part of Figure 12.7 shows scatter plots of the group-specific slopes $\beta_{2,2}$ against $\beta_{1,2}$ for both designs, which are symmetric due to label switching.

As expected from Section 4.3, the posterior draws shown in the upper right-hand part for the generically identified design 2 concentrate around two symmetric modes corresponding to the true values $(2, -2)$ and $(-2, 2)$. Label switching is easily resolved by applying $k$-means clustering to the posterior draws; see the lower right-hand part of Figure 12.7 showing identified posterior draws of the group-specific slopes $\beta_{1,2}$ and $\beta_{2,2}$ for this design.

For design 1, a similar scatter plot of $\beta_{2,2}$ against $\beta_{1,2}$ in the upper left-hand part clearly indicates severe identifiability issues. The posterior draws concentrate around four modes rather than two, with two of them being the symmetric modes corresponding to the true values $(2, -2)$ and $(-2, 2)$. The other two symmetric modes correspond to the second solution $(-3, 3)$ and $(3, -3)$, resulting from generic non-identifiability. When we apply $k$-means clustering to these posterior draws to resolve label switching, we obtain the posterior draws of the group-specific slopes $\beta_{1,2}$ and $\beta_{2,2}$ in the lower left-hand part of

Figure 12.7, showing *intra-component label switching* and indicating identifiability problems for this design. Since $\beta_{g,2}$ switches sign between the two solutions in both groups, it is not possible to recover that "gender" has a strong positive effect on the outcome in one group and a strong negative effect in the other group.

### 12.5.3　Identifiability for mixture of experts models

Hennig (2000) considers mixtures of regression models where the component sizes can arbitrarily depend on covariates and establishes identifiability results in the case where the joint observations of covariates and dependent variable are assumed to be i.i.d. and gives sufficient identifiability conditions for this model. For such a model, the covariates are not assumed fixed or to occur for a fixed design, but random with a specific distribution. In contrast, the ME model is defined conditional on the covariates without specific assumptions concerning their distribution. We will discuss identification for this case.

Consider, as a first example, a simple ME model of $G$ univariate Gaussian distributions for $i = 1, \ldots, n$ outcomes $y_i$ arising from $G$ different groups,

$$y_i | \tilde{x}_i \sim \sum_{g=1}^{G} \eta_g(\tilde{x}_i) \phi(y | \mu_g, \sigma_g^2), \qquad (12.21)$$

where the group weights $\eta_g$ depend on a covariate $\tilde{x}_i$ with group-specific regression parameters, that is,

$$\log \left[ \frac{\eta_g(\tilde{x}_i)}{\eta_{g_0}(\tilde{x}_i)} \right] = \tilde{x}_i \gamma_g, \quad g = 1, \ldots, G, \qquad (12.22)$$

with baseline $g_0$, where $\gamma_{g_0} = 0$. Assume that the component densities differ, that is, $\theta_g \neq \theta'_g$, for $g \neq g'$, where $\theta_g = (\mu_g, \sigma_g^2)$.

For each fixed design point $x = \tilde{x}_i$, (12.21) is a standard finite Gaussian mixture and therefore generically identified. Therefore, if the identity

$$\sum_{g=1}^{G} \eta_g(x) \phi(y | \mu_g, \sigma_g^2) = \sum_{g=1}^{G} \eta_g^\star(x) \phi(y | \mu_g^\star, \sigma_g^{2,\star}) \qquad (12.23)$$

holds, then the two mixtures are related to each other by relabelling, that is, $\mu_g^\star = \mu_{\mathfrak{s}_x(g)}, \sigma_g^{2,\star} = \sigma_{\mathfrak{s}_x(g)}^2$, and $\eta_g^\star(x) = \eta_{\mathfrak{s}_x(g)}(x)$ for $g = 1, \ldots, G$, for some permutation $\mathfrak{s}_x \in \mathfrak{S}(G)$. In contrast to mixtures of regression models, one can show that $\mathfrak{s}_x \equiv \mathfrak{s}_\star$ for all covariate values $x$.

Assume that $\mathfrak{s}_{x_i} \neq \mathfrak{s}_{x_j}$ for two covariates $\tilde{x}_i \neq \tilde{x}_j$, and assume, without loss of generality, that $\mathfrak{s}_{x_i}$ is equal to the identity. Consider first the case of $G = 2$. Then (12.23) implies for $x = \tilde{x}_i$ that

$$\theta_1^\star = \left( \begin{array}{c} \mu_1^\star \\ \sigma_1^{2,\star} \end{array} \right) = \theta_1, \quad \theta_2^\star = \left( \begin{array}{c} \mu_2^\star \\ \sigma_2^{2,\star} \end{array} \right) = \theta_2,$$

whereas for $x = \tilde{x}_j$,

$$\theta_1^\star = \left( \begin{array}{c} \mu_2^\star \\ \sigma_2^{2,\star} \end{array} \right) = \theta_2, \quad \theta_2^\star = \left( \begin{array}{c} \mu_1^\star \\ \sigma_1^{2,\star} \end{array} \right) = \theta_1,$$

contradicting the assumption that $\theta_1 \neq \theta_2$. A similar proof is possible for $G > 2$, where the assumption $\mathfrak{s}_{x_i} \neq \mathfrak{s}_{x_j}$ (assuming again that $\mathfrak{s}_1$ is equal to the identity) implies for $x = \tilde{x}_i$ that $\theta_g^\star = \theta_g$ for all components $g = 1, \ldots, G$, whereas for $x = \tilde{x}_j$ at least one component

$g_i$ exists with $\theta_{g_i}^\star = \theta_{g_j}$, where $g_j = \mathfrak{s}_{x_j}(g_i) \neq g_i$. Hence, $\theta_{g_i} = \theta_{g_j}$, which contradicts the assumption that $\theta_{g_i} \neq \theta_{g_j}$. This implies that $\mathfrak{s}_x \equiv \mathfrak{s}_\star$ for all covariate values $x$.

Therefore, the weight distribution $\eta_1(x), \ldots, \eta_G(x)$ is identified up to relabelling the components, and identification depends on whether $\gamma_g$ can be recovered from the corresponding MNL model (12.22) given the design matrix $X$, constructed rowwise from the covariates $\tilde{x}_i, i = 1, \ldots, n$. Standard conditions for identification in an MNL model imply, for example, that $(X^\top X)^{-1}$ exists (McCullagh & Nelder, 1999). It is well known that identification in a logit, and more generally in an MNL, model fails under complete separation; see, for example, Heinze (2006). Hence, a situation where an ME model is not generically identified occurs if certain clusters do not share covariate values with other clusters; see Hennig (2000, Example 4.2) for illustration. A rather strong condition ensuring generic identifiability for models of this type is an extended coverage condition (Hennig, 2000) requiring that the number of clusters $G$ is exceeded by the minimum number of distinct $q$-dimensional hyperplanes needed to cover the covariate values (excluding the constant) for *each* cluster.

Arguments similar to above apply in general for simple ME models of $G$ probability distributions,

$$y_i \sim \sum_{g=1}^{G} \eta_g(\tilde{x}_i) p(y_i | \theta_g). \tag{12.24}$$

Provided that the parameters in the MNL model (12.22) are identified, it can be shown that an ME model is generically identified if the corresponding standard finite mixture distribution is generically identified. In this case, any other mixture representation (12.24) with parameters $\theta_g^\star$ and $\eta_g^\star(\tilde{x}_i)$ is identified up to (the same) label switching according to a permutation $\mathfrak{s}$ for all possible values $\tilde{x}_i$: $\theta_g^\star = \theta_{\mathfrak{s}(g)}$ and $\eta_g^\star(\tilde{x}_i) = \eta_{\mathfrak{s}(g)}(\tilde{x}_i)$ for $g = 1, \ldots, G$.

It follows that mixtures of experts of multivariate Gaussian distributions (as considered in Section 12.2.2) and Poisson distributions, among many others, are generically identified, provided that the parameters in the MNL model (12.22) are identified. Since a standard finite mixture model is that special case of an ME model where $\tilde{x}_i \equiv 1$ is equal to the intercept, special care must be exercised when the underlying standard finite mixture distribution is generically unidentified, as might be the case when modelling discrete data. It is interesting to note that including $\tilde{x}_i$ in the weight function $\eta_g(\tilde{x}_i)$ in ME models is possible for models where including $\tilde{x}_i$ in the component density $p(y_i | \tilde{x}_i, \theta_g)$ yields a generically non-identified mixture, an example being the regressor $\tilde{x}_i = (1 \; d_i)$, where $d_i$ is a 0/1 dummy variable; see Section 12.5.2.

The situation gets rather complex when covariates $\tilde{x}_i$ (or subsets of these) are included as regressors both in the outcome distribution $p(y_i | \tilde{x}_i, \theta_g)$ and in the weight distribution $\eta_g(\tilde{x}_i)$. The presence of a covariate $\tilde{x}_i$ in $\eta_g(\tilde{x}_i)$ could introduce high discriminative power among the groups and might lead to identification of mixtures of regression models which are not identified, if $\eta_g$ is assumed to be independent of the covariates. To our knowledge, generic identification for general ME models has not been studied systematically and would be an interesting avenue for future research.

As it is, the only way to investigate if the chosen mixture model suffers from identifiability problems is to analyse the results obtained from fitting these models to the data carefully. As the examples in Sections 12.5.1 and 12.5.2 have shown, weird behaviour of the MCMC draws in a Bayesian framework is often a sign of identifiability problems. On the other hand, marginal posterior concentration around pronounced modes, verified for instance through appropriate scatter plots of the MCMC draws for the parameters of interest, indicates that identification might not be an issue for that specific application.

## 12.6  Concluding Remarks

This chapter has outlined the definition, estimation and application of ME models in a number of settings, clearly demonstrating their utility as an analytical tool. Their demonstrated use to cluster observations, and to appropriately capture heterogeneity in cross-sectional data, provides only a glimpse of their potential flexibility and utility in a wide range of settings. The ability of ME models to jointly model response and concomitant variables provides deeper and more principled insight into the relations between such data in a mixture model based analysis.

On a cautionary note, however, when an ME model is employed as an analytic tool, care must be exercised in how and where covariates enter the ME model framework. The interpretation of the analysis fundamentally depends on which of the suite of ME models is invoked. Further, as outlined herein, the identifiability of an ME model must be carefully considered; establishing identifiability for ME models is an outstanding, challenging problem.

# Bibliography

AKAIKE, H. (1973). Information theory and an extension of the maximum likelihood principle. In *2nd International Symposium on Information Theory*, B. N. Petrov & F. Csáki, eds. Budapest: Akadémiai Kiadó, pp. 267–281.

BENAGLIA, T., CHAUVEAU, D., HUNTER, D. & YOUNG, D. (2009). mixtools: An R package for analyzing finite mixture models. *Journal of Statistical Software* **32**, 1–29.

BENTER, W. (1994). Computer-based horse race handicapping and wagering systems: A report. In *Efficiency of Racetrack Betting Markets*, W. T. Ziemba, V. S. Lo & D. B. Haush, eds. San Diego: Academic Press, pp. 183–198.

BISHOP, C. (2006). *Pattern Recognition and Machine Learning*. New York: Springer.

BISHOP, C. M. & SVENSEN, M. (2003). Bayesian hierarchical mixtures of experts. In *Proceedings of the Nineteenth Conference on Uncertainty in Artificial Intelligence*, UAI'03. San Francisco: Morgan Kaufmann.

CHAMROUKHI, F. (2015). Non-normal mixtures of experts. Preprint, arXiv:1506.06707.

CHANDRA, S. (1977). On the mixtures of probability distributions. *Scandinavian Journal of Statistics* **4**, 105–112.

CHIB, S. & GREENBERG, E. (1995). Understanding the Metropolis-Hastings Algorithm. *American Statistician* **49**, 327–335.

COAKLEY, J. & GALLAGHER, M. (2004). *Politics in the Republic of Ireland*. London: Routledge in association with PSAI Press, 4th ed.

DAYTON, C. M. & MACREADY, G. B. (1988). Concomitant-variable latent-class models. *Journal of the American Statistical Association* **83**, 173–178.

DEMPSTER, A. P., LAIRD, N. M. & RUBIN, D. B. (1977). Maximum likelihood from incomplete data via the EM algorithm (with discussion). *Journal of the Royal Statistical Society, Series B* **39**, 1–38.

DESARBO, W. & CRON, W. (1988). A maximum likelihood methodology for clusterwise linear regression. *Journal of Classification* **5**, 248–282.

DIEBOLT, J. & ROBERT, C. P. (1994). Estimation of finite mixture distributions by Bayesian sampling. *Journal of the Royal Statistical Society, Series B* **56**, 363–375.

FRÜHWIRTH-SCHNATTER, S. (2004). Estimating marginal likelihoods for mixture and Markov switching models using bridge sampling techniques. *Econometrics Journal* **7**, 143–167.

FRÜHWIRTH-SCHNATTER, S. (2006). *Finite Mixture and Markov Switching Models*. New York: Springer-Verlag.

FRÜHWIRTH-SCHNATTER, S. (2011a). Dealing with label switching under model uncertainty. In *Mixture Estimation and Applications*, K. Mengersen, C. P. Robert & D. Titterington, eds., chap. 10. Chichester: Wiley, pp. 213–239.

FRÜHWIRTH-SCHNATTER, S. (2011b). Panel data analysis – a survey on model-based clustering of time series. *Advances in Data Analysis and Classification* **5**, 251–280.

FRÜHWIRTH-SCHNATTER, S. (2018). *Applied Bayesian Mixture Modelling. Implementations in MATLAB using the package bayesf Version 4.0.* https://www.wu.ac.at/statmath/faculty-staff/faculty/sfruehwirthschnatter/.

FRÜHWIRTH-SCHNATTER, S. & FRÜHWIRTH, R. (2010). Data augmentation and MCMC for binary and multinomial logit models. In *Statistical Modelling and Regression Structures – Festschrift in Honour of Ludwig Fahrmeir*, T. Kneib & G. Tutz, eds. Heidelberg: Physica-Verlag, pp. 111–132.

FRÜHWIRTH-SCHNATTER, S. & KAUFMANN, S. (2008). Model-based clustering of multiple time series. *Journal of Business & Economic Statistics* **26**, 78–89.

FRÜHWIRTH-SCHNATTER, S., PAMMINGER, C., WEBER, A. & WINTER-EBMER, R. (2012). Labor market entry and earnings dynamics: Bayesian inference using mixtures-of-experts Markov chain clustering. *Journal of Applied Econometrics* **27**, 1116–1137.

FRÜHWIRTH-SCHNATTER, S. & WAGNER, H. (2008). Marginal likelihoods for non-Gaussian models using auxiliary mixture sampling. *Computational Statistics and Data Analysis* **52**, 4608–4624.

GEMAN, S. & GEMAN, D. (1984). Stochastic relaxation, Gibbs distributions and the Bayesian restoration of images. *IEEE Transactions on Pattern Analysis and Machine Intelligence* **6**, 721–741.

GERSHENFELD, N. (1997). Nonlinear inference and cluster-weighted modeling. *Annals of the New York Academy of Sciences* **808**, 18–24.

GEWEKE, J. & KEANE, M. (2007). Smoothly mixing regressions. *Journal of Econometrics* **136**, 252–290.

GORMLEY, I. C. & MURPHY, T. B. (2006). Analysis of Irish third-level college applications data. *Journal of the Royal Statistical Society, Series A* **169**, 361–379.

GORMLEY, I. C. & MURPHY, T. B. (2008a). Exploring voting blocs within the Irish electorate: A mixture modeling approach. *Journal of the American Statistical Association* **103**, 1014–1027.

GORMLEY, I. C. & MURPHY, T. B. (2008b). A mixture of experts model for rank data with applications in election studies. *Annals of Applied Statistics* **2**, 1452–1477.

GORMLEY, I. C. & MURPHY, T. B. (2010a). Clustering ranked preference data using sociodemographic covariates. In *Choice Modelling: The State-of-the-Art and the State-of-Practice*, S. Hess & A. Daly, eds. Bingley: Emerald, pp. 543–569.

GORMLEY, I. C. & MURPHY, T. B. (2010b). A mixture of experts latent position cluster model for social network data. *Statistical Methodology* **7**, 385–405.

GORMLEY, I. C. & MURPHY, T. B. (2018). *MEclustnet: Fitting the mixture of experts latent position cluster model.* R package version 1.0.

GRÜN, B. & LEISCH, F. (2008a). Finite mixtures of generalized linear regression models. In *Recent Advances in Linear Models and Related Areas*, Shalabh & C. Heumann, eds. Heidelberg: Physica-Verlag, pp. 205–230.

GRÜN, B. & LEISCH, F. (2008b). Flexmix version 2: Finite mixtures with concomitant variables and varying and constant parameters. *Journal of Statistical Software* **28**, 1–35.

GRÜN, B. & LEISCH, F. (2008c). Identifiability of finite mixtures of multinomial logit models with varying and fixed effects. *Journal of Classification* **25**, 225–247.

HANDCOCK, M., RAFTERY, A. & TANTRUM, J. M. (2007). Model-based clustering for social networks. *Journal of the Royal Statistical Society, Series A* **170**, 301–354.

HEINZE, G. (2006). A comparative investigation of methods for logistic regression with separated or nearly separated data. *Statistics in Medicine* **25**, 4216–4226.

HENNIG, C. (2000). Identifiability of models for clusterwise linear regression. *Journal of Classification* **17**, 273–296.

HOFF, P. (2009). *A First Course in Bayesian Statistical Methods*. New York: Springer-Verlag.

HOFF, P. D., RAFTERY, A. E. & HANDCOCK, M. S. (2002). Latent space approaches to social network analysis. *Journal of the American Statistical Association* **97**, 1090–1098.

HUERTA, G., JIANG, W. & TANNER, M. A. (2003). Time series modeling via hierarchical mixtures. *Statistica Sinica* **13**, 1097–1118.

HUNTER, D. R. & LANGE, K. (2004). A tutorial on MM algorithms. *American Statistician* **58**, 30–37.

HUNTER, D. R. & YOUNG, D. S. (2012). Semiparametric mixtures of regressions. *Journal of Nonparametric Statistics* **24**, 19–38.

HURN, M., JUSTEL, A. & ROBERT, C. (2003). Estimating mixtures of regressions. *Journal of Compututional and Graphical Statistics* **12**, 1–25.

INGRASSIA, S., PUNZO, A., VITTADINI, G. & MINOTTI, S. C. (2015). The generalized linear mixed cluster-weighted model. *Journal of Classification* **32**, 85–113.

JACOBS, R., JORDAN, M., NOWLAN, S. & HINTON, G. (1991). Adaptive mixtures of local experts. *Neural Computation* **3**, 79–87.

JORDAN, M. & JACOBS, R. (1994). Hierarchical mixtures of experts and the EM algorithm. *Neural Computation* **6**, 181–214.

KASS, R. & RAFTERY, A. (1995). Bayes factors. *Journal of the American Statistical Association* **90**, 773–795.

LANG, J. B., MCDONALD, J. W. & SMITH, P. W. F. (1999). Association-marginal modelling of multivariate categorical responses: A maximim likelihood approach. *Journal of the American Statistical Association* **94**, 1161–71.

LAZEGA, E. (2001). *The Collegial Phenomenon: The Social Mechanisms of Cooperation Among Peers in a Corporate Law Partnership*. Oxford: Oxford University Press.

LI, F., VILLANI, M. & KOHN, R. (2011). Modeling conditional densities using finite smooth mixtures. In *Mixtures: Estimation and Applications*, K. Mengersen, C. Robert & M. Titterington, eds., chap. 6. Wiley, pp. 123–144.

MARSH, M. (1999). The making of the eighth president. In *How Ireland Voted 1997*, M. Marsh & P. Mitchell, eds. Boulder, CO: Westview and PSAI Press, pp. 215–242.

MASOUDNIA, S. & EBRAHIMPOUR, R. (2014). Mixture of experts: A literature survey. *Artificial Intelligence Review* **42**, 275–293.

MAZZA, A., PUNZO, A. & INGRASSIA, S. (2017). *flexCWM: Flexible Cluster-Weighted Modeling*. R package version 1.7.

MCCULLAGH, P. & NELDER, J. A. (1999). *Generalized Linear Models*. London: Chapman & Hall.

MCLACHLAN, G. & PEEL, D. (2000). *Finite Mixture Models*. New York: John Wiley.

MENG, X.-L. & RUBIN, D. B. (1993). Maximum likelihood estimation via the ECM algorithm: A general framework. *Biometrika* **80**, 267–278.

METROPOLIS, N., ROSENBLUTH, A., ROSENBLUTH, M., TELLER, A. & TELLER, E. (1953). Equations of state calculations by fast computing machines. *Journal of Chemical Physics* **21**, 1087–1092.

MUTHÉN, L. K. & MUTHÉN, B. O. (2011). *Mplus User's Guide*. Los Angeles: Muthén and Muthén, 6th ed.

PAMMINGER, C. & FRÜHWIRTH-SCHNATTER, S. (2010). Model-based clustering of categorical time series. *Bayesian Analysis* **5**, 345–368.

PENG, F., JACOBS, R. A. & TANNER, M. A. (1996). Bayesian inference in mixtures-of-experts and hierarchical mixtures-of-experts models with an application to speech recognition. *Journal of the American Statistical Association* **91**, 953–960.

PLACKETT, R. L. (1975). The analysis of permutations. *Applied Statistics* **24**, 193–202.

QUANDT, R. (1972). A new approach to estimating switching regressions. *Journal of the American Statistical Association* **67**, 306–310.

R CORE TEAM (2018). *R: A Language and Environment for Statistical Computing*. Vienna: R Foundation for Statistical Computing.

RAFTERY, A. E., NEWTON, M. A., SATAGOPAN, J. M. & KRIVITSKY, P. (2007). Estimating the integrated likelihood via posterior simulation using the harmonic mean identity (with discussion). In *Bayesian Statistics 8*, J. M. Bernardo, M. J. Bayarri, J. O. Berger, A. P. Dawid, D. Heckerman, A. F. M. Smith, & M. West, eds. Oxford: Oxford University Press, pp. 371–416.

RASMUSSEN, C. E. & GHAHRAMANI, Z. (2002). The infinite mixtures of Gaussian process experts. In *Advances in Neural Information Processing Systems*, vol. 12. Cambridge, MA: MIT Press, pp. 554–560.

SCHWARZ, G. (1978). Estimating the dimension of a model. *Annals of Statistics* **6**, 461–464.

SCOTT, S. L. (2011). Data augmentation, frequentist estimation, and the Bayesian analysis of multinomial logit models. *Statistical Papers* **52**, 87–109.

SINNOTT, R. (1995). *Irish Voters Decide: Voting Behaviour in Elections and Referendums since 1918*. Manchester: Manchester University Press.

SINNOTT, R. (1999). The electoral system. In *Politics in the Republic of Ireland*, J. Coakley & M. Gallagher, eds. London: Routledge & PSAI Press, 3rd ed., pp. 99–126.

STEPHENS, M. (2000). Bayesian analysis of mixture models with an unknown number of components – an alternative to reversible jump methods. *Annals of Statistics* **28**, 40–74.

SUBEDI, S., PUNZO, A., INGRASSIA, S. & MCNICHOLAS, P. D. (2013). Clustering and classification via cluster-weighted factor analyzers. *Advances in Data Analysis and Classification* **7**, 5–40.

TANG, X. & QU, A. (2015). Mixture modeling for longitudinal data. *Journal of Computational and Graphical Statistics* **25**, 1117–1137.

TANNER, M. A. (1996). *Tools for Statistical Inference: Observed Data and Data Augmentation Methods*. New York: Springer-Verlag, 3rd ed.

TEICHER, H. (1961). Identifiability of mixtures. *Annals of Mathematical Statistics* **32**, 244–248.

TEICHER, H. (1963). Identifiability of finite mixtures. *Annals of Mathematical Statistics* **34**, 1265–1269.

TURNER, R. (2014). *mixreg: Functions to fit mixtures of regressions*. R package version 0.0-5.

VERMUNT, J. K. & MAGIDSON, J. (2005). *Latent GOLD 4.0 User's Guide*. Belmont, MA: Statistical Innovations Inc.

VILLANI, M., KOHN, R. & GIORDANI, P. (2009). Regression density estimation using smooth adaptive Gaussian mixtures. *Journal of Econometrics* **153**, 155–173.

VILLANI, M., KOHN, R. & NOTT, D. J. (2012). Generalized smooth finite mixtures. *Journal of Econometrics* **171**, 121–133.

WANG, P., PUTERMAN, M., COCKBURN, I. & LE, N. (1996). Mixed Poisson regression models with covariate dependent rates. *Biometrics* **52**, 381–400.

WATERHOUSE, S., MACKAY, D. & ROBINSON, T. (1996). Bayesian methods for mixtures of experts. In *Advances in Neural Information Processing Systems*. San Mateo, CA: Morgan Kaufmann, pp. 351–357.

WHITE, A. & MURPHY, T. B. (2016). Mixed-membership of experts stochastic blockmodel. *Network Science* **4**, 48–80.

YAKOWITZ, S. J. & SPRAGINS, J. D. (1968). On the identifiability of finite mixtures. *Annals of Mathematical Statistics* **39**, 209–214.

YOUNG, D. S. & HUNTER, D. R. (2010). Mixtures of regressions with predictor-dependent mixing proportions. *Computational Statistics & Data Analysis* **54**, 2253–2266.

# 13

# Hidden Markov Models in Time Series, with Applications in Economics

**Sylvia Kaufmann**

*Study Center Gerzensee, Foundation of the Swiss National Bank, Switzerland*

## CONTENTS

## 13.1 Introduction

Hidden Markov models are mixture models with sequential dependence or persistence in the mixture distribution. For a finite number $G$ of components, persistence in distribution is induced by specifying a latent component indicator which follows a Markov process. The transition probabilities for the Markov process may either be time-invariant or time-varying. In the latter case, hidden Markov models extend mixture of experts models (see Chapter 12) by introducing persistence in the mixtures.

Hidden Markov models in time series econometrics became very popular after the publications of Hamilton (1989, 1990). He transferred earlier regression based approaches like Goldfeld & Quandt (1973) to time series analysis by recognizing their usefulness in capturing asymmetric conditional moments or asymmetric dynamic properties of time series. In Section 13.2 we start by setting out the framework and terminology. In time series analysis, components are usually called states or regimes, and the transition between states is termed regime switch or regime change. This wording will be used in this chapter to be consistent with the econometrics literature. We discuss in separate sections the basic modelling choice of specifying the transition distribution of states. Hamilton (1989, 1990) introduced the model with time-invariant or constant transition distribution, and most of the following literature stayed with this specification.

This is not as restrictive as it may seem at first sight, given that more sophisticated models can be built either by imposing restrictions on the state transition probabilities or by combining multiple latent state indicators in a dynamical or hierarchical way. Change-point models (Chib, 1996; Pesaran et al., 2007; Bauwens et al., 2015) are nested in Markov switching models by imposing appropriate zero restrictions on the transition distribution. Linking multiple latent state indicators dynamically, we can capture many leading/lagging features in multivariate analysis (Phillips, 1991; Paap et al., 2009; Kaufmann, 2010). Linking state indicators hierarchically, we obtain hierarchical Markov mixture models, for example, to disentangle long-term from short-term changing dynamics (Geweke & Amisano, 2011; Bai & Wang, 2011). Nevertheless, constant or exogenous transition distributions do not incorporate an explicit explanation or interpretation of the driving forces underlying the transition distribution.

Including covariate effects in the transition distribution renders it time-varying and yields at least an indication, if not a driving cause, of the regime switches. One of the first proposals was Diebold et al. (1994). Applications followed in business cycle analysis in Filardo (1994) and Filardo & Gordon (1998). Both probit and logit functional forms were used for the transition distribution. Under the assumption of independence between state alternatives, both parameterizations yield essentially the same estimation results. Later on, Koop & Potter (2007) introduced duration-dependent, time-varying probabilities into a change-point model. An interesting alternative is presented in Billio & Casarin (2011), who use a beta autoregressive process to model time-varying transition probabilities.

Against this background, we outline various extensions that are available within the general framework we present. Given that covariates may have state-dependent effects on the transition distribution, we elaborate on various considerations that may flow into the specific parameterization of time-varying transition probabilities. Section 13.2 closes with a discussion of an attractive feature of Markov switching models that has so far, to our knowledge, not been exploited in applied time series modelling. So far, these models have been applied under the assumption that the conditional (i.e. state-dependent) distributions are stationary or, in other words, have finite moments in every period $t$. This need not be the case, however. Many real phenomena are consistent with a process that alternates between a stationary and a non-stationary state-specific distribution. Think of the recent financial crisis, during which dynamics across economic variables may have engaged transitorily on an unsustainable path. Francq & Zakoïan (2001), and more recently Farmer et al. (2009a), derived conditions under which the unconditional distribution of multivariate Markov switching time series processes is stationary and has finite moments. Quite obviously, stationary and non-stationary processes may be combined by a Markov switching process and still yield an unconditionally stationary process, if the non-stationary process does not recur too often and does not prevail for too long. The intriguing thing is that combining state-specific stationary processes is not sufficient to obtain a stationary unconditional process.

In Section 13.3 we outline the estimation of Markov switching models, where the emphasis is on Bayesian estimation. Maximum likelihood estimation and variants of it are based on the EM algorithm, in which the E step takes explicitly into account the state dependence in the mixture to infer about the state indicator (Hamilton, 1990). Extensions to multivariate models followed in Krolzig et al. (2002) and Clements & Krolzig (2003). The forward-filtering backward-sampling (FFBS) algorithm provides the basis for data augmentation in Bayesian estimation (McCulloch & Tsay, 1994; Chib, 1996). Markov chain Monte Carlo methods prove very useful to estimate models with multiple latent variables, like factor models with Markov switching factor mean or factor volatility (Kim & Nelson, 1998).

Hidden Markov models endorse all issues concerning mixture modelling, as comprehensively exposed in Frühwirth-Schnatter (2006). In the present chapter, we therefore discuss in detail the design of state-invariant prior distributions for time-invariant and time-varying transition probabilities (Burgette & Hahn, 2010; Kaufmann, 2015). We then set out the posterior random permutation sampler to obtain draws from the unconstrained, multimodal posterior (Frühwirth-Schnatter, 2001). To sample the parameters of the logit functional form, we borrow from data augmentation algorithms outlined in Frühwirth-Schnatter & Frühwirth (2010) which render the nonlinear, non-Gaussian model in latent utilities linear Gaussian. Parameters are sampled from full conditional distributions rather than by Metropolis–Hastings (Scott, 2011; Holmes & Held, 2006). The approach of Burgette & Hahn (2010) proves very useful to sample parameters of the probit functional form. Instead of normalizing the error covariance of latent utilities with respect to a specific element (McCulloch et al., 2000; Imai & van Dyk, 2005), they propose restricting the trace of the normalized error covariance of the latent utilities, whereby normalization occurs in each iteration of the sampler with respect to a randomly chosen latent state utility. To conclude Section 13.3, we compare estimation time and sampler efficiency between using the logit and the probit functional form to estimate the data generating process of a univariate series driven, respectively, by two and three hidden Markov mixtures. We briefly illustrate that posterior state identification is obtained by post-processing the posterior draws.

In this chapter, we do not discuss prior design and model choice with respect to the number of regimes. The same considerations as outlined in Chapter 7 apply to Markov mixture models in time series analysis, and the interested reader may refer to it. In brief, model choice with respect to the number of regimes can be addressed by means of marginal likelihood (Chib, 1995; Frühwirth-Schnatter, 2004). In the maximum likelihood framework, the issue can only be addressed in a proper statistical way by simulating the test statistic (see Hansen, 1992). The likelihood ratio statistic violates regularity conditions, because models with different numbers of states are not nested within each other. For similar reasons, the widely used information criteria are not an alternative either, or at least should be used with more care than usually done.

In Section 13.4, by discussing informative regime switching we illustrate how to obtain explicit economic interpretations of results from posterior inference. For example, imposing structural restrictions on time-invariant transition probabilities yields explicit interpretations about dynamic relationships across variables. One of the first contributions was Phillips (1991) who analysed country-specific output series in a multivariate setting. Recently, Sims et al. (2008) proposed a general framework to implement and estimate restricted transition distributions in large multiple-equation systems. Including covariate effects in the transition distribution provides an explicit interpretation of the driving factors underlying the latent state indicator. Additionally, prior knowledge may flow into the parameterization of the transition distribution by imposing parameter restrictions (Gaggl & Kaufmann, 2014; Bäurle et al., 2016). In the latter case, this induces a restricted, state-identified prior and may call for some restricted estimation procedures. The list of papers used for illustration is well short of exhaustive and refers mainly to business cycle analysis. Nevertheless, the examples provided are straightforwardly applicable in other areas such as financial econometrics (Hamilton & Susmel, 1994; Bauwens & Lubrano, 1998). Section 13.5 concludes.

The methods discussed in this chapter apply generally to Markov switching models if the dependence on past states is fixed. Models with infinite dependence on past states, as in regime switching generalized ARCH models, are not treated in this chapter. The interested reader may refer to the specific literature (Klaassen, 2002; Gray, 1996; Bauwens et al., 2014) and to Chapter 17 of this volume for an overview. Forecasting is not treated in this chapter either. The interested reader may refer, for example, to Elliott & Timmermann (2005), Pesaran et al. (2007) and Chauvet & Piger (2008). Scenario-based forecasting is used in Kaufmann & Kugler (2010).

## 13.2 Regime Switching: Mixture Modelling over Time

### 13.2.1 Preliminaries and model specification

Hidden Markov models or Markov switching models are mixture models with the typical feature of sequential (time) persistence in the mixture distribution. These models are often applied in time series analysis, where a scalar or a vector of observations is denoted by $y_t$, with $t = 1, \ldots, T$ indexing the observation period.

In a general model with period-specific observation densities

$$y_t \sim f(y_t | x_{1t}, \theta_t), \tag{13.1}$$

persistence is introduced by assuming a time-dependent process for $\theta_t$, $p(\theta_t | x_{2t}, \theta_{t-1})$. In (13.1), $x_{1t}$ denotes covariates which may also include lagged observations of $y_t$, that is, $y_t \sim f(y_t | y_{t-1}, \theta_t)$, which represents the typical time series set-up. The dependence on one lag only is not restrictive, given that any autoregressive process of order $p$ can be represented in its companion form as an autoregressive process of order 1. In the present chapter, $x_{2t}$ denotes covariates that influence the transition distribution of the parameters. In time series analysis, the mixture components are called states, and in hidden Markov models one typically assumes that the set of parameter states is discrete, that is, $\theta_t \in \{\theta_1 \ldots, \theta_G\}$. The latent component indicator $z_t \in \mathcal{G} = \{1, \ldots, G\}$, $G \in \mathbb{N}$, is called the state indicator, and the binary indicator is defined by $z_{tg} = 1$ if and only if $z_t = g$. Conditional on $z_t$, $\theta_t = \sum_{g=1}^G z_{tg}\theta_g = \theta_{z_t}$ and $y_t | z_t \sim f(y_t | x_{1t}, \theta_{z_t})$.

State persistence is introduced by formulating a Markov process for $z_t$:

$$\mathrm{P}(z_t = g | z_{t-1} = g', x_{2t}) = \xi_{t,g'g}, \tag{13.2}$$

with $\sum_{g=1}^G \xi_{t,g'g} = 1$. In the most general specification, covariates $x_{2t}$ render the state transition probabilities $\xi_{t,g'g}$ time-specific or time-varying. Hidden Markov models thus extend mixture of experts models by introducing persistence in the mixture distribution.

In the following, for some derivations it is convenient to gather the transition probabilities ito the matrix

$$\xi_t = \begin{bmatrix} \xi_{t,11} & \cdots & \xi_{t,1G} \\ \vdots & & \vdots \\ \xi_{t,G1} & \cdots & \xi_{t,GG} \end{bmatrix}$$

and likewise into $\xi$ for time-invariant or *homogeneous* transition probabilities. Specific functional forms of state transition call for specific identification restrictions, which are usually set on a reference state. We will denote this reference state by $g_0$. Finally, we define $\mathcal{G}_{-g} = \mathcal{G} \setminus g$.

## 13.2.2 The functional form of state transition

Different functional forms are available to model Markov state transitions $\xi_t$, and each of them needs careful specification. In particular, to extract the information of interest from the data, the researcher has to form expectations about parameterizations to shape the functional form in a sensible way, and, if of interest, to design state-invariant prior distributions. Therefore, in this and other sections, we discuss each functional form in turn.

### 13.2.2.1 Time-invariant switching

The simplest way to parameterize time-invariant switching is to define the $\xi_{g'g}$ directly as transition probabilities:

$$P(z_t = g | z_{t-1} = g') = \xi_{g'g}. \tag{13.3}$$

The condition $0 < \xi_{g'g} < 1$ ensures *ergodicity* of the Markov chain, which means that each state $g$ can be reached from every other state $g'$ in a finite number of steps (equal to 1 under this condition). An ergodic Markov chain has *stationary distribution*

$$\bar{\xi}_g = \lim_{t \to \infty} P(z_t = g) = \lim_{t \to \infty} \xi_{ig}^{(t)} > 0, \quad \forall g \in \mathcal{G}, \, \forall i,$$

where $\xi_{ig}^{(t)}$ is the $(i, g)$th element of the matrix $\xi^t$. In other words, in the matrix of stationary transition probabilities, $\bar{\xi}$, all elements in a column are equal. $\bar{\xi}_g$ is also called *unconditional* or *ergodic* probability.

Persistence probabilities lying strictly between 0 and 1 avoid absorbing states. If absorbing states are present in the data, they should follow non-absorbing states, to be able to identify the state-specific parameters of the latter. In change-point models with a finite and an infinite number of change-points (Chib, 1998; Pesaran et al., 2007; Koop & Potter, 2007), $\xi_{g'g} = 0$ for $g < g'$, that is, the Markov chain is non-ergodic. These models represent a sequence of non-recurrent states, where, after a switch to state $g$, recurrence to states $g' < g$ is no longer allowed. In models with a finite number of change-points, the final state $G$ is an absorbing one, that is,

$$\bar{\xi}_g = \begin{cases} 0, & g = 1, \dots, G-1, \\ 1, & g = G. \end{cases}$$

The logit functional form represents an alternative specification for time-invariant switching:

$$P(z_t = g | z_{t-1} = g') = \xi_{g'g} = \frac{\exp(\gamma_{g'g})}{\sum_{j=1}^{G} \exp(\gamma_{g'j})},$$

where $\gamma_{g'g}$ represents the state-specific constant switching effect. For identification purposes, $\gamma_{g'g_0} = 0$ for some reference state $g_0 \in \mathcal{G}$. No restriction on $\gamma = \{\gamma_{g'g} | g' \in \mathcal{G}, g \in \mathcal{G}_{-g_0}\}$ is needed to ensure that the transition probabilities lie between 0 and 1. In general, working with a functional form also has the advantage that covariates can be included to design time-varying or *informative* regime switching.

### 13.2.2.2   Time-varying switching

To design time-varying switching, we introduce covariates into the transition distribution $\xi_{t,g'g} = \xi(x_{2t}, \gamma)$ with a corresponding vector $\gamma$ of regression coefficients. Depending on the functional form, different restrictions are imposed on $\gamma$ for identification purposes. In the following, we work with a scalar covariate $x_{2t}$. With appropriate adjustments, the generalization to a vector of covariates is straightforward.

The logit functional form with covariates reads

$$\xi_{t,g'g} = \frac{\exp\left(x_{2t}\gamma_{g'g}^x + \gamma_{g'g}\right)}{\sum_{j=1}^{G}\exp\left(x_{2t}\gamma_{g'j}^x + \gamma_{g'j}\right)}, \tag{13.4}$$

in which $\gamma_{g'g}^x$ represents the state-specific covariate effect. For identification purposes, we impose $\gamma_{g'g_0} = 0$ for some reference state $g_0 \in \mathcal{G}$; see, for example, Diebold et al. (1994) for an early contribution.

An alternative is the probit functional form

$$\xi_{t,g'g} = \Phi\left(x_{2t}\gamma_{g'g}^x + \gamma_{g'g}\right), \tag{13.5}$$

where $\Phi(x) = \int_{-\infty}^{x} \phi(u)du$ is the cumulative distribution function with respect to the standard normal density $\phi(\cdot)$. For $G = 2$, the restriction $\gamma_{g'g_0} = -\gamma_{g'g}$ induces identification; see Filardo (1994) and Filardo & Gordon (1998) for early contributions in economics.

A few remarks are in order. First, in both functional forms, we do not fix the reference state to $g_0 = 1$ as is usually done. In estimation, this generalization allows us to apply the permutation sampler to the reference state as well. Second, for $G = 2$, the probit specification has the advantage that a corresponding random utility model exists for $g \neq g_0$ with a latent variable $u_{tg}^*$ following a normal distribution:

$$u_{tg}^* = x_{2t}\gamma_{z_{t-1},g}^x + \gamma_{z_{t-1},g} + \nu_{tg}, \quad \nu_{tg} \sim \mathcal{N}(0,1). \tag{13.6}$$

This renders parameter estimation (e.g. full conditional Gibbs sampling) straightforward.

The situation is more intricate for multi-state regime switching models. Conditional on the state indicators, parameter estimation in a multinomial probit model for $\xi_t$ is not the issue (McCulloch et al., 2000; Imai & van Dyk, 2005; Nobile, 1998; Burgette & Hahn, 2010). However, the state indicators $\mathbf{z} = (z_1, \ldots, z_T)$ are not observed and need to be inferred from the data. This involves the evaluation of the $TG(G-1)$ multivariate integrals

$$\xi_{t,g'g} = \int_{\mathcal{G}_{g'g}} \phi(\nu_t, \Sigma)d\nu_t, \quad g \in \mathcal{G}_{-g_0},$$

where $\phi(\cdot)$ is the density of the $(G-1)$-variate normal distribution with mean 0 and covariance $\Sigma$, and $\nu_t = (\nu_{t1}, \ldots, \nu_{t,g_0-1}, \nu_{t,g_0+1}, \ldots, \nu_{tG})'$ (see also Nobile, 1998). The set $\mathcal{G}_{g'g}$ is given by

$$\mathcal{G}_{g'g} = \bigcap_{j\neq g}\left\{\nu_{tg} - \nu_{tj} > x_{2t}(\gamma_{g'j}^x - \gamma_{g'g}^x) + (\gamma_{g'j} - \gamma_{g'g})\right\} \cap \left\{\nu_{tg} > -(x_{2t}\gamma_{g'g}^x + \gamma_{g'g})\right\}.$$

Various procedures have been proposed to evaluate these integrals (Geweke et al., 1994). They all represent approximations to the transition probabilities, however. Moreover, estimation gets very slow (see Section 13.3.4).

### 13.2.2.3 Nested alternatives

Some alternatives are nested in both the logit and the probit functional forms. If $\gamma_{g'g}^x = 0$, we recover the specification with time-invariant transition probabilities: $\xi_{g'g} = \exp\left(\gamma_{g'g}\right) / \sum_{j=1}^G \exp\left(\gamma_{g'j}\right)$.

State persistence is maintained even if we restrict the covariate effect to be state-independent $\gamma_{g'g}^x = \gamma_g^x$, for all $g'$. If we additionally restrict $\gamma_{g'g} = \gamma_g$, then we obtain a mixture model with time-varying weights:

$$p(y_t|x_{1t}, x_{2t}, \theta) = \sum_{g=1}^G \xi_{tg}(x_{2t}, \gamma_g) f(y_t|x_{1t}, \theta_g),$$

where $\theta$ summarizes all unknown parameters.

Given observations $\mathbf{y} = (y_1, \ldots, y_T)$ and $\mathbf{x}_j = (x_{j1}, \ldots, x_{jT})$, $j = 1, 2$, the relevance of differences across state-dependent parameters or the relevance of covariates can be evaluated using the Savage–Dickey density ratio (Dickey, 1968; Verdinelli & Wasserman, 1995),

$$\mathrm{BF}\left(\mathcal{M}_\mathcal{R}|\mathcal{M}\right) = \left.\frac{p\left(\gamma|\mathbf{y}, \mathbf{x}_1, \mathbf{x}_2\right)}{p\left(\gamma\right)}\right|_{\gamma \in \mathcal{R}}$$

where $p(\gamma)$ is the prior distribution, $\mathcal{M}_\mathcal{R}$ and $\mathcal{M}$ respectively indicate the restricted and the unrestricted model, and $\mathcal{R}$ represents a single or a combination of restrictions on $\gamma$ mentioned above.

### 13.2.3 Generalizations

The framework (13.1) encompasses linear regressions and dynamic models as well, given that $x_{1t}$ may also include lagged observations of $y_t$, that is, $y_t \sim f(y_t|y_{t-1}, \theta_t)$. Self-reinforcing or additional feedback effects may be included by allowing lagged variables to have an effect on the transition distribution, $\theta_t|y_{t-1}, \theta_{t-1}$. The Markov process in (13.2) is of order 1. This is not restrictive, as $p$th-order Markov processes can be reparameterized by defining an encompassing $G^p$ first-order Markov state process with appropriate design of the transition distribution. Likewise, current and (fixed) $p$-lagged state dependence in $f(y_t|\cdot)$ may be reparameterized to current state dependence by defining an encompassing $G^{1+p}$ state variable with appropriately designed transition distribution and enlarging the state-specific parameter to $\theta_t = \{\theta_{z_t}, \ldots, \theta_{z_{t-p}}\}$, with $z_t \in \mathcal{G}$ for all $t$ (see also Hamilton, 1994, Chapter 22).

Model (13.1) is generic in terms of the state dependence of $y_t$ and $\theta_t$, as well as in terms of parameterization of the state indicator. Some elements of $\theta_t$ might be state-independent, as in Frühwirth-Schnatter & Kaufmann (2006), for example. The model encompasses situations where some elements of the appropriately partitioned vector $y_t = (y_{1t}^\top, y_{2t}^\top)^\top$ follow a state-independent distribution, that is, $f(y_t|x_t, \theta_{z_t}) = f(y_{2t}|y_{1t}, x_t, \theta_{2,z_t}) f(y_{1t}|x_t, \theta_{1,z_t})$.

Multiple hidden states $z_t^j$, $j = 1, \ldots, p$, may affect $y_t$. The simplest situation is the case where two independent state processes $z_t^1$ and $z_t^2$ determine the elements in $y_t$, that is, $y_t \sim f(y_{2t}|y_{1t}, x_t, \theta_{z_t^2}) f(y_{1t}|x_t, \theta_{z_t^1})$; see Psaradakis et al. (2005). If observations are independent and driven by independent processes, the models for $y_{1t}$ and $y_{2t}$ might even be analysed separately, with $y_t \sim f(y_{1t}|x_t, \theta_{1,z_t^1}) f(y_{2t}|x_t, \theta_{2,z_t^2})$. Finally, to introduce dependence, the state indicators $z_t^j$, $j = 1, \ldots, p$, may be linked by a dynamic or hierarchical structure (Kaufmann, 2010; Geweke & Amisano, 2011; Bai & Wang, 2011). These models are analysed by defining an encompassing state indicator $z_t^*$, which captures all possible state combinations of the underlying $p$ state indicators;, see Section 13.4 for some examples and references to applications.

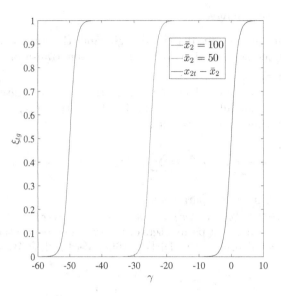

**FIGURE 13.1**
Range of $\gamma$ against $\xi_{tg} = \exp\left(x_{2t}\gamma^x + \gamma\right) / \left(1 + \exp(x_{2t}\gamma^x + \gamma)\right)$, with $\gamma^x = 0.5$, to obtain $\xi_{tg} = 0.5$ when $x_{2t} = \bar{x}_2$, for $\bar{x}_2$ equal to, respectively, 100 and 50 (dashed and solid line). The range of $\gamma$ is scale-invariant (dash-dotted line) if the covariate is mean-adjusted up-front, $x_{2t} - \bar{x}_2$.

### 13.2.4 Some considerations on parameterization

Nothing has been said so far about scaling of covariates $x_{2t}$. In fact, in specification (13.4), the covariate is assumed to be mean-adjusted or normalized, $x_{2t} = \tilde{x}_{2t} - \bar{x}_2$, where $\tilde{x}_{2t}$ and $\bar{x}_2$ are, respectively, the series in levels and the mean or the normalizing level. We call this the *centred* parameterization, in which the time-invariant part of the transition probabilities, $\gamma_{g'g}$, becomes scale-independent. In estimation it scales the range of sensible values for $\gamma_{g'g}$, and in Bayesian inference it allows designing a scale-invariant prior.

To illustrate this, assume that $G = 2$ and the transition probabilities are independent of state $z_{t-1}$:

$$\xi_{tg} = \frac{\exp\left(x_{2t}\gamma^x + \gamma\right)}{\left(1 + \exp(x_{2t}\gamma^x + \gamma)\right)},$$

for $g \neq g_0$ with $\gamma^x = 0.5$, and $\xi_{tg_0} = 1 - \xi_{tg}$. The range of $\gamma$ against $\xi_{tg}$ to obtain $\xi_{tg} = 0.5$ when $\tilde{x}_{2t} = \bar{x}_2$ is not scale-invariant with respect to $\tilde{x}_{2t}$; see the dashed and solid lines in Figure 13.1. Working with the centred version removes the scale-dependence of $\gamma$; see the dash-dotted line in Figure 13.1.

Moreover, it is worthwhile to form expectations about sensible parameter configurations prior to estimation.[1] Assume state-dependent covariate effects, $g_0 = 1$, such that $\gamma_{g'2} = \gamma = (\gamma_1^x, \gamma_2^x, \gamma_1, \gamma_2)$. Figure 13.2 plots the state persistence against the covariate for two settings of $\gamma$. In both settings, state persistence is $\xi_{t,gg} = 0.88$ if $x_{2t} = 0$. If $(\gamma_1^x, \gamma_2^x)$ deviate with equal sign from zero, in the limit, as $x_{2t} \to \pm\infty$, one of the states becomes absorbing. On the other hand, if the parameters differ in sign both states become absorbing if $x_{2t} \to -\infty$, or the indicator switches back and forth between states if $x_{2t} \to \infty$. The conclusion we draw

---

[1]The example follows the one in Kaufmann (2015).

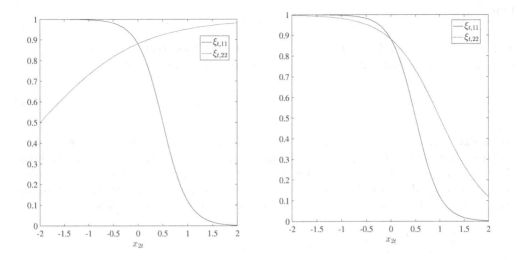

**FIGURE 13.2**

Persistence probabilities $\xi_{t,11} = 1/\left(1 + \exp\left(x_{2t}\gamma_1^x + \gamma_1\right)\right)$ (solid line) and $\xi_{t,22} = \exp\left(x_{2t}\gamma_2^x + \gamma_2\right)/\left(1 + \exp\left(x_{2t}\gamma_{22}^x + \gamma_{22}\right)\right)$ (dashed line). Left: $\gamma = \left(\gamma_1^x, \gamma_2^x, \gamma_1, \gamma_2\right) = (4, 1, -2, 2)$. Right: $\gamma = (4, -2, -2, 2)$.

from this very simple illustration is that if unconstrained covariates have state-dependent effects in the transition distribution, the parameter configuration for $\gamma^x$ should be such that the probability mass is shifted mainly towards one of the states as $x_{2t}$ varies in size.

## 13.2.5 Stability conditions: combining stable and unstable processes

Detailed derivations of stability conditions are beyond the scope of this chapter. Nevertheless, some general conditions for stationarity are re-stated here without proof. From a Bayesian perspective, we might wonder why we should care about stability conditions. In a typical time series set-up, model estimation involves the conditional distribution $f(y_t|z_t, y_{t-1})$, which usually is characterized by finite moments even if some state-specific autoregressive process contains an explosive root. However, from a forecasting perspective we may require the unconditional distribution $\lim_{h\to\infty} f(y_{t+h}|y_t) = f(y_t) = f(y)$ to be stationary and have bounded moments. In the case where some state-specific processes contained explosive roots, the condition would imply a forecasting density with bounded moments in the long run, although forecasting densities for some intermediate forecast horizon could have unbounded moments.

The results provided in this subsection also complement those for the hidden Markov models discussed in Chapter 14. Unlike for those models, time series observations are not independent when we condition on the state indicator $z_t$. This renders the derivation of the unconditional distribution $f(y_t)$ and its moments more involved.

The derivations in Francq & Zakoïan (2001) for multivariate Markov switching autoregressive moving-average processes with homogeneous transition probabilities $\xi$ build on previous results derived for generalized autoregressive processes in Brandt (1986) and Bougerol & Picard (1992). To summarize the relevant results, in what follows $\|\cdot\|$ denotes any matrix operator norm and $\log^+ u = \max\{\log u, 0\}$.

Write a $d$-variate $p$th-order autoregressive Markov switching process in its companion form

$$y_t = c_{z_t} + \rho_{z_t} y_{t-1} + \Sigma_{z_t} \varepsilon_t, \quad \varepsilon_t \sim \mathcal{N}(0, I_d), \tag{13.7}$$

with $I_d$ being the identity matrix and appropriately defined vectors and system matrices $c_{z_t}$, $\rho_{z_t}$ and $\Sigma_{z_t}$. Defining $\omega_{z_t} = c_{z_t} + \Sigma_{z_t}\varepsilon_t$ and solving backwards yields

$$y_t = \omega_{z_t} + \sum_{l=1}^{\infty} \rho_{z_t} \cdots \rho_{z_{t-l+1}} \omega_{z_{t-l}}. \tag{13.8}$$

Given that $\{z_t, \varepsilon_t\}$ is an ergodic process, the process $\{\rho_{z_t}, \omega_{z_t}\}$ is ergodic as well. We can additionally assume $\mathrm{E}\left(\log^+ \|\rho_{z_t}\|\right) < \infty$ and $\mathrm{E}\left(\log^+ \|\omega_{z_t}\|\right) < \infty$. This means respectively that $\rho_{z_t}$ is finite by any matrix norm and shocks are bounded. From the general results in Brandt (1986) and Bougerol & Picard (1992), it follows that the unique strictly stationary solution of (13.7) is given by (13.8), if the top Lyapunov exponent defined by

$$\tilde{\lambda} = \inf_{t \in \mathbb{N}} \left\{ \mathrm{E}\left( \frac{1}{t} \log \|\rho_{z_t} \rho_{z_{t-1}} \cdots \rho_{z_1}\| \right) \right\} \stackrel{a.s.}{=} \lim_{t \to \infty} \frac{1}{t} \log \|\rho_{z_t} \rho_{z_{t-1}} \cdots \rho_{z_1}\|$$

is strictly negative.

**Theorem 13.1** (Francq and Zakoïan, 2001) *Suppose that $\tilde{\lambda} < 0$. Then for all $t \in \mathbb{Z}$, the series*

$$y_t = \omega_{z_t} + \sum_{l=1}^{\infty} \rho_{z_t} \cdots \rho_{z_{t-l+1}} \omega_{z_{t-l}}$$

*converges almost surely (a.s.) and the process $y_t$ is the unique strictly stationary solution of* (13.7).[2]

To establish necessary and sufficient second-order stationarity conditions, we show that $y_t^{(l)} = \rho_{z_t} \cdots \rho_{z_{t-l+1}} \omega_{z_{t-l}}$ converges to zero in $L^2$ at an exponential rate as $l \to \infty$. Ultimately, we obtain $\|y_t\|_2 \leq \sum_{l=0}^{\infty} \|y_t^{(l)}\|_2$, with $y_t^{(0)} = \omega_{z_t}$. Given that $z_t$ and $\varepsilon_t$ are independent, we can write

$$\mathrm{E}\left(\mathrm{vec}\left(y_t^{(l)} y_t^{(l)\top}\right)\right) = \mathrm{E}\left((\rho_{z_t} \otimes \rho_{z_t}) \cdots (\rho_{z_{t-l+1}} \otimes \rho_{z_{t-l+1}}) (\omega_{z_{t-l}} \otimes \omega_{z_{t-l}})\right)$$
$$= \mathrm{E}\left((\rho_{z_t} \otimes \rho_{z_t}) \cdots (\rho_{z_{t-l+1}} \otimes \rho_{z_{t-l+1}}) (\Sigma_{z_{t-l}} \otimes \Sigma_{z_{t-l}})\right) \mathrm{E}(\varepsilon_{t-l} \otimes \varepsilon_{t-l})$$
$$+ \mathrm{E}\left((\rho_{z_t} \otimes \rho_{z_t}) \cdots (\rho_{z_{t-l+1}} \otimes \rho_{z_{t-l+1}}) (c_{z_{t-l}} \otimes c_{z_{t-l}})\right).$$

Define the matrices

$$P = \begin{bmatrix} \xi_{11}(\rho_1 \otimes \rho_1) & \cdots & \xi_{G1}(\rho_1 \otimes \rho_1) \\ \vdots & & \vdots \\ \xi_{1G}(\rho_G \otimes \rho_G) & \cdots & \xi_{GG}(\rho_G \otimes \rho_G) \end{bmatrix},$$

$$S = \begin{bmatrix} \bar{\xi}_1(\Sigma_1 \otimes \Sigma_1) \\ \vdots \\ \bar{\xi}_G(\Sigma_G \otimes \Sigma_G) \end{bmatrix}, \quad C = \begin{bmatrix} \bar{\xi}_1(c_1 \otimes c_1) \\ \vdots \\ \bar{\xi}_G(c_G \otimes c_G) \end{bmatrix},$$

to write compactly

$$\mathrm{E}\left(\mathrm{vec}\left(y_t^{(l)} y_t^{(l)\top}\right)\right) = JP^l \left(S \cdot \mathrm{vec}(I_d) + C\right),$$

---

[2]The presentation in Francq and Zakoïan (2001) has been adjusted to agree with the notation used in this book.

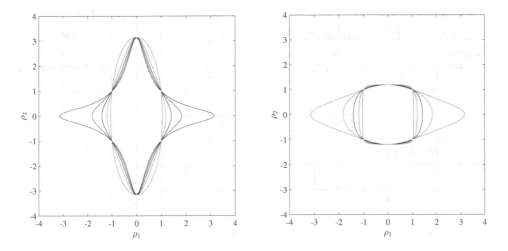

**FIGURE 13.3**
Boundary values for $\rho_1, \rho_2|\xi$ implying a mean-square stable Markov switching univariate autoregressive process. Left: conditional on $\xi_{22} = 0.1$, where $\xi_{11} = \{0.1, 0.3, \ldots, 0.9\}$ from the peaked to the oval boundary. Right: conditional on $\xi_{22} = 0.7$, where $\xi_{11} = \{0.1, 0.3, \ldots, 0.9\}$ from the oval-peaked to the square-like boundary.

where $J = (I_{(dp)^2} \cdots I_{(dp)^2})$ is a matrix of dimension $(dp)^2 \times G(dp)^2$, containing $G$ copies of the identiy matrix $I_{(dp)^2}$.

Given that the matrix norm is sub-multiplicative, we have that

$$\|y_t^{(l)}\|_2 \leq \|JP^l \left(S \cdot \text{vec}(I_d) + C\right)\|^{1/2} \leq \|J\|^{1/2} \|P^l\|^{1/2} \|S \cdot \text{vec}(I_d) + C\|^{1/2}.$$

A sufficient condition for $\|P^l\|$ to converge to zero as $l \to \infty$ is that the maximum eigenvalue of the matrix $P$, $\varrho(P)$, should be less than 1, $\varrho(P) < 1$.

**Theorem 13.2** (Francq and Zakoïan, 2001) *Suppose that $\varrho(P) < 1$. Then, for all $t \in \mathbb{Z}$, the series*

$$y_t = \omega_{z_t} + \sum_{l=1}^{\infty} \rho_{z_t} \cdots \rho_{z_{t-l+1}} \omega_{z_{t-l}}$$

*converges in $L^2$ and the process $y_t$ is the unique non-anticipative second-order stationary solution of (13.7). Then we have*

$$\sum_{l=0}^{\infty} \|JP^l S \cdot \text{vec}(I_d)\| < \infty. \tag{13.9}$$

*Finally, if $c_{z_t} = 0$ in (13.7), a necessary and sufficient condition for the existence of a non-anticipative second-order stationary solution to (13.7) is given by (13.9).*[3]

Non-anticipative solutions are solutions for $y_t$ which are independent of future parameters and shocks, $\{\rho_{z_{t+h}}, \omega_{z_{t+h}}\}$, $h > 0$.

Conditional on $\varrho(P) < 1$, we can derive the expectation and the covariance of a Markov

---

[3]The presentation in Francq and Zakoïan (2001) has been adjusted to agree with the notation used in this book.

switching autoregressive process; see Francq & Zakoïan (2001) for details. Define the vectors $U = (\bar{\xi}_1 \mathrm{E}(y_t^\top | z_t = 1), \ldots, \bar{\xi}_G \mathrm{E}(y_t^\top | z_t = G))^\top$ and $V$ similarly to $U$ by substituting $y_t$ with $\mathrm{vec}(y_t y_t^\top)$. Further, define $c = (\bar{\xi}_1 c_1^\top, \ldots, \bar{\xi}_G c_G^\top)^\top$, $\mathbf{1}_G$ a $1 \times G$ vector of ones and $\iota = (I_d\ 0_{d \times d(p-1)})$ a $d \times dp$ selection matrix. Then the moments of the first $d$ elements in $y_t$, denoted by $y_t^d$, are

$$\mathrm{E}(y_t^d) = (\mathbf{1}_G \otimes \iota)U, \quad \mathrm{vec}\left(\mathrm{E}\left(y_t^d y_t^{d\top}\right)\right) = (\mathbf{1}_G \otimes \iota \otimes \iota)V,$$

replacing $U$ and $V$ with estimates

$$U = (I_{Gdp} - P^*)^{-1} c,$$
$$V = \left(I_{G(dp)^2} - P\right)^{-1} (C + S \cdot \mathrm{vec}(I_d) + DU),$$

in which $P^*$ and $D$ are obtained by replacing $\rho_g \otimes \rho_g$ with, respectively, $\rho_g$ and $c_g \otimes \rho_g + \rho_g \otimes c_g$ in $P$.

In the engineering literature, processes for which the first and second moments converge to bounded limits are called *mean-square stable* (Costa et al., 2004, Farmer et al., 2009b). Clearly, mean-square stability in Markov switching models depends nonlinearly on the state-specific processes $\rho_g$ as well as on the transition distribution $\xi$.

The sufficient condition stated in Theorem 13.1 does not preclude explosive states. In the univariate case, the condition reduces to

$$\tilde{\lambda} = \mathrm{E}(\log \|\rho_{z_t}\|) = \sum_{y=1}^{G} \bar{\xi}_g \log |\rho_g| < 0.$$

It is intuitive that non-stationary states may alternate with stationary ones and yet yield a mean-square stable process if the non-stationary state does not recur too often and prevail for too long. Figure 13.3 plots the boundary values for $(\rho_1, \rho_2)$, conditional on various combinations for $(\xi_{11}, \xi_{22})$, for which a univariate autoregressive process is still mean-square stable. In general, the boundary value for $\rho_g$ decreases the more persistent state $g$ is. We observe that many combinations allow for an explosive root in either state-specific process, even if both states are highly persistent; see, for example, the innermost boundary circle conditional on $(\xi_{11}, \xi_{22}) = (0.9, 0.7)$ in Figure 13.3 (right). By way of illustration, Figure 13.4 plots a simulated time series with $(\rho_1, \rho_2) = (1.1, 0.7)$ and $(\xi_{11}, \xi_{22}) = (0.5, 0.7)$.

However, it is intriguing that stationary state-specific processes are not sufficient to obtain a mean-square stable process. To illustrate this case, we reproduce an example from Farmer et al. (2009b, pp. 1854–1855); see Francq & Zakoïan (2001) for another one. Assume a bivariate two-state Markov switching autoregressive process $y_t$ with state-dependent autoregressive matrices

$$\rho_1 = \begin{bmatrix} 0 & 2 \\ 0 & 0.5 \end{bmatrix}, \quad \rho_2 = \begin{bmatrix} 0.5 & 0 \\ 2 & 0 \end{bmatrix},$$

and set for simplicity $c_{z_t} = 0$ and $\Sigma_{z_t} = I_2$. Both state-specific processes are covariance stationary. Assume that the transition between the two processes is characterized by either of the two transition matrices

$$\xi = \begin{bmatrix} 0.7 & 0.3 \\ 0.4 & 0.6 \end{bmatrix}, \quad \tilde{\xi} = \begin{bmatrix} 0.9 & 0.1 \\ 0.4 & 0.6 \end{bmatrix}.$$

It turns out that in combination with transition probability matrix $\xi$, $\varrho(P)$ is larger than

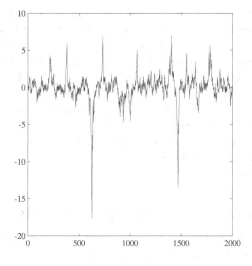

**FIGURE 13.4**
Simulated time series with $\xi_{11} = 0.5$, $\rho_1 = 1.1$, $\xi_{22} = 0.7$, $\rho_2 = 0.7$, $\sigma^2 = 0.1$.

one, while with transition matrix $\tilde{\xi}$, $\varrho(P)$ is smaller than one. The first process is not mean-square stable, although both state-specific processes are stable. Decreasing the transition frequency between states restores mean-square stability.

There are many situations where one or more regimes may be unstable. In economics, for example, periods of hyperinflation are clearly unstable. Also, it is conceivable that economic variables engaged on an unstable path when the last financial crisis broke out. The concept of mean-square stability allows us to combine unstable and stable processes, if, as already stated, the unstable process does not persist for too long and/or does not recur too often. Although this specific feature is appealing for time series and macroeconomic modelling, it has so far been applied very rarely. Davig & Leeper (2007) and Farmer et al. (2009a) derived conditions for unique equilibrium determination in dynamic stochastic general equilibrium (DSGE) models. Foerster et al. (2016) present advances to obtain solutions in Markov switching DSGE models. In time series modelling, most applications estimate Markov switching models with stable state-specific processes. Allowing Markov switching between state-specific stable and unstable explosive processes represents an interesting avenue for future research.

## 13.3 Estimation

As in general for mixture modelling, the latent states are *a priori* not identified in the mixture and the likelihood will be invariant to permutations of states, $\mathfrak{s} = (\mathfrak{s}_1, \ldots, \mathfrak{s}_G)$, $p\left(\mathbf{y}|\mathbf{x}_1, \mathbf{z}, \theta\right) = p\left(\mathbf{y}|\mathbf{x}_1, \mathfrak{s}(\mathbf{z}, \theta)\right)$. Thus, state identification needs ordering restrictions on state-specific parameters $\theta$ or $\xi$. Obviously, these restrictions should be imposed on parameters which indeed differ between states and, if ordering restrictions are known, we may impose them prior to estimation. However, if there is uncertainty about which parameters differ between states, we may first estimate the unidentified model and apply an appropriate state-identifying restriction ex post to obtain a state-identified model.

In the Bayesian framework, knowledge about state-identifying restrictions may be imposed on the prior to estimate the model by restricted sampling. When such restrictions are unknown or uncertain, we design a state-invariant prior to explore the state-invariant posterior $p(\mathbf{z}, \theta, \xi | \mathbf{y}, \mathbf{x}) = p(\mathfrak{s}(\mathbf{z}, \theta, \xi) | \mathbf{y}, \mathbf{x})$. Posterior inference is then obtained by random permutation sampling (Frühwirth-Schnatter, 2001) and state identification by post-processing the posterior output (Frühwirth-Schnatter, 2011).

We discuss estimation conditional on $G$. Testing or model evaluation with respect to $G$ is not discussed, given that methods described in Chapter 7 of this volume can be applied with appropriate adjustments.

### 13.3.1  The complete-data likelihood and the FFBS algorithm

When estimating hidden Markov models, we need to take into account sequential dependence in $z_t$. The complete-data likelihood factorizes into the conditional likelihood:

$$L_c(\theta, \mathbf{z}) = p(\mathbf{y} | \mathbf{x}_1, \mathbf{z}, \theta) = \prod_{t=1}^{T} f(y_t | x_{1t}, \theta_{z_t}), \tag{13.10}$$

where for both frequentist and Bayesian estimation we need an inference on the latent indicator $z_t$. Given the time dependence in $z_t$, the E step in the EM algorithm and the data augmentation step in Bayesian inference are slightly more involved than described in Chapter 2. Both, however, start out with the factorization of the state distribution conditional on all data

$$p(\mathbf{z} | \mathbf{y}, \mathbf{x}_1, \theta, \xi) = p(z_T | \mathcal{I}_T) \prod_{t=1}^{T-1} p(z_t | \mathcal{I}_T, z_{t+1}) p(z_0),$$

where the dependence on $(\theta, \xi)$ is suppressed on the right-hand side for notational convenience. Here $\mathcal{I}_t$ denotes information up to time $t$ and $p(z_0)$ is the starting state probability distribution (Chib, 1996). The factorization includes the typical element

$$p(z_t | \mathcal{I}_T, z_{t+1}) \propto p(z_t | \mathcal{I}_t) p(z_{t+1} | z_t, \mathcal{I}_T), \tag{13.11}$$

where the filter density $p(z_t | \mathcal{I}_t) \propto f(y_t | x_{1t}, \theta_{z_t}) p(z_t | \mathcal{I}_{t-1})$ consists of the likelihood $f(y_t | \cdot)$ and $p(z_t | \mathcal{I}_{t-1}) = \xi_t p(z_{t-1} | \mathcal{I}_{t-1})$ obtained by extrapolation. Based on (13.11), full inference on $p(\mathbf{z} | \mathbf{y}, \mathbf{x}_1, \theta, \xi)$ is obtained by a forward-filtering backward-sampling (FFBS) algorithm, see Algorithm 13.1.

### 13.3.2  Maximum likelihood estimation

Conditional on the smoothed probabilities $P(z_t = g | \mathcal{I}_T)$, $t = 1, \ldots, T$, the observed-data likelihood

$$L_o(\theta, \xi) = p(\mathbf{y} | \mathbf{x}_1, \theta, \xi) = \prod_{t=1}^{T} \sum_{g=1}^{G} P(z_t = g | \mathcal{I}_T) f(y_t | x_{1t}, \theta_g) \tag{13.13}$$

is maximized with respect to the model parameters (the M step). With respect to $\theta$, we solve

$$\sum_{t=1}^{T} \sum_{g=1}^{G} \frac{\partial \, \log f(y_t | x_{1t}, \theta_g)}{\partial \theta'_g} P(z_t = g | \mathcal{I}_T) = 0.$$

**Algorithm 13.1** Forward-filtering backward-smoothing (FFBS) Algorithm

1 Run forwards in time $t = 1, \ldots, T$ to obtain the filter densities $p(z_t|\mathcal{I}_t)$ or the filter probabilities $\mathrm{P}(z_t = g|\mathcal{I}_t)$:

$$\mathrm{P}(z_t = g|\mathcal{I}_{t-1}) = \sum_{g'=1}^{G} \xi_{t,g'g} \mathrm{P}(z_{t-1} = g'|\mathcal{I}_{t-1}),$$

$$\mathrm{P}(z_t = g|\mathcal{I}_t) = \frac{f(y_t|x_{1t}, \theta_g)\mathrm{P}(z_t = g|\mathcal{I}_{t-1})}{\sum_{g'=1}^{G} f(y_t|x_{1t}, \theta_{g'})\mathrm{P}(z_t = g'|\mathcal{I}_{t-1})}.$$

For $t = T$, this yields $p(z_T|\mathcal{I}_T)$.

2 Run backwards in time $t = T - 1, \ldots, 1$ to obtain the smoothed densities $p(z_t|\mathcal{I}_T)$ or smoothed probabilities $\mathrm{P}(z_t = g|\mathcal{I}_T)$:

$$\mathrm{P}(z_t = g|\mathcal{I}_T) = \sum_{g'=1}^{G} \mathrm{P}(z_t = g|\mathcal{I}_t, z_{t+1} = g') \mathrm{P}(z_{t+1} = g'|\mathcal{I}_T), \qquad (13.12)$$

where

$$\mathrm{P}(z_t = g|\mathcal{I}_t, z_{t+1} = g') = \frac{\mathrm{P}(z_t = g|\mathcal{I}_t)\xi_{t+1,gg'}}{\sum_{g=1}^{G} \mathrm{P}(z_t = g|\mathcal{I}_t)\xi_{t+1,gg'}}.$$

---

For example, in a multiple regression set-up with normally distributed error terms, $\theta_g = \{\beta_g, \sigma_g^2\}$, the observation density is

$$f(y_t|x_{1t}, \theta_g) = \frac{1}{\sqrt{2\pi}\sigma_g} \exp\left\{-\frac{1}{2\sigma_g^2}\left(y_t - x_{1t}^\top \beta_g\right)^2\right\},$$

and the maximum likelihood estimate of $\beta_g$ corresponds to a weighted least squares estimate,

$$\hat{\beta}_g = \left(\sum_{t=1}^{T} \tilde{x}_t \tilde{x}_t^\top\right)^{-1} \left(\sum_{t=1}^{T} \tilde{x}_t \tilde{y}_t\right),$$

where $\tilde{x}_t = x_{1t}\sqrt{\mathrm{P}(z_t = g|\mathcal{I}_T)}$ and $\tilde{y}_t = y_t\sqrt{\mathrm{P}(z_t = g|\mathcal{I}_T)}$ are the observations weighted by the square root of the smoothed probabilities. The estimate of the state-dependent variances equals

$$\hat{\sigma}_g^2 = \frac{\sum_{t=1}^{T}\left(\tilde{y}_t - \tilde{x}_t^\top \hat{\beta}_g\right)^2}{\sum_{t=1}^{T} \mathrm{P}(z_t = g|\mathcal{I}_T)}.$$

The estimate of time-invariant Markov transition probabilities (13.3) is given by

$$\hat{\xi}_{g'g} = \sum_{t=2}^{T} \frac{\mathrm{P}(z_t = g, z_{t-1} = g'|\mathcal{I}_T)}{\mathrm{P}(z_{t-1} = g'|\mathcal{I}_T)},$$

where the numerator equals the terms in (13.12) for given $(g, g')$ (Hamilton, 1994, Chapter 22). The first-order conditions with respect to $\gamma$ in the case of time-varying transition probabilities are nonlinear; see, for example, Diebold et al. (1994). To estimate multi-state time-varying transition probabilities one may borrow from recent advances in modelling latent class multinomial logit models (Greene & Hensher, 2003, 2013; Hess, 2014).

Conditional on estimates $(\hat{\theta}, \hat{\xi})$, the maximal value of the observed-data likelihood is

$$L_o(\hat{\theta}, \hat{\xi}) = \prod_{t=1}^{T} \sum_{g=1}^{G} P\left(z_t = g | \mathcal{I}_T\right) f(y_t | x_{1t}, \hat{\theta}_g).$$

### 13.3.3 Bayesian estimation

To make inference on the joint posterior

$$p\left(\mathbf{z}, \theta, \xi | \mathbf{y}, \mathbf{x}\right) \propto p(\mathbf{y} | \mathbf{x}_1, \mathbf{z}, \theta) p(\mathbf{z} | \xi) p(\xi | \mathbf{x}_2) p(\theta), \qquad (13.14)$$

we use data augmentation and the FFBS algorithm (Algorithm 13.1), derived in Section 13.3.1. The first term in (13.14) represents the conditional data likelihood (13.10). The prior $p(\mathbf{z} | \xi) = \prod_{t=1}^{T} p(z_t | z_{t-1}, \xi_t) p(z_0)$ takes into account time dependence in $z_t$.

To sample from the state-invariant posterior, we have to design state-invariant prior distributions $p(\xi | \mathbf{x}_2)$ and $p(\theta)$. A state-invariant prior $p(\theta)$ is often quite straightforward to design, while the design of a state-invariant prior $p(\xi | \mathbf{x}_2)$ is slightly more intricate depending on the level of complexity in the parameterization (see below).

If state-identifying restrictions are known prior to estimation, it may be sensible to impose them on the prior $\theta \sim p(\theta) \mathbb{I}_{\mathcal{R}}$ or on $\xi \sim p(\xi | \mathbf{x}_2) \mathbb{I}_{\mathcal{R}}$, where $\mathbb{I}_{\mathcal{R}}$ is one if the set of restrictions $\mathcal{R}$ is fulfilled. This, however, destroys state invariance of the prior distribution and estimation proceeds with restricted sampling.

#### 13.3.3.1 Prior specifications for the transition distribution

For time-invariant switching (13.3), a conjugate prior is the Dirichlet distribution $(\xi_{g'1}, \ldots, \xi_{g'G}) \sim \mathcal{D}(e_{0,g'1}, \ldots, e_{0,g'G})$, $g' = 1, \ldots, G$. To obtain a state-invariant prior

$$p(\xi) \sim \prod_{g'=1}^{G} \mathcal{D}(e_{0,g'1}, \ldots, e_{0,g'G})$$

we set $e_{0,gg} = \kappa_0$ for all $g$, and $e_{0,g'g} = \kappa_1$ for all $g' \neq g$, where $\kappa_0, \kappa_1 > 0$. An informative prior usually puts more weight on the persistence probabilities (i.e. $\kappa_0 > \kappa_1$). If $G = 2$, the prior distribution is a beta distribution $\mathcal{B}e(e_{0,gg}, e_{0,g'g})$, $g = 1, 2$, $g' \neq g$. Sims et al. (2008) derive a general framework to model Markov switching transition probabilities in large multiple-equation systems, including a framework to design a prior Dirichlet distribution which induces restrictions on the transition probability matrix.

For the logit and the probit functional form (see Section 13.2.2.2), the prior $p(\xi | \mathbf{x}_2)$ is defined by specifying a prior on $\gamma = \{\gamma_g | g \in \mathcal{G}_{-g_0}\}$, for which we assume a normal distribution for each $\gamma_g = \left(\gamma_{1g}^x, \ldots, \gamma_{Gg}^x, \gamma_{1g}, \ldots, \gamma_{Gg}\right)^\top$:

$$p(\gamma) = \prod_{g \in \mathcal{G}_{-g_0}} p(\gamma_g) = \prod_{g \in \mathcal{G}_{-g_0}} \mathcal{N}(e_{0,g}, E_0),$$

with a state-specific mean $e_{0,g}$.

Generally, a state-invariant prior for the logistic functional form is designed in the following way (Kaufmann, 2015). The hyperparameters relating to state persistence are set to $\{e_{0,gg}^x, e_{0,gg}\} = \{\kappa^x, \kappa\}$. Then the hyperparameters referring to transition parameters from the reference state $g_0$ to state $g$ are set to $\{e_{0,g_0g}^x, e_{0,g_0g}\} = \{-\kappa^x, -\kappa\}$ and those referring to transition parameters from other states to state $g$ to zero, $\{e_{0,jg}^x, e_{0,jg}\} = \{0, 0\}$, $j \neq g, g_0$.

Thus, the ordering of the hyperparameters in the prior mean $e_{0,g}$ varies across $g$. When random permutation sampling (see below) includes the reference state $g_0$, hyperparameters $e_{0,g}$ have to be permuted accordingly; see Kaufmann (2015, Appendix). Conditional on $g_0$ (i.e. keeping $g_0$ fixed for estimation), hyperparameters $e_{0,g}$ do not have to be permuted, however.

Under the consideration that relevant parameters in $\gamma_g^x$ should be shifted away from zero in the same direction (as discussed in Section 13.2.4), in fact the only sensible parameterization for a state-invariant prior is $\kappa^x = 0$.

The prior specification proposed in Burgette & Hahn (2010) proves especially useful for applying the permutation sampler when we use a probit functional form for $\xi$. They propose a state-invariant prior which renders normalization independent of a reference state. This means that the parameter $\gamma_{g_0}$ equals the negative of the sum over the parameters governing transitions to the other states, that is, $\gamma_{g_0} = -\sum_{g \neq g_0} \gamma_g$. Additionally, instead of normalizing element $(g_0, g_0)$ in the covariance of the latent utilities, they suggest normalizing the trace of the error covariance

$$z_{tg}^* = x_{2t}\gamma_{z_{t-1},g}^x + \gamma_{z_{t-1},g} + \nu_{tg}, \quad g \in \mathcal{G}_{-g_0}, \tag{13.15}$$

$$\nu_t = (\nu_{t1},\ldots,\nu_{t,g_0-1},\nu_{t,g_0+1},\ldots,\nu_{tG})^\top \sim \mathcal{N}(0,\Sigma^*), \quad \mathrm{tr}\,(\Sigma^*) = G - 1.r$$

To obtain a state-invariant prior, we first set all hyperparameters relating to state persistence to $\{e_{0,gg}^x, e_{0,gg}\} = \{\kappa^x, \kappa\}$, for all $g \neq g_0$, and then all other elements to $\{e_{0,g'g}^x, e_{0,g'g}\} = \{-\kappa^x/(G-1), -\kappa/(G-1)\}$, $g' \neq g$, for all $g', g \neq g_0$. Again, the only sensible parameterization for a state-invariant prior is $\kappa^x = 0$.

### 13.3.3.2 Posterior inference

To obtain a sample from the posterior (13.14), we repeatedly sample from the following conditional distributions:

1. Draw the latent indicators $\mathbf{z}$ from $p(\mathbf{z}|\mathbf{y},\mathbf{x}_1,\theta,\xi)$ by applying the FFBS algorithm (Algorithm 13.1), where the 'BS' step is a backward-sampling step. Given the filter densities $p(z_t = g|\mathcal{I}_t)$, first sample $z_T$ from $p(z_T|\mathcal{I}_T)$. Then, for $t = T-1,\ldots,1$, sample $z_t$ from

$$p(z_t = g|\mathcal{I}_T) \propto p(z_t = g|\mathcal{I}_t)\,\xi_{t+1,g,z_{t+1}}.$$

2. Draw $\xi$ from $p(\xi|\mathbf{x}_2,\mathbf{z})$ or from $p(\xi|\mathbf{x}_2,\mathbf{z})\,\mathbb{I}_\mathcal{R}$ if state-identifying restrictions are imposed on the prior.

3. Draw $\theta$ from $p(\theta|\mathbf{y},\mathbf{x}_1,\mathbf{z})$ or from $p(\theta|\mathbf{y},\mathbf{x}_1,\mathbf{z})\,\mathbb{I}_\mathcal{R}$.

4. If the prior is state-invariant, terminate the sampler with one of the following steps:

   (a) *Random permutation sampler.* Randomly permute the states and state-specific parameters to obtain a sample from the unconstrained multimodal posterior (13.14).

   (b) *Restricted sampler.* Reorder the states and state-specific parameters according to a predefined state-identifying restriction $\mathcal{R}$ to obtain a sample from the constrained posterior.

Sampling from restricted posterior distributions $p(\xi|\mathbf{x}_2,\mathbf{z})\,\mathbb{I}_\mathcal{R}$ and $p(\theta|\mathbf{y},\mathbf{x}_1,\mathbf{z})\,\mathbb{I}_\mathcal{R}$ in steps 2 and 3 requires either restricted or rejection sampling procedures; see Section 13.4 for some examples.

*Posterior sampling of transition parameters*

Under time-invariant switching and a Dirichlet prior $p(\xi)$, the posterior $p(\xi|\mathbf{x}_2, \mathbf{z}) = p(\xi|\mathbf{z})$ in step 2 is also Dirichlet,

$$p(\xi|\mathbf{z}) \sim \prod_{g'=1}^{G} \mathcal{D}(e_{g'1}, \ldots, e_{g'G}),$$

with $e_{g'g} = e_{0,g'g} + n_{g'g}$, where the prior hyperparameter is updated by $n_{g'g} = \sum_{t=1}^{T} \mathbb{I}(z_t = g, z_{t-1} = g')$, the number of times state $g$ is preceded by state $g'$.

To obtain $p(\xi|\mathbf{x}_2, \mathbf{z})$ under the logit or probit functional form, we derive $p(\gamma|\mathbf{x}_2, \mathbf{z})$. To sample under the logit functional form, we introduce latent state-specific random utilities $u_{tg}^*$ for all but the reference state $g_0$ (McFadden, 1974):

$$u_{tg}^* = w_t^\top \gamma_g + \nu_{tg}, \quad \forall g \in \mathcal{G}_{-g_0}. \tag{13.16}$$

$$\nu_{tg} \text{ i.i.d. Type I extreme value.}$$

where $w_t^\top = (x_{2t}z_{t-1,1}, \ldots, x_{2t}z_{t-1,G}, z_{t-1,1}, \ldots, z_{t-1,G})$. Given that maximum utility is obtained for the observed state, we use a partial representation of the model, in which the latent utilities are expressed in differences from the maximum utility of all other states:

$$u_{tg} = u_{tg}^* - u_{t,-g}^* = w_t^\top \gamma_g - \log\left(\sum_{j \in \mathcal{G}_{-g}} \exp\left(w_t^\top \gamma_j\right)\right) + \epsilon_{tg}, \quad \forall g \in \mathcal{G}_{-g_0}, \tag{13.17}$$

$$\epsilon_{tg} \text{ i.i.d. logistic,}$$

where $u_{t,-g}^* = \max_{j \in \mathcal{G}_{-g}} u_{tj}^*$, $\epsilon_{tg} = \nu_{tg} - \nu_{t,-g}$ and the constant term $\log(\cdot)$ is independent of $\gamma_g$.

The latent utilities model (13.17) is linear in the coefficients $\gamma_g$, with non-normal errors, however. Hence, introducing $\mathbf{u}_g = (u_{1g}, \ldots, u_{Tg})$ as a first layer of data augmentation renders the model linear in $\gamma_g$, and posterior sampling of $\gamma_g$ may be based on a Metropolis–Hastings step; see Scott (2011) and Holmes & Held (2006). This approach has also been used in economic applications; see Hamilton & Owyang (2012) and Owyang et al. (2015).

Frühwirth-Schnatter & Frühwirth (2007) show that sampling efficiency is considerably improved by introducing a second layer of data augmentation. This is achieved by approximating the logistic distribution of $\epsilon_{tg}$ by a mixture of normals with $K$ (typically 10) components, introducing the mixture components $\mathbf{r}_g = (r_{1g}, \ldots, r_{Tg})$, $r_{tg} \in \{1, \ldots, K\}$, as latent variables. Conditional on $\mathbf{u}_g$ and $\mathbf{r}_g$, we obtain a normal posterior for $\gamma_g$,

$$\gamma_g|\mathbf{z}, \mathbf{x}_2, \mathbf{u}_g, \mathbf{r}_g \sim \mathcal{N}(e_g(\mathbf{u}_g, \mathbf{r}_g), E_g(\mathbf{r}_g)),$$

with moments $e_g(\cdot)$ and $E_g(\cdot)$ explicitly derived in Kaufmann (2015, Appendix). In addition, Kaufmann (2015) shows that higher sampling efficiency is achieved when using the partial representation (13.17) rather than specification (13.16).

When working with the multinomial probit functional form, data augmentation and parameter estimation can generally proceed as in Albert & Chib (1993), McCulloch et al. (2000) and Nobile (1998). The (normalized) model (13.15) for the latent random utilities is linear Gaussian and the posterior $p(\gamma|\mathbf{z}, \mathbf{x}_2, \mathbf{u}, \Sigma^*)$ is normal. Burgette & Hahn (2010) propose a sampler that is particularly useful to apply with random permutation sampling. Given their detailed description, we do not reproduce the sampler here.

**TABLE 13.1**
Time in minutes for 1,=000 draws and inefficiency factors for posterior samples retaining all draws and only every fifth draw

| | Logit | | | | Probit | | | |
|---|---|---|---|---|---|---|---|---|
| **$G = 2$** | | | | | | | | |
| | $g = 2$ | | | | $g = 2$ | | | |
| Inefficiency | all | 5th | | | all | 5th | | |
| $\gamma_{1g}^x$ | 80.83 | 22.31 | | | 86.64 | 18.95 | | |
| $\gamma_{2g}^x$ | 53.65 | 9.60 | | | 44.59 | 9.64 | | |
| $\gamma_{1g}$ | 39.85 | 10.00 | | | 36.37 | 6.00 | | |
| $\gamma_{2g}$ | 41.07 | 8.19 | | | 17.61 | 3.17 | | |
| Time | 3.2 | | | | 3.4 | | | |
| **$G = 3$** | | | | | | | | |
| | $g = 2$ | | $g = 3$ | | $g = 2$ | | $g = 3$ | |
| Inefficiency | all | 5th | all | 5th | all | 5th | all | 5th |
| $\gamma_{1g}^x$ | 19.85 | 2.60 | 12.59 | 2.76 | 3.66 | 1.68 | 2.28 | 0.87 |
| $\gamma_{2g}^x$ | 13.19 | 3.83 | 19.43 | 2.37 | 2.15 | 0.65 | 3.44 | 1.13 |
| $\gamma_{3g}^x$ | 16.95 | 2.78 | 19.53 | 4.30 | 3.49 | 1.03 | 2.86 | 1.08 |
| $\gamma_{1g}$ | 9.12 | 2.19 | 16.56 | 2.81 | 2.94 | 0.84 | 2.65 | 1.05 |
| $\gamma_{2g}$ | 14.82 | 3.85 | 21.53 | 4.13 | 3.01 | 1.22 | 3.17 | 0.94 |
| $\gamma_{3g}$ | 22.05 | 5.76 | 9.86 | 2.38 | 2.55 | 0.60 | 3.13 | 0.61 |
| Time | 3.4 | | | | 38.6 | | | |

## 13.3.4 Sampler efficiency: logit versus probit

By way of illustration, we simulate a time series $y_t$ of length $T = 500$ with autoregressive processes subject to, respectively, $G = 2$ and $G = 3$ regimes with time-varying transition probabilities:

$$y_t = \rho_{z_t} y_{t-1} + \varepsilon_t, \quad \varepsilon_t \sim \mathcal{N}(0, \sigma_{z_t}^2), \tag{13.18}$$

with parameter settings

$$(\rho_{z_t}, \sigma_{z_t}^2) = \begin{cases} (0.2, 0.1), & z_t = 1, \\ (0.8, 0.01), & z_t = 2, \\ (0, 1), & z_t = 3. \end{cases}$$

We simulate $z_t$ using the logit functional form for the transition probabilities, using $g_0 = 1$ as baseline, and assume the following parameters for $\gamma$ for $G = 2$,

$$\gamma^\top = \begin{pmatrix} \gamma_{12}^x & \gamma_{22}^x & \gamma_{12} & \gamma_{22} \end{pmatrix} = \begin{pmatrix} 4 & 1 & -2 & 2 \end{pmatrix};$$

and for $G = 3$,

$$\gamma_g^\top = \begin{pmatrix} \gamma_{1g}^x & \gamma_{2g}^x & \gamma_{3g}^x & \gamma_{1g} & \gamma_{2g} & \gamma_{3g} \end{pmatrix} = \begin{cases} \begin{pmatrix} 4 & 3 & 0 & -1 & 1 & 0 \end{pmatrix}, & g = 2, \\ \begin{pmatrix} 0 & 2 & 4 & -1 & 0 & 2 \end{pmatrix}, & g = 3. \end{cases}$$

For the covariate $x_{2t}$, we assume a persistent process, $x_{2t} = 0.8x_{2,t-1} + \epsilon_t$, $\epsilon_t \sim \mathcal{N}(0, 1)$.

We estimate the model using either the logit or probit functional form for the transition probabilities and evaluate estimation time and sampler efficiency. For estimation, we include a state-specific constant $c_g$ in (13.18). We use independent prior specifications: $c_g \sim \mathcal{N}(0, 0.0225)$, $\rho_g \sim \mathcal{N}(0, 0.21)$, $\sigma_g^2 \sim \mathcal{IG}(5, 0.25)$. For $\gamma$, we use $E_0 = I_{2G}$, where $I_{2G}$ is the $2G \times 2G$ identity matrix, as well as $(\kappa^x, \kappa) = (0, 2)$ and $(\kappa^x, \kappa) = (0, 0)$ when working

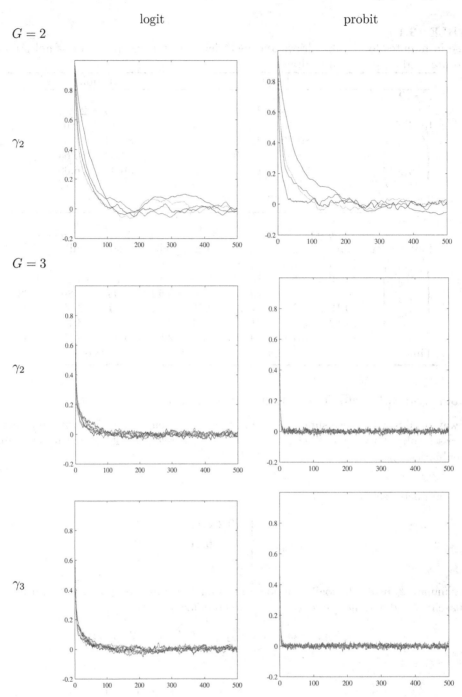

**FIGURE 13.5**
Autocorrelation of state-identified posterior draws for $\gamma_g$, $G = \{2, 3\}$. Each line corresponds to the autocorrelation for a specific parameter $\gamma_{g'g}^x$ or $\gamma_{g'g}$, $g' \in \{1, \ldots, G\}$.

with the logit and probit functional form, respectively. When we use the probit functional form, the trace-restricted inverse Wishart prior for $\Sigma^*$ is parameterized according to Burgette & Hahn (2010).

We draw $M = 30{,}000$ samples from the posterior and retain the last $15{,}000$ draws for posterior inference. In Table 13.1 we report the time to obtain 1000 draws from the random permutation sampler and the inefficiency factors for the state-identified posterior draws of $\gamma^{(m)}$, $m = 1, \ldots, M$. Obviously, there is a trade-off. For $G = 2$, estimation time and efficiency are comparable across both functional forms. By thinning out the posterior sample, we reduce inefficiency roughly by a factor of four. For $G = 3$, however, estimation time remains nearly unchanged when working with the logit functional form, and there is a tenfold increase in estimation time when working with the probit functional form. Obviously, this is due to the fact that in each iteration, $TG(G-1)$ multivariate integrals have to be evaluated to compute the transition probabilities $\xi_t$. On the other hand, the probit sampler of Burgette & Hahn (2010) performs much better in terms of efficiency. This is also reflected in Figure 13.5, in which we depict the autocorrelation function of the state-identified posterior draws for $\gamma_g^{(m)}$, $m = 1, \ldots, M$, for $g = \{2, 3\}$. While the autocorrelation functions drop at about the same rate for $G = 2$, the autocorrelations drop very quickly to zero when working with the probit functional form for $G = 3$. Given these results, the researcher may use the logit functional form to save computation time, and thin out considerably the posterior sample to adjust for the relative inefficiency.

### 13.3.5 Posterior state identification

We illustrate posterior state identification based on the posterior output of model $G = 3$ estimated with the logit functional form (that obtained using the probit functional form is identical). The top row of Figure 13.6 reproduces scatter plots of the unsorted posterior draws. In panel (a), $\rho_{g'}^{(m)}$, $g \neq g'$ is plotted against $\rho_g^{(m)}$, while in panel (b), $\log \sigma_g^{2\,(m)}$ is plotted against $\rho_{g'}^{(m)}$. Panel (a) reflects for $\rho_g$ as well as $\rho_{g'}$ the three modes of the unidentified posterior output of the random permutation sampler. On the other hand, panel (b) reveals that a restriction imposed jointly on $(\rho_g^{(m)}, \sigma_g^{2\,(m)})$ identifies the states well. We reorder each draw $\left( \rho_1^{(m)}, \rho_2^{(m)}, \rho_3^{(m)} \right)$ according to a permutation $\mathfrak{s}^{(m)} = \left( \mathfrak{s}_1^{(m)}, \mathfrak{s}_2^{(m)}, \mathfrak{s}_3^{(m)} \right)$ of $\{1, 2, 3\}$ which fulfils the restriction $\rho_{\mathfrak{s}_1^{(m)}}^{(m)} < \rho_{\mathfrak{s}_2^{(m)}}^{(m)} < \rho_{\mathfrak{s}_3^{(m)}}^{(m)}$ and permute accordingly the states and state-dependent parameters:

$$z_t^{(m)} = \mathfrak{s}^{(m)} \left( z_t^{(m)} \right), \quad t = 1, \ldots, T,$$

$$\left( c_g^{(m)}, \sigma_g^{2\,(m)} \right) = \left( c_{\mathfrak{s}_g^{(m)}}^{(m)}, \sigma_{\mathfrak{s}_g^{(m)}}^{2\,(m)} \right), \quad g = 1, 2, 3,$$

$$\gamma_g^{(m)} = \left( \gamma_{\mathfrak{s}_g^{(m)}, \mathfrak{s}_g^{(m)}}^{x\,(m)}, \gamma_{\mathfrak{s}_g^{(m)}, \mathfrak{s}_g^{(m)}}^{(m)} \right) \quad \text{with } \gamma_{g_0} = 0, \quad g = 1, 2, 3,$$

$$\gamma_g^{(m)} = \gamma_g^{(m)} - \gamma_1^{(m)}, \quad g = 2, 3, \quad \text{to normalize to } g_0 = 1.$$

For posterior inference we retain all ordered draws $m \in M_0 \subseteq \{1, \ldots, M\}$, which fulfil a joint restriction on $(\rho_g^{(m)}, \sigma_g^{2\,(m)})$, $g = 1, 2, 3$:

$$M_0 = \left\{ m : \mathbb{I}_{\{\rho_1 < \rho_2 < \rho_3\}}(\rho_1^{(m)}, \rho_2^{(m)}, \rho_3^{(m)}) \cdot \mathbb{I}_{\{\sigma_1^2 > \sigma_2^2 > \sigma_3^2\}}(\sigma_1^{2\,(m)}, \sigma_2^{2\,(m)}, \sigma_3^{2\,(m)}) = 1 \right\}.$$

That is, we retain those draws which show the smallest $\rho$ and largest $\sigma^2$ in the first state, intermediate values for $\rho$ and $\sigma^2$ in the second state and the largest $\rho$ and smallest $\sigma^2$ in

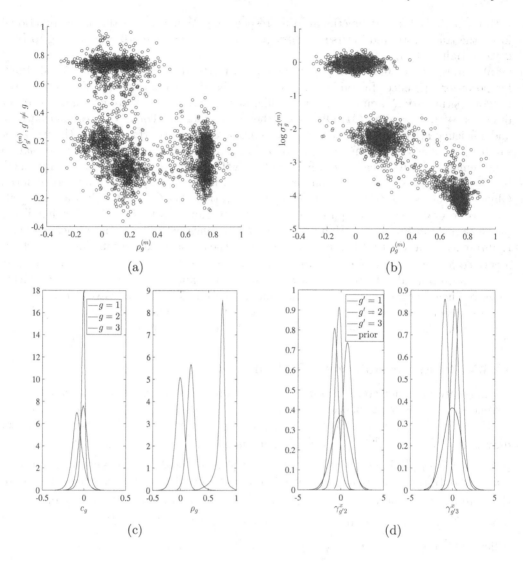

**FIGURE 13.6**
Posterior draws for $G = 3$. Scatter plots of unsorted posterior draws in panels (a) and (b). Panels (c) and (d) are based on state-identified posterior draws, with panel (c) showing the marginal posterior distributions of $c_g$ and $\rho_g$ and panel (d) posterior versus prior distributions for $\gamma_{g'j}^x$, $j = 2, 3$.

the third state. For illustration of the identified output, panels (c) and (d) in Figure 13.6 reproduce the state-identified marginal posterior distributions of $c_g$, $\rho_g$ and $\gamma_{g'g}^x$, $g' = 1, 2, 3$.

## 13.4 Informative Regime Switching in Applications

In this section we discuss various possibilities for introducing information into the transition distribution. The usual critique of Markov switching models with time-invariant transition distributions is that the switches of the state indicator remain unexplained. In economic analysis, the usual approach to giving an interpretation to the state indicator is to relate ex-post estimated state-specific periods to some statistical measures of the investigated series, like state-specific means or volatilities. Another possibility is to relate the estimated state of the hidden Markov chain to some officially released indicator, such as the business cycle turning point dates released by the US National Bureau of Economic Research.

To address this critique, one can directly include information into the transition distribution. Being more specific on the design of the transition distribution, one can obtain informative switching for time-invariant specification. Including explicitly covariates which affect the transition probabilities renders the transition distribution informative and time-varying. A further possibility for including information is to impose restrictions on the parameters of the transition distribution.

### 13.4.1 Time-invariant switching

Consider the time-invariant specification of the transition probabilities $P(z_t = g|z_{t-1} = g') = \xi_{g'g}$, and collect the probabilities in the transition matrix

$$
\xi = \begin{bmatrix} \xi_{11} & \cdots & \xi_{1G} \\ \vdots & & \\ \xi_{G1} & \cdots & \xi_{GG} \end{bmatrix}. \tag{13.19}
$$

Information can be included by explicitly designing or imposing restrictions on the transition matrix. In this sense, the time-invariant transition distribution becomes informative. In the following we discuss various examples. In Sims et al. (2008) the interested reader can find an encompassing framework for imposing and analysing restrictions on Markov transition probabilities in large multiple-equation models.

#### 13.4.1.1 Unconditional switching

Imposing the restriction $\xi_{g'g} = \xi_g$ for all $g' = 1, \ldots, G$ in (13.19) renders state switching unconditional. The Markov mixture model simplifies to a simple mixture model

$$
p(\mathbf{y}|\theta, \xi) = \prod_{t=1}^{T} \sum_{g=1}^{G} \xi_g f(y_t|\theta_g), \quad P(z_t = g|\xi) = \xi_g.
$$

#### 13.4.1.2 Structured Markov switching

The general set-up (13.19) introduces state persistence with expected duration $(1 - \xi_{gg})^{-1}$ for state $g$. It does not ensure a minimum time duration, which is defined for some economic features such as a recession. A recession is usually defined as two consecutive quarters (half a year) of negative gross domestic product (GDP) growth. When working with quarterly GDP data, we might include this minimum cycle duration by designing an encompassing state indicator $z_t^*$ with transition matrix:

$$
\xi^* = \begin{bmatrix} \xi_{11} & \xi_{12} & 0 & 0 \\ 0 & 0 & 0 & 1 \\ 1 & 0 & 0 & 0 \\ 0 & 0 & \xi_{21} & \xi_{22} \end{bmatrix} \quad \begin{array}{l} z_t^* = 1, z_t = 1 \text{ recession} \\ z_t^* = 2, z_t = 1 \text{ trough} \\ z_t^* = 3, z_t = 2 \text{ peak} \\ z_t^* = 4, z_t = 2 \text{ expansion} \end{array}
$$

which imposes a minimum cycle duration from peak to peak or from trough to trough of five quarters (Artis et al., 2004). Generalizing to longer state durations or to state-specific cycle durations is straightforward.

In multivariate analysis, sub-vectors of data in $y_t = (y_{1t}^\top, y_{2t}^\top)^\top$ may be affected by different state indicators $z_t^1$ and $z_t^2$. The simplest set-up is the case where both indicators follow independent transition distributions $\xi_1$ and $\xi_2$ and $f(y_t|x_t, z_t, \theta, \xi) = f(y_{2t}|y_{1t}, x_t, z_t^2, \theta, \xi_2) f(y_{1t}|x_t, z_t^1, \theta, \xi_1)$. Taking business cycle analysis again as an example, a common feature in macroeconomic data is that a group of variables is perceived as leading the business cycle, while another group of variables moves contemporaneously with GDP. To include this feature in the model, we may impose that the states of $z_t^2$ should lead the states of $z_t^1$ by designing the encompassing state $z_t^*$ with transition matrix:

$$
\xi^* = \begin{bmatrix} \xi_{11}^* & \xi_{12}^* & 0 & 0 \\ 0 & \xi_{22}^* & 0 & \xi_{24}^* \\ \xi_{31}^* & 0 & \xi_{33}^* & 0 \\ 0 & 0 & \xi_{43}^* & \xi_{44}^* \end{bmatrix} \quad \begin{array}{l} z_t^* = 1, \ z_t^1 = 1 \text{ and } z_t^2 = 1 \\ z_t^* = 2, \ z_t^1 = 1 \text{ and } z_t^2 = 2 \\ z_t^* = 3, \ z_t^1 = 2 \text{ and } z_t^2 = 1 \\ z_t^* = 4, \ z_t^1 = 2 \text{ and } z_t^2 = 2 \end{array} \tag{13.20}
$$

where $\xi_{g'g}^*$ are appropriately scaled convolutions of the transition probabilities of the underlying states $z_t^1$ and $z_t^2$. The restrictions impose a minimum duration of five periods for a full cycle. They also impose that the leading state indicator $z_t^2$ can only switch across states when $z_t^1$ has reached the same state. The approach was used by Phillips (1991) to model international data. Kaufmann (2010) uses the set-up to cluster a large panel of macroeconomic data into a group of series leading the business cycle and a group of series contemporaneously moving with GDP growth. The posterior evaluation of (13.20) provides additional interpretations. For example, the expected lead of $z_t^2$ into a recovery ($z_t^1 = z_t^2 = 2$) is $1/(1 - \xi_{22}^*)$. Probabilistic forecasts about, for example, the probability of reaching the trough ($z_{T+2}^* = 2$) within the next half year conditional on being in recession currently ($z_T^* = 1$) are available from the forecast density $p(z_{T+h}^*|y_T, x_T, z_T^*)$.

Dynamic structure between states can also be designed by using varying time leads for transition. Paap et al. (2009) define asymmetric leads across two-state business cycle phases ($z_t \in \{0, 1\}$ in their set-up):

$$
z_t^2 = \begin{cases} \prod_{j=\kappa_1}^{\kappa_2} z_{t+j}^1, & \text{if } \kappa_1 \leq \kappa_2, \\ 1 - \prod_{j=\kappa_2}^{\kappa_1} \left(1 - z_{t+j}^1\right), & \text{if } \kappa_1 > \kappa_2, \end{cases}
$$

where for specific values such as $\kappa_1 = 8$ and $\kappa_2 = 5$ one may even obtain state dynamics with overlapping phases of different cycles. In Paap et al. (2009), $\kappa_1$ and $\kappa_2$ are also part of model estimation.

Yet another example is found in Kim et al. (2005) who link hierarchically two binary state variables $z_t^1$ and $z_t^2$ to a three-state indicator $z_t$ with restricted transition probability matrix

$$
\xi = \begin{bmatrix} \xi_{11} & 1 - \xi_{11} & 0 \\ 0 & \xi_{22} & 1 - \xi_{22} \\ 1 - \xi_{33} & 0 & \xi_{33} \end{bmatrix},
$$

and define $z_t^1 = 1$ if $z_t = 2$ and $z_t^2 = 1$ if $z_t = 3$. They specify an autoregressive process for the mean-adjusted GDP growth,

$$\rho(L)\left(y_t - \theta_0 - \theta_1 z_t^1 - \theta_2 z_t^2 - \theta_3 \sum_{j=1}^{p} z_{t-j}^1\right) = \varepsilon_t,$$

where $\rho(L) = 1 - \rho_1 L - \rho_2 L^2 - \dots$ represents the autoregressive polynomial in the lag operator $Ly_t = y_{t-1}$. By including the sum over $p$ lagged values of $z_t^1$ (the last term in parentheses) they capture potential bounce-back effects in GDP growth after recessions.

## 13.4.2    Time-varying switching

### 13.4.2.1    Duration dependence and state-identifying restrictions

Covariates affecting the transition distribution render it informative in the sense that they explain what drives the latent state. Early contributions, mainly in business cycle analysis, are Diebold et al. (1994), Filardo (1994), McCulloch & Tsay (1994) and Filardo & Gordon (1998). Duration dependence is obtained when the persistence of states depends on the number of periods that the current regime has been prevailing. In change-point modelling, a time-varying extension is presented in Koop & Potter (2007), where regime duration is modelled by a Poisson distribution.

A multi-country, multi-state extension in business cycle analysis has recently been proposed in Billio et al. (2016), where the authors include weighted information on lagged business cycle states of all euro area countries and the United States into the transition distribution. The covariate $x_{2t}$ is a weighted average of lagged country-specific state indicators $z_{it} \in \{1, 2, 3\}$, $x_{2t} = \sum_{i=1}^{n} w_{it}\mathbb{I}(z_{i,t-1} = g)$.

Gaggl & Kaufmann (2014) work with a panel of 21 groups of US occupation data to analyse the phenomenon of jobless recoveries that characterizes the US labor market since the early 1990s. They formulate a dedicated factor model for occupational growth with a latent four-state indicator process in the mean factor growth rate with transition matrix

$$\xi_t = \begin{bmatrix} \xi_{t,11} & \xi_{t,12} & 0 & \xi_{t,14} \\ \xi_{t,21} & \xi_{t,22} & \xi_{t,23} & 0 \\ 0 & 0 & \xi_{t,33} & \xi_{t,34} \\ 0 & 0 & \xi_{t,43} & \xi_{t,44} \end{bmatrix}.$$

Interpreting states 1 and 3 as recessions and states 2 and 4 as expansions, the zero restrictions imply a one-time change in the business cycle phase-specific growth rates. The change can only occur at a turning point, that is, when exiting or falling into a recession.

The parameterization for the transition probabilities defines the reference state to be $g_0 = 1$ and reads

$$\xi_{t,12} = \frac{\exp\left(\gamma_{12,0} + \gamma_{12,1} x_{2t} + \gamma_{12,2} t\right)}{1 + \sum_{g=2,4} \exp\left(x_t^\top \gamma_{1g}\right)}, \quad \xi_{t,14} = \frac{\exp\left(\gamma_{14,2} t\right)}{1 + \sum_{g=2,4} \exp\left(x_t^\top \gamma_{1g}\right)},$$

with $x_t^\top = (1, x_{2t}, t)$, where $x_{2t}$ is GDP growth, and similar specifications for $\xi_{t,22}$ and $\xi_{t,23}$. The restrictions $\gamma_{j2,2} \leq 0$ and $\gamma_{j4,2} \leq 0$ identify states 2 and 4 as expansions, given that positive GDP growth $x_{2t} > 0$ increases the probability of switching to state 2 or 4. Time $t$ helps identifying the break point. Therefore $\gamma_{14,2} \geq 0$ and $\gamma_{23,2} \geq 0$.

The results document that since the recession in the early 1990s, routine jobs have experienced stronger job losses during recessions, while non-routine jobs have experienced weaker job growth during expansions.

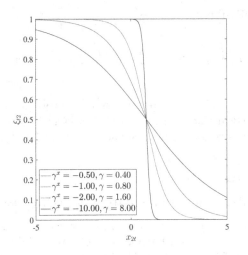

**FIGURE 13.7**
State probabilities $\xi_{t2}$, implemented to keep the threshold at $\tilde{\gamma} = 0.8$ for increasing covariate effect.

### 13.4.2.2 Shape restrictions

For some applications, for example when only a few observations are expected to be assigned to a state or to be available to estimate a transition to a state, it may be useful to impose restrictions on parameters in $\gamma$. This results in formulating a restricted prior $p(\gamma)\mathbb{I}_{\mathcal{R}}$. For example, Bäurle et al. (2016) analyse changing dynamics in a vector autoregressive model for Swiss macroeconomic variables when the interest rate approaches the zero lower bound. The model specification allows for two states ($G = 2$), and sets $g_0 = 1$. The state probabilities are assumed to be state-independent,

$$\xi_{t2} = \frac{\exp\left(\gamma^x x_{2t} + \gamma\right)}{1 + \exp\left(\gamma^x x_{2t} + \gamma\right)},$$

where $x_{2t}$ is the lagged level of the interest rate. By restricting $\gamma^x < 0$ the probability of state 2 is increasing as the interest rate approaches the zero lower bound.

Implicitly, the parameters $\gamma^x \neq 0$ and $\gamma$ define a threshold value for $x_{2t}$, $\tilde{\gamma} = -\gamma/\gamma^x$, that is, the value at which $\xi_{t2} = 0.5$. If we have some idea about an upper bound $\overline{\gamma}$ (which should not be too high) and a lower bound $\underline{\gamma}$ (which could, for example, be between 0 and 1) for the threshold $\tilde{\gamma}$, then

$$\underline{\gamma} < \tilde{\gamma} \leq \overline{\gamma}$$
$$\underline{\gamma} < -\gamma/\gamma^x \leq \overline{\gamma} \text{ or } -\gamma^x\underline{\gamma} < \gamma \leq -\gamma^x\overline{\gamma}.$$

These restrictions may be imposed on the prior distribution for $\gamma^x$ and $\gamma$:

$$p\left(\gamma^x, \gamma\right)\mathbb{I}_{\mathcal{R}} = \mathcal{N}\left(g_0, G_0\right)\mathbb{I}\left(\gamma^x < 0\right)\mathbb{I}\left(-\gamma^x\underline{\gamma} < \gamma \leq -\gamma^x\overline{\gamma}\right),$$

where $\gamma^x < 0$ is a state-identifying restriction. These restrictions render $\gamma$ and $\gamma^x$ highly correlated, which may be taken into account when specifying the prior moments. Figure 13.7 plots $\xi_{t2}$ against values for $x_{2t}$. To implement a prior threshold value at $\tilde{\gamma} = 0.8$, stronger effects of the covariate imply a higher value for $\gamma$. Depending on the informativeness of the data, we may be more or less informative about $\gamma^x$. For example, in Bäurle et al. (2016) only a few observations for $x_{2t}$, the interest rate, are available near the zero lower bound to estimate the transition to state 2. Therefore, the authors use an informative prior on $\gamma^x$. Posterior draws are obtained by sampling from conditional constrained posterior distributions $p(\gamma^x, \gamma | x_{2t}, z_t) = p(\gamma | \gamma^x, x_{2t}, z_t) p(\gamma^x | x_{2t}, z_t) \mathbb{I}(\gamma^x < 0) \mathbb{I}(-\gamma^x \underline{\gamma} < \gamma \le -\gamma^x \overline{\gamma})$; see Bäurle et al. (2016) for details.

## 13.5 Concluding Remarks

In time series analysis, hidden Markov models introduce persistence into the mixture distribution. The persistence is induced by defining a latent state process, which evolves according to a Markov transition distribution. This distribution may be parameterized as either time-invariant or time-varying. We discussed the parameterization of the logit and probit functional form to model time-varying transition distributions. Emphasis is put on Bayesian estimation. We discussed in detail the design of state-invariant prior distributions, in particular for the parameters of the transition distribution. We described the random permutation sampler with which we obtain a sample from the unconstrained posterior distribution. The evaluation of estimation time and sampler efficiency when using the logit or the probit functional form reveals a strong trade-off. While estimation time does not increase significantly with the number of latent states when working with the logit functional form, there is a tenfold increase in estimation time for a three-state model when working with the probit functional form. On the other hand, draws from the probit symmetric sampler of Burgette & Hahn (2010) are as efficient as from the thinned out logit posterior. The researcher may therefore opt to work with the logit functional form to save estimation time for models with more than two latent states, and to thin out considerably the posterior draws to reduce sampling inefficiency.

We illustrate how explicit economic interpretation may be obtained from posterior inference by imposing structural or dynamic restrictions on the transition distribution of the latent state process. Prior knowledge or expectations may also flow into the specification of the prior transition distribution. A restricted, state-identified prior then calls for restricted posterior sampling procedures.

An interesting feature of Markov switching models has not been applied so far in time series analysis. Based on general results for dynamical systems, we summarize conditions under which Markov switching processes with time-invariant transition probabilities are stationary and obtain a unique stationary solution. We provide conditions for second-order stationarity and report first and second moments of Markov switching processes. Results show that stationarity conditions do not rule out state-specific processes with explosive roots. Thus stationary state-specific processes can be combined with non-stationary, explosive processes, as long as the latter do not recur too often or persist for too long. However, the intriguing feature is that combining only stationary state-specific processes is not sufficient to obtain a stationary unconditional process. Exploiting these characteristics represents an interesting avenue for future research.

## Acknowledgements

The author thanks Klaus Neusser and Markus Pape for valuable comments and suggestions.

# Bibliography

ALBERT, J. H. & CHIB, S. (1993). Bayesian analysis of binary and polychotomous response data. *Journal of the American Statistical Association* **88**, 669–679.

ARTIS, M. J., MARCELLINO, M. & PROIETTI, T. (2004). Dating business cycles: A methodological contribution with an application to the euro area. *Oxford Bulletin of Economics and Statistics* **66**, 537–565.

BAI, J. & WANG, P. (2011). Conditional Markov chain and its application in economic time series analysis. *Journal of Applied Econometrics* **26**, 715–734.

BÄURLE, G., KAUFMANN, D., KAUFMANN, S. & STRACHAN, R. W. (2016). Changing dynamics at the zero lower bound. Working Paper 16.02, Study Center Gerzensee.

BAUWENS, L., DE BACKER, B. & DUFAYS, A. (2014). A Bayesian method of change-point estimation with recurrent regimes: Application to GARCH models. *Journal of Empirical Finance* **29**, 207–229.

BAUWENS, L., KOOP, G., KOROBILIS, D. & ROMBOUTS, J. V. (2015). The contribution of structural break models to forecasting macroeconomic series. *Journal of Applied Econometrics* **30**, 596–620.

BAUWENS, L. & LUBRANO, M. (1998). Bayesian inference on GARCH models using the Gibbs sampler. *Econometrics Journal* **1**, C23–C46.

BILLIO, M. & CASARIN, R. (2011). Beta autoregressive transition Markov-switching models for business cycle analysis. *Studies in Nonlinear Dynamics & Econometrics* **15**, 1–32.

BILLIO, M., CASARIN, R., RAVAZZOLO, F. & VAN DIJK, H. K. (2016). Interactions between eurozone and US booms and busts: A Bayesian panel Markov-switching VAR model. *Journal of Applied Econometrics* **31**, 1352–1370.

BOUGEROL, P. & PICARD, N. (1992). Strict stationarity of generalized autoregressive processes. *Annals of Probability* **20**, 1714–1730.

BRANDT, A. (1986). The stochastic equation $y_{n+1} = a_n y_n + b_n$ with stationary coefficients. *Advances in Applied Probability* **18**, 211–220.

BURGETTE, L. F. & HAHN, P. R. (2010). Symmetric Bayesian multinomial probit models. Working Paper 10-26, Department of Statistical Science, Duke University.

CHAUVET, M. & PIGER, J. (2008). A comparison of the real-time performance of business cycle dating methods. *Journal of Business & Economic Statistics* **26**, 42–49.

CHIB, S. (1995). Marginal likelihood from the Gibbs output. *Journal of the American Statistical Association* **90**, 1313–1321.

CHIB, S. (1996). Calculating posterior distributions and modal estimates in Markov mixture models. *Journal of Econometrics* **75**, 79–97.

CHIB, S. (1998). Estimation and comparison of multiple change-point models. *Journal of Econometrics* **86**, 221–241.

CLEMENTS, M. & KROLZIG, H.-M. (2003). Business cycle asymmetries: Characterization and testing based on Markov-switching autoregressions. *Journal of Business & Economic Statistics* **21**, 196–211.

COSTA, O., FRAGOSO, M. & MARQUES, R. (2004). *Discrete-Time Markov Jump Linear Systems*. New York: Springer.

DAVIG, T. & LEEPER, E. M. (2007). Generalizing the Taylor principle. *American Economic Review* **97**, 607–635.

DICKEY, J. M. (1968). Three multidimensional integral identities with Bayesian applications. *Annals of Statistics* **39**, 1615–1627.

DIEBOLD, F. X., LEE, J.-H. & WEINBACH, G. C. (1994). Regime switching with time-varying transition probabilities. In *Nonstationary Time Series Analysis and Cointegration*, C. P. Hargreaves, ed. Oxford: Oxford University Press, pp. 283–302.

ELLIOTT, G. & TIMMERMANN, A. (2005). Optimal forecast combination under regime switching. *International Economic Review* **46**, 1081–1102.

FARMER, R. E., WAGGONER, D. F. & ZHA, T. (2009a). Indeterminacy in a forward-looking regime switching model. *International Journal of Economic Theory* **5**, 69–84.

FARMER, R. E., WAGGONER, D. F. & ZHA, T. (2009b). Understanding Markov-switching rational expectations models. *Journal of Economic Theory* **144**, 1849–1867.

FILARDO, A. J. (1994). Business-cycle phases and their transitional dynamics. *Journal of Business & Economic Statistics* **12**, 299–308.

FILARDO, A. J. & GORDON, S. F. (1998). Business cycle durations. *Journal of Econometrics* **85**, 99–123.

FOERSTER, A., RUBIO-RAMÍREZ, J., WAGGONER, D. F. & ZHA, T. (2016). Perturbation methods for Markov-switching dynamic stochastic general equilibrium models. *Quantitative Economics* **7**, 637–669.

FRANCQ, C. & ZAKOÏAN, J.-M. (2001). Stationarity of multivariate Markov switching ARMA models. *Journal of Econometrics* **102**, 339–364.

FRÜHWIRTH-SCHNATTER, S. (2001). MCMC estimation of classical and dynamic switching and mixture models. *Journal of the American Statistical Association* **96**, 194–209.

FRÜHWIRTH-SCHNATTER, S. (2004). Estimating marginal likelihoods for mixture and Markov switching models using bridge sampling techniques. *Econometrics Journal* **7**, 143–167.

FRÜHWIRTH-SCHNATTER, S. (2006). *Finite Mixture and Markov Switching Models*. New York: Springer.

FRÜHWIRTH-SCHNATTER, S. (2011). Dealing with label switching under model uncertainty. In *Mixtures: Estimation and Applications*, K. L. Mengersen, C. P. Robert & D. M. Titterington, eds. Chichester: John Wiley, pp. 213–240.

FRÜHWIRTH-SCHNATTER, S. & FRÜHWIRTH, R. (2007). Auxiliary mixture sampling with applications to logistic models. *Computational Statistics and Data Analysis* **51**, 3509–3528.

FRÜHWIRTH-SCHNATTER, S. & FRÜHWIRTH, R. (2010). Data augmentation and MCMC for binary and multinomial logit models. In *Statistical Modelling and Regression Structures – Festschrift in Honour of Ludwig Fahrmeir*, T. Kneib & G. Tutz, eds. Heidelberg: Physica-Verlag, pp. 111–132.

FRÜHWIRTH-SCHNATTER, S. & KAUFMANN, S. (2006). How do changes in monetary policy affect bank lending? An analysis of Austrian bank data. *Journal of Applied Econometrics* **21**, 275–305.

GAGGL, P. & KAUFMANN, S. (2014). The cyclical component of labor market polarization and jobless recoveries in the U.S. Working Paper 14.03, Study Center Gerzensee.

GEWEKE, J. & AMISANO, G. (2011). Hierarchical Markov normal mixture models with applications to financial asset returns. *Journal of Applied Econometrics* **26**, 1–29.

GEWEKE, J., KEANE, M. & RUNKLE, D. (1994). Alternative computational approaches to inference in the multinomial probit model. *Review of Economics & Statistics* **76**, 609–632.

GOLDFELD, S. M. & QUANDT, R. E. (1973). A Markov model for switching regressions. *Journal of Econometrics* **1**, 3–16.

GRAY, S. (1996). Modelling the conditional distribution of interest rates as a regime-switching process. *Journal of Financial Economics* **42**, 27–62.

GREENE, W. H. & HENSHER, D. A. (2003). A latent class model for discrete choice analysis: Contrasts with mixed logit. *Transportation Research Part B* **37**, 681–698.

GREENE, W. H. & HENSHER, D. A. (2013). Revealing additional dimensions of preference heterogeneity in a latent class mixed multinomial logit model. *Applied Economics* **45**, 1897–1902.

HAMILTON, J. D. (1989). A new approach to the economic analysis of nonstationary time series and the business cycle. *Econometrica* **57**, 357–384.

HAMILTON, J. D. (1990). Analysis of time series subject to changes in regime. *Journal of Econometrics* **45**, 39–70.

HAMILTON, J. D. (1994). *Time Series Analysis*. Princeton, NJ: Princeton University Press.

HAMILTON, J. D. & OWYANG, M. T. (2012). The propagation of regional recessions. *Review of Economics & Statistics* **94**, 935–947.

HAMILTON, J. D. & SUSMEL, R. (1994). ARCH and changes in regime. *Journal of Econometrics* **45**, 307–333.

HANSEN, B. E. (1992). The likelihood ratio test under nonstandard conditions: Testing the Markov switching model of GNP. *Journal of Applied Econometrics* **7**, S61–S82.

HESS, S. (2014). Latent class structures: Taste heterogeneity and beyond. In *Handbook of Choice Modelling*, S. Hess & A. Daly, eds. Cheltenham: Edward Elgar, pp. 311–329.

HOLMES, C. C. & HELD, L. (2006). Bayesian auxiliary variable models for binary and multinomial regression. *Bayesian Analysis* **1**, 145–168.

IMAI, K. & VAN DYK, D. A. (2005). A Bayesian analysis of the multinomial probit model using marginal data augmentation. *Journal of Econometrics* **124**, 311–334.

KAUFMANN, S. (2010). Dating and forecasting turning points by Bayesian clustering with dynamic structure: A suggestion with an application to Austrian data. *Journal of Applied Econometrics* **25**, 309–344.

KAUFMANN, S. (2015). *K*-state switching models with time-varying transition distributions – does loan growth signal stronger effects of variables on inflation? *Journal of Econometrics* **187**, 82–94.

KAUFMANN, S. & KUGLER, P. (2010). A monetary real-time forecast of euro area inflation. *Journal of Forecasting* **29**, 388–405.

KIM, C.-J., MORLEY, J. & PIGER, J. (2005). Nonlinearity and the permanent effects of recessions. *Journal of Applied Econometrics* **20**, 291–309.

KIM, C.-J. & NELSON, C. R. (1998). Business cycle turning points, a new coincident index, and tests of duration dependence based on a dynamic factor model with regime-switching. *Review of Economics & Statistics* **80**, 188–201.

KLAASSEN, F. (2002). Improving GARCH volatility forecasts with regime-switching GARCH. *Empirical Economics* **27**, 363–394.

KOOP, G. & POTTER, S. M. (2007). Estimation and forecasting in models with multiple breaks. *Review of Economics & Statistics* **74**, 763–789.

KROLZIG, H.-M., MARCELLINO, M. & MIZON, G. (2002). A Markov-switching vector equilibrium correction model of the UK labour market. *Empirical Economics* **27**, 233–254.

MCCULLOCH, R. & TSAY, R. (1994). Statistical analysis of economic time series via Markov switching models. *Journal of Time Series Analysis* **15**, 523–539.

MCCULLOCH, R. E., POLSON, N. G. & ROSSI, P. E. (2000). A Bayesian analysis of the multinomial probit model with fully identified parameters. *Journal of Econometrics* **99**, 173–193.

MCFADDEN, D. (1974). Conditional logit analysis of qualitative choice behaviour. In *Frontiers of Econometrics*, P. Zarembka, ed. New York: Adacemic Press, pp. 105–142.

NOBILE, A. (1998). A hybrid Markov chain for the Bayesian analysis of the multinomial probit model. *Statistics and Computing* **8**, 229–242.

OWYANG, M. T., PIGER, J. M. & WALL, H. J. (2015). Forecasting national recessions using state level data. *Journal of Money, Credit and Banking* **47**, 847–866.

PAAP, R., SEGERS, R. & VAN DIJK, D. (2009). Do leading indicators lead peaks more than troughs? *Journal of Business & Economic Statistics* **27**, 528–543.

PESARAN, M. H., PETTENUZZO, D. & TIMMERMAN, A. (2007). Forecasting time series subject to multiple structural breaks. *Review of Economics & Statistics* **74**, 763–789.

PHILLIPS, K. L. (1991). A two-country model of stochastic output with changes in regime. *Journal of International Economics* **31**, 121–142.

PSARADAKIS, Z., RAVN, M. O. & SOLA, M. (2005). Markov switching causality and the money-output relationship. *Journal of Applied Econometrics* **20**, 665–683.

SCOTT, S. L. (2011). Data augmentation, frequentistic estimation, and the Bayesian analysis of multinomial logit models. *Statistical Papers* **52**, 87–109.

SIMS, C. A., WAGGONER, D. F. & ZHA, T. (2008). Methods for inference in large multiple-equation Markov-switching models. *Journal of Econometrics* **146**, 255–274.

VERDINELLI, I. & WASSERMAN, L. (1995). Computing Bayes factors using a generalization of the Savage–Dickey density ratio. *Journal of the American Statistical Association* **90**, 614–618.

# 14

# Mixtures of Nonparametric Components and Hidden Markov Models

**Elisabeth Gassiat**

*Laboratoire de Mathématiques d'Orsay, Université Paris-Sud, CNRS, and Université Paris-Saclay, France*

## CONTENTS

## 14.1 Introduction

The topic of this chapter is statistical inference of nonparametric finite mixtures. The latent variables (and thus the observations) will be mostly taken to be independent and identically distributed (i.i.d.), but in some cases they will be possibly non-independently distributed. For each observation, the corresponding latent variable indicates from which population the observation comes. In particular, when the latent variables form a Markov chain, the observation process will come from a nonparametric hidden Markov model (HMM) with finite state space. We would like to emphasize the fact that nonparametric modelling will only cover the conditional distribution of the observations given the latent variables, excluding the mixing distribution. Nonparametric modelling of the mixing distribution (with possibly infinitely denumerable or continuous support) is considered in Chapter 6.

To fix ideas, assume that a random variable $X$ follows a distribution

$$P = \sum_{g=1}^{G} \eta_g F_g. \tag{14.1}$$

In many problems, inference for the distributions $F_g$ is of interest in itself, for instance in genomic applications, signal analysis, or econometric modelling; see references in Bonhomme et al. (2016b) or Gassiat et al. (2016). The distribution $P$ may have a density given by

$$p(x) = \sum_{g=1}^{G} \eta_g f_g(x). \tag{14.2}$$

Parametric modelling of the distributions $F_g$ (or the densities $f_g$) constrains the distributions to have prescribed shapes. Because nonparametric modelling leads to more flexibility, methods to deal with nonparametric models have been investigated in several applied papers; see, for instance, the references in Gassiat et al. (2016) or De Castro et al. (2016) for HMMs. One major obstacle in using nonparametric modelling seems to be the very basic question of identifiability (apart from the specific aspect related to label switching or the identification of the number of hidden states).

However, it should be emphasized that nonparametric estimation of the distribution $P$ of the observations (or its density $p$) is always possible, even if the model is not identifiable. What will be of interest for us is the estimation of the weights $\eta_g$ and of the so-called emission distributions $F_g$ (or the emission densities $f_g$) for $g = 1, \ldots, G$. If the model is not identifiable, inference on the weights and the emission distributions is hopeless. But although identifiability is impossible to achieve in the widest generality, it has been shown recently that it is possible to obtain identifiability for particular classes of models. The aim of this chapter is to review situations where identifiability has been proved, and where inference thus can be meaningful.

Let us proceed with some general ideas about the situations we will consider. If the observation is one-dimensional, it is obvious that, to obtain identifiability of the weights and the emission distributions from the marginal distribution $P$, one needs to put some restrictions on the emission distributions. This will be the subject of Sections 14.2 and 14.3. In Section 14.2 we will consider mixtures of two populations (i.e. $G = 2$) under specific restrictions on the emission densities. In Section 14.3 we will consider mixtures of translated densities. A first hint on the link between HMMs and multidimensional mixtures appears in this section. If one wants to achieve fully nonparametric modelling of the translated distribution, then one requires a block of two observations to be dependent, which is the case for consecutive observations in HMMs.

Section 14.4 deals with multivariate mixtures. It appears that when the observations are at least three-dimensional and the coordinates are conditionally independent, that is, the emission distributions are the tensor products of the marginal distributions of the coordinates (or of three blocks of coordinates), then identifiability holds under a simple linear independence assumption that will be discussed in detail in Section 14.4.1. Using the fact that, conditionally on the present state, past and future states of a Markov chain are independent, it is possible to write the distribution of three consecutive observations in an HMM as a multidimensional mixture where the coordinates of the three-dimensional observations (consisting of the consecutive observations) are conditionally independent, and identifiability is obtained for HMMs. One way to prove identifiability is constructive, by applying spectral methods to the tensor of the three-dimensional distribution. We subsequently present nonparametric inference based on spectral methods in Section 14.4.2. Once the model is proven to be identifiable, nonparametric estimation methods may be proposed based on frequentist model selection ideas or based on Bayesian methodology. Model selection methods usually lead to oracle inequalities for the risk of the estimation of the distribution of the observations. One then has to go back from the distribution of the observations to the weights and the emission densities. Development of results on these ideas is the aim of Section 14.4.3. A specific section, Section 14.4.4, is dedicated to HMMs.

We conclude the chapter by discussing some related questions and extensions to other nonparametric mixture models in Section 14.5.

## 14.2 Mixtures with One Known Component

In this section we consider mixtures of two populations on the real line. Obviously, any distribution may be split into any mixture of itself, so that one has to specify some assumptions on the emission densities. One way to do this is to fix one of the components, so that for a known fixed density $g$,

$$p(x) = (1 - \eta)g(x) + \eta f(x), \quad x \in \mathbb{R}. \tag{14.3}$$

Apart from the knowledge of the first component, one still has to restrict the set of possible densities $f$ for the other component. Indeed, one has, for instance, $(1 - \eta)g(x) + \eta f(x) = (1 - \eta/2)g(x) + (\eta/2)[g(x) + 2f(x)]$ so that, with the proportion $\eta/2$ and the second component equal to $g(x) + 2f(x)$, one gets the same mixture as (14.3).

### 14.2.1 The case where the other component is symmetric

One possibility for obtaining consistency is to assume that the second component is symmetric around some unknown value $\mu$, so that

$$p(x) = (1 - \eta)g(x) + \eta f(x - \mu), \tag{14.4}$$

where $f$ is a symmetric density, that is, for all $x \in \mathbb{R}$, $f(-x) = f(x)$. The parameter to be estimated is $(f, \theta)$, with $\theta = (\eta, \mu) \in [0, 1] \times \mathbb{R}$. As soon as one has an estimator $\widehat{\theta} = (\widehat{\eta}, \widehat{\mu})$ of $\theta$, one can construct an estimator of the unknown $f$ from a nonparametric estimator $\widehat{p}$ of $p$ by taking a symmetrized version of $\widehat{p}(\cdot + \widehat{\mu})/\widehat{\eta} - (1/\widehat{\eta} - 1)g(\cdot)$.

Bordes et al. (2006a) prove, under a condition on the existence of moments, that identifiability holds when constraining $f(\cdot)$ to symmetry only and almost everywhere on $\theta$. They propose a minimum distance estimator for the parameters $\theta$, based on a symmetrization-based distance, and prove that this estimator is consistent. Bordes & Vandekerkhove (2010) then established that the estimator is $\sqrt{n}$-consistent and asymptotically normal. Further, Hohmann & Holzmann (2013a) studied the situation where the known component $g$ is symmetric with a known centre of symmetry. Identifiability is proved by comparing the tails of (and thus under assumptions on) the characteristic functions. The authors construct asymptotically normal estimators. Xiang et al. (2014) also proposed an asymptotically normal estimator based on the minimum profile Hellinger distance.

### 14.2.2 Mixture of a uniform and a non-decreasing density

Model (14.3) is often considered in applications when dealing with multiple testing problems. When the data considered are the $p$-values, the model becomes

$$p(x) = (1 - \eta)u(x) + \eta f(x), \quad x \in [0, 1], \tag{14.5}$$

where the random variables take values in $[0, 1]$ and the first component $u(x)$ of the mixture is the density of the uniform distribution $\mathcal{U}[0, 1]$ on $[0, 1]$.

Nguyen & Matias (2014b) investigate efficient estimation of the proportion $\eta$ when the nonparametric component is assumed to be in $\mathcal{F}_\delta$, the set of non-increasing probability densities that are positive on $[0, 1 - \delta)$ and zero on $[1 - \delta, 1]$. When $\delta > 0$, it is possible to compute the efficient Fisher information that gives a lower bound of the asymptotic variance of $\sqrt{n}$-consistent estimators. The authors prove that a histogram-based estimator of $\eta$ is $\sqrt{n}$-consistent and conjecture that there exist no asymptotically efficient estimators that can achieve the lower bound given by the efficient Fisher information. This lower bound explodes when $\delta$ goes to 0, which implies that, when $\delta = 0$, the quadratic risk of any estimator cannot converge to a finite value at a parametric rate.

When one has at hand a preliminary estimator of the unknown proportion $\eta$ and a non-parametric estimator $\widehat{p}$ of the mixture density, it is possible to estimate the nonparametric component $f$ from equation (14.5), after plugging in the estimators for $\eta$ and $p$. However, although this estimator may have good asymptotic properties in theory, it does not behave well in practice. Nguyen & Matias (2014a) propose two different estimators for $f$ that exhibit better performance in simulation studies. The first estimator is a randomly weighted kernel estimator, while the second estimator is a maximum smoothed likelihood estimator.

## 14.3    Translation Mixtures

A case of particular interest is the situation where the components of the mixture are shifted versions of one (unknown) distribution, that is, when, for some unknown distribution $F$, and some unknown parameters $\mu_1, \ldots, \mu_G$, the mixture distribution is

$$P(\cdot) = \sum_{g=1}^{G} \eta_g F(\cdot - \mu_g). \tag{14.6}$$

When the observations $X_1, \ldots, X_n$ are i.i.d. with distribution $P$, modelling the distribution as (14.6) is not enough to get identifiability, and one has to add some assumption on the unknown distribution $F$. This is the subject of Section 14.3.1. However, when the observations are no longer independent, it is possible to obtain identifiability without any assumption on $F$, as will be discussed in Section 14.3.2.

### 14.3.1    Translation of a symmetric density

The usual assumption for obtaining identifiability is that $F$ is restricted to be a symmetric cumulative function in the sense that, for all $x \in \mathbb{R}$, $F(-x) + F(x) = 1$. This situation was studied independently by Bordes et al. (2006b) and Hunter et al. (2007).

Let $\mathcal{F}$ be the set of symmetric cumulative functions. Denote by $\Omega_G$ the set of weights $(\eta_g)_{1 \leq g \leq G}$ and translation parameters $(\mu_g)_{1 \leq g \leq G}$ in $\mathbb{R}^G$. The weights $(\eta_g)_{1 \leq g \leq G}$ have to be such that $\eta_g \geq 0$, $g = 1, \ldots, G$, and $\sum_{g=1}^{G} \eta_g = 1$. Denote by $\Omega_G^\star$ the subset of identifiable parameters of $\Omega_G$, that is, the parameters $(\eta_g)_{1 \leq g \leq G}$, $(\mu_g)_{1 \leq g \leq G}$ such that, for any $F \in \mathcal{F}$, one may recover the parameters from $\sum_{g=1}^{G} \eta_g F(\cdot - \mu_g)$.

The main identifiability result (see Hunter et al., 2007, Theorem 1) says that the following two statements are equivalent:

(a) The set of parameters $(\eta_g)_{1 \leq g \leq G}$, $(\mu_g)_{1 \leq g \leq G}$ is in $\Omega_G^\star$.

(b) For all $(\eta'_g)_{1 \leq g \leq G}$, $(\mu'_g)_{1 \leq g \leq G}$, the convolution $\sum_{g=1}^{G} \eta_g \delta_{\mu_g} \star \sum_{g=1}^{G} \eta'_g \delta_{-\mu'_g}$ is symmetric if and only if $\sum_{g=1}^{G} \eta'_g \delta_{-\mu'_g} = \sum_{g=1}^{G} \eta_g \delta_{-\mu_g}$.

From this result, Hunter et al. (2007) deduce that, when $G = 2$, identifiability holds if and only if $\eta_1 \notin \{0, 1/2, 1\}$, which is equivalent to the assumption that the mixture $\eta_1 F(\cdot - \mu_1) + (1 - \eta_1)F(\cdot - \mu_2)$ is not symmetric.

Bordes et al. (2006b) propose an iterative estimation procedure for which they prove that the resulting estimator of the parameters is $n^{-1/4+\alpha}$-consistent for any positive $\alpha$. Hunter et al. (2007) build on their symmetry considerations to develop an estimator which is proven to be $\sqrt{n}$-consistent and asymptotically normal under technical assumptions. Butucea & Vandekerkhove (2014) propose a new class of M-estimators for these parameters based on a Fourier approach, and prove that these estimators are $\sqrt{n}$-consistent under mild regularity conditions.

### 14.3.2 Translation of any distribution and hidden Markov models

We now consider observations with marginal distribution (14.6) that are not independent. This section mainly builds on Gassiat & Rousseau (2016). To obtain identifiability, we need two dependent consecutive observations. Then to use identifiability for building estimators, we need repetitions of such consecutive observations. We may have independent repetitions of a block of two non-independent variables, or we may have a stationary sequence of random variables. Thus, HMMs fits within this setting.

Consider a pair $(X_1, X_2)$ of random variables such that the marginal distribution of $X_1$ ($X_2$) is (14.6), and the latent variables $(Z_1, Z_2)$ have a distribution given by the $G \times G$ matrix $\xi$ on $\{1, \ldots, G\}^2$, that is, for all $g, g'$, $\xi_{gg'} = \mathrm{P}(Z_1 = g, Z_2 = g')$. It is obvious that, by translating the distribution $F$ and translating all the $\mu_g$ in the reverse direction, one obtains the same distribution. We thus fix arbitrarily $\mu_1$ to 0, and consider the set of parameters $\Theta_G$ for the matrix $\xi$ and the translation parameters $\mu_g$, $g = 1, \ldots, G$, such that $\mu_1 = 0 < \mu_2 < \ldots < \mu_G$ and such that $\xi$ is full rank.

The main identifiability result (Theorem 1) of Gassiat & Rousseau (2016) is the following. If the parameters lie in $\Theta_G$ for some (possibly unknown) $G$ and if $F$ is any probability distribution, then one may recover $G$, $\xi$, $\mu_g$, $g = 1, \ldots, G$, and $F$ from the distribution of $(X_1, X_2)$. This is a strong result. Indeed, it requires very weak assumptions. In particular, no assumption is made about $F$. The assumption that the translation parameters are distinct is obviously a basic assumption (the order constraint just fixes the label switching). The only structural assumption is that $\xi$ has full rank. Note that, with only two latent states (i.e. $G = 2$), assuming that $\xi$ has full rank is the same as assuming that the variables $X_1$ and $X_2$ are *not* independent.

The proof of the identifiability result uses characteristic functions and relies on complex analysis arguments. Let us explain the main ideas. Let $\theta$ be a parameter vector, containing the parameters in the matrix $\xi$ as well as the translation parameters $(\mu_1, \ldots, \mu_G)$, and let $F$ be a probability distribution. One may rewrite the model as

$$X_i = \mu_{Z_i} + U_i, \quad i = 1, 2, \tag{14.7}$$

where $U_1$ and $U_2$ are independent variables with distribution $F$ and independent of $(Z_1, Z_2)$ which has distribution $\xi$. Then the characteristic functions of the variables $X_i$ are the product of the characteristic functions of the corresponding variables $\mu_{Z_i}$ and the characteristic functions of the corresponding variables $U_i$. Consider $(\theta_1, F_1)$ and $(\theta_2, F_2)$ such that the distribution of the pair $(X_1, X_2)$ is the same under both sets of parameters. Using the fact that the characteristic functions of $(X_1, X_2)$, of $X_1$, and of $X_2$, are the same for both sets of parameters, one can see that it is possible to separate what comes from $F_1$, $F_2$ and what comes from $\theta_1$, $\theta_2$. This leads to the following equation, for all $(s,t)$ in a neighbourhood of $(0,0)$:

$$\Phi_{\theta_1}(s,t)\phi_{\theta_2}(s)\phi_{\theta_2}(t) = \Phi_{\theta_2}(s,t)\phi_{\theta_1}(s)\phi_{\theta_1}(t), \tag{14.8}$$

where, for any $\theta$, $\Phi_\theta$ is the characteristic function of $(\mu_{Z_1}, \mu_{Z_2})$ under $\theta$, and $\phi_\theta$ is the characteristic function of $\mu_{Z_1}$ (or $\mu_{Z_2}$) under $\theta$. The identifiability proof is completed by studying the set of zeros of those functions, using properties of the entire function. The fact that $\xi$ has full rank is used to obtain that $\Phi_\theta$ is not the null function.

The milestone of the work in Gassiat & Rousseau (2016) is that, as soon as (14.8) holds in a neighbourhood of $(0,0)$, then $\theta_1 = \theta_2$. Let $\theta_1$ be the true (unknown) parameter and let $\theta_2$ be any possible $\theta$. Taking the modulus of the difference of both sides of (14.8) and integrating in a neighbourhood of $(0,0)$, one obtains a contrast function $M(\theta)$ which is non-negative, and zero if and only if $\theta$ is the true (unknown) value.

Since characteristic functions may be estimated empirically, this allows us to build an empirical contrast function $M_n(\theta)$ that estimates $M(\theta)$ well enough to define an estimator $\widehat\theta$ as a minimizer of $M_n(\theta)$. It is proved in Gassiat & Rousseau (2016) that such an estimator has good properties (parametric rate $\sqrt n$ of convergence and asymptotic Gaussian distribution).

Once an estimator of $\theta$ is given, one may use a model selection approach to estimate the distribution $F$. One possibility is described in Gassiat & Rousseau (2016) and works as follows. Assume that possible distributions are dominated, and have a density $f \in \mathcal{F}$, where $\mathcal{F}$ is an infinite-dimensional set of densities. Assume a given a collection $(\mathcal{F}_k)_{k\geq 1}$ of finite-dimensional approximating sets of $\mathcal{F}$ (the larger the value of $k$, the better is the approximation and the larger is the dimension of $\mathcal{F}_k$). For instance, $\mathcal{F}_k$ may be the set of stepwise densities defined by a partition which is refined when $k$ gets larger, or $\mathcal{F}_k$ may be the finite-dimensional space spanned by the first $k$ elements of a basis. One may define, for any density function $f$,

$$\ell_n(f) = \frac{1}{n}\sum_{i=1}^n \log\left[\sum_{g=1}^G \widehat\eta_g f(X_i - \widehat\mu_g)\right],$$

which could be seen as the log likelihood if the variables were independent (which is not the case) and where $\theta$ is replaced by a preliminary estimator. For any $k$, define $\widehat f_k$ to be the maximizer of $\ell_n(f)$ over $\mathcal{F}_k$. As usual one has to make a trade-off between complexity and variance so that the estimator will be chosen by selecting $k$ using a penalized criterion such as

$$D_n(k) = -\ell_n\left(\widehat f_k\right) + \text{pen}(k,n),$$

where $\text{pen}(k,n)$ is some penalty term that has to be chosen. Then the estimator is defined by $\widehat f = \widehat f_{\widehat k}$, where $\widehat k$ is a minimizer of $D_n$. In this way, adaptivity results are proved in Gassiat & Rousseau (2016) on the estimation of the marginal density of $X_1$ when the penalty is adequately chosen. Furthermore, when it is possible to go back from the risk on $p(\cdot) = \sum_{g=1}^G \eta_g f(\cdot - \mu_g)$ to that on $f$, adaptivity results are proved for the estimation of $f$.

Let us briefly describe how this works. Following model selection theory, one gets an oracle inequality for the square of the Hellinger distance $h^2(p,\widehat p)$ from which adaptivity for the estimation of $p$ is deduced. Here, $\widehat p(\cdot) = \sum_{g=1}^G \widehat\eta_g \widehat f(\cdot - \widehat\mu_g)$, and for any pair of densities $p$ and $q$ with respect to some measure $\nu$, $h^2(p,q) = \int(\sqrt p - \sqrt q)^2 d\nu(x)$. The problem is now to bound $h(p,\widehat p)$ from below by some distance between $f$ and $\widehat f$. This may be done using the $L_1(\nu)$-distance, when $\sup_g \eta_g > 1/2$. Indeed, one has $h(p,\widehat p) \geq \|\widehat p - p\|_1$, and then, using the triangular inequality, the fact that the $L_1$-norm of a density is 1 and that the weights add up to 1, we have, on the one hand,

$$\left\| \sum_g \widehat{\eta}_g \widehat{f}(\cdot - \widehat{\mu}_g) - \eta_g f(\cdot - \mu_g) \right\|_1 \leq \sum_g |\widehat{\eta}_g - \eta_g| + \sup_g \| f(\cdot - \widehat{\mu}_g) - f(\cdot - \mu_g) \|_1.$$

On the other hand, using the triangle inequality iteratively, one obtains that

$$\left\| \sum_g \eta_g (\widehat{f}(\cdot - \widehat{\mu}_g) - f(\cdot - \widehat{\mu}_g)) \right\|_1 \geq \left( 2 \max_g \eta_g - 1 \right) \| \widehat{f} - f \|_1.$$

Provided that (a) one has parametric rates for the estimation of $\eta_g$ and $\mu_g$, $g = 1, \ldots, G$, (b) $f$ satisfies a Lipschitz property, and (c) $\max_g \eta_g > 1/2$ (which is a weak assumption when $G = 2$), one can transfer adaptive results from $h^2(p, \widehat{p})$ to adaptive results on $\| \widehat{f} - f \|_1$.

This is a first example of how model selection methods lead to adaptive estimators: one obtains an oracle inequality for the estimation of the density of the observed variables and transfers it to the nonparametric part of the model if one can prove an inequality linking both risks. Note that doing so requires one to have a preliminary estimator with parametric rate. Indeed, one could use the model selection strategy to estimate $\theta$ and $f$ together, but the usual methods do not directly give a parametric rate for the parametric part alone, so that it does not seem easy to go back to adaptive rates for the nonparametric part. Finally, Bayesian methods can also be used; see Vernet (2015b).

## 14.4 Multivariate Mixtures

Let us now consider multidimensional mixtures for which the coordinates may be partitioned into at least $d \geq 3$ blocks that constitute random variables that are conditionally independent given the component. That is, the observation has distribution $P$ given by

$$P = \sum_{g=1}^{G} \eta_g \otimes_{j=1}^{d} F_{g,j}, \tag{14.9}$$

where, for each $g = 1, \ldots, G$, $F_{g,1}, \ldots, F_{g,d}$ are probability distributions on $d$ spaces (that may have different dimensions). When the spaces are equal and $F_{g,1} = \ldots = F_{g,d}$ for all $g = 1, \ldots, G$, this may be seen as modelling independently repeated measurements in unknown several populations. We shall first consider the situation where the observations $X_1, \ldots, X_n$ are i.i.d., then we will exhibit a structural link with HMMs via the fact that, when $(Z_t)_{t \geq 1}$ is a Markov chain, conditionally on $Z_t$, $Z_{t-1}$ and $Z_{t+1}$ are independent, so that what has been understood for independent variables will be used to understand HMMs with finite state space nonparametric in Section 14.4.4.

### 14.4.1 Identifiability

In the statistical literature, the first results may be found in Hall & Zhou (2003) where the case $G = 2$ and $d = 2$ or 3 is addressed, and in Hall et al. (2005) who do not fix $G$ and discuss $d = 2$ and $d > 2$. General insights on identifiability for various latent structure models are developed in Allman et al. (2009). Here, the authors point out a fundamental algebraic result by Kruskal (1977), which may be stated as an identifiability result of model (14.9) when $d = 3$ and the probability distributions are on finite sets.

Building upon this result, Allman et al. (2009) prove that, when $d \geq 3$, model (14.9) is identifiable as soon as, for $j = 1, \ldots, d$, the probability measures $F_{1,j}, \ldots, F_{G,j}$ are linearly independent (which reduces to $F_{1,j}$ and $F_{2,j}$ being distinct when $G = 2$). Results in the parametric literature about spectral methods such as Anandkumar et al. (2012) and Song et al. (2014) may be used to get the same result; see also Bonhomme et al. (2016a,b).

Let us present the spectral methods argument of Anandkumar et al. (2012) in more detail. For the sake of simplicity, we assume that $d = 3$ and that the $F_{g,j}$ are distributions on $\mathbb{R}$ so that an observation takes the form

$$X = \begin{pmatrix} X_1 \\ X_2 \\ X_3 \end{pmatrix} \in \mathbb{R}^3.$$

Let $\phi_1, \ldots, \phi_M$ be $M$ real-valued functions and denote by $A^{(j)}$, for $j = 1, 2, 3$, the $M \times G$ matrix such that $A^{(j)}_{l,g} = \int \phi_l dF_{g,j}$, $l = 1, \ldots, M$, $g = 1, \ldots, G$. For instance, when the $\phi_l$ are indicator functions of a partition of $\mathbb{R}$, then $A^{(j)}$ has the conditional distributions of the associated discretized coordinate $X_j$ as columns. More generally, when the $\phi_l$ are such that $(\phi_l)_{l \geq 1}$ forms a basis of the space of densities or of the space of distributions, then for large enough $M$, $A^{(j)}$ has rank $M$ as soon as $F_{1,j}, \ldots, F_{G,j}$ are linearly independent, which we now assume for $j = 1, 2, 3$. Let $D$ be the diagonal $G \times G$ matrix having the $\eta_g$ on the diagonal and denote by $S$ the $M \times M$ matrix such that $S_{l,m} = \mathrm{E}[\phi_l(X_1)\phi_m(X_2)]$. Then

$$S = A^{(1)} D (A^{(2)})^\top,$$

so that as soon as $A^{(1)}$ and $A^{(2)}$ have full rank, which occurs for large enough $M \geq G$, one has $\mathrm{rank}(S) = G$. Thus, $G$ is identifiable based on the joint distribution of $(X_1, X_2)$. We now fix $M$ such that $A^{(1)}$, $A^{(2)}$ and $A^{(3)}$ have full rank.

Let $T$ be the $M \times M \times M$ tensor such that

$$T(l_1, l_2, l_3) = \mathrm{E}[\phi_{l_1}(X_1)\phi_{l_2}(X_2)\phi_{l_3}(X_3)],$$

and let $U_1$ and $U_2$ be $M \times G$ matrices such that $U_1^\top S U_2$ is invertible (such matrices may be found by a singular value decomposition of $S$). Let $V$ be a vector in $\mathbb{R}^M$, and define $T[V]$ to be the $M \times M$ matrix given by applying the tensor $T$ to $V$, that is,

$$T[V]_{l,m} = \mathrm{E}[\phi_l(X_1)\phi_m(X_2)\langle V, \Phi(X_3)\rangle], \quad l, m = 1, \ldots, M,$$

where $\Phi(X_3) = (\phi_h(X_3))_{1 \leq h \leq M}$. Define now, for all $V$, the matrix $B(V)$ (which may be computed as soon as one knows the distribution of $X$) by

$$B(V) = (U_1^\top T[V] U_2)(U_1^\top S U_2)^{-1}.$$

Then, denoting by $\Delta(V)$ the diagonal $G \times G$ matrix with the coordinates of $(A^{(3)})^\top V$ on the diagonal, one has

$$B(V) = (U_1^\top A^{(1)}) \Delta(V) (U_1^\top A^{(1)})^{-1}.$$

Thus, all matrices $B(V)$ have the same eigenvectors, and their eigenvalues are the coordinates of $(A^{(3)})^\top V$. This means that, by exploring various vectors $V$, one may recover $A^{(3)}$. The eigenvectors also stay the same when permuting coordinates 2 and 3 of the observed variable, so that one may recover $A^{(2)}$, and thus also $A^{(1)}$. Recovering $D$ is then also possible. Finally, by taking $M$ to infinity, one may recover the distributions $F_{1,g}$, $F_{2,g}$ and $F_{3,g}$, $g = 1, \ldots, G$.

### 14.4.2 Estimation with spectral methods

As seen in the spectral proof of identifiability, for large enough $M$, one may recover all parameters by spectral analysis (singular value decompositions and eigenvalue decompositions) using the matrix $S$ and the tensor $T$. Given a sample of observations of $X$, $S$ and $T$ may be estimated empirically by taking empirical means as estimators of the expectations involved. Thus, by spectral analysis using the estimators $\widehat{S}$ of $S$ and $\widehat{T}$ of $T$, one obtains estimators of $A^{(1)}$, $A^{(2)}$, $A^{(3)}$, and $D$ for fixed $M$. Such an algorithm is studied in Anandkumar et al. (2012) and Song et al. (2014).

To achieve the right rate of convergence for the estimators of the nonparametric part, that is, the distributions $F_{1,g}$, $F_{2,g}$ and $F_{3,g}$, $g = 1, \ldots, G$, one has to choose $M$ appropriately as a function of the number of observations $n$, typically in a way that depends on the smoothness of the densities $f_{1,g}$, $f_{2,g}$ and $f_{3,g}$, $g = 1, \ldots, G$, of the distributions. This is studied in Bonhomme et al. (2016b), for repeated measurements, and in Bonhomme et al. (2016a), where asymptotic results are given at the minimax asymptotic rate. However, choosing the right $M$ when using spectral methods for estimation requires prior knowledge on the smoothness of densities.

### 14.4.3 Estimation with nonparametric methods

When identifiability holds, one may use model selection methods for the estimation of the parameters of the model (parametric weights and nonparametric probability distributions), as described in Section 14.3.2. This leads to oracle inequalities for the risk of the estimator of the density of the observed variables, and one can deduce results for the risk on the parameters if it is possible to relate both risks. We shall describe some possible ways of using such ideas in multivariate mixtures below.

Let us first mention that Hall & Zhou (2003) propose an estimator of the weight and of the repartition functions of the distributions that are $\sqrt{n}$- consistent for $d$-dimensional mixtures with $d \geq 3$ and $G = 2$. The method is to minimize some distance between the empirical $d$-dimensional repartition function and the set of repartition functions belonging to model (14.9) with $G = 2$, and then to take the minimizer as an estimator. Benaglia et al. (2009) propose an EM-type algorithm for semi- and nonparametric estimation in multivariate mixtures, but do not provide theoretical properties for the estimator obtained.

We now describe possible model selection methods such as penalized maximum likelihood and penalized least squares. Assume that on each of the $d$ spaces, possible probability distributions are dominated and denote by $\mathcal{F}_j$ the (nonparametric) sets of possible densities, $j = 1, \ldots, d$. For each $j = 1, \ldots, d$, denote by $(\mathcal{F}_{j,k})_{k \geq 1}$ a collection of finite-dimensional approximating sets of $\mathcal{F}_j$. The log likelihood is given by

$$\ell_n \left( (\eta_g)_{1 \leq g \leq G}, (f_{g,j})_{1 \leq g \leq G, 1 \leq j \leq d} \right) = \frac{1}{n} \sum_{i=1}^{n} \log \left[ \sum_{g=1}^{G} \eta_g \prod_{j=1}^{d} f_{g,j} \left( (X_i)_j \right) \right],$$

and for any $k = (k_1, \ldots, k_d) \in (\mathbb{N}^*)^d$, define $((\widehat{\eta}_g)_{1 \leq g \leq G}, (\widehat{f}_{g,j,k})_{1 \leq g \leq G, 1 \leq j \leq d})$ as the maximizer of $\ell_n(\cdot)$ for $(\eta_g)_{1 \leq g \leq G}$ and for $f_{g,j} \in \mathcal{F}_{j,k_j}$, $1 \leq g \leq G, 1 \leq j \leq d$. To achieve a trade-off between complexity and variance, the estimator will be chosen by selecting $\widehat{k}$ as minimizing a penalized criterion such as

$$D_n(k) = -\ell_n \left( (\widehat{\eta}_g)_{1 \leq g \leq G}, (\widehat{f}_{g,j,k})_{1 \leq g \leq G, 1 \leq j \leq d} \right) + \mathrm{pen}(k_1, \ldots, k_d, n).$$

Let $p$ denote the density of $X_1 = ((X_1)_j)_{1 \leq j \leq d}$ and $\widehat{p}$ the estimator of $p$ obtained with $\widehat{k}$.

Then, following Massart (2007), it may be possible to prove some oracle inequality with a statement as follows. Assume that the penalty is larger than some quantity related to the complexity of the model times $\log n/n$. Then the risk $\mathrm{E}\left[h^2\left(\widehat{p}, p\right)\right]$ for the estimation of $p$ may be upper-bounded, up to some universal constant $C$, by the best (over all approximation spaces) Kullback–Leibler divergence of $p$ to its best approximation plus the penalty, plus some negligible term.

Instead of using maximum likelihood, one may use least squares by minimizing, on each approximating space, the contrast function $\gamma_n$ given by

$$\gamma_n\left((\eta_g)_{1\leq g\leq G}, (f_{g,j})_{1\leq g\leq G, 1\leq j\leq d}\right) = \int p^2(x)dx - \frac{2}{n}\sum_{i=1}^{n} p\left(X_i\right)$$

where, for $x = (x_j)_{1\leq j\leq d}$,

$$p(x) = \sum_{g=1}^{G} \eta_g \prod_{j=1}^{d} f_{g,j}(x_j),$$

and by selecting $k$ with a penalized criterion as before. Then the oracle inequality that may be proved is now based on the $L_2$-risk $\mathrm{E}\|\widehat{p} - p\|_2^2$.

A very important feature of such model selection methods is that the choice of the approximation space is data-driven, and often leads to adaptive minimax rates of estimation. Moreover, more practically, one may apply the so-called slope heuristic to calibrate the penalty; see Baudry et al. (2012) and Chapter 7, Section 7.2.2.2, above for details.

If one is interested in error evaluations for the parameters of the mixture, one has to follow the same route as in Section 14.3.2. First, find a preliminary estimator of the weights $\eta_g$, $g = 1, \ldots, G$, with convergence rate $\sqrt{n}$, and use model selection to obtain an estimator for the nonparametric part by plugging the estimator of the weights into the estimation criterion. Second, obtain an oracle inequality (this requires slightly more elaborate analysis due to the plug-in estimator in the criterion). Third, go back from the risk on the density of the observations to the risk of the emission densities.

Let us show an example of such inequality in the simple case of repeated measurements models, that is, when, for all $g$, the $f_{g,j}$ are the same, $f_{g,j} = f_g$, $j = 1, \ldots, d$. We assume that all the $f_g$ are in $L^2(I)$ for some subset $I$ of $\mathbb{R}$. In what follows, norms of functions are $L_2$-norms and norms of vectors are Euclidian norms. We denote by $\langle \cdot, \cdot \rangle$ the $L_2$-inner product between functions. For $\eta = (\eta_1, \ldots, \eta_G)$ and $\mathbf{f} = (f_1, \ldots, f_G)$, denote

$$p_{\eta, \mathbf{f}}(x_1, x_2, x_3) = \sum_{g=1}^{G} \eta_g f_g(x_1) f_g(x_2) f_g(x_3).$$

Let $\mathcal{H}$ be a closed bounded subset of $L^2(I)$. Let $\mathcal{F}$ be the subspace spanned by $f_1, \ldots, f_G$, and define, for $\mathbf{q} = (q_1, \ldots, q_G) \in \mathbb{R}^G$ and $\mathbf{u} = (u_1, \ldots, u_G) \in \mathcal{F}^G$,

$$D(\mathbf{q}, \mathbf{u}) = 3\sum_{g,j} q_g q_j \langle f_g, f_j \rangle^2 \langle u_g, u_j \rangle + 6\sum_{g,j} q_g q_j \langle f_g, f_j \rangle \langle u_g, f_j \rangle \langle f_g, u_j \rangle.$$

When writing $u_1, \ldots, u_G$ in the basis $f_1, \ldots, f_G$, $D(\mathbf{q}, \mathbf{u})$ is a quadratic form in the coordinates of $u_1, \ldots, u_G$, with $G^2 \times G^2$ matrix $A(\mathbf{q}, \mathbf{f})$. To obtain the inequality, we shall assume that the determinant $|A(\eta, \mathbf{f})| \neq 0$. Note that $|A(\mathbf{q}, \mathbf{f})|$ is a polynomial in the $q_j$ and the $\langle f_g, f_j \rangle$. Thus, if it is not the null function, the set of zeros of $|A(\mathbf{q}, \mathbf{f})|$ is negligible. But by taking functions $f_g$ with non-intersecting supports, we easily obtain that

$$D(\mathbf{q}, \mathbf{u}) \geq 3\sum_{g=1}^{G} q_g^2 \|f_g\|^4 \|u_g\|^2,$$

so that for such $\mathbf{f}$, $|A(\mathbf{q}, \mathbf{f})| \neq 0$ and $|A(\mathbf{q}, \mathbf{f})|$ is not the null function. The assumption $|A(\eta, \mathbf{f})| \neq 0$ is thus generically satisfied. Moreover, by continuity, under this assumption, there exists a neighbourhood of $\eta$ such that, for $\mathbf{q}$ in this neighbourhood, $|A(\mathbf{q}, \mathbf{f})| \neq 0$. We now state our theorem.

**Theorem 14.1** *Assume $\eta_g > 0$, $g = 1, \ldots, G$, and that $f_1, \ldots, f_G$ are linearly independent. Assume, moreover, that $|A(\eta, \mathbf{f})| \neq 0$. Let $\mathcal{K}$ be a compact neighbourhood of $\eta$ in $\mathbb{R}^G$ such that if $\mathbf{q} = (q_1, \ldots, q_G) \in \mathcal{K}$, then $|A(\mathbf{q}, \mathbf{f})| \neq 0$ and $q_g > 0$, $g = 1, \ldots, G$. Then there exists a constant $c(\mathcal{K}, \mathbf{f}) > 0$ such that, for all $\mathbf{h} \in \mathcal{H}^G$ and all $\mathbf{q} = (q_1, \ldots, q_G) \in \mathcal{K}$,*

$$\|p_{\mathbf{q}, \mathbf{f}+\mathbf{h}} - p_{\mathbf{q}, \mathbf{f}}\|^2 \geq c(\mathcal{K}, \mathbf{f}) \left( \sum_{g=1}^{G} \|h_g\|^2 \right).$$

Note that for large enough $n$ the estimator of $\eta$ will be in $\mathcal{K}$ with large probability.

Let us now prove Theorem 14.1. Denote

$$N(\mathbf{q}, \mathbf{h}) = \|p_{\mathbf{q}, \mathbf{f}+\mathbf{h}} - p_{\mathbf{q}, \mathbf{f}}\|^2.$$

Note first that the identifiability proof uses only spectral arguments, so that it may be extended to obtain that

$$\mathbf{h} \in \mathcal{H}^G \text{ is such that } N(\mathbf{q}, \mathbf{h}) = 0 \iff \mathbf{h} = 0.$$

One may compute

$$\begin{aligned} N(\mathbf{q}, \mathbf{h}) = \quad & \textstyle\sum_{g,j} q_g q_j \langle f_g + h_g, f_j + h_j \rangle^3 - q_g q_j \langle f_g, f_j + h_j \rangle^3 \\ & -q_g q_j \langle f_g + h_g, f_j \rangle^3 + q_g q_j \langle f_g, f_j \rangle^3. \end{aligned}$$

Let $\mathcal{F}^{\perp}$ be the orthogonal of $\mathcal{F}$. For $g = 1, \ldots, G$, let $u_g$ be the projection of $h_g$ on $\mathcal{F}$ and $h_g^{\perp}$ its projection on $\mathcal{F}^{\perp}$. Then

$$N(\mathbf{q}, \mathbf{h}) = N(\mathbf{q}, \mathbf{u}) + M(\mathbf{q}, \mathbf{u}, \mathbf{h}^{\perp}),$$

where

$$\begin{aligned} M(\mathbf{q}, \mathbf{u}, \mathbf{h}^{\perp}) = \quad & \textstyle\sum_{g,j} q_g q_j \left\{ \langle h_g^{\perp}, h_j^{\perp} \rangle^3 + 3 \langle h_g^{\perp}, h_j^{\perp} \rangle^2 \langle f_g + u_g, f_j + u_j \rangle \right. \\ & \left. +3 \langle h_g^{\perp}, h_j^{\perp} \rangle \langle f_g + u_g, f_j + u_j \rangle^2 \right\}. \end{aligned}$$

One may write $N(\mathbf{q}, \mathbf{u})$ as a finite sum of homogeneous functions in the $q_g$ and the $u_g$. The constant and the linear terms are zero, and denote the homogeneous term of degree 2 by $D(\mathbf{q}, \mathbf{u})$. Using

$$N(\mathbf{q}, \mathbf{u}) = D(\mathbf{q}, \mathbf{u}) + o \left( \sum_{i=1}^{G} \|h_i\|^2 \right),$$

we easily obtain that, for all $\mathbf{q} \in \mathcal{K}$ and all $\mathbf{u} \in \mathcal{F}^G$, $D(\mathbf{q}, \mathbf{u}) \geq 0$.

Denote $\|\mathbf{h}\|^2 = \sum_{g=1}^{G} \|h_g\|^2$ and define

$$c_1(\mathcal{K}, \mathbf{f}) := \inf_{\mathbf{q} \in \mathcal{K}, \mathbf{u} \in (\mathcal{F} \cap \mathcal{H})^G} \frac{N(\mathbf{q}, \mathbf{u})}{\|(\mathbf{q}, \mathbf{u})\|^2}.$$

Let $(\mathbf{q}_n, \mathbf{u}_n)_n$ be a sequence realizing $c_1(\mathcal{K}, \mathbf{f})$. By compactness (closed and bounded subset in a finite-dimensional space), $(\mathbf{q}_n, \mathbf{u}_n)_n$ has a limit point $(\bar{\mathbf{q}}, \bar{\mathbf{h}})$. If $(\bar{\mathbf{q}}, \bar{\mathbf{h}}) \neq 0$, one has $N(\bar{\mathbf{q}}, \bar{\mathbf{h}}) \neq 0$ and we obtain $c_1(\mathcal{K}, \mathbf{f}) > 0$. Otherwise

$$c_1(\mathcal{K}, \mathbf{f}) = \lim_{n \to +\infty} \frac{D(\mathbf{q_n}, \mathbf{u_n})}{\|(\mathbf{q}_n, \mathbf{u}_n)\|^2}.$$

But using the fact that $D(\mathbf{q}, \mathbf{u})$ is non-negative and non-degenerate by the assumption $|A(\mathbf{f})| \neq 0$, we obtain that $c_1(\mathcal{K}, \mathbf{f}) > 0$.

Now using Schur's theorem (which says that the Hadamard product of two positive matrices gives a positive matrix) and the fact that Gram matrices are non-negative, we easily obtain

$$M(\mathbf{q}, \mathbf{u}, \mathbf{h}^{\perp}) \geq 3\lambda_{\min}(G(\mathbf{f} + \mathbf{u})) \left( \sum_{g=1}^{G} q_g^2 \|h_g^{\perp}\|^2 \right),$$

where $G(\star)$ denotes the Gram matrix of the function $\star$ and $\lambda_{\min}(\square)$ is the minimum eigenvalue of $\square$. Then

$$
\begin{aligned}
N(\mathbf{q}, \mathbf{h}) &\geq 3\lambda_{\min}(G(\mathbf{f} + \mathbf{u})) \left( \sum_{g=1}^{G} q_g^2 \|h_g^{\perp}\|^2 \right) + c_1(\mathcal{K}, \mathbf{f}) \left( \sum_{g=1}^{G} \|u_g\|^2 \right) \\
&\geq 3(\lambda_{\min}(G(\mathbf{f} + \mathbf{u}))(\inf_{\mathbf{q} \in \mathcal{K}} q_g^2) \sum_{g=1}^{G} \|h_g^{\perp}\|^2 + c_1(\mathcal{K}, \mathbf{f}) \left( \sum_{g=1}^{G} \|u_g\|^2 \right).
\end{aligned}
$$

Now let $(\mathbf{q}_n, \mathbf{h}_n)_n$ be such that

$$c(\mathcal{K}, \mathbf{f}) := \inf_{(\mathbf{q}, \mathbf{h})} \frac{N(\mathbf{q}, \mathbf{h})}{\|(\mathbf{q}, \mathbf{h})\|^2} = \lim_{n \to +\infty} \frac{N(\mathbf{q}_n, \mathbf{h}_n)}{\|(\mathbf{q}_n, \mathbf{h}_n)\|^2}.$$

If $c(\mathcal{K}, \mathbf{f}) = 0$, and since $c_1(\mathcal{K}, \mathbf{f}) > 0$, we have that $\mathbf{u}_n$ tends to 0. But using the fact that $\lambda_{\min}(G(\mathbf{f} + \mathbf{u}))$ is a continuous function of $\mathbf{u}$ (in the finite-dimensional space $\mathcal{F}^G$), we obtain the contradiction

$$0 = c(\mathcal{K}, \mathbf{f}) \geq \left( \lambda_{\min}(G(\mathbf{f}))(\inf_{\mathbf{q} \in \mathcal{K}} q_g^2) \wedge c_1(\mathcal{K}, \mathbf{f}) \right) > 0,$$

so that we may conclude that $c(\mathcal{K}, \mathbf{f}) > 0$.

### 14.4.4 Hidden Markov models

We shall now consider stationary finite state space nonparametric HMMs. In this section observations $(X_t)_{t \geq 1}$ are independent conditionally on the latent variables $(Z_t)_{t \geq 1}$, and the conditional distribution of $X_t$ depends only on $Z_t$. The latent variables $(Z_t)_{t \geq 1}$ form a stationary Markov chain on the finite set $\{1, \ldots, G\}$. We shall denote by $\xi$ the transition matrix of the chain, and assume that it is irreducible and aperiodic. Let $H_g$ denote the distribution of $X_t$ conditional on $Z_t = g$, $g = 1, \ldots, G$. As discussed in Chapter 13, finite state space HMMs are widely used as extensions of finite mixture models to model dependent variables coming from different populations. If $(\eta_g)_{g=1,\ldots,G}$ denotes the stationary probability mass function of $\xi$, then the marginal distribution of each variable $X_t$ is the finite mixture

$$\sum_{g=1}^{G} \eta_g H_g.$$

Now let us exhibit the structural link between HMMs and multivariate mixtures given by (14.9). Suppose $X$ to be the vector of three consecutive observations $X_{t-1}, X_t, X_{t+1}$. One may write the probability distribution of $X$ as

$$\sum_{g=1}^{G} \left( \sum_{g_1=1}^{G} \eta_{g_1} \xi_{g_1,g} H_{g_1} \right) \otimes H_g \otimes \left( \sum_{g_3} \xi_{g,g_3} H_{g_3} \right),$$

which, since all weights $\eta_g$ are positive, is the same as

$$\sum_{g=1}^{G} \eta_g \left( \sum_{g_1=1}^{G} \frac{\eta_{g_1} \xi_{g_1,g}}{\eta_g} H_{g_1} \right) \otimes H_g \otimes \left( \sum_{g_3} \xi_{g,g_3} H_{g_3} \right),$$

and may thus be seen as coming from the fact that, when $(Z_t)_{t \geq 1}$ is a Markov chain, the past $Z_{t-1}$ and the future $Z_{t+1}$ are independent conditional on the present $Z_t$. Thus the distribution of $X$ is a three-dimensional mixture given by (14.9) with

$$F_{g,1} = \sum_{g_1=1}^{G} \frac{\eta_{g_1} \xi_{g_1,g}}{\eta_g} H_{g_1}, \quad F_{g,2} = H_g, \quad F_{g,3} = \sum_{g_3} \xi_{g,g_3} H_{g_3},$$

and one may apply the identifiability results of Section 14.4.1. This is what is done in Hsu et al. (2012) for parametric HMMs where the observations can take finitely many values, and in Gassiat et al. (2016) for nonparametric HMMs.

Moreover, if $\xi$ is of full rank and the probability distributions $H_1, \ldots, H_G$ are linearly independent, then $F_{1,1}, \ldots, F_{G,1}$ are linearly independent, $F_{1,3}, \ldots, F_{G,3}$ are linearly independent, and the finite state space nonparametric HMM model is identifiable. A more general identifiability result is proven in Alexandrovich et al. (2016) by using Kruskal's algebraic result and the methods of Allman et al. (2009). They obtain that for a large enough number of consecutive variables (more than three), depending on $G$, one may identify the HMM as soon as $\xi$ is of full rank and the distributions $H_1, \ldots, H_G$ are distinct. However, the proof is not constructive, in contrast to proofs using spectral methods.

Gassiat et al. (2016) propose several estimation methods and present real data results to support the conclusion that clustering using nonparametric HMMs may lead to better results than by using conventional parametric HMM algorithms. Spectral algorithms are proposed in Hsu et al. (2012) and their application in a nonparametric setting is investigated in De Castro et al. (2017) where theoretical results on the rates of convergence are given. Vernet (2015b) gives assumptions under which Bayesian methods lead to consistent posterior distributions, amd the rates of convergence are studied in Vernet (2015a).

De Castro et al. (2016) propose model selection, based on penalized least squares estimators for the emission distributions, which is statistically optimal and practically tractable. They prove a non-asymptotic oracle inequality for the nonparametric estimator of the emission distributions $H_g$, $g = 1, \ldots, G$. This requires an inequality similar to Theorem 14.1 which is proved to hold under a very weak assumption. A consequence is that this estimator is rate minimax adaptive up to a logarithmic term. The results hold under the assumption that the transition matrix $\xi$ is of full rank and in settings where the emission distributions have square integrable densities that are linearly independent. Simulations are given that show the improvement obtained when applying the least squares minimization consecutively to the spectral estimation. Lehéricy (2016) uses simulations ato study the sensitivity of the method to the linear independence assumption. He shows that when the smallest eigenvalue of the Gram matrix of the scalar products of the densities is very small, more observations are needed for good performance.

## 14.5 Related Questions

### 14.5.1 Clustering

Identification of the parameters of a mixture may be the first step for model-based clustering; see Section 8.1 above. De Castro et al. (2017) consider the filtering and smoothing recursions in nonparametric HMMs when the parameters of the model are unknown and replaced by estimators. They provide an explicit and time-uniform control of the filtering and smoothing errors in total variation norm as a function of the parameter estimation errors, so that performance on estimation may be transferred to performance in clustering via *a posteriori* probabilities.

### 14.5.2 Order estimation

In a parametric context, the question of order estimation has been discussed in Chapter 7 above. General principles for order identification as described in Gassiat (2018, Section 4.1) remain valid, but are far more difficult to apply. Kasahara & Shimotsu (2014) develop a procedure to estimate consistently a lower bound on the number of components in a multivariate finite mixture with conditionally independent coordinates. Lehéricy (2016) provides two different methods to estimate consistently the number of hidden states in a nonparametric HMM, one using a thresholding method on the singular values of the estimated distribution of two consecutive variables, the other using model selection techniques such as developed in De Castro et al. (2016). The theoretical results are completed by a very interesting simulation study to compare the methods.

### 14.5.3 Semi-parametric estimation

As may be understood and proved, for instance, using spectral methods, the weights $\eta_g$, $g = 1, \ldots, G$, of the mixture may be estimated at a parametric $\sqrt{n}$ rate. Computing the efficient Fisher information throws light on the loss occurring due to the fact that the nonparametric emission distributions are unknown. In the case of finite mixtures of multidimensional distributions which are tensor products of at least three one-dimensional distributions, using step functions to approximate the densities, one obtains parametric models in which the weights may be asymptotically efficiently estimated. Of course, the finer the approximation, the better the asymptotic efficient variance of the estimator. However, choosing the right degree of approximation with a finite number of observations is a non-trivial problem. Such questions are investigated in Gassiat et al. (2018).

### 14.5.4 Regressions with random (observed or non-observed) design

One may consider situations where the model is a mixture of regression models with unknown regression functions. The regressor variable may be random or not random, observed or not observed. Finite mixtures of regressions with observed design are discussed in Kasahara & Shimotsu (2009), Hunter & Young (2012), Hohmann & Holzmann (2013b) and Vandekerkhove (2013); see also Chapter 12 above.

When the regressor variable is random and observed with noise, one can relate the model to the so-called errors-in-variables model. But when a regression model with random design is considered and the regressor variable is not observed, then one is faced with a model which is a mixture of regressions. When the design follows a continuous distribution, the mixture is no longer finite, and difficult identifiability problems occur. A recent situation where identifiability can be solved may be found in Dumont & Le Corff (2017).

## 14.6   Concluding Remarks

In this chapter we have presented several mixture models where identifiability is verified with nonparametric modelling of the population distribution. In such cases, one may use nonparametric strategies such as model selection or nonparametric Bayesian methods, with provable guarantees. There is still a lot to investigate on both the applied and the theoretical side. It is, for instance, fascinating that mixture models for which identifiability was obviously not true become identifiable when considering that the observations are dependent variables.

# Bibliography

ALEXANDROVICH, G., HOLZMANN, H. & LEISTER, A. (2016). Nonparametric identification and maximum likelihood estimation for hidden Markov models. *Biometrika* **103**, 423–434.

ALLMAN, E. S., MATIAS, C. & RHODES, J. A. (2009). Identifiability of parameters in latent structure models with many observed variables. *Annals of Statistics* **37**, 3099–3132.

ANANDKUMAR, A., HSU, D. & KAKADE, S. M. (2012). A method of moments for mixture models and hidden Markov models. In *Proceedings of Machine Learning Research*, S. Mannor, N. Srebro & R. C. Williamson, eds., vol. 23. PMLR.

BAUDRY, J.-P., MAUGIS, C. & MICHEL, B. (2012). Slope heuristics: Overview and implementation. *Statistics and Computing* **22**, 455–470.

BENAGLIA, T., CHAUVEAU, D. & HUNTER, D. R. (2009). An EM-like algorithm for semi- and nonparametric estimation in multivariate mixtures. *Journal of Computational and Graphical Statistics* **18**, 505–526.

BONHOMME, S., JOCHMANS, K. & ROBIN, J.-M. (2016a). Estimating multivariate latent-structure models. *Annals of Statistics* **44**, 540–563.

BONHOMME, S., JOCHMANS, K. & ROBIN, J.-M. (2016b). Nonparametric estimation of finite mixtures from repeated measurements. *Journal of the Royal Statistical Society, Series B* **78**, 211–229.

BORDES, L., DELMAS, C. & VANDEKERKHOVE, P. (2006a). Semiparametric estimation of a two-component mixture model where one component is known. *Scandinavian Journal of Statistics* **33**, 733–752.

BORDES, L., MOTTELET, S. & VANDEKERKHOVE, P. (2006b). Semiparametric estimation of a two-component mixture model. *Annals of Statistics* **34**, 1204–1232.

BORDES, L. & VANDEKERKHOVE, P. (2010). Semiparametric two-component mixture model with a known component: An asymptotically normal estimator. *Mathematical Methods of Statistics* **19**, 22–41.

BUTUCEA, C. & VANDEKERKHOVE, P. (2014). Semiparametric mixtures of symmetric distributions. *Scandinavian Journal of Statistics* **41**, 227–239.

DE CASTRO, Y., GASSIAT, E. & LACOUR, C. (2016). Minimax adaptive estimation of nonparametric hidden Markov models. *Journal of Machine Learning Research* **17**, Paper No. 111.

DE CASTRO, Y., GASSIAT, E. & LE CORFF, S. (2017). Consistent estimation of the filtering and marginal smoothing distributions in nonparametric hidden Markov models. *IEEE Transactions in Information Theory* **63**, 4758–4777.

DUMONT, T. & LE CORFF, S. (2017). Nonparametric regression on hidden $\Phi$-mixing variables: Identifiability and consistency of a pseudo-likelihood based estimation procedure. *Bernoulli* **23**, 990–1021.

GASSIAT, E. (2018). Springer Monographs in Mathematics. Cham: Springer International Publishing.

GASSIAT, E., CLEYNEN, A. & ROBIN, S. (2016). Inference in finite state space non parametric hidden Markov models and applications. *Statistics and Computing* **26**, 61–71.

GASSIAT, E. & ROUSSEAU, J. (2016). Non parametric finite translation hidden Markov models and extensions. *Bernoulli* **22**, 193–212.

GASSIAT, E., ROUSSEAU, J. & VERNET, E. (2018). Efficient semiparametric estimation and model selection for multidimensional mixtures. *Electronic Journal of Statistics* **12**, 703–740.

HALL, P., NEEMAN, A., PAKYARI, R. & ELMORE, R. (2005). Nonparametric inference in multivariate mixtures. *Biometrika* **92**, 667–678.

HALL, P. & ZHOU, X.-H. (2003). Nonparametric estimation of component distributions in a multivariate mixture. *Annals of Statistics* **31**, 201–224.

HOHMANN, D. & HOLZMANN, H. (2013a). Semiparametric location mixtures with distinct components. *Statistics* **47**, 348–362.

HOHMANN, D. & HOLZMANN, H. (2013b). Two-component mixtures with independent coordinates as conditional mixtures: Nonparametric identification and estimation. *Electronic Journal of Statistics* **7**, 859–880.

HSU, D., KAKADE, S. M. & ZHANG, T. (2012). A spectral algorithm for learning hidden Markov models. *Journal of Computational System Science* **78**, 1460–1480.

HUNTER, D. R., WANG, S. & HETTMANSPERGER, T. P. (2007). Inference for mixtures of symmetric distributions. *Annals of Statistics* **35**, 224–251.

HUNTER, D. R. & YOUNG, D. S. (2012). Semiparametric mixtures of regressions. *Journal of Nonparametric Statistics* **24**, 19–38.

KASAHARA, H. & SHIMOTSU, K. (2009). Nonparametric identification of finite mixture models of dynamic discrete choices. *Econometrica* **77**, 135–175.

KASAHARA, H. & SHIMOTSU, K. (2014). Nonparametric identification and estimation of the number of components in multivariate mixtures. *Journal of the Royal Statistical Society, Series B* **76**, 97–111.

KRUSKAL, J. B. (1977). Three-way arrays: Rank and uniqueness of trilinear decompositions, with application to arithmetic complexity and statistics. *Linear Algebra and Applications* **18**, 95–138.

LEHÉRICY, L. (2016). Consistent order estimation for nonparametric hidden Markov models. Preprint, arXiv:1606.00622.

MASSART, P. (2007). *Concentration Inequalities and Model Selection*, vol. 1896 of *Lecture Notes in Mathematics*. Berlin: Springer-Verlag.

NGUYEN, V. H. & MATIAS, C. (2014a). Nonparametric estimation of the density of the alternative hypothesis in a multiple testing setup. Application to local false discovery rate estimation. *ESAIM: Probability and Statistics* **18**, 584–612.

NGUYEN, V. H. & MATIAS, C. (2014b). On efficient estimators of the proportion of true null hypotheses in a multiple testing setup. *Scandinavian Journal of Statistics* **41**, 1167–1194.

SONG, L., ANANDKUMAR, A., DAI, B. & XIE, B. (2014). Nonparametric estimation of multiview latent variable models. In *Proceedings of the 31th International Conference on Machine Learning, ICML 2014, Beijing, China, 21-26 June 2014*.

VANDEKERKHOVE, P. (2013). Estimation of a semiparametric mixture of regressions model. *Journal of Nonparametric Statistics* **25**, 181–208.

VERNET, E. (2015a). Non parametric hidden Markov models with finite state space: Posterior concentration rates. Preprint, arXiv:1511.08624.

VERNET, E. (2015b). Posterior consistency for nonparametric hidden Markov models with finite state space. *Electronic Journal of Statistics* **9**, 717–752.

XIANG, S., YAO, W. & WU, J. (2014). Minimum profile Hellinger distance estimation for a semiparametric mixture model. *Canadian Journal of Statistics* **42**, 246–267.

# Part III

# Selected Applications

# 15

## Applications in Industry

**Kerrie Mengersen, Earl Duncan, Julyan Arbel, Clair Alston-Knox, Nicole White**

*ACEMS Queensland University of Technology, Australia; ACEMS Queensland University of Technology, Australia; Laboratoire Jean Kuntzmann, Université Grenoble Alpes and INRIA Grenoble—Rhône-Alpes, France; Griffith University, Australia; Institute for Health and Biomedical Innovation, Queensland University of Technology, Australia*

## CONTENTS

## 15.1 Introduction

There are various definitions of the term *industry*, ranging from a traditional focus on manufacturing enterprises, to a slightly more relaxed inclusion of general trade, to a very broad umbrella of dedicated work. In this chapter we take the middle ground and include activities that have a commercial focus. This definition embraces an alphabet of fields, spanning agriculture, business and commerce, defence, engineering, fisheries, gas and oil, health, and so on.

A very wide range of commonly encountered problems in these industries are amenable to statistical mixture modelling and analysis. These include process monitoring or quality control, efficient resource allocation, risk assessment, prediction, and so on. Commonly articulated reasons for adopting a mixture approach include the ability to describe non-standard outcomes and processes, the potential to characterize each of a set of multiple outcomes or processes via the mixture components, the concomitant improvement in interpretability of the results, and the opportunity to make probabilistic inferences such as component membership and overall prediction.

In this chapter, we illustrate the wide diversity of applications of mixture models to problems in industry, and the potential advantages of these approaches, through a series of case studies. The first of these focuses on the iconic and pervasive need for process monitoring, and reviews a range of mixture approaches that have been proposed to tackle complex multimodal and dynamic or online processes. The second study reports on mixture approaches to resource allocation, applied here in a spatial health context but applicable more generally. The next study provides a more detailed description of a multivariate Gaussian mixture approach to a biosecurity risk assessment problem, using big data in the form of satellite imagery. This is followed by a final study that again provides a detailed description of a mixture model, this time using a nonparametric formulation, for assessing an industrial impact, notably the influence of a toxic spill on soil biodiversity.

## 15.2   Mixtures for Monitoring

Process monitoring is an iconic problem in many industrial settings, also known as statistical process control, quality control or health monitoring (Ge et al., 2013). Historically, statistical tools developed for monitoring these processes were based on the assumption that the associated data were generated from a single distribution, representative of the population under study. It has since been acknowledged that many of these processes may be sufficiently heterogeneous to warrant the use of a mixture distribution. From their general formulation in Chapter 1, a variety of distributions can be considered as components in a mixture model, which represent interpretable features of the population or allow for flexible modelling of non-standard data. An example of such an approach, and the currency of interest in this problem, is given by Sindhu et al. (2015), who describe Bayesian estimation of Gumbel mixture models and the development of associated cumulative quality control charts.

Fault detection is another example of a common industrial problem where mixture models have been successfully applied for purposes of monitoring. Faults can be due to a range of factors, such as ageing of equipment, drifting of sensors or reactions, or modifications to the underlying process (Xie & Shi, 2012). Since multiple faults may be considered simultaneously, each with different operating conditions, it is natural to consider a mixture model of some form to describe the process. A $d$-dimensional Gaussian mixture model is commonly assumed,

$$y_i \sim \sum_{g=1}^{G} \eta_g \mathcal{N}_d \left( \mu_g, \Sigma_g \right), \tag{15.1}$$

where $y_i$ denotes an observed process output or a transformed output, for example, principal components. The resulting model is then combined with other statistics to construct an overall index for fault detection. An example of work in this area is provided by Yu & Quin (2008), who proposed a Gaussian finite mixture model for multimode chemical process monitoring. In this application, each mixture component described an operating mode, with the resulting model estimated by the EM algorithm. Following model estimation, the authors proposed a Bayesian approach for subsequent inference on fault detection, by first calculating posterior probabilities of component membership for each observation. These probabilities were then combined with local, component-specific Mahalanobis distances to construct an overall index for fault detection. The authors argue, through examples, that the mixture approach is superior to the more traditional multivariate process monitoring methods such as principal components and partial least squares, both of which inadequately assume the process follows a unimodal Gaussian distribution. In related work, Wen et al. (2015) propose the mixture canonical variate analysis model, in which Gaussian mixture components are again used to describe different operating modes and singular value decomposition of the covariance matrix is employed for each cluster. Monitoring indices are then formed based on local statistics derived from the canonical variates for each cluster. These authors and others have proposed a range of variations for the use of mixture models for complex process monitoring, such as multiphase batch processes (Yu & Quin, 2009; Chen & Zhang, 2010), nonlinear multimode non-Gaussian processes (Yu, 2012a) and other dynamic or online processes (Yu, 2012b; Xie & Shi, 2012; Lin et al., 2013).

Mixtures can also be applied to a wide range of other monitoring problems. In ecology, for example, Neubauer et al. (2013) adopt a Dirichlet process mixture of multivariate Gaussians for species distribution estimation and source identification based on observed geochemical signatures. Ecological monitoring of abundance can also be framed as a binomial mixture model (Wu et al., 2015). Here, the authors cast the mixture in a Bayesian hierarchical framework in order to monitor spatially referenced replicated count data with characteristic unbalanced sampling and overdispersion. The use of mixture models to describe spatial variation in geographic monitoring has been explored by a large number of authors; one example of this is detailed in a later section of this chapter.

Estimation of species abundance is another area where mixture models offer an appealing solution. In particular, mixtures have seen wide application in dealing with zero inflation, often encountered in abundance data, see for instance Korner-Nievergelt et al. (2015) and McCarthy (2007, p. 264). The zero-inflated Poisson model (Lambert, 1992) is a special case of a mixture model, with components comprising a Poisson distribution and a Dirac mass at zero; see also Chapter 9, Section 9.2.3, above. Membership in each component is modelled by a Bernoulli distribution with unknown probability $\eta(x_i)$, which in turn characterizes the underlying zero generating process. These models take the general form

$$p(y_i) = \begin{cases} \eta(x_i)\,\mathbb{I}_0(y_i) + (1 - \eta(x_i))\mathcal{P}(\mu(x_i)), & \text{if } y_i = 0, \\ (1 - \eta(x_i))\mathcal{P}(\mu(x_i)), & \text{if } y_i > 0, \end{cases}$$

where covariate effects, $x_i$, are often included for predicting both the probability of a non-zero count, $\eta(x_i)$, and the Poisson rate parameter, $\mu(x_i)$. Rhodes et al. (2008) adopt a similar modelling strategy for monitoring the long-term impact of chemical pollution on aquatic environments. Here, the authors formulate a zero-inflated Poisson mixed effects model to estimate the effect of copper exposure on the reproductive output of a copepod (a small crustacean found in nearly every marine and freshwater body) over three generations. In addition to covariate effects, random effects are included in each component of the proposed model, to account for correlation among observations taken from the same experimental unit (a single female copepod). In a different study, Lyashevska et al. (2016) developed a zero-inflated Poisson model with spatially correlated random effects, with an application to abundance estimation of *Macoma balthica*, an invertebrate found in the Wadden Sea. Similar to applications already presented, the motivation for the use of mixtures in these studies included the ability to describe the non-standard distributions of the measures considered and to make more subtle inferences. A nonparametric approach to monitoring similar industrial impacts is described in more detail in a later section of this chapter.

A final example of an industry in which monitoring is a fundamental tool is target tracking and recognition. Bayesian and non-Bayesian mixture models for clustering, classification and signal separation in this context have been used for more than a decade (see, for example, Sadjadi, 2001; Vigneron et al., 2010) and have become more sophisticated with adaptation, real-time capabilities and ability to better distinguish foreground, background and trajectories (e.g. KaewTraKulPong & Bowden, 2002). Mixture models for human–computer interactions are also popular; for example, Pietquin (2004) discusses early ideas for unsupervised learning of dialogue strategies, which have been refined and expanded markedly in the last ten years.

## 15.3   Health Resource Usage

The health industry is a very large industry, incorporating activities associated with drug development, diagnosis, surgery, rehabilitation, and other activities relating to healthcare. Good management of financial, clinical, administrative and other resources is required to provide high-quality healthcare. Therefore, any models which may assist managers of health-care resources are a valuable tool. Two case studies are presented below which demonstrate the use of mixture models in the health industry and how they can provide management with insights into the practices regarding health resource usage.

### 15.3.1   Assessing the effectiveness of a measles vaccination

The goal of an immunization program is to target certain groups of individuals with a vaccine so as to achieve herd immunity in the population. The success of an immunization program can be assessed by studying the number of individuals who are deemed immune, as determined by high levels of antibodies, before and after administration of the vaccine. In the case of measles, individuals have traditionally been classified as immune if a serum sample indicates that their concentration of measles antibodies is greater than some threshold. However, this approach tends to have poor test sensitivity, and does not account for different degrees of immunity.

Del Fava et al. (2012) illustrate how a mixture model can be used to estimate age-specific population prevalences and more accurately classify individuals' state of immunity. The case study in Del Fava et al. (2012) involves a national measles immunization program in Tuscany, Italy, targeted at children less than 15 years old, conducted in 2004–2005. Serum samples were collected pre- and post-immunization (2003 and 2005–2006, respectively) from individuals aged up to 49 years, and the concentration of antibodies was recorded. To account for the heterogeneity in the levels of antibody concentration, each sample of antibody concentration is modelled by a (univariate) Gaussian mixture,

$$y_i \sim \sum_{g=1}^{G} \eta_g \mathcal{N}\left(\mu_g, \sigma_g^2\right), \quad i = 1, \ldots, n,$$

where $\mu_g$ and $\sigma_g^2$ are the component-specific mean and variance, and are given non-informative priors:

$$
\begin{aligned}
\mu_1 &\sim \mathcal{U}\left(y_{\min}, y_{\max}\right) \mathbb{I}_{(-\infty, \mu_2)}, \\
\mu_g &\sim \mathcal{U}\left(y_{\min}, y_{\max}\right) \mathbb{I}_{(\mu_{g-1}, \mu_{g+1})}, \quad g = 2, \ldots, G-1, \\
\mu_G &\sim \mathcal{U}\left(y_{\min}, y_{\max}\right) \mathbb{I}_{(\mu_{G-1}, \infty)}, \\
\sigma_g^2 &\sim \mathcal{G}^{-1}\left(0.01, 0.01\right).
\end{aligned}
$$

The latent indicator is modelled by the multinomial distribution

$$z_i \sim \mathcal{M}\left(1, \eta(a_i)\right),$$

where the age-specific mixture probabilities $\eta(a_i) = (\eta_{1,a_i}, \ldots \eta_{G,a_i})$ are given the non-informative, conjugate Dirichlet prior,

$$(\eta_{1,a_i}, \ldots \eta_{G,a_i}) \sim \mathcal{D}(1, \ldots, 1), \quad i = 1, \ldots, n,$$

and $a_i$ is the age of the $i$th individual. The number of components is unknown, but is unlikely to be very large in this context. However, contrary to the conventional dichotomous approach to classification of immunity, the number of components is most likely larger than two, even though individuals will ultimately be classified as either susceptible or immune. In the Tuscany measles immunization program case study, separate models were fitted to the pre- and post-immunization sample data, and three and four subpopulations were identified by the models, respectively. The "optimal" number of mixture components was determined by refitting the models with different numbers of components, and selecting between the two models by goodness of fit.

Two simplifications to this model are discussed by Del Fava et al. (2012). Firstly, the component-specific variance $\sigma_g^2$ may be considered homogeneous (i.e. $\sigma_g^2 = \sigma^2$ for all $g$). Secondly, the mixture probabilities $\eta = (\eta_1, \ldots \eta_G)$ can be made age-independent, following

$$(\eta_1, \ldots \eta_G) \sim \mathcal{D}(1, \ldots, 1).$$

The results in Del Fava et al. (2012) are based on models using homogeneous variance and age-dependent mixture probabilities.

By modelling the antibody concentration using a mixture model, different levels of immunity may be observed (see Figure 15.1). For example, in the measles case study, four subpopulations were identified by the model fitted to the post-immunization data. The component with the smallest mean represented individuals who were the most susceptible and were most likely unvaccinated, while the components with the two largest means represented individuals who exhibited high degrees of immunity, where the high level of antibody concentration may be the result of the vaccination or natural infection. The remaining component represented a distinctly separate group of individuals who exhibited some degree of immunity, but with comparatively lower levels of antibody concentration.

In the case of the age-dependent model, useful inferences can be drawn from a simple time series plot of the proportion of individuals belonging to each mixture component (i.e. the prevalence) against the age of the individuals. Such a plot may help analyse the effects of initial vaccine uptake and immunization boosts, and identify age groups which are the most susceptible and thus perhaps should be targeted in future immunization programs.

## 15.3.2 Spatio-temporal disease mapping: identifying unstable trends in congenital malformations

In most disease mapping studies, the data are collected over a long period of time, but the temporal effect on the relative risk is often ignored. Abellan et al. (2008) demonstrate how a Bayesian spatio-temporal mixture model can be used to simultaneously estimate the relative risk for each area at each time point, and identify areas which exhibit a temporal pattern with "substantial and distinctive variability". Being able to detect such variability, or instability, may be of interest to medical researchers, health practitioners, and government bodies because it helps identify potential, sudden fluctuations in reported cases, either due to changes in risk factors or healthcare practices, necessitating further investigation.

The case study in Abellan et al. (2008) involves data on the annual number of non-chromosomal congenital malformations in England, observed across 970 artificially constructed square areas over 16 years. Due to the low rate of malformations, a binomial

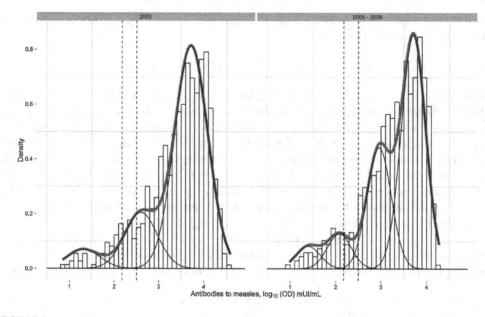

**FIGURE 15.1**
Histogram of the antibody concentration for the pre- and post-immunization samples, over-laid by the estimated mixture component densities. The dashed lines represent traditional cut-off values for classifying individuals as either susceptible or immune. Reproduced with permission from Del Fava et al. (2012).

random variable is assumed to model the count data,

$$y_{it} \sim \mathcal{B}\left(n_{it}, \pi_{it}\right),$$

where $n_{it}$ is the total number of births (including stillbirths) and $\pi_{it}$ is the relative risk of congenital malformations in area $i = 1, \ldots, 970$, and year $t = 1, \ldots, 16$. The relative risk is decomposed into four parameters associated by the logistic link function,

$$\log(\pi_{it}) = \mu + \beta_i + \beta_t + \beta_{it}, \tag{15.2}$$

where $\mu$ is the estimated overall risk, $\beta_i$ and $\beta_t$ are the spatial and temporal random effects respectively, and $\beta_{it}$ is a space-time interaction parameter. Intrinsic conditional autoregress-ive models are used in the specification of the priors for $\beta_i$ and $\beta_t$ to account for spatial and temporal heterogeneity. To account for heterogeneity of the space-time interaction, a two-component Gaussian mixture model is used,

$$\beta_{it} \sim \eta \mathcal{N}\left(0, \sigma_1^2\right) + (1 - \eta)\mathcal{N}\left(0, \sigma_2^2\right),$$

where the latent indicator

$$z_{it} \sim \mathcal{M}\left(1, \eta, 1 - \eta\right)$$

determines whether the variance $\sigma_{z_{it}}^2$ is small or large:

$$\sigma_1^2 \sim \mathcal{N}\left(0, 0.01\right) \mathbb{I}_{(0,+\infty)},$$
$$\sigma_2^2 \sim \mathcal{N}\left(0, 100\right) \mathbb{I}_{(0,+\infty)}.$$

Including additional covariate information in the regression equation (15.2) should be straightforward. For the full model specification, we refer the reader to Abellan et al. (2008).

By analysing the posterior probabilities that $\beta_{it}$ has a large variance for a given area, one can characterize the stability of the spatial patterns of the relative risk in area $i$. Given the spatial nature of the data in such studies, a choropleth map may be useful for identifying whether the areas with unstable temporal trends are spatially correlated. In the congenital malformations case study, 125 (13%) of the 970 areas were identified as having unstable temporal patterns. A map of England discretized into the 970 square areas supcrimposed with the 125 unstable areas indicated no signs of spatial correlation.

Additional analysis of the data corresponding to the mixture component representing unstable temporal trends may provide more insight. For example, in the malformations case study, the 125 unstable areas were further classified into one of five clusters according to the similarity of their estimated space-time interaction effects. One cluster, consisting of only two areas, indicated a large spike in the risk of malformations in a particular year. Further investigations revealed that the number of kidney malformations reported in that year had increased as a result of changes to recording practices.

## 15.4 Pest Surveillance

Exotic and native pests in plague proportions can cause major social and economic problems in a range of industries such as agriculture and eco-tourism. The case study presented here uses Bayesian mixture models to assist managers in pre-season assignment of areas of habitat that are potentially valuable to search for fire ant infestation.

South American fire ants (*Solenopsis invicta*) were first discovered in the Port of Brisbane, Australia, in 2001. The seriousness of this threat in terms of social, environmental and economic consequences was recognized across all levels of government and affected industry, and, as a result, the National Red Imported Fire Ant Eradication Program was established in late 2001.

Since this time, data have been collected on the location of each colony that has been found, with a total of 5027 locations spanning the years 2001–2010 being used in this analysis. However, as eradication was seen as a viable outcome, very few details of the colonies were documented or collated, resulting in the only available data being the year of discovery and the longitude and latitude of the infestation. Given this limitation, it was decided to use Landsat images taken in the winter months of each year of discovery, along with 18 potential habitat indices which may fit with the theories and observations formed by long-term biological control officers within the program. The use of Landsat imagery in this analysis attempts to enhance the prospects of surveillance at a relatively low cost. As discussed in Spring & Cacho (2015), the appropriate use of remote surveillance, in combination with ground surveillance, can increase the likelihood of either containment or eradication, which is the purpose of this analysis.

### 15.4.1 Data and models

The aim of the analysis is to cluster areas of the Brisbane region that may be useful in identifying suitable habitat for the invasive species. An appealing starting point is to use a multivariate normal mixture model as a soft clustering technique. For computational reasons, the model was first developed without reference to spatial dependence, with the spatial component considered in a secondary analysis, following Alston et al. (2009). The

---

**Algorithm 15.1** Gibbs sampler for estimation of parameters in a normal mixture model

1 Update $z_i \sim \mathcal{M}(1, \tau_{i1}, \tau_{i2}, \ldots, \tau_{iG})$, where

$$\tau_{ig} = \frac{\eta_g \phi(y_i | \mu_g, \Sigma_g)}{\sum_{j=1}^{G} \eta_j \phi(y_i | \mu_j, \Sigma_j)}.$$

2 Update $\eta | \mathbf{y}, \mathbf{z} \sim \mathcal{D}(e_0 + n_1, e_0 + n_2, \ldots, e_0 + n_G)$, where $n_g = \sum_{i=1}^{n} \mathbb{I}(z_i = g)$.

3 Update $\Sigma_g^{-1} | \mathbf{y}, \mathbf{z} \sim \mathcal{W}(c_g, C_g)$, where $c_g = c_0 + n_g$ and

$$C_g = C_0 + W_g + \frac{n_g N_0}{N_0 + n_g} (\bar{y}_g - b_0)(\bar{y}_g - b_0)^{\top},$$

where

$$\bar{y}_g = \frac{1}{n_g} \sum_{i=1}^{n} \mathbb{I}(z_i = g) y_i, \quad W_g = \sum_{i=1}^{n} \mathbb{I}(z_i = g)(y_i - \bar{y}_g)(y_i - \bar{y}_g)^{\top}.$$

4 Update $\mu_g | \Sigma_g, \mathbf{y}, \mathbf{z} \sim \mathcal{N}(b_g, B_g)$, where $B_g = \Sigma_g / (N_0 + n_g)$ and

$$b_g = b_0 \cdot \frac{N_0}{N_0 + n_g} + \bar{y}_g \cdot \frac{n_g}{N_0 + n_g}.$$

---

non-spatial model is usually represented as in equation (15.1) which can also be represented as

$$Y_i | z_i \sim \sum_{g=1}^{G} \mathbb{I}(z_i = g) \cdot \mathcal{N}_d(y_i | \mu_g, \Sigma_g),$$

where $\mathcal{N}_d$ denotes a multivariate normal density with mean $\mu_g$ and variance–covariance matrix $\Sigma_g$. As described in Chapter 1, a vector $\mathbf{z} = \{z_1, z_2, \ldots, z_n\}$ can be associated with this sample. It consists of a set of unobserved indicator vectors denoting component membership, which are estimated as another parameter in the model using Gibbs sampling. As in earlier chapters, $\eta_g$ represents the weight (proportion) of each component in the model and determines the prior probability $P(z_i = g | \eta_g) = \eta_g$. Denote $\eta = (\eta_1, \ldots, \eta_G)$.

We use standard conjugate priors in this mixture model. Specifically,

$$\Sigma_g^{-1} \sim \mathcal{W}(c_0, C_0),$$
$$\mu_g | \Sigma_g \sim \mathcal{N}\left(b_0, \frac{1}{N_0} \Sigma_g\right),$$
$$\eta \sim \mathcal{D}_G(e_0).$$

Frühwirth-Schnatter (2006, pp. 192–193) provides an extensive discussion on the choice of hyperparameters for these prior distributions. In this analysis, we have followed the recommendations of Robert (1996) for hyperparameter selection. The resulting algorithm for estimating the parameters of this mixture model is well known. It is a two-stage Gibbs sampler, outlined in Algorithm 15.1.

As the number of pixels in an image was quite large, in excess of 2 million, to assist our clients implement these models, a library was developed in PyMCMC (Strickland et al., 2012), which includes a Gibbs sampler to estimate the mixture model parameters, as per this algorithm. In this analysis, step 1 of the algorithm is the most laborious, and as such

---

**Algorithm 15.2** Label switching reordering scheme used in Gibbs sampler for estimation of parameters of the normal mixture model.

---

1 Perform mixture model updating for a specified burn-in period.

2 During the next 10,000 iterations, calculate the likelihood of the mixture at every 100th iteration, based on the current parameter estimates.

3 From these calculated likelihoods, choose the set of parameters that maximizes the likelihood. This parameter set then becomes the base ordering, from which any label switching is detected. Note that there is no mean or variance ordering imposed on this set.

4 For each iteration thereafter, test the updated parameters for the latent variable against this base ordering. The possibility of switched labels is then determined. This is computed by comparing the current allocations of $z$ to the allocations of $z$ during the chosen optimal iteration. A misclassification matrix, $C$, is constructed, and the cost of misclassifications determined using the Munkres algorithm (also called the Hungarian algorithm or the Kuhn–Munkres algorithm; see Kuhn, 1955).

---

was programmed with the option of parallel computation for the end user. Additionally, rather than store all the iterates, we wished to compute posterior estimates "on the run", saving on storage space. To enable this, we needed to deal with label switching within the mixture, and this was achieved using Algorithm 15.2, which is a single-chain variation on that presented by Cron & West (2011).

Technically, there is no limit on the number of dimensions (variables) we could fit in this mixture model. So potentially, we could include all 18 variables in the analysis. However, from a pragmatic viewpoint, there is little to be gained, and a large computational burden to incur, by fitting variables that have no influence on habitat suitability. For this reason, we decided to pre-determine which of the 18 variables were modelled via mixtures. Using expert opinion and CART analysis, it was decided to use Landsat band 3 (visible red), Landsat band 6 (mid infrared) and a soil brightness index to assess the probability that the area associated with each pixel is habitable terrain for inclusion in the upcoming surveillance season. The multivariate analysis allowed managers to create meaningful clusterings that reflect the sometimes complex combinations of conditions that form habitat suitability, rather than relying on single derived indices. Figure 15.2 illustrates the difference in these three variables between known infestation sites and non-infested areas.

### 15.4.2 Resulting clusters

To test the potential of the model, we estimated the parameters of a multivariate normal mixture model on a Landsat image from 2011 based on Landsat bands 3 and 6, and the soil brightness index. The component estimates for the component weights, means and covariance matrices are given in Table 15.1. As the density estimates in Figure 15.2 indicate that positive centred values of these pixels are dominant in regions of known *S. invicta* colonies, it can be seen that components 3–6 are of most interest for suitable habitat. Component 2 is of some interest, as each mean is positive but the distribution straddles the negative-valued regions, and we would consider areas of reasonable size which are allocated to this component. Component 1, which has negative-valued means, is considered not suitable for *S. invicta* habitat in terms of surveillance, due to the initial findings of the classification and regression tree (CART) analysis. Component 1 is associated with over 85% of the area in the Landsat image, so this is helpful in terms of surveillance planning in that the tracts of land associated with component 1 can be largely discounted for surveillance if budgetary restraints are an issue.

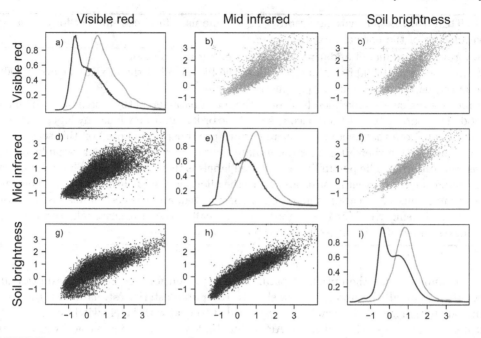

**FIGURE 15.2**
Density and scatter plots of the three variables of interest in habitat modelling. (a) Density of visible red in areas of previously detected fire ant colonies (light grey) and areas not currently infested (dark gray). (b) Scatter plot of visible red against mid infrared in areas of known infestation. (c) Scatter plot of visible red against soil brightness in areas of known infestation. (d) Scatter plot of visible red against mid infrared in non-infested areas. (e) Density of mid infrared in areas of previously detected fire ant colonies (light grey) and areas not currently infested (dark grey). (f) Scatter plot of mid infrared against soil brightness in areas of known infestation. (g) Scatter plot of visible red against soil brightness in non-infested areas. (h) Scatter plot of mid infrared against soil brightness in non-infested areas. (i) Density of soil brightness in areas of previously detected fire ant colonies (light grey) and areas not currently infested (dark grey).

A big advantage of viewing the region in terms of this mixture model, from a management point of view, is the ability to create posterior probability maps which will predict the likelihood of areas belonging to suitable habitat, and combining this with expert knowledge from the ground workers, previous finds and budgetary restraints to make effective decisions in terms of where to concentrate the surveillance effort in any year. Figure 15.3 illustrates the range of estimated posterior probabilities for each pixel belonging to components 2–6. These values are converted into a map within ArcGIS software and layered with other relevant input to form a surveillance strategy. The probability aids managers to make decisions about prioritizing areas.

Generally, we found that this method highlighted new areas of urban growth, which anecdotally were developments where experienced staff expected the possibility of fire ant infestations. Fire ant colonies that were missed using this technique tended to be roadside infestations caused by transportation via tyres, and, realistically, these are not habitats in which we would wish to perform surveillance, relying instead on public education to detect infestations within habitats best defined by component 1 of the mixture.

**TABLE 15.1**

Estimated component means, variances and weights from the mixture model for Landsat band 3 ($LS_3$), Landsat band 6 ($LS_6$) and soil brightness (SB). Associated credible intervals given in parentheses.

| Cluster | Variable | Mean | Covariance matrix | | | |
|---|---|---|---|---|---|---|
| 1 | $LS_3$ | −0.305 | 26.80 | | | |
| | | (−0.307,−0.303) | (26.6, 27.0) | | | |
| | $LS_6$ | −0.260 | −25.50 | 34.5 | | |
| | | (−0.262, −0.258) | (−25.7, −25.4) | (34.4, 34.7) | | |
| | SB | −0.193 | 7.57 | −14.7 | 9.02 | |
| | | (−0.195, −0.191) | (7.52, 7.62) | (−14.8, −14.6) | (8.97, 9.06) | |
| Weight | 0.8510 | | | | | |
| | (0.8490, 0.8520) | | | | | |
| 2 | $LS_3$ | 1.35 | 7.83 | | | |
| | | (1.34, 1.36) | (7.71, 7.95) | | | |
| | $LS_6$ | 1.37 | 0.92 | 5.99 | | |
| | | (1.36, 1.37) | (0.82, 1.01) | (5.91, 6.06) | | |
| | SB | 1.01 | −8.59 | −6.18 | 19.10 | |
| | | (1.00, 1.02) | (−8.75, −8.44) | (−6.30, −6.07) | (18.80, 19.30) | |
| Weight | 0.1030 | | | | | |
| | (0.1020, 0.1040) | | | | | |
| 3 | $LS_3$ | 1.85 | 9.5 | | | |
| | | (1.80, 1.90) | (9.20, 9.82) | | | |
| | $LS_6$ | 0.978 | 3.3 | 8.64 | | |
| | | (0.920, 1.04) | (3.05, 3.56) | (8.17, 9.08) | | |
| | SB | 0.597 | −11.3 | −11.5 | 21.7 | |
| | | (0.560, 0.638) | (−11.8, −10.8) | (−12.0, −11.1) | (20.9, 22.4) | |
| Weight | 0.0308 | | | | | |
| | (0.0302, 0.0314) | | | | | |
| 4 | $LS_3$ | 2.62 | 3.630 | | | |
| | | (2.51, 2.74) | (3.29, 3.99) | | | |
| | $LS_6$ | 2.92 | 0.95 | 0.96 | | |
| | | (2.85, 2.98) | (0.75, 1.15) | (0.81, 1.11) | | |
| | SB | 2.58 | −4.66 | −1.15 | 8.54 | |
| | | (2.51, 2.65) | (−5.13, −4.22) | (−1.40, −0.89) | (7.62, 9.50) | |
| Weight | 0.0009 | | | | | |
| | (0.0008, 0.0010) | | | | | |
| 5 | $LS_3$ | 3.92 | 12.8 | | | |
| | | (3.82, 4.03) | (12.0, 13.8) | | | |
| | $LS_6$ | 3.19 | 5.78 | 4.24 | | |
| | | (3.13, 3.25) | (5.24, 6.28) | (3.85, 4.64) | | |
| | SB | 2.46 | −20.5 | −10.6 | 34.7 | |
| | | (2.38, 2.55) | (−22.0, −19.1) | (−11.5, −9.6) | (32.1, 37.4) | |
| Weight | 0.0132 | | | | | |
| | (0.0119, 0.0144) | | | | | |
| 6 | $LS_3$ | 8.08 | 0.73 | | | |
| | | (7.97, 8.19) | (0.66, 0.80) | | | |
| | $LS_6$ | 4.83 | 0.41 | 0.95 | | |
| | | (4.74, 4.91) | (0.37, 0.45) | (0.88, 1.02) | | |
| | SB | 5.71 | −1.29 | −1.45 | 3.23 | |
| | | (5.55, 5.86) | (−1.40, −1.19) | (−1.56, −1.34) | (2.94, 3.50) | |
| Weight | 0.0012 | | | | | |
| | (0.0011, 0.0013) | | | | | |

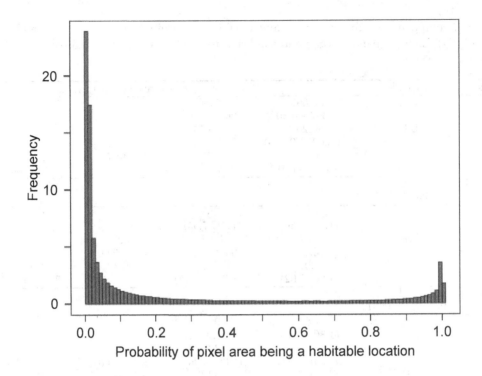

**FIGURE 15.3**
Histogram representing posterior probability of individual pixels in 2011 Landsat image belonging to habitat suitable for fire ant infestation surveillance.

## 15.5 Toxic Spills

Toxic spills are externalities of industrial activity that affect soils. Their impact on the biodiversity is one of many detrimental consequences of toxic spills on an ecosystem. Biodiversity is a measure of the variety of organisms present in an ecosystem. Since mixture models are probability models for representing the presence of subpopulations within an overall population, it is reasonable to use mixtures for modelling biodiversity. We will pursue this approach here, under a Bayesian nonparametric (BNP) framework; see Chapter 6 for a comprehensive review of BNP mixture models.

The data of the present case study consist of microbial communities, or groups of species, observed as counts at locations in the soil, or sites, along with a toxic contaminant measurement. The composition of species may differ among the sites, and the main inferential interest amounts to understanding the contaminant impact on species diversity.

In contrast to most of the mixture models presented in this handbook, the identity of mixture components is actually observed in the data. As we shall see, this feature leads to inferential and computational techniques contrasting with other approaches.

Note that the exposition is in terms of species due to the actual application to toxic spills. Nevertheless, the approach is not limited to species sampling problems. This application is based on Arbel et al. (2015, 2016).

## 15.5.1 Data and model

The data consist of a soil microbial data set acquired across a hydrocarbon contamination gradient at the location of a fuel spill at Australia's Casey Station in East Antarctica (110° 32′ E, 66° 17′ S), along a transect at 22 locations. Microbes are classified as operational taxonomic units, which we also generically refer to as species. Species measurements are paired with a contaminant called total petroleum hydrocarbon (TPH), suspected to impact diversity. We refer to Snape et al. (2015) for a complete account of the data set acquisition.

The state space is the set of species. We label them with positive integers, and order them in overall (over all sites) decreasing abundance. Probabilities of presence, say $\eta_g$ for species $g$, are a set of self-evident parameters in this kind of species sampling models. They are positive and sum to one. When the total number of species is fixed, say $G$, the presence probabilities are elements of the unit simplex of dimension $G - 1$. Numerous community summaries of interest to ecologists are described in terms of probabilities of presence, such as diversity, richness, evenness, to name just a few. Instances of predominant indices in ecology include the Shannon index $-\sum_g \eta_g \log \eta_g$ (or entropy), the Simpson index (or Gini index) $1 - \sum_g \eta_g^2$, and the Good index $-\sum_g \eta_g^\alpha (\log \eta_g)^\beta$, $\alpha, \beta \geq 0$, which generalizes both.

The data can be further described as follows. To each site $i = 1, \ldots, I$ corresponds a covariate value $x_i \in \mathcal{X}$, where the space $\mathcal{X}$ is a subset of $\mathbb{R}$, for instance $[0, \infty)$ in the present case. Individual observations $y_{n,i}$ at site $i$ are indexed by $n = 1, \ldots, n_i$, where $n_i$ denotes the total abundance, or number of observations, at site $i$. Observations $y_{n,i}$ take on positive natural numbers $g \in \{1, \ldots, G_i\}$ where $G_i$ denotes the number of distinct species observed at site $i$. We denote by $(\mathbf{x}, \mathbf{y})$ the observations over all sites, where $\mathbf{x} = (x_i)_{i=1,\ldots,I}$, $\mathbf{y} = (y_i^{n_i})_{i=1,\ldots,I}$ and $y_i^{n_i} = (y_{n,i})_{n=1,\ldots,n_i}$. The abundance of species $g$ at site $i$ is denoted by $n_{ig}$, representing the number of times that $y_{n,i} = g$ with respect to index $n$. The relative abundance satisfies $\sum_{g=1}^{G_i} n_{ig} = n_i$.

The multinomial distribution provides a natural framework when the sampling process consists of independent and identically distributed (i.i.d.) observations of a fixed number of species, say $G$ (see applications in ecology such as Fordyce et al., 2011; De'ath, 2012; Holmes et al., 2012). Within this framework, individual observations follow a categorical distribution, which is a generalization of the Bernoulli distribution when the sample space is a set of $G > 2$ items. Namely, the probability mass function of an individual observation $y_n$, which can take on a value of a species in $\{1, \ldots, G\}$, is $\mathrm{P}(y_n = g) = \eta_g$. A tantamount notation, also more reminiscent of mixture models, is

$$y_n \overset{\text{ind}}{\sim} \sum_{g=1}^{G} \eta_g \delta_g, \tag{15.3}$$

where $\delta_g$ denotes a Dirac point mass at $g$. We shall adopt an extension of model (15.3) in two respects. First, we will not assume that the total number of species $G$ is known. We let the number of species take arbitrarily large values, hence we replace the finite sum in (15.3) by a countable infinite sum, leading us to adopt a nonparametric approach. As a consequence, the weights $(\eta_1, \eta_2, \ldots)$ become infinite vectors. Their elements sum to one and belong to the simplex of infinite dimension. Second, due to the site-by-site nature of the data, we formulate sitewise independent, but not identically distributed, generative models. Each site $i$ is parameterized by a vector of presence probabilities denoted by $\eta(x_i) = (\eta_1(x_i), \eta_2(x_i), \ldots)$, and we denote by $\eta = (\eta(x_1), \ldots, \eta(x_I))$ the full parameter. Taking into account these desiderata, we end up with the following data model which is a mixture model with a countable infinite number of components:

$$y_{n,i} \mid \eta(x_i), x_i \overset{\text{ind}}{\sim} \sum_{g=1}^{\infty} \eta_g(x_i)\delta_g, \tag{15.4}$$

for $i = 1, \dots, I$, $n = 1, \dots, n_i$. We assume a fixed design, meaning that the covariates $x_i$ are not randomized. Additionally, we adhere to a Bayesian viewpoint and need to endow the parameter $\eta$ with a prior distribution.

The initial motivation for the prior distribution on the presence probabilities $\eta$ stems from dependent Dirichlet processes introduced by MacEachern (1999). The Dirichlet process (Ferguson, 1973) is a popular BNP distribution for species modelling which conveys an interesting natural clustering mechanism. Dependent Dirichlet processes were proposed by MacEachern in order to extend Dirichlet processes to multiple-site situations, and to allow for borrowing of strength across the sites; see also Section 6.2.3 above. Dirichlet processes (and their dependent version) are (almost surely) discrete random probability measures, hence they consist of random weights and random locations. Of interest for us is the distribution of the random weights which shall constitute the prior distribution on the presence probabilities. We first describe the distribution of the weights of the Dirichlet process, also known as the Griffiths–Engen–McCloskey (GEM) distribution, used as a prior for $\eta(x)$ for any given covariate value $x$, and then turn to describe the distribution of the weights of the dependent Dirichlet process, which we call dependent GEM, used as a joint prior for $\eta = (\eta(x_1), \dots, \eta(x_I))$.

The marginal prior distribution on $\eta(x)$, "marginal" meaning at a given covariate value $x$, is defined in a constructive way, called *stick-breaking*, as follows (Sethuraman, 1994). Let us introduce i.i.d. beta random variables

$$V_g(x) \overset{iid}{\sim} \mathcal{B}e(1, \alpha), \quad g \geq 1,$$

where $\alpha > 0$. Then the prior distribution induced on $\eta(x)$ by setting $\eta_1(x) = V_1(x)$ and, for $g > 1$,

$$\eta_g(x) = V_g(x) \prod_{l < g} (1 - V_l(x)), \tag{15.5}$$

is called the GEM distribution, and $\alpha$ is called the precision parameter.

For an exhaustive description of the prior distribution on $\eta$, the marginal description (15.5) needs be complemented by specifying a distribution for stochastic processes $(V_g(x), x \in \mathcal{X})$, for any positive integer $g$. Since (15.5) requires i.i.d. beta marginals, natural candidates are i.i.d. beta processes. A simple yet effective construct to obtain a beta process is to transform a Gaussian process by the inverse cumulative distribution function (cdf) transform as follows. Denote by $z \sim \mathcal{N}(0, \sigma^2)$ a Gaussian random variable, by $\Phi_\sigma$ its cdf and by $F_\alpha$ the cdf of a $\mathcal{B}e(1, \alpha)$ random variable. Then $F_\alpha^{-1}(u) = 1 - (1 - u)^{1/\alpha}$, and the random variable

$$V = h_{\sigma,\alpha}(z) = F_\alpha^{-1} \circ \Phi_\sigma(z) \tag{15.6}$$

is $\mathcal{B}e(1, \alpha)$ distributed. The idea of including a transformed Gaussian process within a stick-breaking process is used in previous articles (see, for instance, Rodriguez & Dunson, 2011).

Of the full path of a Gaussian process on $\mathcal{X}$, only the values at observed covariates $\mathbf{x} = (x_1, \dots, x_I)$ are used. For any positive integer $g$, denote these values by $z_g = (z_g(x_1), \dots, z_g(x_I))$, and let $\mathbf{z} = (z_1, z_2, \dots)$ be a set of i.i.d. copies. Then the inverse cdf transform (15.6) maps $\mathbf{z}$ on a set of i.i.d. copies of beta vectors $\mathbf{V} = (V_1, V_2, \dots)$.

In turn, the stick-breaking construction (15.5) maps V on the set of presence probabilities $\eta = (\eta_1, \eta_2, \ldots)$. Hence, the prior distribution on the presence probabilities $\eta$ is the distribution induced under the composition of transforms (15.6) and (15.5). We denote it by

$$\eta \sim \text{Dep-GEM}(\alpha, \lambda, \sigma^2), \tag{15.7}$$

where $\lambda$ and $\sigma^2$ are two parameters of the Gaussian processes that we describe now.

A Gaussian process is fully specified by a mean function, which we take equal to zero, and a covariance function $\Sigma$ defined by

$$\Sigma(x_i, x_l) = \text{Cov}\big(z_g(x_i), z_g(x_l)\big). \tag{15.8}$$

We control the overall variance of $z_g$ by a positive pre-factor $\sigma^2$ and write $\Sigma = \sigma^2 \tilde{\Sigma}$, where $\tilde{\Sigma}$ is normalized in the sense that $\tilde{\Sigma}(x_i, x_i) = 1$ for all $i$. Possible choices of covariance structures include the squared exponential, Ornstein–Uhlenbeck, and rational quadratic covariance functions; see Rasmussen & Williams (2006) for more details. We work equally with one of these three options, without trying to learn the covariance structure, but only a parameter $\lambda$ involved in all three called the length scale of the process. It tunes how far apart two points in the space $\mathcal{X}$ have to be for the process to change significantly. The shorter $\lambda$ is, the rougher are the paths of the process. We stress the dependence on $\lambda$ by denoting $\Sigma = \sigma^2 \tilde{\Sigma}_\lambda$. The prior distribution of $z_g$ is the following multivariate normal:

$$p(z_g | \mathbf{x}, \sigma^2, \lambda) \propto \big(\sigma^{2I} |\tilde{\Sigma}_\lambda|\big)^{-1/2} \exp\left(-\frac{1}{2\sigma^2} z_g^\top \tilde{\Sigma}_\lambda^{-1} z_g\right).$$

The prior distribution is complemented by specifying the prior distribution $p(\sigma^2, \lambda, \alpha)$ over the hyperparameters. We adopt independent prior distributions, that is, $p(\sigma^2, \lambda, \alpha)$ takes the form $p(\sigma^2)p(\lambda)p(\alpha)$. More specifically, gamma prior distributions are used for all three hyperparameters: the inverse variance $1/\sigma^2$, the inverse length scale $1/\lambda$ and the precision parameter $\alpha$. These prior distributions are also common choices in the absence of dependence since they turn out to be conjugate.

---

The complete Bayesian model is given by the following mixture model:

$$y_{n,i} \,|\, \eta(x_i), x_i \overset{\text{ind}}{\sim} \sum_{g=1}^{\infty} \eta_g(x_i)\delta_g,$$

$$i = 1, \ldots, I,\ n = 1, \ldots, n_i,$$

$$\eta \sim \text{Dep-GEM}(\alpha, \lambda, \sigma^2),$$

$$\sigma^2 \sim \mathcal{IG}(c_\sigma, C_\sigma),$$

$$\lambda \sim \mathcal{IG}(c_\lambda, C_\lambda),$$

$$\alpha \sim \mathcal{G}(c_\alpha, C_\alpha),$$

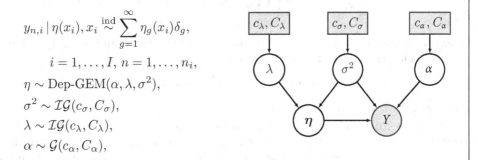

with fixed shape parameters $c_\sigma, c_\lambda, c_\alpha$ and scale parameters $C_\sigma, C_\lambda, C_\alpha$. In the graphical representation of the model, rectangles indicate fixed parameters, circles indicate random variables, and filled-in shapes indicate known values.

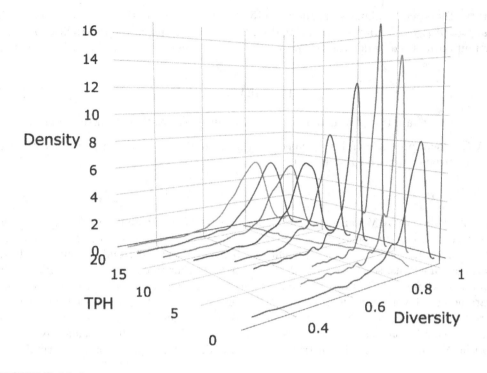

**FIGURE 15.4**
Posterior distributions of diversity at varying pollution levels from 0 to 20 TPH units, along with the posterior mean of the diversity.

### 15.5.2   Posterior sampling and summaries

Here we expose posterior sampling in terms of the Gaussian processes $\mathbf{z}$, keeping in mind that the main parameters of interest, the presence probabilities $\eta$, can be recovered through the composition of the inverse cdf transform (15.6) and the stick-breaking construction (15.5). The likelihood can be easily obtained from these two transforms as

$$p(\mathbf{y}|\mathbf{z}, \mathbf{x}, \sigma^2, \alpha) = \prod_{g=1}^{G} \prod_{i=1}^{I} h_{\sigma,\alpha}(z_g(x_i))^{n_{ig}} (1 - h_{\sigma,\alpha}(z_g(x_i)))^{\bar{n}_{i,g+1}},$$

where we denote by $\bar{n}_{i,g+1} = \sum_{l>g} n_{il}$ the sum of abundances of species $\{g+1, g+2, \ldots\}$ at site $i$. The posterior distribution is then

$$p(\mathbf{z}, \lambda, \sigma^2, \alpha|\mathbf{y}, \mathbf{x}) \propto p(\mathbf{y}|\mathbf{z}, \mathbf{x}, \sigma^2, \alpha)p(\mathbf{z}|\mathbf{x}, \sigma^2, \lambda)p(\sigma^2)p(\lambda)p(\alpha),$$

where $p(\mathbf{z}|\mathbf{x}, \sigma^2, \lambda) = \prod_{g=1}^{G} p(z_g|\mathbf{x}, \sigma^2, \lambda)$.

Sampling from the posterior distribution of $(\mathbf{z}, \sigma^2, \lambda, \alpha)$ in the Dep-GEM model is performed by a Markov chain Monte Carlo algorithm comprising Gibbs and Metropolis–Hastings steps. It proceeds by sequentially updating each parameter $\mathbf{z}$, $\sigma^2$, $\lambda$ and $\alpha$ via its conditional distribution as follows.

(a) Conditional for $\mathbf{z}$: we use $G$ independent Metropolis algorithms with Gaussian proposals. The covariance matrix for the proposal is set proportional to the prior covariance matrix $\tilde{\Sigma}_\lambda$. For any $g \in \{1, \ldots, G\}$, the target distribution is

$$p(z_g|\cdot) \propto p(\mathbf{y}|\mathbf{z}, \mathbf{x}, \sigma^2, \alpha)p(z_g|\mathbf{x}, \sigma^2, \lambda).$$

(b) Conditional for $\sigma^2$: Metropolis–Hastings algorithm with a Gaussian proposal left-truncated to 0 with target distribution

$$p(\sigma^2|\cdot) \propto p(\mathbf{y}|\mathbf{z},\mathbf{x},\sigma^2,\alpha)p(\mathbf{z}|\mathbf{x},\sigma^2,\lambda)p(\sigma^2).$$

(c) Conditional for $\lambda$: Metropolis–Hastings algorithm with a Gaussian proposal left-truncated to 0 with target distribution

$$p(\lambda|\cdot) \propto p(\mathbf{z}|\mathbf{x},\sigma^2,\lambda)p(\lambda).$$

(d) Conditional for $\alpha$: Metropolis algorithm with a Gaussian proposal left-truncated to 0 with target distribution

$$p(\alpha|\cdot) \propto p(\mathbf{y}|\mathbf{z},\mathbf{x},\sigma^2,\alpha)p(\alpha).$$

The posterior sample allows for a probabilistic evaluation of various quantities of interest for ecologists. Figure 15.4 provides posterior distributions of diversity at varying pollution levels from 0 to 20 TPH units. This shows a better posterior precision around 5 TPH units. The posterior mean of the diversity first increases with the pollution level with a maximum at 4 TPH units, and then decreases. Such a variation may depict a hormetic effect, a dose-response phenomenon characterized by favourable responses to low exposures to pollutant. In addition to diversity, so-called *effective concentrations* are highly relevant criteria in determining guidelines for protection of an ecosystem. The effective concentration at level $\ell$, denoted by $EC_\ell$, is the concentration of a contaminant that causes $\ell\%$ effect on the population relative to the baseline community (e.g. Newman, 2012). For example, the $EC_{50}$ is the median effective concentration and represents the concentration of a contaminant which induces a response halfway between the control baseline and the maximum after a specified exposure time. Estimation of both diversity and effective concentrations is straightforward from a posterior sample. See Arbel et al. (2015) for detailed results.

## 15.6  Concluding Remarks

This chapter has showcased a number of industrial applications where mixture modelling has been shown to be a powerful tool for inference. While the applications presented were diverse, spanning manufacturing, ecology and health, each shared the common motivation of characterizing heterogeneity by multiple components, with each component describing a key feature of the population or process under study. The breadth of applications also demonstrated the flexibility of mixture specification, with models varying in terms of distributional assumptions for individual components, the form of the mixing proportions and the use of mixtures as generative models for the observed data and as a prior distribution for unknown parameters.

A review of monitoring studies strengthened the case for mixture models as an elegant solution for explaining observed heterogeneity, in cases where the use of a single distribution is inadequate for reliable inference. In studies of fault detection, the utility of mixture models was twofold: (i) as a way of characterizing different operating modes of a process; and (ii) to construct a meaningful, global summary of the process, taking into account component membership uncertainty. In other examples presented, focused in ecology, the flexibility of mixtures in terms of component specification was highlighted. This flexibility extended to different types of data (continuous, count) and mixture component specification to account for zero inflation, for improved predictions of species abundance.

The two case studies pertaining to health resource usage showed how mixture models can account for heterogeneity relating to tangible characteristics, such as antibody concentration, and technical constructs, such as a random effect for space-time interaction. This ability to address different sources of heterogeneity is a result of applying the mixture model in different ways. In the first case study, the likelihood was represented by the mixture model directly, as is typical in many applications. In the second case study, however, the mixture model was applied to a random effect parameter, or, more specifically, its prior distribution. In both case studies, the mixture models were able to identify emerging patterns and guide the management of health resources.

In the pest surveillance example, the multivariate Bayesian mixture model when applied to data drawn from standard Landsat images, is shown to be an effective method of estimating the probability of land in an urban space being suitable habitat for an invasive pest. Additionally, these models can be used to either cluster image pixels into either a most probable component or the probability of a pixel being suitable habitat, if suitability is encompassed by several clusters. The model flexibility, combined with the Bayesian approach using Markov chain Monte Carlo, allows researchers and managers to value add and extend scenario testing with relative ease.

The flexibility of the Bayesian mixture model is likely to make this modelling procedure extensible to other types of land use (e.g. rural) and other forms of imagery (e.g. aerial and drone-captured snapshots, ASTER and Spot 5). This is an active area of research globally, as technology becomes more readily available. Estimation of bare earth is interesting in many applications on a worldwide scale, with reports such as Weber et al. (2010) indicating that this research is ongoing in the quest to detect this land use pattern using remote sensing data.

The last case study instantiated how infinite mixtures and Bayesian nonparametric methods can be successfully applied to industry-related issues, namely the influence of a toxic spill on soil biodiversity. Mixtures were demonstrated to adequately model covariate-dependent species data, allowing one to make useful inference on biodiversity, effective concentrations, etc., as well as to provide predictions outside the range of observed covariates.

# Bibliography

ABELLAN, J. J., RICHARDSON, S. & BEST, N. (2008). Use of space-time models to investigate the stability of patterns of disease. *Environmental Health Perspectives* **116**, 1111–1119.

ALSTON, C. L., MENGERSEN, K. L. & GARDNER, G. E. (2009). A new method for calculating the volume of primary tissue types in live sheep using computed tomography scanning. *Animal Production Science* **49**, 1035–1042.

ARBEL, J., MENGERSEN, K. L., RAYMOND, B., WINSLEY, T. & KING, C. (2015). Application of a Bayesian nonparametric model to derive toxicity estimates based on the response of Antarctic microbial communities to fuel contaminated soil. *Ecology and Evolution* **5**, 2633–2645.

ARBEL, J., MENGERSEN, K. L. & ROUSSEAU, J. (2016). Bayesian nonparametric dependent model for partially replicated data: The influence of fuel spills on species diversity. *Annals of Applied Statistics* **10**, 1496–1516.

CHEN, T. & ZHANG, J. (2010). On-line multivariate statistical monitoring of batch processes using Gaussian mixture model. *Computers and Chemical Engineering* **34**, 500–507.

CRON, A. J. & WEST, M. (2011). Efficient classification-based relabeling in mixture models. *American Statistician* **65**, 16–20.

DE'ATH, G. (2012). The multinomial diversity model: Linking Shannon diversity to multiple predictors. *Ecology* **93**, 2286–2296.

DEL FAVA, E., SHKEDY, Z., BECHINI, A., BONANNI, P. & MANFREDI, P. (2012). Towards measles elimination in Italy: Monitoring herd immunity by Bayesian mixture modelling of serological data. *Epidemics* **4**, 124–131.

FERGUSON, T. (1973). A Bayesian analysis of some nonparametric problems. *Annals of Statistics* **1**, 209–230.

FORDYCE, J. A., GOMPERT, Z., FORISTER, M. L. & NICE, C. C. (2011). A hierarchical Bayesian approach to ecological count data: A flexible tool for ecologists. *PLoS ONE* **6**, e26785.

FRÜHWIRTH-SCHNATTER, S. (2006). *Finite Mixture and Markov Switching Models*. New York: Springer-Verlag.

GE, Z., SONG, Z. & GAO, F. (2013). Review of recent research on data-based process monitoring. *Industrial Engineering and Chemical Research* **52**, 3542–3562.

HOLMES, I., HARRIS, K. & QUINCE, C. (2012). Dirichlet multinomial mixtures: Generative models for microbial metagenomics. *PloS One* **7**, e30126.

KAEWTRAKULPONG, P. & BOWDEN, R. (2002). An improved adaptive background mixture model for real-time tracking with shadow detection. In *2nd European Workshop on Advanced Video Based Surveillance Systems, 2001*.

KORNER-NIEVERGELT, F., ROTH, T., VON FELTEN, S., GUÉLAT, J., ALMASI, B. & KORNER-NIEVERGELT, P. (2015). *Bayesian Data Analysis in Ecology Using Linear Models with R, BUGS and STAN*. London: Academic Press.

KUHN, H. (1955). The Hungarian method for the assignment problem. *Naval Research Logistics Quarterly* **2**, 83–97.

LAMBERT, D. (1992). Zero-inflated Poisson regression, with an application to defects in manufacturing. *Technometrics* **34**, 1–14.

LIN, Y., CHEN, M. & ZHOU, D. (2013). Online probabilistic operational safety assessment of multi-mode engineering systems using Bayesian methods. *Reliability Engineering and System Safety* **119**, 150–157.

LYASHEVSKA, O., BRUS, D. & MEER, J. (2016). Mapping species abundance by a spatial zero-inflated Poisson model: A case study in the Wadden Sea, the Netherlands. *Ecology and Evolution* **6**, 532–543.

MACEACHERN, S. N. (1999). Dependent nonparametric processes. In *ASA Proceedings of the Section on Bayesian Statistical Science*. Alexandria, VA: American Statistical Association, pp. 50–55.

MCCARTHY, M. A. (2007). *Bayesian Methods for Ecology*. Cambridge: Cambridge University Press.

NEUBAUER, P., SHIMA, J. S. & SWEARER, S. E. (2013). Inferring dispersal and migrations from incomplete geochemical baselines: Analysis of population structure using Bayesian infinite mixture models. *Methods in Ecology and Evolution* **4**, 836–845.

NEWMAN, M. C. (2012). *Quantitative Ecotoxicology*. Boca Raton, FL: CRC Press.

PIETQUIN, O. (2004). *A Framework for Unsupervised Learning of Dialogue Strategies*. Ph.D. thesis, Faculté Polytechnique de Mons, TCTS Lab, Belgium.

RASMUSSEN, C. E. & WILLIAMS, C. K. I. (2006). *Gaussian Processes for Machine Learning*. Cambridge, MA: MIT Press.

RHODES, J. R., GRIST, E. P. M., KWOK, K. W. H. & LEUNG, K. M. Y. (2008). A Bayesian mixture model for estimating intergeneration chronic toxicity. *Environmental Science and Technology* **42**, 8108–8114.

ROBERT, C. P. (1996). Mixtures of distributions: Inference and estimation. In *Markov Chain Monte Carlo in Practice*, W. R. Gilks, S. Richardson & D. Spiegelhalter, eds. London: Chapman & Hall, pp. 441–464.

RODRIGUEZ, A. & DUNSON, D. B. (2011). Nonparametric Bayesian models through probit stick-breaking processes. *Bayesian Analysis* **6**, 145–177.

SADJADI, F. A., ed. (2001). *Automated Target Recognition*, vol. 11 of *SPIE Conference Volume*, Bellingham, WA. SPIE.

SETHURAMAN, J. (1994). A constructive definition of Dirichlet priors. *Statistica Sinica* **4**, 639–650.

SINDHU, T. N., RIAZ, M., ASLAM, M. & AHMED, Z. (2015). Bayes estimation of Gumbel mixture models with industrial applications. *Transactions of the Institute of Measurement and Control* **38**, 201–214.

SNAPE, I., SICILIANO, S. D., WINSLEY, T., VAN DORST, J., MUKAN, J., PALMER, A. S. & LAGEREWSKIJ, G. (2015). *Operational Taxonomic Unit (OTU) Microbial Ecotoxicology data from Macquarie Island and Casey Station: TPH, Chemistry and OTU Abundance Data.* Australian Antarctic Data Centre.

SPRING, D. & CACHO, O. J. (2015). Estimating eradication probabilities and trade-offs for decision analysis in invasive species eradication programs. *Biological Invasions* **17**, 191–204.

STRICKLAND, C. M., DENHAM, R. J., ALSTON, C. L. & MENGERSEN, K. L. (2012). A Python package for Bayesian estimation using Markov chain Monte Carlo. In *Case Studies in Bayesian Statistical Modelling and Analysis*, C. L. Alston, K. L. Mengersen & A. N. Pettitt, eds. Chichester: John Wiley, pp. 421–460.

VIGNERON, V., ZARZOSO, V., MOREAU, E., GRIBONVAL, R. & VINCENT, E., eds. (2010). *Latent Variable Analysis and Signal Separation: 9th International Conference, LVA/ICA 2010, St. Malo, France, September 27-30, 2010*, vol. 6365, Berlin. Springer.

WEBER, K., GLENN, N. & TIBBITTS, J. (2010). Investigation of potential bare ground modeling techniques using multispectral satellite imagery. Tech. Rep. NNG06GD82G. Section within Final Report: Forecasting Rangeland Condition with GIS in Southeastern Idaho.

WEN, Q., GE, Z. & SONG, Z. (2015). Multimode dynamic process monitoring based on mixture canonical variate analysis model. *Industrial and Engineering Chemical Research* **54**, 1605–1614.

WU, G., HOLAN, S. H., NILON, C. H. & WIKLE, C. K. (2015). Bayesian binomial mixture models for estimating abundance in ecological monitoring studies. *Annals of Applied Statistics* **9**, 1–26.

XIE, X. & SHI, H. (2012). Dynamic multimode process modeling and monitoring using adptive Gaussian mixture models. *Industrial and Engineering Chemical Research* **51**, 5497–5505.

YU, J. (2012a). A nonlinear kernel Gaussian mixture model based inferential monitoring approach for fault detection and diagnosis of chemical processes. *Chemical Engineering Science* **68**, 506–519.

YU, J. (2012b). A particle filter driven dynamic Gaussian mixture model approach for complex process monitoring and fault diagnosis. *Journal of Process Control* **22**, 778–788.

YU, J. & QUIN, S. J. (2008). Multimode process monitoring with Bayesian inference-based finite Gaussian mixture models. *AIChE Journal* **54**, 1811–1829.

YU, J. & QUIN, S. J. (2009). Multiway Gaussian mixture model based multiphase batch process monitoring. *Industrial and Engineering Chemical Research* **48**, 8585–8594.

# 16

# Mixture Models for Image Analysis

**Florence Forbes**

*Univ. Grenoble Alpes, Inria, CNRS, Grenoble INP\*, LJK, 38000 Grenoble, France*
*\* Institute of Engineering Univ. Grenoble Alpes*

## CONTENTS

## 16.1 Introduction

Image analysis includes a variety of tasks such as image restoration, segmentation, registration, visual tracking, retrieval, texture modelling, classification and sensor fusion. Important application domains are medical imaging, remote sensing and computer vision. Problems involving incomplete data, where part of the data is missing or unobservable, are common, and mixture models can be used in many of these tasks directly or indirectly. The aim may be to recover an original image which is hidden and has to be estimated from a noisy or blurred version (restoration). More generally, the observed and hidden data are not necessarily of the same nature (segmentation). The observations may represent measurements, for example, multidimensional variables recorded at each pixel of an image, while the hidden data could consist of an unknown class assignment to be estimated at each pixel. To give an idea of the variety of uses, mixture models have been used for image restoration (e.g. Niknejad et al., 2015), for image registration (Gerogiannis et al., 2009), visual tracking (Karavasilis et al., 2012), image retrieval (Beecks et al., 2015), texture modelling (Blanchet & Forbes, 2008), classification (Bouveyron et al., 2007) and sensor fusion (Gebru et al., 2016), to name only a few of the relevant papers. However, the most typical and direct use relates to image segmentation which can be recast straightforwardly into a clustering task. More generally, in this chapter we will focus on problems that can be posed as labelling or clustering problems in which the solution is a set of labels assigned to image pixels or features.

In the context of statistical image segmentation, choosing the probabilistic model that best accounts for the observations is an important first step for the quality of the subsequent estimation and analysis. Hidden Markov random field (HMRF) models were revealed to be a powerful tool for image segmentation (Geman & Geman, 1984; Besag, 1986). They are very useful in accounting for spatial dependencies between the different pixels of an image, but these spatial dependencies are also responsible for a typically large amount of computation. Markov model-based segmentation requires estimation of the model parameters. A common approach involves alternately restoring the unknown segmentation (labelling or clustering) based on a maximum *a posteriori* rule and then estimating the model parameters using the observations and the restored data. This is the case, for instance, in the popular iterated conditional mode (ICM) algorithm of Besag (1986) which makes use of the pseudo-likelihood approximation (Besag, 1974). This combination usually provides reasonable segmentations but is known to lead to biased parameter estimates, essentially due to the restoration step. Because of the missing data structure of the task, the expectation-maximization (EM) algorithm provides another justifiable formalism for such an alternating scheme. It has the advantage of dealing with conditional probabilities instead of committing to suboptimal restorations of the hidden data.

In this chapter, we first present how HMRF models generalize standard mixture models (Section 16.2). We propose an inference procedure using variational approximation (Section 16.3) and illustrate the framework with two real medical image applications (Section 16.4).

## 16.2   Hidden Markov Model Based Clustering

Hidden structure models, and more specifically Gaussian mixture models, are among the most statistically mature methods for clustering. A clustering or labelling problem is specified in terms of a set of sites $S$ and a set of labels $\mathcal{G}$. A site often represents an item, a point or a region in Euclidean space such as an image pixel or an image feature. A set of sites may be categorized in terms of their regularity. Sites on a lattice are considered as spatially regular (e.g. the pixels of a two-dimensional image). Sites which do not present spatial regularity are considered as irregular. This is the usual case when sites represent geographic locations (Green & Richardson, 2002) or features extracted from images at a more abstract level, such as *interest points* (see Lowe, 2004; Blanchet & Forbes, 2008). It can also be that the sites correspond to items (e.g. genes) that are related to each other through a distance or dissimilarity measure (Vignes & Forbes, 2009) or simply to a collection of independent items.

A label is an event that may happen to a site. We will consider only the case where a label assumes a discrete value in a set of $G$ labels. In the following developments, it is convenient to consider $\mathcal{G}$ as the set of $G$-dimensional indicator vectors $\mathcal{G} = \{e_1, \ldots, e_G\}$, where each $e_g$ has all its components being 0, except the $g$th which is 1. The labelling problem is to assign a label from a label set $\mathcal{G}$ to each of the sites. If there are $n$ sites, the set $\mathbf{z} = \{z_1, \ldots, z_n\}$, with $z_i \in \mathcal{G}$ for all $i \in S$, is called a labelling of the sites in $S$ in terms of the labels in $\mathcal{G}$. We consider cases where the data naturally divide into observed data $\mathbf{y} = \{y_1, \ldots, y_n\}$ and unobserved or missing membership data $\mathbf{z} = \{z_1, \ldots, z_n\}$. They are considered as random variables denoted by $\mathbf{Y} = \{Y_1, \ldots, Y_n\}$ and $\mathbf{Z} = \{Z_1, \ldots, Z_n\}$ with domain $\mathcal{Z}$ being equal to $\mathcal{Z} = \mathcal{G}^n$.

For image analysis or spatial data clustering, dependencies or contextual information can be taken into account using HMRF models, which can be seen as a generalization of standard mixture models.

### 16.2.1 Mixture models

Before introducing the HMRF model, we recall the underlying mixture modelling corresponding to independent $Z_i$. The model reduces to a standard mixture model that we will refer to as an *independent mixture*; see also Chapter 1. The distribution of $(\mathbf{Y}, \mathbf{Z})$ is defined by

$$p(\mathbf{z}) = \prod_{i \in S} p(z_i), \tag{16.1}$$

$$p(\mathbf{y}|\mathbf{z}) = \prod_{i \in S} p(y_i|z_i). \tag{16.2}$$

Equation (16.1) means that the hidden variables $Z_i$ are independent, while equation (16.2) is sometimes referred to as the *independent noise assumption*. Under (16.1) and (16.2), the $Y_i$ are also independent variables. To recover the standard mixture definition, we need to assume that the $Z_i$ are identically distributed according to a multinomial distribution with parameters $\eta = \{\eta_1, \ldots, \eta_G\}$. Similarly, the conditional distribution for class $g$, $p(\cdot|Z_i = e_g) = f(\cdot|\theta_g)$, is assumed not to depend on $i$ but only on some parameter $\theta_g$. Different choices are possible for $f(\cdot|\theta_g)$. The most commonly encountered in applications are multivariate Gaussians (Celeux & Govaert, 1995; Banfield & Raftery, 1993), multivariate Student (McLachlan & Peel, 2000; Gerogiannis et al., 2009), and Poisson distributions for count data (Green & Richardson, 2002; Forbes et al., 2013; Karlis & Meligkotsidou, 2007); see also Chapter 8 above for a review of mixture modelling of count data.

### 16.2.2 Markov random fields: Potts model and extensions

When the $Z_i$ are not independent, the interrelationship between sites can be modelled by a so-called neighbourhood system usually defined through a graph. Two neighbouring sites correspond to two nodes of the graph linked by an edge. The dependencies between neighbouring $Z_i$ are then modelled by further assuming that the joint distribution of $Z_1, \ldots, Z_n$ is a discrete Markov random field (MRF) on this specific graph defined by

$$p(\mathbf{z}) = W^{-1} \exp(-H(\mathbf{z})), \tag{16.3}$$

where $W$ is a normalizing constant and $H$ is a function assumed to be of the following form (we restrict to pairwise interactions):

$$H(\mathbf{z}) = \sum_{i \sim j} V_{ij}(z_i, z_j) + \sum_{i \in S} V_i(z_i), \tag{16.4}$$

where the $V_{ij}$ ($V_i$) are functions referred to as pair (singleton) potentials. We write $i \sim j$ when sites $i$ and $j$ are neighbours on the graph, so that the sum above is only over neighbouring sites.

A simple model is the so-called Ising model where the $Z_i$ are binary variables representing spin orientations. The more general Potts model allows the $Z_i$ to take $G$ values that correspond to $G$ classes with $G > 2$.

The singleton potentials $V_i(z_i)$ impact the probability of assigning site $i$ to label or class $z_i$. When these potentials depend on $i$ and not only on $z_i$, they are referred to as a non-stationary external field. Such non-stationarity can be useful to account for *a priori* knowledge that may vary with the site. This is typically the case when introducing probabilistic atlases in brain MRI data analysis (Forbes et al., 2011). Often, however, we restrict to a stationary external field that can be then denoted by $V_i(z_i) = -\alpha_{z_i}$. The potentials are then defined by a $G$-dimensional vector $\alpha = \{\alpha_1, \ldots, \alpha_G\}$ using the vector notation $V_i(z_i) = -\langle z_i, \alpha \rangle$, where $\langle z_i, \alpha \rangle$ denotes the scalar product between $z_i$ and $\alpha$. The latter notation has the advantage that it still makes sense when the vectors are arbitrary and not necessarily indicators. This will be useful when describing the algorithms of Section 16.3.

The singleton potentials are linked to the standard mixture proportions $\eta$. When the $V_{ij}$ are zero in (16.4), $p(\mathbf{z})$ in (16.3) reduces to a standard mixture model up to the reparameterization

$$p(\mathbf{z}) = \prod_{i \in S} \frac{\exp(\langle z_i, \alpha \rangle)}{\sum_{g'=1}^{G} \exp(\alpha_{g'})} . \tag{16.5}$$

From (16.5), we can identify the link between $\alpha$ and $\eta$ as, for all $g = 1, \ldots, G$,

$$\eta_g = \frac{\exp(\alpha_g)}{\sum_{g'=1}^{G} \exp(\alpha_{g'})} .$$

The pair potentials allow us to model the dependence between $Z_i$ and $Z_j$ at sites $i$ and $j$. We consider pair potentials $V_{ij}$ that depend on $z_i$ and $z_j$ but also possibly on $i$ and $j$. Since the $z_i$ can only take a finite number of values, for each $i$ and $j$, we can define a $G \times G$ matrix $\mathbb{V}_{ij} = (\mathbb{V}_{ij}(k,l))_{1 \le k,l \le G}$ and write without lost of generality $V_{ij}(z_i, z_j) = \mathbb{V}_{ij}(k,l)$ if $z_i = e_k$ and $z_j = e_l$, or, using the indicator vector notation, $V_{ij}(z_i, z_j) = -\langle z_i, \mathbb{V}_{ij} z_j \rangle$.

If, for all $i$ and $j$, $\mathbb{V}_{ij} = \beta \times I_G$ where $\beta$ is a scalar and $I_G$ is the $G \times G$ identity matrix, then the pair potentials reduce to a single scalar interaction parameter $\beta$ and we get the Potts model traditionally used for image segmentation (Besag, 1986). Note that this model is appropriate most of the time for segmentation since, for positive $\beta$, it tends to favour neighbours that are in the same class.

In practice, these parameters can be tuned according to experts or *a priori* knowledge or they can be estimated from the data. In the latter case, the part to be estimated is usually assumed independent of the indices $i$ and $j$. In what follows, the Markov model parameters will reduce to a single matrix $\mathbb{V}$. Note that, formulated as such, the model is not identifiable in the sense that different values of the parameters, namely $\mathbb{V}$ and $\mathbb{V} + c\mathbb{1}$ (where $\mathbb{1}$ denotes the $G \times G$ matrix with all its components being 1 and $c$ an arbitrary scalar value) lead to the same probability distribution. This issue is generally easily handled by imposing some additional constraint such as $\mathbb{V}(k,l) = 0$ for one of the components $(k,l)$.

### 16.2.3    Hidden Markov field with independent noise

The independent noise assumption (16.2) underlying standard mixture models is also crucial in the more general hidden Markov random field. When the goal is to estimate $\mathbf{z}$ from the observed $\mathbf{Y} = \mathbf{y}$, most approaches fall into two categories. The first ones focus on finding the best $\mathbf{z}$ using a Bayesian decision principle such as maximum *a posteriori* or maximum posterior mode rules. This explicitly involves the use of $p(\mathbf{z}|\mathbf{y})$ and uses the fact that the conditional field denoted by $\mathbf{Z}|\mathbf{Y} = \mathbf{y}$ is a Markov field. This includes methods such as ICM (Besag, 1986) and simulated annealing (Geman & Geman, 1984) which differ in the way they deal with the intractable $p(\mathbf{z}|\mathbf{y})$ and use its Markovianity. A second type of approach is related to a missing-data point of view, for which the focus is on estimating parameters when some of the data are missing (the $z_i$ here). The reference algorithm in such cases is the EM algorithm (Dempster et al., 1977). In addition to estimating the parameters, the EM algorithm provides also a segmentation $\mathbf{z}$ by offering the possibility of restoring the missing data; see Chapter 2 above for a detailed review of the EM algorithm.

However, when applied to hidden Markov fields, the algorithm is not tractable and requires approximations. It follows a number of procedures including the Gibbsian EM of Chalmond (1989), the MCEM algorithm and a generalization of it (Qian & Titterington, 1991), the PPL-EM algorithm of Qian & Titterington (1991) and various mean-field-like approximations of EM (Celeux et al., 2003). Such approximations are also all based on the Markovianity of $\mathbf{Z}|\mathbf{Y} = \mathbf{y}$. This property is a critical requirement for any further developments.

When $\mathbf{Z}$ is Markovian, a simple way to guarantee the Markovianity of $\mathbf{Z}|\mathbf{Y} = \mathbf{y}$ is the independent noise assumption (16.2). Indeed, equations (16.2) and (16.3) imply that the conditional field $(\mathbf{Y}, \mathbf{Z})$ is a Markov random field, which implies that $\mathbf{Z}|\mathbf{Y} = \mathbf{y}$ is an MRF too. This standard and widely used situation is referred to in Benboudjema & Pieczynski (2005) as the hidden Markov field with independent noise (HMF-IN) model. Equation (16.2) is a conditional independence and non-correlated noise condition. Denoting by $\theta = \{\theta_1, \ldots, \theta_G\}$ the class-dependent distribution parameters, the HMF-IN parameters are denoted by $\Psi = (\theta, \alpha, \mathbb{V})$. In the one-dimensional Gaussian case, $\theta_g = (\mu_g, \sigma_g^2)$, the mean and variance parameters of the Gaussian distribution.

Like standard mixture models, hidden Markov (random) fields can then be used for a number of segmentation or clustering tasks. Many applications are related to image analysis, but other examples include population genetics (François et al., 2006) and bioinformatics (Vignes & Forbes, 2009). The fact that $\mathbf{Z}$ is Markovian is not strictly necessary. However, in a segmentation or clustering context, it has the advantage of providing some insight into and control of the segmentation regularity through a meaningful and easy-to-understand parametric model, but it also somewhat reduces the modelling capabilities of the approach (see Blanchet & Forbes, 2008). More general approaches involve so-called couple MRF or triplet MRF (Benboudjema & Pieczynski, 2005; Blanchet & Forbes, 2008) but will not be described in this chapter.

## 16.3   Markov Model Based Segmentation via Variational EM

The model complexity of a hidden Markov random field is greater than that of standard mixtures and makes the EM algorithm intractable. Solutions have been proposed which associate the pseudo-likelihood approximation (Besag, 1974) and Monte Carlo simulations (Chalmond, 1989), but the corresponding algorithms are time-consuming. In this section we present a variational approximation approach. In a number of complex real imaging applications, it has been observed as a competitive alternative to Markov chain Monte Carlo approaches, in terms of the quality of the results, with a great gain in terms of computation time (for a comparison in functional MRI analysis, see, for example, Chaari et al., 2013). The variational approximation relates to the so-called mean field approximation in statistical physics (Chandler, 1987). For a hidden Markov field model, the likelihood of $(\mathbf{Y}, \mathbf{Z})$ is called the complete (or complete-data) likelihood and is given by

$$p(\mathbf{y}, \mathbf{z}|\Psi) = p(\mathbf{y}|\mathbf{z}, \theta)\, p(\mathbf{z}|\alpha, \mathbb{V}). \tag{16.6}$$

The conditional field $\mathbf{Z}$ given $\mathbf{Y} = \mathbf{y}$ is also a Markov field, with energy function $H(\mathbf{z}; \alpha, \mathbb{V}) - \log p(\mathbf{y}|\mathbf{z}, \theta)$. Henceforth, we will refer to the Markov fields $\mathbf{Z}$ and $\mathbf{Z}$ given $\mathbf{Y} = \mathbf{y}$ as the marginal and the conditional fields, respectively. Recovering the unknown $\mathbf{Z}$ requires values for the vector parameter $\Psi = (\theta, \alpha, \mathbb{V})$. If unknown, the parameters are often estimated from the maximum likelihood perspective as

$$\hat{\Psi} = \arg\max_{\Psi} \log p(\mathbf{y}|\Psi), \qquad (16.7)$$

where $p(\mathbf{y}|\Psi)$ is the incomplete (also called observed-data) likelihood. This optimization is usually solved by the iterative EM procedure (Dempster et al., 1977); see also Chapter 2 above. Each iteration may be formally decomposed into two steps. Given the current value of the parameter $\Psi^{(s)}$ at iteration $s$, the E step involves computing the expectation of the complete log likelihood knowing the observations $\mathbf{y}$ and the current estimate $\Psi^{(s)}$. In the M step, the parameter is then updated by maximizing this expected complete log likelihood,

$$\Psi^{(s+1)} = \arg\max_{\Psi} \sum_{\mathbf{z} \in \mathcal{Z}} \log p(\mathbf{y}, \mathbf{z}|\Psi) \, p(\mathbf{z}|\mathbf{y}, \Psi^{(s)}). \qquad (16.8)$$

It is known that, under mild regularity conditions, EM converges to the set of stationary points of the incomplete likelihood $\Psi \mapsto p(\mathbf{y}|\Psi)$ (Wu, 1983). As discussed in Csiszar & Tusnady (1984) and Neal & Hinton (1998), denoting by $\mathcal{D}$ the set of distributions on missing data, EM can be viewed as an alternating maximization procedure of a function $F$ defined, for any probability distribution $q \in \mathcal{D}$, by

$$F(q, \Psi) = \sum_{\mathbf{z} \in \mathcal{Z}} \log \left( \frac{p(\mathbf{y}, \mathbf{z}|\Psi)}{q(\mathbf{z})} \right) q(\mathbf{z}). \qquad (16.9)$$

Starting from current values $(q^{(s)}, \Psi^{(s)})$, set

$$q^{(s+1)} = \arg\max_{q \in \mathcal{D}} F(q, \Psi^{(s)}) \qquad (16.10)$$

and

$$\Psi^{(s+1)} = \arg\max_{\Psi} F(q^{(s+1)}, \Psi) \qquad (16.11)$$

$$= \arg\max_{\Psi} \sum_{\mathbf{z} \in \mathcal{Z}} \log p(\mathbf{y}, \mathbf{z}|\Psi) \, q^{(s+1)}(\mathbf{z}).$$

The first optimization (16.10) has an explicit solution $q^{(s+1)} = p(\cdot|\mathbf{y}, \Psi^{(s)})$, so that the solutions of (16.8) and (16.11) are the same. Hence the "marginal" sequence $\{\Psi^{(s)}\}_s$ of the sequence $\{(q^{(s)}, \Psi^{(s)})\}_s$ produced by the alternating maximization procedure of $F$ is an EM path. The maximization (16.11) can also be understood as the minimization of a Kullback–Leibler divergence, up to some convention on $p(\mathbf{y})$, thus justifying the name of alternating minimization procedure (e.g. Csiszar & Tusnady, 1984; Byrne & Gunawardana, 2005).

There exist different generalizations of EM when the M step (16.8) is intractable; it can be relaxed by requiring just an increase rather than an optimum. This yields generalized EM (GEM) procedures (McLachlan & Krishnan, 2008; see also Boyles, 1983, for a convergence result). Unfortunately, EM (or GEM) is not appropriate for solving the optimization problem (16.7) in HMRFs due to the complex structure of the hidden variables $\mathbf{Z}$. The distribution $p(\mathbf{z}|\alpha, \mathbb{V})$ is only known up to its normalizing constant $W$ (the partition function) which depends upon the parameters of interest $\mathbb{V}$. The domain $\mathcal{Z}$ is too large, so that the E step is intractable.

Alternative approaches have been proposed and they can be understood as generalizations of the alternating maximization procedures mentioned above: the optimization (16.10) is solved over a restricted class of probability distributions $\tilde{\mathcal{D}}$ on $\mathcal{Z}$, and the M step (16.11) remains unchanged. This yields the variational EM (VEM) algorithms (Jordan et al., 1998). For a convex optimization justification see also Wainwright & Jordan (2003, 2005). Byrne & Gunawardana (2005) proved that, under mild regularity conditions, VEM converges to the set $\mathcal{L}$ of the stationary points of the function $F$ in $\tilde{\mathcal{D}}$. Here again, generalizations of VEM can be defined by requiring an increase rather than an optimum in the M step (16.11), thus defining generalized VEM procedures. These relaxation methods are part of the generalized alternating minimization procedures (Byrne & Gunawardana, 2005). The most popular form of VEM occurs when $\tilde{\mathcal{D}}$ is the set of independent probability distributions on $\mathcal{Z}$ so that $q^{(s+1)}(\mathbf{z})$ is a factorized distribution $\prod_{i \in S} q_i^{(s+1)}(z_i)$. Then optimizing (16.10) with respect to $q_i^{(s+1)}(e_k)$ leads to a fixed point equation for all $i \in S$ and for all $e_k \in V$:

$$\log q_i^{(s+1)}(e_k) = c_i + \sum_{\mathbf{z} \in \mathcal{Z}} \log p(\mathbf{z}|\mathbf{y}, \Psi^{(s)}) \mathbb{I}(z_i = e_k) \prod_{j \neq i} q_j^{(s+1)}(z_j), \qquad (16.12)$$

where $c_i$ is the normalizing constant. The Markov property implies that the right-hand side of the equation only involves the probability distributions $q_j$ for $j$ in the neighbourhood of $i$ that we will denote by $j \in \mathcal{N}(i)$. Another equivalent form of (16.12) is to update in turn, for each $i$ in $S$,

$$q_i^{(s+1)}(z_i) \propto \exp(\mathrm{E}_{q_{\backslash i}^{(s+1)}}[\log p(z_i|\mathbf{y}, \mathbf{Z}_{\backslash i}, \Psi^{(s)})]) \qquad (16.13)$$

where the expectation is taken with regard to

$$q_{\backslash i}^{(s+1)}(\mathbf{z}_{\backslash i}) = \prod_{j \in \mathcal{N}(i)} q_j^{(s+1)}(z_j).$$

See Chaari et al. (2013, Appendix) for a straightforward way to derive (16.13) using the Kullback–Leibler divergence properties.

In practice, when developing the right-hand side of (16.13), the terms that do not depend on $z_i$ can be omitted. The latter are part of the normalizing constant that can be deduced (e.g. in the exponential family case as explained in Beal & Ghahramani, 2003) or computed afterwards. In the HMF-IN case, it becomes

$$q_i^{(s+1)}(z_i) \propto \exp(\mathrm{E}_{q_{\backslash i}^{(s+1)}}[\log p(y_i|z_i, \theta^{(s)}) + \log p(z_i|\mathbf{Z}_{\backslash i}, \alpha^{(s)}, \mathbb{V}^{(s)})]).$$

For a pairwise potential MRF, it becomes

$$q_i^{(s+1)}(z_i) \propto p(y_i|z_i, \theta^{(s)}) \exp\left(\left\langle z_i, \sum_{j \in \mathcal{N}(i)} \mathbb{V}_{ij}^{(s)} \mathrm{E}_{q_j^{(s+1)}}(Z_j) + \alpha_i^{(s)} \right\rangle\right). \qquad (16.14)$$

By way of illustration, let us consider a two-class Potts model for which $z_i \in \{e_1, e_2\}$ and the potentials are defined by $\alpha = 0$ and $V_{ij}(z_i, z_j) = \beta \langle z_i, z_j \rangle$, that is, $\mathbb{V}_{ij} = \mathbb{V} = \beta I_2$ where $I_2$ is the $2 \times 2$ identity matrix. It follows that $\mathrm{E}_{q_j^{(s+1)}}(Z_j) = (q_j^{(s+1)}(e_1), q_j^{(s+1)}(e_2))^\top$. Then equation (16.14) for $z_i = e_1$ reads

$$q_i^{(s+1)}(e_1) \propto p(y_i|z_i = e_1, \theta_1^{(s)}) \exp\left(\beta \sum_{j \in \mathcal{N}(i)} q_j^{(s+1)}(e_1)\right),$$

which after normalization leads to

$$
q_i^{(s+1)}(e_1) = \left( 1 + \frac{p(y_i|z_i = e_2, \theta_2^{(s)})}{p(y_i|z_i = e_1, \theta_1^{(s)})} \exp\left\{ \beta \sum_{j \in \mathcal{N}(i)} (q_j^{(s+1)}(e_2) - q_j^{(s+1)}(e_1)) \right\} \right)^{-1},
$$

$$
q_i^{(s+1)}(e_2) = 1 - q_i^{(s+1)}(e_1). \tag{16.15}
$$

The fixed point equations (16.15) must be solved iteratively, updating each site in turn.

Equation (16.14) can also be recovered from a different point of view. The idea when considering a particular site $i$ is to neglect the fluctuations of the sites interacting with $i$ so that the resulting system behaves as one composed of independent variables. More specifically, for all $j$ different from $i$, the $z_j$ are fixed at their current conditional mean value $\mathrm{E}(Z_j|\mathbf{y}, \Psi^{(s)}])$. However, these mean values are unknown and it is the goal of the approximation to compute them. Therefore, the approximation depends on a self-consistency condition: the mean values that can be computed from the approximate distribution must be equal to the mean values used to define this approximate distribution. Then replacing the exact conditional mean values by the mean values in the approximation leads to a fixed point equation involving these mean values (see Celeux et al., 2003, for more details). Existence and uniqueness of a solution to (16.12) are properties that have not yet been fully understood and will not be discussed here. We refer to Tanaka (2001) for a better insight into the properties of the (potentially multiple) solutions of the mean field equations. Such solutions are usually computed iteratively (see Ambroise & Govaert, 1998, Zhang, 1996, and an erratum in Fessler, 1998).

Despite the relaxation which may make the summation of the VEM E step explicit for a convenient choice of $\tilde{\mathcal{D}}$ (i.e. the computation of $F(q^{(s+1)}, \Psi)$ in (16.11)), VEM remains intractable for hidden Markov random fields. From (16.6) and (16.11), $\theta$ and $(\alpha, \mathbb{V})$ are updated independently, given $q^{(s+1)}$. Under additional commonly used assumptions on $p(\mathbf{y}|\mathbf{z}, \theta)$, $\theta^{(s+1)}$ is computed in closed form (see, for example, Section 16.4). The issue is the update of $(\alpha, \mathbb{V})$ since it requires an explicit expression of the partition function or some related quantities (its gradient, for example).

To overcome this difficulty, different approaches have been proposed. The *mean field* and *simulated field* algorithms proposed in Celeux et al. (2003) are alternatives to VEM that propagate the approximation $q^{(s+1)}$ of $p(\mathbf{z}|\mathbf{y}, \Psi^{(s)})$ to $p(\mathbf{z}|\alpha, \mathbb{V})$. The MCVEM approach (Forbes & Fort, 2007) differs from the previous one in that the approximation method does not lead to a simple valid model but appears as a succession of approximations to overcome successive computational difficulties. Similar ideas have been used successfully to estimate $\mathbb{V}$ in various applications, for example, in Chaari et al. (2013) and Forbes et al. (2013). Another common solution is to fixed $\mathbb{V}$ to a sequence of values using an annealing scheme; see, for example, Scherrer et al. (2009). The parameter $\alpha$ is often set to zero, although it can be added to the set of unknown parameters to be estimated without much difficulty (Celeux et al., 2004).

### 16.3.1 Links with the iterated conditional mode and the Gibbs sampler

As presented in Besag (1974), the iterated conditional mode algorithm involves updating in turn a solution $\mathbf{z}^*$ that satisfies

$$
z_i^{*(s+1)} = \arg\max_{z_i} p(y_i|z_i, \theta^{(s)}) p(z_i|\mathbf{z}_{\mathcal{N}(i)}^{*(s)})
$$

$$
= \arg\max_{z_i} p(y_i|z_i, \theta^{(s)}) \exp\left( \beta \left\langle z_i, \sum_{j \in \mathcal{N}(i)} z_i^{*(s)} \right\rangle \right).
$$

As noted in Celeux et al. (2003), it can be seen as a modal version of the variational mean field in the sense that the fixed point equations are similar, with the mean operator replaced by the mode (max) operator. Similarly, the Gibbs sampler as presented in Geman & Geman (1984) is recovered by sampling

$$z_i^{*(s+1)} \quad \sim \quad p(y_i|z_i, \theta^{(s)}) \exp\left(\beta \left\langle z_i, \sum_{j \in \mathcal{N}(i)} z_i^{*(s)} \right\rangle\right),$$

where $\sim$ indicates a simulation according to the distribution defined on the right-hand side. The Gibbs sampler can be seen as a simulated version of the mean field approximation.

## 16.4 Illustration: MRI Brain Scan Segmentation

We illustrate how the modelling and estimation scheme presented could provide general guidelines to deal with complex joint processes in medical image analysis. We provide two applications, both involving brain MRI data, but in different contexts and illustrating different capabilities of the models presented. Section 16.4.1 deals with image data where each pixel is associated with a univariate observation (a single MR sequence). The emphasis is on a sophisticated use of the external field or singleton potential parameters ($\alpha$). In Section 16.4.2 multivariate observations are considered. Multiple MR sequences are segmented simultaneously. The emphasis is put on the design of the pair potential parameters ($\mathbb{V}$).

### 16.4.1 Healthy brain tissue and structure segmentation

The analysis of MR brain scans is a complex task that requires several sources of information to be taken into account and combined. The analysis is frequently based on segmentations of tissues and of subcortical structures performed by human experts. For automatic segmentation, difficulties arise from the presence of various artefacts such as noise or intensity non-uniformities (see Figure 16.1(a) and (c)). For structures, the segmentation requires in addition the use of prior information usually encoded via a pre-registered atlas. Interest has been growing (see, for example, Ashburner & Friston, 2005; Pohl et al., 2006) in tackling this complexity by allowing the possibility of introducing mutual interactions between components of a model. Such a coupling can be naturally expressed in a statistical framework via the definition of a joint distribution that performs a number of essential tasks. The statistical framework illustrated in this section allows (1) for tissue segmentation using *local* HMRF models, (2) for MRF segmentation of structures and (3) for *local affine* registration of an atlas. All tasks are linked, and completing each one of them can help in refining the others. We specify a joint model from which conditional models are derived. As a result, cooperation between tissues and structures and interaction between the segmentation and registration steps are easily introduced. An explicit joint formulation has the advantage of providing a strategy to construct more consistent or complete models that are open to incorporation of new tasks. Estimation is then carried out using a variational EM framework (see Scherrer et al., 2009, and Forbes et al., 2011, for details). The evaluation performed on both phantoms and real 3 tesla brain scans shows good results and demonstrates the clear improvement provided by coupling the registration step to tissue and structure segmentation.

#### 16.4.1.1   A Markov random field approach to segmentation and registration

We consider a finite set $S$ of $n$ voxels on a regular three-dimensional grid. The observed data $\mathbf{y} = \{y_1, \ldots, y_n\}$ are the intensity values observed respectively at each voxel and the missing data $\mathbf{z} = (\mathbf{t}, \mathbf{s})$ is made up of two sets: the tissue classes $\mathbf{t} = \{t_1, \ldots, t_n\}$ and the subcortical structure classes $\mathbf{s} = \{s_1, \ldots, s_n\}$. The $t_i$ take their values in $\{e_1, e_2, e_3\}$, which represents the three tissues cerebro-spinal fluid (CSF), grey matter and white matter (see Figure 16.1(b)). For the subcortical structure (see Figure 16.1(c)) segmentation we consider $L$ structures, the $s_i$ taking their values in $\{e'_1, \ldots, e'_L, e'_{L+1}\}$ where $e'_{L+1}$ corresponds to an additional background class. Tissues and structures are linked and we denote by $T^{s_i}$ the tissue of structure $s_i$ at voxel $i$. The model parameters $\Psi = (\theta, \mathcal{R})$ include both the intensity distribution parameters $\theta$ and the registration parameters $\mathcal{R}$. We consider them in a Bayesian framework as realizations of random variables. The MRF parameters will be considered here as fixed (see below).

To capture interactions between the various fields $\mathbf{y}$, $\mathbf{t}$, $\mathbf{s}$ and $\Psi$ we adopt a *conditional random field* approach which involves specifying a conditional model $p(\mathbf{t}, \mathbf{s}, \Psi | \mathbf{y})$. We define $p(\mathbf{t}, \mathbf{s}, \Psi | \mathbf{y})$ as a Gibbs measure with energy function $H(\mathbf{t}, \mathbf{s}, \Psi | \mathbf{y})$,

$$p(\mathbf{t}, \mathbf{s}, \Psi | \mathbf{y}) \propto \exp(H(\mathbf{t}, \mathbf{s}, \Psi | \mathbf{y})),$$

where the energy is decomposed into the following terms. We denote by $f(y_i | t_i, s_i, \theta_i)$ positive functions of $y_i$ and consider the decomposition

$$
\begin{aligned}
H(\mathbf{t}, \mathbf{s}, \Psi | \mathbf{y}) = {} & H_T(\mathbf{t}) + H_S(\mathbf{s}) + H_{T,S}(\mathbf{t}, \mathbf{s}) + H_{T,\mathcal{R}}(\mathbf{t}, \mathcal{R}) + H_{S,\mathcal{R}}(\mathbf{s}, \mathcal{R}) \\
& + H_\theta(\theta) + H_\mathcal{R}(\mathcal{R}) + \sum_{i \in S} \log f(y_i | t_i, s_i, \theta_i) .
\end{aligned}
\tag{16.16}
$$

In what follows, we discuss a number of essential tasks and show the terms in (16.16) can be specified so that the model performs the tasks listed below.

*Robust-to-noise segmentation*

Robust-to-noise segmentation is generally addressed via MRF modelling. It introduces local spatial dependencies between voxels, providing a labelling regularization. For tissue and structure segmentations, we use the standard Potts model setting

$$H_T(\mathbf{t}) = \sum_{i \in S} \sum_{j \in \mathcal{N}(i)} \beta_T \langle t_i, t_j \rangle,$$

$$H_S(\mathbf{s}) = \sum_{i \in S} \sum_{j \in \mathcal{N}(i)} \beta_S \langle s_i, s_j \rangle,$$

where $\langle t_i, t_j \rangle$ denotes the scalar product, $\mathcal{N}(i)$ represents the voxels neighbouring $i$, and $\beta_T$ and $\beta_S$ are additional interaction strength parameters.

*Local approach to deal with non-uniformity*

Tissue intensity models are generally estimated globally through the entire volume and then suffer from imperfections at a local level. We adopt a local segmentation alternative. The principle is to locally compute the tissue models in various subvolumes of the initial volume. These models better reflect local intensity distributions and are likely to handle different sources of intensity non-uniformity.

We consider intensity models that depend on the tissue class $k$ but also on the voxel localization, so that $\theta$ decomposes into $\theta = \{\theta_i, \ i \in S\}$ where $\theta_i = (\theta_i^1, \theta_i^2, \theta_i^3)^\top$. Although

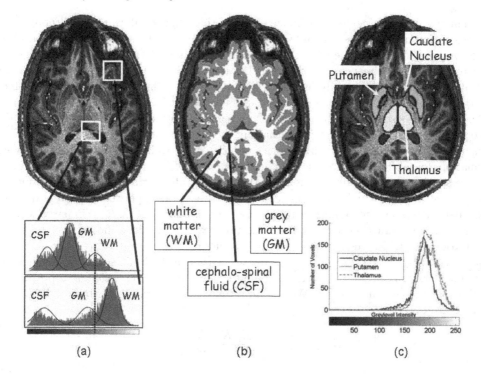

**FIGURE 16.1**

Obstacles to accurate segmentation of MR brain scans. Image (a) illustrates spatial intensity variations: two local intensity histograms (bottom) in two different subvolumes (top) are shown with their corresponding Gaussians fitted using three-component mixture models for the three brain tissues considered. The vertical line corresponds to some intensity value labelled as grey matter or white matter depending on the subvolume. Image (b) illustrates a segmentation into three tissues: white matter, grey matter and cerebro-spinal fluid. Image (c) shows the largely overlapping intensity histograms (bottom) of three grey matter structures segmented manually (top): the putamen, the thalamus and the caudate nuclei.

possible in our Bayesian framework, this general setting results in too many parameters which could not be estimated accurately. The local approach provides an intermediate efficient solution where the $\theta_i$ are first considered as constant over subvolumes. Let $\mathcal{C}$ be a regular cubic partitioning of the volume $S$ into a number of non-overlapping subvolumes $\{V_c, \ c \in \mathcal{C}\}$. We write $\theta = \{\theta_c, \ c \in \mathcal{C}\}$, where $\theta_c = (\theta_c^1, \theta_c^2, \theta_c^3)^\top$ is the common value of all $\theta_i$ for $i \in V_c$. In addition, to ensure consistency and spatial regularity between the local estimations of the $\theta_c$s, we consider an MRF prior $p(\theta) \propto \exp(H_\theta(\theta))$. When Gaussian intensity distributions are considered, this corresponds to assigning *auto-normal* Markov priors to the mean parameters. Apart from the issue of estimating $\theta$, having voxel dependent $\theta_i$ is not a problem. We can easily return to this case from the estimated $\theta_c$s, by using a cubic splines interpolation step.

*Incorporating* a priori *knowledge via local affine atlas registration*

The *a priori* knowledge required for structure segmentation is classically provided via a global non-rigid atlas registration. Most methods first register the prior information to the medical image and then segment the image based on that aligned information. Although reliable registration methods are available, it is still important in the subsequent segmentation task to overcome biases caused by commitment to the initial registration. Also segmentation results provide information that can be used for feedback on registration. Global registration approaches generally lead to a high-dimensional minimization problem which is computationally greedy and subject to a large number of local optima.

We instead choose a hierarchical local affine registration model as in Pohl et al. (2006). We consider (a) a global affine transformation given by parameters $\mathcal{R}^G$, which describes the non-structure-dependent deformations, and (b) one local affine structure-dependent deformation for each structure, defined in relation to $\mathcal{R}^G$ and capturing the residual structure-specific deformations. It follows that $L + 2$ affine transformation parameters $\mathcal{R} = (\mathcal{R}^G, \mathcal{R}_1^S, \ldots, \mathcal{R}_{L+1}^S)$ have to be estimated. Interactions between labels and registration parameters are introduced through $H_{T,\mathcal{R}}(\mathbf{t}, \mathcal{R})$ and $H_{S,\mathcal{R}}(\mathbf{s}, \mathcal{R})$. Similarly to Pohl et al. (2006), the interaction between the structure classes $\mathbf{s}$ and $\mathcal{R}$ is chosen so as to favour configurations for which the segmentation of a structure $l$ is aligned on its prior atlas. We denote by $\zeta_S = \{\zeta_S^l, l = 1, \ldots, L+1\}$ the statistical atlas of the brain subcortical structures under consideration and by $\mathfrak{f}(\mathcal{R}^G, \mathcal{R}_l^S, i)$ the interpolation function assigning a position in the atlas space to the image space. We compute the spatial *a priori* distribution $f_S^l(\mathcal{R}, \cdot)$ of one structure $l$ as

$$f_S^l(\mathcal{R}, i) = \frac{\zeta_S^l(\mathfrak{f}(\mathcal{R}^G, \mathcal{R}_l^S, i))}{\sum_{l'=1}^{L+1} \zeta_S^{l'}(\mathfrak{f}(\mathcal{R}^G, \mathcal{R}_{l'}^S, i))}.$$

The normalization across all structures is necessary as $\mathcal{R}_l^S$ are structure-dependent parameters and multiple voxels in the atlas space could be mapped to one location in the image space. Although some atlas is potentially available for tissues, in our setting we build $f_T$, the spatial *a priori* distribution of the $K = 3$ tissues, from the $f_S^l$:

$$f_T^k(\mathcal{R}, i) = \sum_{l:T^l=k} f_S^l(\mathcal{R}, i) + \frac{1}{K} f_S^{L+1}(\mathcal{R}, i).$$

Agreement between structure segmentation and the atlas is then favoured by setting

$$H_{S,\mathcal{R}}(\mathbf{s}, \mathcal{R}) = \sum_{i \in S} \langle s_i, \log(f_S(\mathcal{R}, i) + \epsilon) \rangle,$$

with the vectorial notation $f_S = (f_S^1, \ldots, f_S^{L+1})^\top$. The logarithm and a positive scalar $\epsilon$ are introduced respectively for homogeneity between probabilities and energies, and to ensure the existence of the logarithm. We choose $\epsilon = 1$, making in addition $H_{S,\mathcal{R}}(\mathbf{s}, \mathcal{R})$ positive, but the overall method does not seem to be sensitive to its exact value. Similarly, we define the interaction between $\mathbf{t}$ and $\mathcal{R}$ by

$$H_{T,\mathcal{R}}(\mathbf{t}, \mathcal{R}) = \sum_{i \in V} \langle t_i, \log(f_T(\mathcal{R}, i) + \epsilon) . \rangle$$

Then the term $H_{\mathcal{R}}(\mathcal{R})$ can be used to introduce *a priori* knowledge to favour estimation of $\mathcal{R}$ close to some average registration parameters computed from a training data set if available. In our case, no such data set were available and we set $H_{\mathcal{R}}(\mathcal{R}) = 0$.

### Cooperative tissue and structure segmentations

Tissues and structures are linked: a structure is made up of a specific tissue and knowledge on structures, and locations provide information for tissue segmentation. Inducing cooperation between tissue and structure segmentations can be done through the term $H_{T,S}(\mathbf{t},\mathbf{s})$. We set

$$H_{T,S}(\mathbf{t},\mathbf{s}) = \sum_{i \in S} \langle t_i, e_{T^{s_i}} \rangle$$

so as to favour situations for which the tissue $T^{s_i}$ of structure $s_i$ is the same as the tissue given by $t_i$. Cooperation between tissue and structure labels also appears via the energy data term $\sum_{i \in S} f(y_i|t_i, s_i, \theta_i)$. Considering Gaussian intensity distributions, we denote by $\mathcal{N}(\cdot|\mu, \lambda^{-1})$ the Gaussian distribution with mean $\mu$ and precision $\lambda$ (i.e. the inverse of the variance). Denoting $\theta_i^k = \{\mu_i^k, \lambda_i^k\}$, we see $\theta_i$ as a three-dimensional vector, so that when $t_i = e_k$, $\mathcal{N}(y_i|\langle t_i, \theta_i \rangle)$ denotes the Gaussian distribution with mean $\mu_i^k$ and precision $\lambda_i^k$. To account for both tissue and structure information, we set

$$f(y_i|t_i, s_i, \theta_i) = \mathcal{N}(y_i|\langle t_i, \theta_i \rangle)^{\frac{1+\langle s_i, e'_{L+1}\rangle}{2}} \mathcal{N}(y_i|\langle e_{T^{s_i}}, \theta_i \rangle)^{\frac{1-\langle s_i, e'_{L+1}\rangle}{2}} .$$

When tissue and structure segmentations contain the same information at voxel $i$, that is, either $t_i = e_{T^{s_i}}$ or $s_i = e'_{L+1}$, the expression for $f$ above reduces to the usual $\mathcal{N}(y_i|\langle t_i, \theta_i \rangle)$. When this is not the case, the expression for $f$ above leads to $\mathcal{N}(y_i|\langle t_i, \theta_i \rangle)^{1/2} \mathcal{N}(y_i|\langle e_{T^{s_i}}, \theta_i \rangle)^{1/2}$, which is a more appropriate compromise.

This achieves the definition of the hierarchical model that can then be fitted to data using a VEM approach as specified in Scherrer et al. (2009) and Forbes et al. (2011).

### 16.4.1.2 Experiments: Joint tissue and structure segmentation

We consider both phantoms and real 3 T brain scans. We use the normal 1 mm³ BrainWeb phantoms database from the McConnell Brain Imaging Center (Collins et al., 1998). These phantoms are generated from a realistic brain anatomical model and an MRI simulator that simulates MR acquisition physics, in which different values of non-uniformity and noise can be added. Because these images are simulated we can quantitatively compare our tissue segmentation to the underlying tissue generative model to evaluate the segmentation performance.

We perform a quantitative evaluation using the Dice similarity metric (Dice, 1945). This metric measures the overlap between a segmentation result and the gold standard. Denoting by $\text{TP}_k$ the number of true positives for class $k$, $\text{FP}_k$ the number of false positives and $\text{FN}_k$ the number of false negatives, the Dice metric is given by

$$d_k = \frac{2\text{TP}_k}{2\text{TP}_k + \text{FN}_k + \text{FP}_k}.$$

It takes values in $[0, 1]$, where 1 represents perfect agreement. Since BrainWeb phantoms contain only tissue information, three subcortical structures were manually segmented by three experts: the left caudate nucleus, the left putamen and the left thalamus. The results we report are for eight BrainWeb phantoms, for 3%, 5%, 7% and 9% of noise with 20% and 40% of non-uniformity for each noise level. Regarding real data, we then evaluate our method on real 3 T MR brain scans (T1 weighted sequence) coming from the Grenoble Institute of Neuroscience.

We then evaluate the performance of the joint tissue and structure segmentation. We consider two cases: our combined approach with fixed registration parameters (LOCUSB-TS) and with estimated registration parameters (LOCUSB-TSR). Table 16.1 shows the evaluation on BrainWeb images using our reference segmentation of the three structures. The table shows the means and standard deviations of the Dice coefficient values obtained for the eight BrainWeb images. It also shows the means and standard deviations of the relative

**TABLE 16.1**
Mean Dice coefficient values obtained on three structures using LOCUSB-TS and LOCUSB-TSR for BrainWeb images, over eight experiments for different values of noise (3%, 5%, 7%, 9%) and non-uniformity (20%, 40%). The corresponding standard deviations are shown in parentheses. The second column shows the results when registration is done as a pre-processing step (LOCUSB-TS). The third columns shows the results with our full model including iterative estimation of the registration parameters (LOCUSB-TSR). The last column shows the relative Dice coefficient improvement for each structure.

| Structure | LOCUSB-TS | LOCUSB-TSR | Relative Improvement |
|---|---|---|---|
| Left thalamus | 91% (0) | 94% (1) | 4% (1) |
| Left putamen | 90% (1) | 95% (0) | 6% (1) |
| Left caudate nucleus | 74% (0) | 91% (1) | 23% (1) |

improvements between the two models LOCUSB-TS and LOCUSB-TSR. In particular, a significant improvement of 23% is observed for the caudate nucleus for the latter model.

The three structure segmentations improve when registration is combined. In particular, in LOCUSB-TS the initial global registration of the caudate nucleus is largely suboptimal, but it is then corrected in LOCUSB-TSR. More generally, for the three structures we observe a stable gain for all noise and inhomogeneity levels.

Figure 16.2 shows the results obtained with LOCUSB-T, and LOCUSB-TSR on a real 3 T brain scan. The structures emphasized in image (c) are the two lateral ventricles, the caudate nuclei, the putamens and the thalamus. Figure 16.2(e) shows in addition a 3D reconstruction of 17 structures segmented with LOCUSB-TSR. The results with LOCUSB-TS are not shown because the differences with LOCUSB-TSR were not visible at this graphical resolution.

We therefore observe a gain in combining tissue and structure segmentation, in particular through the improvement of tissue segmentation for areas corresponding to structures such as the putamens and thalamus. The additional integration of a registration parameter estimation step also provides some significant improvement. It allows for an adaptive correction of the initial global registration parameters and a better registration of the atlas locally.

### 16.4.2 Brain tumor detection from multiple MR sequences

The previous subsection described a possible model for healthy brain segmentation using three normal tissues. When considering brain damage, the number of extra tissues to take into account can vary with the pathology. In this section, we illustrate the possibility of modelling interactions between these tissues via the pair potential parameters.

#### 16.4.2.1 Tissue interaction modelling

A fully automatic algorithm is now proposed to segment glioma MR sequences, by availing of the additional information provided by multiple MR sequences. We adopt a data model comprising five normal tissue classes; white matter, grey matter, ventricular CSF, extra-ventricular CSF, and other. The glioma is modelled by a further four classes representing the diseased tissue state: oedema, non-enhancing, enhancing and necrotic. As illustrated in the previous section, the standard Potts model is often appropriate for clustering since it tends to favour neighbours that are in the same class. However, this model penalizes pairs that have different classes with the same penalty, regardless of the tissues they represent. In practice, it may be more appropriate to encode higher penalties when the tissues are known

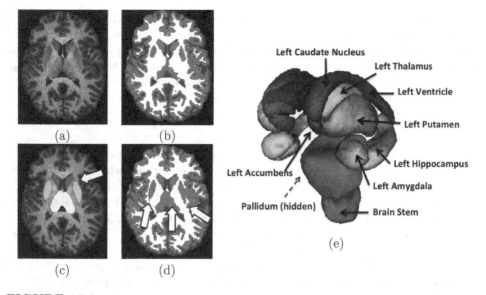

(a)  (b)

(c)  (d)

(e)

**FIGURE 16.2**
Evaluation of LOCUSB-TSR on a real 3 T brain scan shown in image (a). For comparison, the tissue segmentation obtained with LOCUSB-TS is given in image (b). The results obtained with LOCUSB-TSR are shown in the second row. Major differences between tissue segmentations (images (c) and (d)) are indicated using arrows. Image (e) shows the corresponding 3D reconstruction of 17 structures segmented using LOCUSB-TSR. The names of the left structures (use symmetry for the right structures) are indicated in the image.

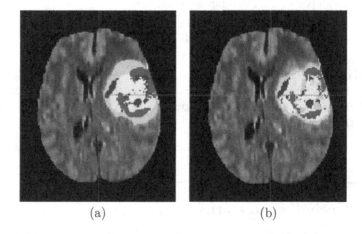

(a)  (b)

**FIGURE 16.3**
Evaluation of P-LOCUS in image (b) on a real 3 T brain scan. The ground truth is shown in image (a).

to be unlikely neighbours. For example, the penalty for a white matter and extraventricular CSF pair is expected to be greater than that of a grey matter and extraventricular CSF pair, as these two classes are more likely to form neighbourhoods. This models the undesirability of abrupt changes in neighbouring tissues. In practice, the interaction matrix $\mathbb{V}$ can be tuned according to experts' *a priori* knowledge, or can be estimated from the data. In the absence of sufficient data to robustly and accurately estimate a full free $\mathbb{V}$ with $G = 9$, further constraints are imposed on the matrix. The four glioma classes are considered a single structure, whose interaction with the normal tissue classes is not dependent on the specific glioma tissue state. Letting $\tau$ be the set of classes comprising the glioma structure and $\bar{\tau}$ the set of healthy tissues, we propose to use for $\mathbb{V}$ a matrix defined by

$$
\mathbb{V}(g, g') = \left\{ \begin{array}{ll} \beta, & \text{for all } (g, g') \text{ such that } g \in \tau \text{ and } g' \in \bar{\tau}, \\ \beta_{g, g'}, & \text{otherwise.} \end{array} \right.
$$

More generally, when prior knowledge indicates that, for example, two given classes are likely to be next to each other, this can be encoded in the matrix with a higher entry for this pair. Conversely, when there is enough information in the data, a full free $\mathbb{V}$ matrix can be estimated and will reflect the class structure (i.e. which class is next to which as indicated by the data) and will then mainly serve as a regularizing term to encode additional spatial information. The fine design of $\mathbb{V}$ may be important in such a case. For another illustration of a non-standard $\mathbb{V}$, see also Forbes et al. (2013).

For the distribution of the observed variables $\mathbf{y}$ given the classification $\mathbf{z}$, the usual conditional independence assumption is made. It follows that the conditional probability of the hidden field $\mathbf{z}$ given the observed field $\mathbf{y}$ is

$$
p(\mathbf{z}|\mathbf{y}, \theta, \mathbb{V}) = W(\mathbb{V})^{-1} \exp \left( -H_{\mathbf{z}}(\mathbf{z}, \mathbb{V}) + \sum_{i \in S} \log f(y_i|z_i, \theta) \right).
$$

For simplicity, no external field $\alpha$ is specified here, but it can be used in practice to account for prior knowledge via anatomical or vascular atlases (see Kabir et al., 2007, or the Appendix in Menze et al., 2015 for more details).

### 16.4.2.2 Experiments: Lesion segmentation

The algorithm referred to as P-LOCUS, for "Pathological LOCUS", was tested on real-patient data from the BRATS 2013 data set. A more complete description of the model used and the results is given in (Menze et al., 2015, Appendix). As an illustration, Figure 16.3 shows the correspondence between the ground truth corresponding to a manual segmentation and the P-LOCUS result.

## 16.5  Concluding Remarks

In this chapter we focused on image segmentation as a typical image processing task that can benefit from a mixture modelling approach. Regarding the specific brain MR application we described, the framework can be adapted to other applications. It provides a strategy and guidelines for dealing with complex joint processes involving more than one identified subprocess. It is based on the idea that defining conditional models is usually more straightforward and captures more explicitly cooperative aspects, including cooperation with external knowledge.

The Bayesian formulation provides additional flexibility such as the possibility of dealing, in a well-based manner, with some sort of non-stationarity in the parameters (like that due to intensity non-uniformities in our MRI example). Of course, depending on the application in mind, more complex energy functions than the one given in our MRI illustration may be necessary. In particular, for our example, it was enough to consider separately cooperation between label sets and spatial interactions. However, one useful extension, to be investigated in future work, would be to add a spatial component in the cooperation mechanisms themselves.

Beyond this illustration, images can be considered in a broad sense meaning that the observed data do not need to be made up of a set of 2D or 3D pixels but could correspond to more general graph structures. The material in this chapter can be more generally applied to *dependent data clustering* as illustrated in Green & Richardson (2002), Vignes & Forbes (2009), and Forbes et al. (2013). In addition, we have not discussed a number of common complications that can occur in the measurement process. This includes issues such as the high dimensionality of the observations, missing observations and heterogeneous observations. Solutions exist in such cases. Some procedures using the EM approach are implemented in the SpacEM$^3$ software (Vignes et al., 2011) available at http://spacem3.gforge.inria.fr. Finally, since image analysis is a vast domain in terms of both methodology and applications, many important contributions are not cited or mentioned in this chapter.

# Bibliography

AMBROISE, C. & GOVAERT, G. (1998). Convergence proof of an EM-type algorithm for spatial clustering. *Pattern Recognition Letters* **19**, 919–927.

ASHBURNER, J. & FRISTON, K. J. (2005). Unified segmentation. *NeuroImage* **26**, 839–851.

BANFIELD, J. D. & RAFTERY, A. E. (1993). Model-based Gaussian and non-Gaussian clustering. *Biometrics* **49**, 803–821.

BEAL, M. J. & GHAHRAMANI, Z. (2003). The variational Bayesian EM algorithm for incomplete data: With application to scoring graphical model structures. In *Bayesian Statistics 7*, J. M. Bernardo, J. O. Berger, A. P. Dawid & A. F. M. Smith, eds. Oxford: Oxford University Press.

BEECKS, C., UYSAL, M. S. & SEIDL, T. (2015). Content-based image retrieval with Gaussian mixture models. In *MultiMedia Modeling: 21st International Conference, MMM 2015, Sydney,NSW, Australia, January 5-7, 2015, Proceedings, Part I*. Cham: Springer, pp. 294–305.

BENBOUDJEMA, D. & PIECZYNSKI, W. (2005). Unsupervised image segmentation using triplet Markov fields. *Computer Vision and Image Understanding* **99**, 476–498.

BESAG, J. (1974). Spatial interaction and the statistical analysis of lattice systems. *Journal of the Royal Statistical Society, Series B* **36**, 192–236.

BESAG, J. (1986). On the statistical analysis of dirty pictures. *Journal of the Royal Statistical Society, Series B* **48**, 259–302.

BLANCHET, J. & FORBES, F. (2008). Triplet Markov fields for the classification of complex structure data. *IEEE Transactions on Pattern Analysis Machine Intelligence* **30**, 1055–1067.

BOUVEYRON, C., GIRARD, S. & SCHMID, C. (2007). High-dimensional discriminant analysis. *Communications in Statistics Theory and Methods* **36**, 2607–2623.

BOYLES, R. (1983). On the convergence of EM algorithms. *Journal of the Royal Statistical Society, Series B* **45**, 47–50.

BYRNE, W. & GUNAWARDANA, A. (2005). Convergence theorems of generalized alternating minimization procedures. *Journal of Machine Learning Research* **6**, 2049–2073.

CELEUX, G., FORBES, F. & PEYRARD, N. (2003). EM procedures using mean field-like approximations for model-based image segmentation. *Pattern Recognition* **36**, 131–144.

CELEUX, G., FORBES, F. & PEYRARD, N. (2004). Modèle de Potts avec champ externe et algorithme de type EM pour la segmentation d'images. In *RFIA, 14th French Meeting AFRIF-AFIA Reconnaissance des Formes & Intelligence Artificielle*. Toulouse.

CELEUX, G. & GOVAERT, G. (1995). Gaussian parsimonious clustering models. *Pattern Recognition* **28**, 781–793.

CHAARI, L., VINCENT, T., FORBES, F., DOJAT, M. & CIUCIU, P. (2013). Fast joint detection-estimation of evoked brain activity in event-related fMRI using a variational approach. *IEEE Transactions on Medical Imaging* **32**, 821–837.

CHALMOND, B. (1989). An iterative Gibbsian technique for reconstruction of $m$-ary images. *Pattern Recognition* **22**, 747–761.

CHANDLER, D. (1987). *Introduction to Modern Statistical Mechanics*. New York: Oxford University Press.

COLLINS, D. L., ZIJDENBOS, A. P., KOLLOKIAN, V., SLED, J. G., KABANI, N. J., HOLMES, C. J. & EVANS, A. C. (1998). Design and construction of a realistic digital brain phantom. *IEEE Transactions on Medical Imaging* **17**, 463–468.

CSISZAR, I. & TUSNADY, G. (1984). Information geometry and alternating minimization procedures. *Statistics & Decisions* **1**, 205–237.

DEMPSTER, A. P., LAIRD, N. M. & RUBIN, D. B. (1977). Maximum likelihood from incomplete data via the EM algorithm. *Journal of the Royal Statistical Society, Series B* **39**, 1–38.

DICE, L. R. (1945). Measures of the amount of ecologic association between species. *Ecology* **26**, 297–302.

FESSLER, J. (1998). Comments on "The convergence of mean field procedures for MRF's". *IEEE Transactions on Image Processing* **7**, 917.

FORBES, F., CHARRAS-GARRIDO, M., AZIZI, L., DOYLE, S. & ABRIAL, D. (2013). Spatial risk mapping for rare disease with hidden Markov fields and variational EM. *Annals of Applied Statistics* **7**, 1192–1216.

FORBES, F. & FORT, G. (2007). Combining Monte Carlo and mean field like methods for inference in hidden Markov random fields. *IEEE Transactions on Image Processing* **16**, 824–837.

FORBES, F., SCHERRER, B. & DOJAT, M. (2011). Bayesian Markov model for cooperative clustering: Application to robust MRI brain scan segmentation. *Journal de la Société Française de Statistique* **152**.

FRANÇOIS, O., ANCELET, S. & GUILLOT, G. (2006). Bayesian clustering using hidden Markov random fields in spatial genetics. *Genetics* **174**, 805–816.

GEBRU, I., ALAMEDA-PINEDA, X., FORBES, F. & HORAUD, R. (2016). EM algorithms for weighted-data clustering with application to audio-visual scene analysis. *IEEE Transactions on Pattern Analysis and Machine Intelligence* **38**, 2402–2415.

GEMAN, S. & GEMAN, D. (1984). Stochastic relaxation, Gibbs distributions and the Bayesian restoration of images. *IEEE Transactions on Pattern Analysis and Machine Intelligence* **6**, 721–741.

GEROGIANNIS, D., NIKOU, C. & LIKAS, A. (2009). The mixtures of Student's $t$-distributions as a robust framework for rigid registration. *Image Vision Computing* **27**, 1285–1294.

GREEN, P. J. & RICHARDSON, S. (2002). Hidden Markov models and disease mapping. *Journal of American Statistical Association* **97**, 1055–1070.

JORDAN, M. I., GHAHRAMANI, Z., JAAKKOLA, T. S. & SAUL, L. K. (1998). An introduction to variational methods for graphical models. In *Learning in Graphical Models*, M. I. Jordan, ed. Dordrecht: Kluwer Academic Publishers, pp. 105–162.

KABIR, Y., DOJAT, M., SCHERRER, B., FORBES, F. & GARBAY, C. (2007). Multimodal MRI segmentation of ischemic stroke lesions. In *29th Annual International Conference of the IEEE Engineering in Medicine and Biology Society (EMBC)*. Lyon, France.

KARAVASILIS, V., BLEKAS, K. & NIKOU, C. (2012). A novel framework for motion segmentation and tracking by clustering incomplete trajectories. *Computer Vision and Image Understanding* **116**, 1135–1148.

KARLIS, D. & MELIGKOTSIDOU, L. (2007). Finite mixtures of multivariate Poisson distributions with application. *Journal of Statistical Planning and Inference* **137**, 1942–1960.

LOWE, D. (2004). Distinctive image features from scale-invariant keypoints. *International Journal of Computer Vision* **60**, 91–110.

MCLACHLAN, G. & KRISHNAN, T. (2008). *The EM Algorithm and Extensions*. New York: Wiley. Second Edition.

MCLACHLAN, G. & PEEL, D. (2000). Robust mixture modelling using the t distribution. *Statistics and Computing* **10**, 339–348.

MENZE, B., REYES, M. & VAN LEEMPUT, K. F. A. (2015). The multimodal brain tumorimage segmentation benchmark (BRATS). *IEEE Transactions on Medical Imaging* **34**, 1993–2024.

NEAL, R. M. & HINTON, G. E. (1998). A view of the EM algorithm that justifies incremental, sparse, and other variants. In *Learning in Graphical Models*, M. I. Jordan, ed. Dordrecht: Kluwer Academic Publishers, pp. 355–368.

NIKNEJAD, M., RABBANI, H. & BABAIE-ZADEH, M. (2015). Image restoration using Gaussian mixture models with spatially constrained patch clustering. *IEEE Transactions on Image Processing* **24**, 3624–3636.

POHL, K. M., FISHER, J., GRIMSON, E., KIKINIS, R. & WELLS, W. M. (2006). A Bayesian model for joint segmentation and registration. *NeuroImage* **31**, 228–239.

QIAN, W. & TITTERINGTON, M. (1991). Estimation of parameters in hidden Markov models. *Philosophical Transactions of the Royal Society London Series A* **337**, 407–428.

SCHERRER, B., FORBES, F. & DOJAT, M. (2009). A conditional random field approach for coupling local registration with robust tissue and structure segmentation. In *Medical Image Computing and Computer-Assisted Intervention – MICCAI 2009: 12th International Conference, London*, G.-Z. Yang, D. Hawkes, D. Rueckert, A. Noble & C. Taylor, eds. Berlin: Springer-Verlag.

TANAKA, T. (2001). Information geometry of mean-field approximation. In *Advanced Mean Field Methods*, M. Opper & D. Saad, eds., chap. 17. Cambridge, MA: MIT Press.

VIGNES, M., BLANCHET, J., LEROUX, D. & FORBES, F. (2011). SpaCEM3, a software for biological module detection when data is incomplete, high dimensional and dependent. *Bioinformatics* **27**, 881–882.

VIGNES, M. & FORBES, F. (2009). Gene clustering via integrated Markov models combining individual and pairwise features. *IEEE Transaction on Computational Biology and Bioinformatics* **6**, 260–270.

WAINWRIGHT, M. & JORDAN, M. (2003). Graphical models, exponential families, and variational inference. Tech. Rep. 649, UC Berkeley, Department of Statistics.

WAINWRIGHT, M. & JORDAN, M. (2005). A variational principle for graphical models. In *New Directions in Statistical Signal Processing*, S. Haykin, T. Principe, T. Sejnowski & J. McWhirter, eds., chap. 11. Cambridge, MA: MIT Press.

WU, C. F. J. (1983). On the convergence properties of the EM algorithm. *Annals of Statistics* **11**, 95–103.

ZHANG, J. (1996). The convergence of mean field procedures for MRF's. *IEEE Transactions on Image Processing* **5**, 1662–1665.

# 17

---

## Applications in Finance

**John M. Maheu and Azam Shamsi Zamenjani**

*McMaster University, Canada; University of New Brunswick, Canada*

## CONTENTS

---

## 17.1 Introduction

The financial literature has recognized for some time (Fama, 1965; Mandelbrot, 1997) that return distributions exhibit fat tails and asymmetry which are significantly different than the normal distribution. This phenomenon is more evident in daily and higher-frequency returns and has led many researchers to use different models to explain skewness and excess kurtosis. Mixture models provide a flexible approach to capture these features of the return distribution.

Mixture models can be divided into two categories in finance. Continuous mixtures have a mixing variable which follows a continuous distribution. The stochastic volatility model of Clark (1973) and the subsequent literature (Taylor, 1994) is a good example of this specification. The second class of models are discrete mixtures. These can be finite or countably infinite and have a mixing variable that follows a discrete distribution. Examples include a finite mixture of normals and Markov switching models. This chapter will focus on discrete mixture models used in finance. Section 17.2 considers finite mixture models, whereas Section 17.3 deals with (countably) infinite mixture models. Section 17.4 concludes.

## 17.2    Finite Mixture Models

We begin with independent and identically distributed (i.i.d.) mixtures combined with a conditional variance function.

### 17.2.1    i.i.d. mixture models with volatility dynamics

Time-series data at quarterly frequencies and higher display pronounced changes and strong persistence in the conditional variance. Two standard models to capture this are generalized autoregressive conditional heteroscedasticity (GARCH) and stochastic volatility (SV) models. Discrete mixture specifications can be incorporated with volatility dynamics to improve the fit of the conditional distribution.

As a simple example consider the model in Alexander & Lazar (2006) which combines a finite mixture with GARCH. The density of $y_t$ given $y^{t-1}$ after de-meaning the data is

$$
\begin{aligned}
y_t &= \varepsilon_t, \\
p(\varepsilon_t | y^{t-1}) &= \sum_{g=1}^{G} \eta_g \phi_g(\varepsilon_t | \mu_g, \sigma_{gt}^2), \quad \sum_{g=1}^{G} \eta_g = 1, \quad \sum_{g=1}^{G} \eta_g \mu_g = 0, \\
\sigma_{gt}^2 &= \omega_g + \alpha_g \varepsilon_{t-1}^2 + \beta_g \sigma_{g,t-1}^2, \quad g = 1, \dots, G.
\end{aligned}
$$

In this model, the effect of past shocks, $\varepsilon_{t-1}^2$, does not depend on the past state, that is, is not state-dependent. This model provides a wide range of values for kurtosis. Estimation is conducted using maximum likelihood. Alexander & Lazar (2006) show that a mixture of two normal-GARCH models describes exchange rate returns better than the GARCH model with symmetric and asymmetric Student $t$ innovations. Similar results are found in Bai et al. (2003). Extensions to the multivariate setting with multivariate GARCH processes directing the conditional covariance are discussed in Bauwens et al. (2007) and Galeano & Ausín (2010).

Mixture models have also been incorporated into SV specifications. Durham (2007) models the distribution of daily S&P 500 index data as a mixture of normals with conditional variances following a stochastic volatility process. The model is

$$
\begin{aligned}
y_t &= \mu + \exp(h_{t-1}/2)\, u_t, \\
h_t &= \alpha + \delta h_{t-1} + \sigma_v v_t, \\
v_t &= \rho u_t + (1 - \rho)\epsilon_t, \quad \epsilon_t \sim \mathcal{N}(0, 1),
\end{aligned}
$$

where $u_t$ follows a mixture of normals and $u_t$ and $\epsilon_t$ are uncorrelated (i.e. $\mathrm{Corr}(u_t, \epsilon_t) = 0$). The model presented here differs from the usual specification since $u_t$ follows a mixture of normals and is non-Gaussian while the leverage effect still operates through $\mathrm{Corr}(u_t, v_t) = \rho$. Durham shows that the mixture normal-SV model considerably improves upon several affine-jump models in capturing the features of the data and the shape of the distribution.

### 17.2.2    Markov switching models

A finite Markov switching (MS) model or hidden Markov model (HMM) assumes that the dynamics of a data series, $y_t$, depend on a discrete latent variable, $z_t$, postulated to follow a Markov chain with realizations in $\{1, \dots, G\}$. This model was popularized by Hamilton (1988, 1989) in economics where it was applied to interest rate regimes and expansion and recessions in real growth; see also Chapter 13 of this handbook.

An important contribution of these early papers was to provide a filter that gives inference on the state $z_t$ given time $t$ information, $p(z_t|y^t)$, and facilities construction of the likelihood function. The latent variable $z_t$ is referred to as the state or regime of the time series at time $t$ and provides a natural descriptor to many financial concepts such as low and high volatility or bull and bear stock market states. Evolution of the state is governed by the transition probabilities $\xi_{gh} = \mathrm{P}(z_t = h|z_{t-1} = g)$ with transition matrix

$$\xi = \begin{bmatrix} \xi_{11} & \xi_{12} & \cdots & \xi_{1G} \\ \xi_{21} & \xi_{22} & \cdots & \xi_{2G} \\ \cdots & \cdots & \cdots & \cdots \\ \xi_{G1} & \xi_{G2} & \cdots & \xi_{GG} \end{bmatrix},$$

where each row sums to 1. For a discussion of the properties of a Markov chain, see Hamilton (1994, Chapter 22).

Given a conditional data density $f(y_{t+1}|\theta_{z_{t+1}})$ with state-dependent parameter $\theta_{z_{t+1}}$, the predictive density one period ahead conditional on $z_t$ is the mixture

$$p(y_{t+1}|z_t) = \sum_{z_{t+1}=1}^{G} \xi_{z_t, z_{t+1}} f(y_{t+1}|\theta_{z_{t+1}}). \tag{17.1}$$

Unlike the i.i.d. mixture discussed in most other chapters, the predictive density mixture weights $\xi_{z_t, z_{t+1}}$ depend on $z_t$ and allow for time variation in the mixture. This makes it possible not only to capture changes in the conditional mean and conditional variance, but also changes in the shape of the density such as tail behaviour and asymmetry.

By far the most common approach to estimation is likelihood based. Maximum likelihood estimation (Hamilton, 1989) and Gibbs sampling (Chib, 1996) for these models are now standard. Hamilton (1990) presents an EM algorithm for estimation of MS models which exploits the smoothed estimates of the states. Chib (1996) augments the parameter space to include the latent variables and simulates the latent variables jointly through the forward-filter backward sampler (FFBS); see also Algorithm 13.1 in Chapter 13 of this volume.

Early applications of the model were to the equity market risk premium and bull and bear markets. Turner et al. (1989) consider a two-state economy, characterized by high and low volatility, in which the variance of a portfolio's excess return is assumed to be state-dependent. They model the risk premium of an asset as

$$\mu_t = \alpha + \gamma\, \mathrm{P}(z_t = 1|y^{t-1}),$$

with $z_t \in \{0, 1\}$, where $\alpha$ represents the asset's risk premium in the low-variance state and $\mathrm{P}(z_t = 1|y^{t-1})$ captures the agent's learning about the state.

Pagan & Schwert (1990) compare Hamilton's two-state MS model for US monthly stock return volatility with several models, including GARCH and exponential GARCH models. Maheu & McCurdy (2000a) model bull and bear markets by defining a duration-dependent two-state autoregressive model to account for the nonlinearity in both the conditional mean and the conditional variance of stock returns:

$$y_t = \mu_{z_t} + \sum_{j=1}^{l} \varphi_j (y_{t-j} - \mu_{z_{t-j}}) + \sqrt{h_{z_t}}\, v_t, \quad v_t \sim \mathcal{N}(0, 1), \tag{17.2}$$

where $l$ is the number of lags, and $h_{z_t}$ follows an ARCH structure. Duration, the length of time in the same state, is defined as

$$D(z_t) = \begin{cases} D(z_{t-1}) + 1, & \text{if } z_t = z_{t-1}, \\ 1, & \text{otherwise.} \end{cases}$$

Duration $D(z_t)$ is the length of a run of consecutive states $z_t$ and enters the transitions probabilities as

$$p(z_t = j | z_{t-1} = j, D(z_{t-1}) = d) = (1 + \exp(-\gamma_{1j} - \gamma_{2j} d))^{-1}$$

for parameters $\gamma_{1j}, \gamma_{2j}$, $j = 1, 2$. They find negative duration dependence (the probability of exiting the state decreases with duration) with $\gamma_{2j} > 0$ in both states and extend this model to allow the state duration to affect the conditional mean and conditional variance.

Rydén et al. (1998) show that many of the stylized facts from a long time series of daily S&P 500 returns can be captured by a MS model with the exception of the strong persistence properties of the absolute value of returns. Billio & Pelizzon (2000) adopt a two-state MS model to estimate the value-at-risk[1] of a portfolio and show that it outperforms the GARCH and independent mixture models. Guidolin & Timmermann (2007) consider asset allocation problems in a four-moment international capital asset pricing model framework, and find distinguishable bull and bear regimes.

In the exchange rate context, Engel (1994) applies an MS model to several quarterly exchange rates and shows its superiority at predicting the direction of change compared to a random walk or the forward rate. Bollen et al. (2000) assess both in-sample and out-of-sample performance of MS models with state-dependent means and variances in explaining dynamics of foreign exchange rates. Beine et al. (2003) link the dynamics (regime switches) of returns and volatility of exchange rates with official interventions.

What is clear from this literature is that MS models are well suited to capturing lower-frequency changes in the distribution of returns. With some exceptions these models cannot account for the autocorrelation patterns of daily volatility proxies. One method of obtaining a richer autocorrelation structure is to model a higher-order Markov chain as in Maheu & McCurdy (2000b) through duration dependence. Another approach that is consistent with long memory in volatility is the Markov switching multifractal model of Calvet & Fisher (2004). In this model returns follow

$$
\begin{aligned}
y_t &= \sigma_t v_t, \quad v_t \overset{i.i.d.}{\sim} \mathcal{N}(0,1), \\
\sigma_t &= \sigma(M_{1,t} M_{2,t} \ldots, M_{G,t})^{1/2},
\end{aligned}
$$

where $G$ is the number of volatility frequencies, $\sigma$ is a positive constant, and $M_{g,t}$, $g = 1, \ldots, G$, are volatility components with marginal distribution $M$ such that $M_{g,t} \geq 0$ and $\mathrm{E}(M_{g,t}) = 1$. The multipliers $M_{1,t}, M_{2,t}, \ldots, M_{G,t}$ at a given time $t$ are statistically independent. The vector $M_t = (M_{1,t}, M_{2,t}, \ldots, M_{G,t})^\top$ is latent, and must be inferred recursively by Bayesian updating. $M_t$ follows a first-order Markov process:

$$
M_{g,t} = \begin{cases} \text{a new draw from distribution } M, & \text{with probability } \pi_g, \\ M_{g,t-1}, & \text{with probability } 1 - \pi_g, \end{cases}
$$

where $\pi_g = 1 - (1 - \pi_1)^{b^{g-1}}$. Calvet & Fisher (2004) show that even a simple two-state distribution for $M$ can capture long memory in volatility as well as sudden volatility movements. The unknown parameters are $\sigma > 0$, $\pi_1 \in (0,1)$, $b \in (1, \infty)$, and the hyperparameters in distribution $M$.

Another approach to capturing the strong persistence in volatility is to combine MS dynamics with conventional GARCH and SV models.

---

[1] Value-at-risk indicates the potential loss associated with an unfavourable movement in market prices over a given time period at a certain confidence level.

## 17.2.3  Markov switching volatility models

The earlier literature explored combining ARCH with Markov switching dynamics (Cai, 1994; Hamilton & Susmel, 1994; Maheu & McCurdy, 2000a; Susmel, 2000; Kaufmann & Frühwirth-Schnatter, 2002). The Markov switching GARCH (MS-GARCH) model is more challenging to estimate because of its intractable likelihood function stemming from path dependence. Consider a simple example of a $G$-state MS-GARCH$(1,1)$ model in which only the intercept in the conditional variance is subject to parameter change,[2]

$$
\begin{aligned}
y_t &= \mu + \varphi y_{t-1} + \varepsilon_t, \\
\varepsilon_t &= \sigma_{z_t} v_t, \quad v_t \sim \mathcal{N}(0,1), \quad z_t \in \{1, \ldots, G\}, \\
\sigma_t^2(z^t) &= \omega_{z_t} + \alpha \varepsilon_{t-1}^2 + \beta \sigma_{t-1}^2(z^{t-1}),
\end{aligned}
\tag{17.3}
$$

with $\omega_{z_t} > 0$, $\alpha \geq 0$ and $\beta \geq 0$. In this formulation $\sigma_1^2(z_1)$ is a function of $z_1$ through $\omega_{z_1}$ which feeds into $\sigma_2^2(z^2)$ which is a function of both $\omega_{z_2}$ and $\sigma_1^2(z_1)$ and so on. This process continues so that the entire history $z^t$ of states up to time $t$ affects the conditional variance $\sigma_t^2(z^t)$ at time $t$. At any $t$, there are $G^t$ paths that the volatility process can take. These must be integrated out of the likelihood for maximum likelihood estimation.

One practical, albeit *ad hoc*, way to get around this path dependence is to truncate the dependence. Gray (1996) replaces (17.3) with two equations,

$$
\sigma_t^2(z_t) = \omega_{z_t} + \alpha \varepsilon_{t-1}^2 + \beta \sigma_{t-1}^2,
$$

$$
\sigma_{t-1}^2 = \sum_{g=1}^{G} \mathrm{P}(z_{t-1} = g | y^{t-2}) \sigma_{t-1}^2(z_{t-1} = g),
$$

where $\mathrm{P}(z_{t-1} = g | y^{t-2})$ is the probability of state $z_{t-1}$ given time $t-2$ information. Integrating over the states in $\sigma_{t-1}^2$ removes the path dependence problem. Applying this model to short-term interest rate provides a better fit to the data compared to existing competitive models.

Taking a similar approach, Dueker (1997) considers a two-state MS-GARCH model with Student $t$ innovations, and shows that adding MS to the pure GARCH model improves the predictions of the option-implied volatility of stocks. Klaassen (2002) uses the same idea as in Gray (1996), after substituting $p(z_{t-1} | y^{t-2})$ with $p(z_{t-1} | y^{t-1})$ and therefore using more information, and shows that applying the model to data on three US dollar exchange rates results in a significant improvement in out-of-sample volatility forecasts.

One criticism of Gray's (Gray (1996)) approach, raised by Haas et al. (2004), is that the GARCH parameters within each regime are not clearly associated. The solution suggested by Haas et al. (2004) is to keep track of $G$ different GARCH processes using the $G$ state-dependent parameters. An example of their model is

$$
\begin{aligned}
y_t &= \varepsilon_t, \\
\varepsilon_t &= \sigma_{z_t} v_t, \quad v_t \sim \mathcal{N}(0,1), \quad z_t \in \{1, \ldots, G\}, \\
\sigma_t^2(z_t) &= \omega_{z_t} + \alpha \varepsilon_{t-1}^2 + \beta \sigma_{t-1}^2(z_t).
\end{aligned}
$$

Note that $\sigma_t^2(z^t)$ is replaced by $\sigma_t^2(z_t)$. They apply this analytically tractable model to several exchange rate return series and show that the model fits the exchange rate data and also captures the dynamic properties of the data better than a GARCH model with Student $t$ innovations. Haas & Mittnik (2008) extend this framework to a multivariate setting.

---

[2]It is straightforward to allow $\alpha$ and $\beta$ to be state-dependent, but empirically $\omega$ is the most important parameter to allow to change.

Ang & Chen (2002) apply several models including MS and MS-GARCH models, following the Gray (1996) approach, to examine the asymmetric correlations between stocks (correlations between stocks are much greater for downside moves than for upside moves.). Francq & Zakoïan (2005) study a MS-GARCH model, and present the necessary and sufficient conditions for the existence of moments.

A different approach to dealing with the path dependence in the MS-GARCH model is to integrate it out in the estimation process. This is consistent with the model assumptions but computationally intensive. Bauwens et al. (2010) propose a Gibbs sampling algorithm for estimating the MS-GARCH model which circumvents the path dependence problem by augmenting the parameter space to include the state variables. Bauwens et al. (2011) approach the problem by particle MCMC (combination of sequential Monte Carlo and Markov chain Monte Carlo) which allows them to sample the state variables jointly and improve the model performance. Henneke et al. (2011) estimate an MS-ARMA-GARCH model by developing an MCMC algorithm, and apply the model to returns from the New York Stock Exchange. He & Maheu (2010) pursue a Bayesian particle filtering approach in an MS-GARCH model designed to detect structural changes in volatility.

Billio et al. (2016) propose a number of Bayesian sampling approaches for the latent state including a state-space representation of the MS-GARCH model estimated by an approximate Kalman filter. Augustyniak (2014) develops a framework exploiting both a Monte Carlo expectation maximization algorithm and importance sampling.

Stochastic volatility modelling has also been augmented with Markov switching dynamics. Due to the complications in estimating an MS-SV, which includes two levels of the latent variables, most of the work uses Bayesian simulation methods. A version of the MS-SV model, considered in So et al. (1998), is formulated as

$$
\begin{aligned}
y_t &= \mu + \exp(h_t/2)u_t, \quad u_t \sim \mathcal{N}(0,1), \\
h_t &= \alpha_{z_t} + \delta h_{t-1} + \sigma_v v_t, \quad v_t \sim \mathcal{N}(0,1),
\end{aligned}
\tag{17.4}
$$

where

$$
\alpha_{z_t} = \gamma_1 + \sum_{g=2}^{G} \gamma_g \mathbb{I}(z_t > g),
$$

and all $\gamma_g$ for $g \geq 2$ are negative; hence, the first regime corresponds to the highest-level state. Note that in this formulation the drift of the latent variable $h_t$ is determined by the state variable $z_t$. Working in a Bayesian framework, So et al. (1998) sample $h_t$ by viewing model (17.4) as a partial non-Gaussian state space model (Shephard, 1994). They jointly sample the state variables using data augmentation and applying the filter introduced in Carter & Kohn (1994) and Chib (1996). They estimate a model with three states for returns on the S&P 500. Working with the same model, Carvalho & Lopes (2007) propose a customized auxiliary particle filtering approach by which they sequentially learn about states and parameters. Therefore, there is no need to rerun the whole algorithm after the arrival of a new observation, as opposed to the MCMC approach employed in So et al. (1998).

Billio & Cavicchioli (2013) consider the MS-SV model

$$
\begin{aligned}
y_t &= \mu + \exp(h_t/2)u_t, \quad u_t \sim \mathcal{N}(0,1), \\
h_t &= \alpha_{z_t} + \delta_{z_t} h_{t-1} + \sigma_{z_t} v_t, \quad v_t \sim \mathcal{N}(0,1),
\end{aligned}
$$

where $h_t$ denotes the log variance. After linearizing (Harvey et al., 1994), they apply an approximate Kalman filter (Kim & Nelson, 1999) to the resulting state space model, and show the feasibility of the approach using one-month US Treasury bill rates.

### 17.2.4 Jumps

Another area of the financial literature that features discrete mixtures is models of jumps. Since Press (1967), jumps have become important in capturing unusual return events as well as reconciling pricing models with derivative prices.

Consider the following SV-jump model from Lopes & Polson (2010). After an Euler discretization we have the following model for log returns $y_t$:

$$
\begin{aligned}
y_t &= \exp(h_t/2)\varepsilon_t + J_t v_t, \qquad\qquad\qquad (17.5)\\
h_t &= \alpha + \beta h_{t-1} + \sigma u_t, \quad u_t \sim \mathcal{N}(\mu_u, \sigma_u^2),\\
J_t &\sim \mathcal{B}er(p), \quad v_t \sim \mathcal{N}(\mu_v, \sigma_v^2).
\end{aligned}
$$

$\varepsilon_t$ is normally distributed and $h_t$ denotes the unobserved log volatility. $J_t \in \{0, 1\}$ is a jump indicator and governed by a Bernoulli distribution with parameter $p$, while $v_t$ is the jump size. $y_t$ given $h_t$ is a mixture of normal and jump innovations, $\varepsilon_t$ and $v_t$. This model can be estimated by MCMC or particle filtering methods as done in Lopes & Polson (2010). Often the focus is to estimate jump times $\{t|J_t = 1\}$ and the associated jump sizes.

The previous model is an example of jumps that impact prices only. Recent research has focused on volatility jumps. Eraker et al. (2003) incorporate jumps in both returns and volatility in a bivariate continuous-time stochastic volatility model. Using S&P 500 and Nasdaq 100 index returns, they show that both of these components (jump in return and jump in volatility) have a strong impact on option pricing and generate more realistic implied volatility curves. Todorov & Tauchen (2011) analyse the movements of high-frequency data on the VIX index and suggest that jumps are an important channel for generating leverage effects since jumps in returns and volatility mostly occur at the same time but in opposite directions.

Bates (2000) compares a continuous SV-jump model with a pure SV model to explain option prices, and shows that the negative correlation between return and volatility is not sufficient to explain the negative skewness implicit in the post-1987 data, and that during this period the SV-jump model outperforms the SV model in fitting the S&P 500 option prices. Eraker (2004) considers a continuous SV-jump model in which jumps to volatility and prices follow the same Poisson process. This model of stochastic volatility with correlated jumps is written as

$$
\begin{aligned}
\frac{dS_t}{S_t} &= a\,dt + \sqrt{V_t}\,dW_t^S + dJ_t^S,\\
dV_t &= \kappa(\theta - V_t)\,dt + \sigma_V\sqrt{V_t}\,dW_t^V + dJ_t^V,
\end{aligned}
$$

where $S_t$ denotes the price, $V_t$ is the volatility process (variance), and $dW_t^S$ and $dW_t^V$ are Brownian processes such that $\mathrm{E}(dW_t^S, dW_t^V) = \rho$. Eraker (2004) also allows the jump intensity to depend on volatility. He finds that this model fits the option prices in-sample, but that adding jumps to the model does not improve the performance significantly.

In a closely related literature, the SV component in the model above is replaced with a GARCH process (Vlaar & Palm, 1993). This model is extended to a time-varying jump intensity by Chan & Maheu (2002) with

$$
\begin{aligned}
y_t &= \mu + \sigma_t u_t + \sum_{g=1}^{n_t} v_{t,g}, \quad u_t \sim \mathcal{N}(0, 1),\\
\sigma_t^2 &= \omega + \alpha\varepsilon_{t-1}^2 + \beta\sigma_{t-1}^2, \quad \varepsilon_t = y_t - \mu,\\
n_t | y^{t-1} &\sim \mathcal{P}(\lambda_t), \quad v_{t,g} \sim \mathcal{N}(\mu_{v,t}, \sigma_{v,t}^2),
\end{aligned}
$$

where $n_t$ is the number of jumps that arrive between $t-1$ and $t$ (allowing for multiple jumps during a time interval) governed by a Poisson distribution, and $v_{t,g}$ is the conditionally normal jump size. $\mu_{v,t}$ and $\sigma_{v,t}^2$ can be set to constants or considered to be functions of past returns. The jump intensity $\lambda_t$ is time-varying and follows

$$
\begin{aligned}
\lambda_t &= \lambda_0 + \rho\lambda_{t-1} + \gamma q_{t-1}, \\
q_{t-1} &= \mathrm{E}(n_{t-1}|y^{t-1}) - \mathrm{E}(n_{t-1}|y^{t-2}),
\end{aligned}
$$

where $q_{t-1}$ is a derived jump innovation. The model for $\lambda_t$ explicitly allows for jump clustering over time through autocorrelation when $\rho \neq 0$. Chan & Maheu (2002) apply the model to capture jump dynamics in stock market returns, and infer the number of jumps. Allowing the jump intensity to be time-varying improves the volatility forecasts of the stock market. Maheu & McCurdy (2004) apply a similar GARCH-jump model to both individual stocks and indices, but allow inferred jumps to affect the GARCH volatility component. They find that the GARCH-jump mixture model accurately identifies jumps and provides superior out-of-sample forecasts of conditional variance compared to the asymmetric GARCH model with fat-tailed innovations. Maheu et al. (2013) extend Chan & Maheu (2002) and Maheu & McCurdy (2004), and show that jump risk is priced in the market index and contributes to the equity premium.

A closely related model is found in Christoffersen et al. (2012). They estimate four nested models of GARCH-jump models with correlated jumps in returns and volatility and time-varying jump intensity which together deliver a better fit to the returns and improve option valuation. The general model is formulated as

$$
\begin{aligned}
y_t &= \mu + (\lambda_u - \tfrac{1}{2})h_{u,t} + (\lambda_v - \zeta)h_{v,t} + u_t + v_t, \quad u_t \sim \mathcal{N}(0, h_{u,t}), \\
h_{u,t} &= \omega_u + b_u h_{u,t-1} + \frac{a_u}{h_{u,t-1}}(u_{t-1} + v_{t-1} - c_u h_{u,t-1})^2, \\
h_{v,t} &= \omega_v + b_v h_{v,t-1} + \frac{a_v}{h_{v,t-1}}(u_{t-1} + v_{t-1} - c_v h_{v,t-1})^2,
\end{aligned}
$$

where $h_{u,t}$ is the conditional variance of $u_t$. $v_t$ is conditionally distributed as compound Poisson $J(h_{v,t}, \theta, \delta^2)$ where $h_{v,t}$ is the jump intensity, $\theta$ the mean jump size, and $\delta^2$ the jump size variance.

Duan et al. (2006) model correlated jumps in asset prices and volatility in a related GARCH-jump model, provide the risk-neutral dynamics, and show that the discrete-time GARCH-jump models converge to their continuous-time counterparts.

Working with a bivariate GARCH-jump model with time-varying common jump intensity, Chan (2008) models the joint density of currency spot rates and futures, and offers a dynamic hedging strategy which results in a lower portfolio risk for foreign currencies.

## 17.3  Infinite Mixture Models

### 17.3.1  Dirichlet process mixture model

Countably infinite mixtures are a popular approach in Bayesian nonparametric methods. A typical model for $y_t$ is

$$
p(y_t) = \int f(y_t|\theta)dH(\theta), \tag{17.6}
$$

where $f(\cdot|\cdot)$ is a kernel density, $H$ is almost surely discrete,

$$H(\cdot) = \sum_{g=1}^{\infty} \eta_g \delta_{\theta_g}(\cdot), \tag{17.7}$$

and $\eta = (\eta_1, \eta_2, \ldots)$ is a vector of the weights such that $\eta_g \geq 0$ for all $g$, $\sum_{g=1}^{\infty} \eta_g = 1$, and $\delta_x(\cdot)$ denotes the distribution degenerate at $x$. $\theta_1, \theta_2, \ldots$ are assumed to be draws from some known distribution. $H$ is the unknown discrete mixing distribution and from a Bayesian perspective requires a prior assumption; see also Chapter 6 for a comprehensive review of Dirichlet process mixtures (DPMs) and more general nonparametric Bayesian mixture models.

An infinite mixture model of this form can flexibly represent a wide range of continuous distributions. This nonparametric model is often embedded in a richer time series model since it is able to capture the fat tails and skewness existing in financial data, but it is not able to capture time dependencies.

Before discussing the model in (17.6)–(17.7), we discuss selection of a prior for $H$ and some of its properties. By far the most popular prior for $H$ is the Dirichlet process (DP) prior. The DP, introduced by Ferguson (1973), is a distribution of probability measures over a measurable space. A draw from a DP, say $H$, is a discrete distribution with a countable number of atoms. The DP is indexed by two parameters: a positive scalar $\alpha$, called the concentration parameter, and a base distribution $H_0$. The draw $H \sim \mathcal{DP}(\alpha, H_0)$ is centred on the base distribution $H_0$ which serves as a prior guess. The concentration parameter $\alpha$ controls how close we believe the unknown $H$ is to $H_0$. The larger the value of $\alpha$, the stronger the belief in $H_0$ and the more distinct atoms we have with significant mass. This prior is convenient in that we can centre it around a well-known family of parametric distributions such as the Gaussian, and impose some measure of parsimony on the predictive density of $y_t$ through $\alpha$.

A key feature of the DP is that for any finite partition $\{A_1, \ldots, A_G\}$ of the parameter space,

$$H(A_1), \ldots, H(A_G)|\alpha, H_0 \sim \mathcal{D}(\alpha H_0(A_1), \ldots, \alpha H_0(A_G)), \tag{17.8}$$

where $\mathcal{D}(a_1, \ldots, a_G)$ denotes a Dirichlet distribution with parameter vector $(a_1, \ldots, a_G)$. This property implies an expectation and variance of, respectively, $E(H(A_i)) = H_0(A_i)$ and $V(H(A_i)) = H_0(A_i)(1 - H_0(A_i))/(1 + \alpha)$.

There are several different representations for the DP, namely the Chinese restaurant representation, the Pólya urn process, and the stick-breaking representation. A *Chinese restaurant process* (CRP), illustrated by Figure 17.1, is a distribution over partitions. Imagine a restaurant with an infinite number of tables that have a unique dish $\theta_g \sim H_0$ that is eaten at each table $g$. Assume that $G$ tables are currently occupied by $n - 1$ customers. A new customer $\phi_n$ (the $n$th customer) chooses an empty table with probability $\alpha/(n-1+\alpha)$ which is served the new dish $\theta_{G+1} \sim H_0$ or decides to sit at one of the occupied tables that have dish $\theta_g$, $g = 1, \ldots, G$, with probability $n_g/(n-1+\alpha)$ where $n_g$ is the number of people sitting at that table.

Another closely related representation for a draw from $H \sim \mathcal{DP}(\alpha, H_0)$, is the *Pólya urn scheme*, also discussed in Chapter 6. Imagine an urn with $\alpha$ black balls. Whenever a black ball is drawn, a new colour is generated from $H_0$ and the new colour ball along with the black ball is placed back in the urn. If the drawn ball is non-black, it is placed back together with a ball with the same colour. Thus, the $n$th new ball that we put in the urn is the same colour as one of the existing colours, say colour $g$, in the urn with probability $n_g/(n-1+\alpha)$ where $n_g$ is the number of existing balls with colour $g$, or has a new colour with probability $\alpha/(n-1+\alpha)$.

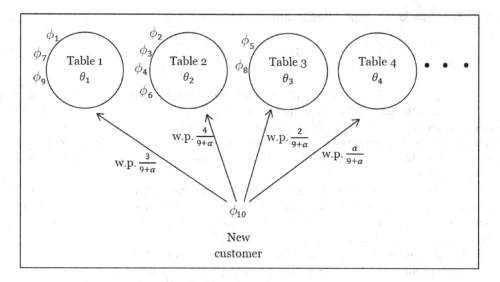

**FIGURE 17.1**
Chinese restaurant process.

Blackwell & MacQueen (1973) prove that the distribution of colours after $n$ draws converges to a discrete distribution that follows a DP. Let $z_n$ be an indicator and record the colour of the $n$th ball. If the first ball is $z_1 \sim H_0$, thereafter ball colours are generated as

$$z_{n+1}|z^n \sim \sum_{g=1}^{G} \frac{n_g}{\alpha + n} \delta_{z_{n+1},g} + \frac{\alpha}{\alpha + n} \delta_{z_{n+1},G+1},$$

where $\delta_{h,g}$ denotes the Kronecker delta and $G$ is the current number of distinct colours.

The observations from a Pólya urn are exchangeable, and this key point means we can always rearrange our sample to assume we are dealing with the last observation and apply the Pólya urn (conditional distribution) results. This is the basis of the Gibbs sampler we discuss below.

Another, constructive, definition of the DP is due to Sethuraman (1994) and often referred to as a *stick-breaking representation* due to the construction of the weights. A draw from the DP, $H \sim \mathcal{DP}(\alpha, H_0)$, almost surely has the form

$$H(\cdot) \quad = \quad \sum_{g=1}^{\infty} \eta_g \delta_{\theta_g}(\cdot), \tag{17.9}$$

$$\eta_1 \quad = \quad v_1, \quad \eta_g = v_g \prod_{l=1}^{g-1}(1 - v_l), \quad g \geq 2, \tag{17.10}$$

$$v_g \quad \overset{i.i.d.}{\sim} \quad \mathcal{B}e(1, \alpha), \quad g = 1, 2, \dots, \tag{17.11}$$

$$\theta_g \quad \overset{i.i.d.}{\sim} \quad H_0, \quad g = 1, 2, \dots. \tag{17.12}$$

This formulation and the CRP and Pólya urn provide different avenues to posterior sampling.

Note that any draw $H$ from a $\mathcal{DP}(\alpha, H_0)$ is a discrete probability measure (Ferguson, 1973), and therefore it is possible for a sample from $H$ to have repeated elements. Moreover, the DP cannot be used as a prior for distributions of continuous variables. To deal with this issue, Lo (1984) introduces the Dirichlet process mixture model and suggests working with a hierarchical model. The DPM model takes the following form

$$y_t|\phi_t \quad \sim \quad F(\phi_t), \quad t = 1, \ldots, T, \tag{17.13}$$

$$\phi_t \quad \overset{i.i.d.}{\sim} \quad H, \tag{17.14}$$

$$H|\alpha, H_0 \quad \sim \quad \mathcal{DP}(\alpha, H_0), \tag{17.15}$$

where $F(\phi_t)$ is the distribution associated with density $f(y_t|\phi_t)$. If $F(\cdot)$ is a continuous distribution then so is the DPM model. The parameters $\phi_t$ are assumed to be distributed according to an unknown distribution $H$ which follows the DP prior.

Equivalently, the model in the form of (17.6) with $H$ from (17.9)–(17.12) becomes

$$p(y_t) = \sum_{g=1}^{\infty} \eta_g f(y_t|\theta_g), \tag{17.16}$$

where $\theta_g$ represent the unique draws from $H_0$, assuming that $H_0$ is a continuous distribution, while the $\phi_t$ in (17.14) represent draws from $H$ which may have repeats.

With the advent of Markov chain Monte Carlo methods, Escobar (1994) and Escobar & West (1995) provide tractable approaches to posterior simulation. This is the first generation of posterior samplers for DPM models. With some exceptions they tend to be restrictive and suffer from some drawbacks. First, the Gibbs sampler approach relies on a conjugate prior $H_0$.[3] Second, $H$, the unknown distribution, is integrated out and therefore inference on this object is not possible.

We now discuss the Gibbs sampling method for the model defined in (17.13)–(17.15). Denote the set of distinct values of all the $\phi_t$ by $\theta = \{\theta_1, \theta_2, \ldots, \theta_G\}$ of size $G \leq T$. Conditional on $G$, we introduce indicators $z_t = g$ if $\phi_t = \theta_g$ so that, given $z_t = g$ and $\theta$, $y_t \sim F(\theta_g)$. The configuration set $\mathbf{z} = \{z_1, \ldots, z_T\}$ partitions the data $\mathbf{y} = \{y_1, \ldots, y_T\}$ into $G$ distinct groups so that the $n_g = \#\{t : z_t = g\}$ observations in group $g$ have the same parameter value $\theta_g$. Also define $I_g = \{t : z_t = g\}$ for the set of indices of observations in group $g$ and $\mathbf{y}_g = \{y_t : z_t = g\}$ as the corresponding group of observations. Define $\phi_{-t} = \{\phi_1, \ldots, \phi_{t-1}, \phi_{t+1}, \ldots, \phi_T\}$, $\mathbf{z}_{-t} = \{z_1, \ldots, z_{t-1}, z_{t+1}, \ldots, z_T\}$, $G^{(t)}$ as the number of distinct values in $\phi_{-t}$ and $n_g^{(t)} = \#\{z_t \in \mathbf{z}_{-t} : z_t = g\}$. Escobar et al. (1994) show that the Gibbs sampling steps sample from the corresponding densities,

(a)  $\phi_t|\alpha, \phi_{-t}, \mathbf{y}, t = 1, \ldots, T,$

(b)  $\theta|\alpha, \mathbf{z}, \mathbf{y},$

(c)  $\alpha|\mathbf{z}.$

The first step is a consequence of samples from the DP being exchangeable and is based on the following result:

$$(\phi_t|\phi_{-t}, \mathbf{z}_{-t}, G^{(t)}, \mathbf{y}) \sim \frac{\alpha}{T-1+\alpha} h(y_t) H(d\theta|y_t) + \frac{1}{T-1+\alpha} \sum_{g=1}^{G^{(t)}} n_g^{(t)} f(y_t|\theta_g),$$

where $h(y_t) = \int f(y_t|\theta) dH_0(\theta)$ is the predictive density derived from the prior evaluated at

---

[3]There are exceptions such as Neal (2000) for non-conjugate models, but these are not likely to perform well in high-dimensional settings.

$y_t$ and $H(d\theta|y_t) \propto f(y_t|\theta)dH_0(\theta)$ is the posterior distribution based on one observation. This stage generates a new configuration by sequentially sampling indicators from the posteriors

$$
z_t = \begin{cases} g, & \text{with probability } \dfrac{n_g^{(t)}}{T-1+\alpha}f(y_t|\theta_g), \quad g = 1,\dots,G^{(t)}, \\ G^{(t)}+1, & \text{with probability } \dfrac{\frac{\alpha}{1}}{T-1+\alpha}h(y_t), \end{cases}
$$

for any index $t$. If $z_t = G^{(t)} + 1$, draw a new $\phi_t$ from $H(d\theta|y_t)$. The second sampling step $\theta|\alpha, \mathbf{z}, \mathbf{y}$, is not necessary, but it tends to improve the mixing of the Markov chain. The first step only changes one parameter at a time conditional on all others, while the second step allows for an update of all unique values $\theta$ at one time. The final sampling step of $\alpha$ follows the Gibbs step from Escobar & West (1995).

Escobar et al. (1994) derive the conditional predictive density for future data $y_{T+1}$,

$$
p(y_{T+1}|\theta, \mathbf{z}, \alpha, \mathbf{y}) = \frac{\alpha}{\alpha+T} \int f(y_{T+1}|\theta_{G+1})dH_0(\theta_{G+1}) + \sum_{g=1}^{G} \frac{n_g}{\alpha+T}f(y_{T+1}|\theta_g), \quad (17.17)
$$

where $\theta_{G+1}$ is a new independent draw from $H_0$. Note that (17.17) represents a potentially infinite mixture model since for each new observation there is a possibility (proportional to $\alpha$) to introduce a new parameter $\theta$.

The final estimate of the predictive density is obtained by integrating out the parameter uncertainty. Given $M$ Gibbs draws of $\phi^{(m)}, \mathbf{z}^{(m)}$ and $\alpha^{(m)}$, the estimate is

$$
p(y_{T+1}|\mathbf{y}) \approx \frac{1}{M}\sum_{m-1}^{M} p(y_{T+1}|\theta^{(m)}, \mathbf{z}^{(m)}, \alpha^{(m)}, \mathbf{y}). \quad (17.18)
$$

There are important alternatives to the Gibbs sampler. Ishwaran & James (2001) provide the details on the stick-breaking notion of constructing infinite-dimensional priors (see also Chapters 6 and 7 in this volume). Equation (17.9) represents the DP as a countably infinite sum of atomic measures, which leads to the infinite mixture for the density of $y_t$ in (17.16). The second generation of posterior sampling methods allows for inference on $H$ and uses (17.16) directly. One approach is to truncate this infinite mixture to a finite number large enough to approximate the true model. Ishwaran & James (2001) provide details on the approximation and block sampling methods. A second approach to deal with the infinite number of parameters in (17.16) is the slice sampler method of Walker (2007). He introduces a latent variable that randomly truncates the model to a finite mixture model, but whose marginal distribution preserves the original model. This turns an infinite-dimensional sampling problem into a finite-dimensional one.

*An application of the DPM to stock returns*

To illustrate how an infinite mixture model is able to capture the fat tails in a data series, we nonparametrically estimate the density of gold returns, applying the DPM model and using the Gibbs sampler discussed above. The data are the monthly log returns from January 1970 to November 2012 with $T = 515$ observations. Table 17.1 displays summary statistics of the data set. This table shows that monthly returns of gold display skewness and excess kurtosis.

In this example, $\theta = (\mu, h)$ and the base measure, $H_0$, is assumed to have a conjugate prior of

$$
H_0(\mu_t, h_t) \equiv \mathcal{NG}\left(\mu_t, h_t|\mu_0, \rho, \tfrac{\alpha_0}{2}, \tfrac{\beta_0}{2}\right),
$$
$$
\mu_0 = 0, \quad \rho = 4, \quad \alpha_0 = 4, \quad \beta_0 = 20,
$$

**TABLE 17.1**

Summary statistics of gold returns. The data set is the monthly returns from January 1970 to November 2012

| Total number | Mean | Variance | Skewness | Excess kurtosis |
|---|---|---|---|---|
| 515 | 0.327 | 4.6 | 8.1 | 1.26 |

where $\mathcal{NG}(\cdot)$ is the normal-gamma distribution,[4] while $\alpha \sim \mathcal{G}(a,b)$ where $\mathcal{G}(a,b)$ is the gamma distribution.[5] Using the results above, we have

$$h(y_t) \quad \propto \quad t_{\alpha_0}\left(y_t|\mu_0, \frac{\beta_0(1+\rho)}{\alpha_0\,\rho}\right),$$

$$H(d\theta|y_t) \quad \propto \quad \mathcal{NG}\left(\mu, h|\bar{\mu}_t, (1+\rho)^{-1}, \frac{\alpha_0+1}{2}, \frac{\omega_t^*}{2}\right),$$

in which

$$\bar{\mu}_t = \frac{y_t + \rho\mu_0}{1+\rho}, \quad \omega_t^* = -(1+\rho)\bar{\mu}_t^2 + y_t^2 + \rho\mu_0^2 + \beta_0.$$

$t_\nu(\mu, \sigma^2)$ denotes a Student $t$ density with location $\mu$, scale parameter $\sigma^2$ and degree of freedom $\nu$, and $t_\nu(x|\mu, \sigma^2)$ is the associated density function evaluated at $x$.

The predictive density given parameter draws is a mixture of normals and Student $t$ distributions,

$$p(y_{T+1}|\theta, \mathbf{z}, \alpha)$$

$$= \frac{\alpha}{\alpha+T} \int f(y_{T+1}|\mu_{G+1}, h_{G+1})dH_0(\mu_{G+1}, h_{G+1}) + \sum_{g=1}^{G} \frac{n_g}{\alpha+T} f(y_{T+1}|\mu_g, h_g)$$

$$= \frac{\alpha}{\alpha+T} t_{\alpha_0}\left(y_{T+1}|\mu_0, \frac{\beta_0(1+\rho)}{\alpha_0\,\rho}\right) + \sum_{g=1}^{G} \frac{n_g}{\alpha+T} \phi(y_{T+1}|\mu_g, h_g),$$

where $G$ is the current number of mixture components. The predictive density is obtained by averaging this result over the MCMC draws as in (17.18).

We run the MCMC algorithm for $M = 7000$ iterations. After dropping the first $M_0 = 1000$ draws as burn-in, we estimate the predictive density. Figure 17.2 shows the estimated nonparametric predictive density of monthly returns with $\alpha \sim \mathcal{G}(1, 24)$ (top) as well as its logarithmic scale plot (bottom) in comparison to the normal distribution with mean and variance equal to sample mean and sample variance, respectively. The nonparametric model captures significant deviations from the normal distribution by using approximately five components, on average. The log density of the nonparametric model displays thicker upper tails, meaning large increases in gold prices are more likely than large drops. The posterior mean of the precision parameter $\alpha$ is 0.17.

---

[4]$(X, T) \sim \mathcal{NG}(\mu, \rho, a, b)$ with density

$$f(x, \tau|\mu, \rho, a, b) = \frac{b^a\sqrt{\rho}}{\Gamma(a)\sqrt{2\pi}}\tau^{a-1}e^{-\tau b}e^{\frac{-\rho\tau(x-\mu)^2}{2}}.$$

[5]$X \sim \mathcal{G}(a, b)$ with density

$$f(x|a, b) = \frac{b^a}{\Gamma(a)}x^{a-1}e^{-xb},$$

then $E(x) = \frac{a}{b}$.

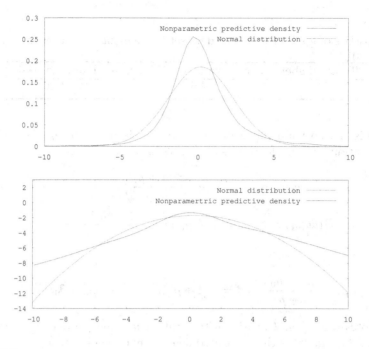

**FIGURE 17.2**
Top: a comparison of the nonparametric predictive density with the normal distribution.
Bottom: the associated log densities.

Pooling data into different clusters is a distinct advantage of the DPM model. For instance, Jensen & Fisher (2018) use the DPM in the context of a linear factor structure to classify the skill of mutual funds into subpopulations.

The DPM model is suitable for modelling an unknown i.i.d. distribution such as the unconditional distribution of returns. Without modification it cannot deal with the pronounced volatility dynamics in return data. For this reason the DPM is often embedded in a more sophisticated time series specification of returns. We turn to some examples next.

## 17.3.2   GARCH-DPM and SV-DPM

To make the model richer and more suitable for financial data, recent research combines the DPM model with a time series model of volatility such as GARCH or stochastic volatility.

*GARCH*

The first class of studies incorporates the GARCH (Bollerslev, 1986) functional form into the infinite mixture model. These models impose an ARMA structure on squared innovations along with a parametric normal, Student $t$, or in some cases (Bauwens et al., 2007; Galeano & Ausín, 2010) a finite mixture of normals for the innovation density. By extending GARCH models to a Bayesian semi-parametric setting, the functional form of the conditional density of the data can be estimated nonparametrically in the same framework as the DPM model (17.13)–(17.15), but with a GARCH factor entering into each of the components of this mixture. For general reviews on multivariate GARCH (MGARCH) models, see Bauwens et al. (2006) and Virbickaite et al. (2015).

Kalli et al. (2013) define an infinite mixture model with GARCH components to estimate a unimodal and asymmetric conditional return distribution. The unknown distribution $H$ has a stick-breaking prior (SBP) with a standard exponential distribution as the base distribution $H_0$. The SBP is a more general case of the DP for which there exist less tractable results. With the SBP, the rate of weight decay is controlled by a potentially infinite number of parameters, instead of only one parameter, $\alpha$, which is the case for the DP.

Kalli et al. (2013) look at the daily log returns of three stock indices and find evidence in favour of their semi-parametric model against GARCH, EGARCH, and GJR-GARCH (Glosten et al., 1993) models. Both the EGARCH and the GJR-GARCH model allow for an asymmetric response to past volatility shocks. Other works on Bayesian nonparametric GARCH models are Lau & Cripps (2012) and Ausín et al. (2014).

Jensen & Maheu (2013) propose a Bayesian nonparametric modelling approach for the MGARCH model,

$$y_t | \mu_t, B_t, \Sigma_t \sim \mathcal{N}_d(\mu_t, \Sigma_t^{1/2} B_t^{-1}(\Sigma_t^{1/2})^\top), \quad t = 1, \ldots, T,$$

$$\Sigma_t = \Gamma_0 + \Gamma_1 \odot y_{t-1} y_{t-1}^\top + \Gamma_2 \odot \Sigma_{t-1}, \tag{17.19}$$

$$\mu_t, B_t | H \overset{i.i.d.}{\sim} H, \tag{17.20}$$

$$H | \alpha, H_0 \sim \mathcal{DP}(\alpha, H_0),$$

$$H_0(\mu_t, B_t) \equiv \mathcal{N}_d(\mu_0, V_0) \times \mathcal{W}(c + d - 1, C), \quad c \geq 1, \tag{17.21}$$

where $y_t = (y_{1t}, \ldots, y_{dt})^\top$ is the $d$-dimensional vector of returns and the symbol $\odot$ denotes the Hadamard product. They assume a parametric model for the dynamics of the conditional variance matrix of returns in (17.19) proposed by Ding & Engle (2001), and a DPM prior for the multivariate distribution of the returns in (17.20). In the Sethuraman (1994) representation of the DPM model, this model can be cast as an infinite mixture of multivariate normals with mixing over both the location and scale matrix of the normal components,

$$p(y_t | \Sigma_t) = \sum_{g=1}^{\infty} \eta_g \phi(y_t | \mu_g, \Sigma_t^{1/2} B_g^{-1}(\Sigma_t^{1/2})^\top), \quad t = 1, \ldots, T, \tag{17.22}$$

$$\eta_1 = v_1, \quad \eta_g = v_g \prod_{l=1}^{g-1}(1 - v_l), \quad g \geq 2, \tag{17.23}$$

$$v_g \overset{i.i.d.}{\sim} \mathcal{B}(1, \alpha), \quad g = 1, 2, \ldots,$$

$$\theta_g \equiv (\mu_g, B_g) \overset{i.i.d.}{\sim} H_0, \quad g = 1, 2, \ldots.$$

This nests the Gaussian case when $\eta_1 = 1$ and $\eta_g = 0$, $g > 1$. If $\mu_g = \mu$, for all $g$ and $\alpha \to \infty$, then $H \to H_0$ and we obtain a Student $t$ distribution.

Using the conditionally conjugate priors (normal and Wishart) as the base measure, $H_0$, slice sampling methods from Walker (2007) and Kalli et al. (2009) can be used for posterior simulation. In contrast to the previous example that was based on the Pólya urn and integrated the unknown $H$ out, slice sampling works directly with the infinite mixture representation in (17.22). In the following sampling steps, $\mathbf{z} = \{z_1, \ldots, z_T\}$ is the configuration set where $z_t = g$ if the $t$th observation uses the $g$th component's parameters $(\mu_g, B_g)$, and $n_g = \#\{t : z_t = g\}$. $\mathbf{u} = (u_1, \ldots, u_T)$ are the auxiliary variables introduced by Walker (2007) to make the sampling more tractable, while $\Sigma = (\Sigma_1, \ldots, \Sigma_T)$.

**Step 1** The posterior distribution of $(\mu_g, B_g)$, $g = 1, \ldots, G$: Using the transformation $y_t^\star =$

$\Sigma_t^{-1/2} y_t$, and making use of conditionally conjugate prior, we have

$$
\begin{aligned}
B_g | \mathbf{y}, \mathbf{z}, \mu_g, \Sigma &\sim \mathcal{W}(n_g + d - 1 + c, C_{gT}), \\
\mu_g | \mathbf{y}, \mathbf{z}, B_g, \Sigma &\sim \mathcal{N}_d(\mu_{gT}, V_{gT}),
\end{aligned}
$$

in which

$$
C_{gT} = \left( C^{-1} + \sum_{t:z_t=g} (y_t^\star - \Sigma_t^{-1/2} \mu_g)(y_t^\star - \Sigma_t^{-1/2} \mu_g)^\top \right)^{-1},
$$

$$
V_{gT} = \left( \sum_{t:z_t=g} (\Sigma_t^{-1/2})^\top B_g \Sigma_t^{-1/2} + V_0^{-1} \right)^{-1},
$$

$$
\mu_{gT} = V_{gT} \left( \sum_{t:z_t=g} (\Sigma_t^{-1/2})^\top B_g y_t^\star + V_0^{-1} \mu_0 \right).
$$

**Step 2** Updating $v_g$, $g = 1, \ldots, G$: by the conjugacy of the generalized Dirichlet distribution to multinomial sampling (Ishwaran & James, 2001) we have

$$
v_g | \mathbf{z} \sim \mathcal{B}e \left( 1 + \sum_{t=1}^{T} \mathbb{I}(z_t = g), \alpha + \sum_{t=1}^{T} \mathbb{I}(z_t > g) \right).
$$

Then update $\eta$ according to (17.23).

**Step 3** Updating $u_t$, $t = 1, \ldots, T$ (Walker, 2007):

$$
u_t | \eta, \mathbf{z} \sim \mathcal{U}(0, \eta_{z_t}).
$$

Then we update $G$ such that $\sum_{g=1}^{G} \eta_g > 1 - \min\{u_t\}_{t=1}^{T}$. Additional $\eta_g$ and $(\mu_g, B_g)$ will need to be generated from the priors if $G$ is incremented.

**Step 4** Updating $\mathbf{z}$ (Walker, 2007): for each $t = 1, \ldots, T$,

$$
\mathrm{P}(z_t = g | \mathbf{y}, \mu_g, B_g, \Sigma) \propto \mathbb{I}(\eta_g > u_t) \phi(y_t | \mu_g, \Sigma_t^{1/2} B_g^{-1} (\Sigma_t^{1/2})^\top), \quad g = 1, \ldots, G.
$$

**Step 5** Updating the GARCH parameters $\Gamma = (\Gamma_0, \Gamma_1, \Gamma_2)$: assuming a prior $p(\Gamma)$ for $\Gamma$, the posterior reads

$$
p(\Gamma | \mu, B, \mathbf{z}, \mathbf{y}, \Sigma) \propto p(\Gamma) \times \prod_{t=1}^{T} \phi(y_t | \mu_{z_t}, \Sigma_t^{1/2} B_{z_t}^{-1} (\Sigma_t^{1/2})^\top),
$$

which is not of standard form, and the Metropolis–Hastings sampler can be applied.

Given a large number of draws from the sampling steps above, the predictive density of $y_{T+1}$ given $\mathbf{y}$ can be approximated using $M$ draws of the posterior as follows:

$$
p(y_{T+1} | \mathbf{y}) \approx \frac{1}{M} \sum_{m=1}^{M} \phi(y_{T+1} | \mu_{z_{T+1}^{(m)}}^{(m)}, (\Sigma_{T+1}^{(m)})^{1/2} (B_{z_{T+1}^{(m)}}^{(m)})^{-1} ((\Sigma_{T+1}^{(m)})^{1/2})^\top),
$$

where $\theta^{(m)}$ is the $m$th posterior draw of parameter $\theta$, and $z_{T+1}^{(m)}$ at each iteration $m$ is one of the $G^{(m)}$ components, say component $g$, with probability $\eta_g^{(m)}$, or is a new component with probability $1 - \sum_{g=1}^{G^{(m)}} \eta_g^{(m)}$. Note that we are able to compute $\Sigma_{T+1}^{(m)}$ at each iteration of the algorithm from

$$\Sigma_{T+1}^{(m)} = \Gamma_0^{(m)} + \Gamma_1^{(m)} \odot y_T (y_T)^\top + \Gamma_2^{(m)} \odot \Sigma_T^{(m)}.$$

This is recursively computed from $t = 1$ to $t = T + 1$.

Jensen & Maheu (2013) consider two data sets, equity return and foreign exchange rate, to estimate the model employing Pólya urn and stick-breaking sampling schemes. Comparison of the Bayes factors and density forecasts with parametric GARCH models (Gaussian and Student $t$ innovations) supports the flexible semi-parametric approach, particularly in the case of asymmetric distributions and during highly volatile periods.

Working with a univariate version of the model, Ausín et al. (2014) estimate the return density of the Hang Seng and Bombay Stock Exchange indices and carry out Bayesian prediction of the value-at-risk. They compare the results of the semi-parametric model with a Gaussian, a Student $t$, and a mixture of two zero-mean Gaussian distributions, and find significant differences in the return predictive distribution, particularly in the tails.

*Stochastic volatility*

The second class of studies combines stochastic volatility with a DPM model. The main difference between the SV and GARCH models is that for the former, conditional on time $t$ information, the conditional variance is stochastic and can be thought of as the impact of an unobserved news flow process.

Jensen & Maheu (2010) extend the standard SV specification that has parametric return innovations to the following semi-parametric (SV-DPM) setting:

$$y_t | \mu_t, \lambda_t^2 \sim \mathcal{N}(\mu_t, \lambda_t^{-2} \exp(h_t)),$$
$$h_t | h_{t-1} \sim \mathcal{N}(\delta h_{t-1}, \sigma_v^2) \text{ and } h_t \perp y_t, \quad |\delta| < 1,$$
$$(\mu_t, \lambda_t^2) \sim H, \tag{17.24}$$
$$H \sim \mathcal{DP}(\alpha, H_0), \tag{17.25}$$
$$H_0(\mu_t, \lambda_t^2) \equiv \mathcal{N}(m, (\tau \lambda_t^2)^{-1}) \times \mathcal{G}(a_0/2, b_0/2). \tag{17.26}$$

The latent log volatility $h_t$ follows a parametric, stationary, first-order autoregressive (AR) process defined with the AR parameter $\delta$, but the rest of the model is nonparametric inasmuch no assumption is made about the underlying distribution of return innovations. Note that, assuming independence, $h_t \perp y_t$, Jensen & Maheu (2010) remove any leverage effect (Jacquier et al., 2004). Equations (17.24) and (17.25) assume the mixture's probabilities and parameters follow the DP prior. The base distribution, $H_0$, is a conjugate conditional normal-gamma distribution. The Sethuraman (1994) representation for this semi-parametric model is

$$y_t \sim \sum_{g=1}^{\infty} \eta_g \mathcal{N}(\mu_g, \lambda_g^{-2} \exp(h_t)), \tag{17.27}$$

with the mixture weights distributed as $\eta_1 = v_1$, $\eta_g = v_g \prod_{l=1}^{g-1}(1 - v_l)$, $g > 1$, where $v_g \sim \mathcal{Be}(1, \alpha)$. The mixture parameters $(\mu_g, \lambda_g^2)$ have the $\mathcal{NG}$ prior distribution (17.26). Jensen & Maheu (2010) construct an MCMC sampling method for the model, apply it to a long series of daily market returns and find strong deviations from both normal and Student $t$ distributions.

To take into consideration the correlation between returns and volatility innovations

(Jacquier et al., 2004; Nakajima & Omori, 2009), Jensen & Maheu (2014) extend the univariate algorithm of the semi-parametric stochastic volatility model above by lifting the independence assumption $y_t \perp h_t$. To model the unknown joint distribution of return and volatility innovations ($u_t$ and $v_t$ in the following expressions), they choose an asymmetric SV-DPM model (ASV-DPM) as follows:

$$y_t = \mu + \exp(h_t/2)u_t,$$
$$h_t = \delta h_{t-1} + v_t,$$
$$(u_t, v_t)^\top \sim \mathcal{N}_2(0, \Lambda_t),$$
$$\Lambda_t \sim H,$$
$$H \sim \mathcal{DP}(\alpha, H_0),$$
$$H_0 \equiv \mathcal{W}^{-1}(c, C).$$

They compare the ASV-DPM model with the parametric SV model by daily predictive likelihoods.[6] The empirical experiment provides evidence favouring the nonparametric asymmetric stochastic volatility model more often than the parametric version.

Delatola & Griffin (2013) extend the parametric, linearized stochastic volatility model of Omori et al. (2007) to capture the leverage effect in an infinite mixture model. They include a constant leverage effect and nonparametrically model the distribution of the log squared returns. The semi-parametric stochastic volatility model with leverage effect in this case is specified as

$$y_t^* = h_t + z_t^*,$$
$$h_t = \delta h_{t-1} + d_t \rho \sigma_v \exp(\mu_t/2)[a^* + b^*(y_t^* - h_t - \mu_t)] + \sigma_v(1 - \rho^2)\epsilon_t^*,$$
$$z_t^*|\mu_t \sim \mathcal{N}(\mu_t, \delta\sigma_z^2), \quad \epsilon_t^* \sim \mathcal{N}(0, 1),$$
$$\mu_t \sim H,$$
$$H \sim \mathcal{DP}(\alpha, H_0),$$

where $a^* = \exp(\delta\sigma_z^2/8)$, $b^* = 0.5a^*$, $d_t = \text{sign}(y_t)$, and, as in Nakajima & Omori (2009) and Jacquier et al. (2004), $y_t^* = \log(y_t^2 + c)$ and $z_t^*$ follows an unknown distribution that is approximated by an infinite mixture of normals.

Kalli & Griffin (2015) define a Bayesian nonparametric model for the cross-sectional aggregation of AR(1) models (Robinson, 1978; Granger, 1980) to account for long memory in volatility. Suppose that we have $d$ time series, $\{h_{i1}, \ldots, h_{iT}\}_{i=1}^d$, where each time series follows an AR(1) process with persistence parameter $\phi_i \overset{i.i.d.}{\sim} F_\phi$. The finite aggregate process is written as

$$h_t = \frac{1}{d}\sum_{i=1}^d h_{it}, \quad t = 1, \ldots, T. \tag{17.28}$$

---

[6]The predictive likelihood is defined as

$$p(y_L, \ldots, y_T|y^{L-1}, \mathcal{M}) = \prod_{t=L}^T p(y_t|y^{t-1}, \mathcal{M}),$$

where $\mathcal{M}$ denotes the particular model, and $L > 1$ is chosen to eliminate the influence of the priors. The one-step-ahead predictive likelihoods $p(y_t|y^{t-1}, \mathcal{M})$, can be estimated by computing the sample mean of the likelihood using MCMC draws of the unknown parameters and latent variables conditional on the data history $y^{t-1}$.

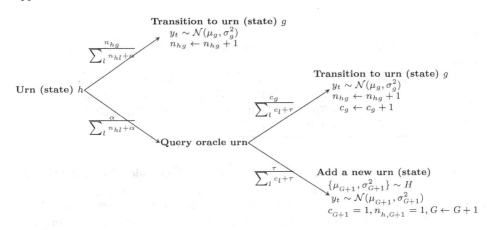

**FIGURE 17.3**

Pólya urn for infinite hidden Markov model. The oracle urn has parameter $\tau$. There are a countably infinite number of transition urns (states) with parameter $\alpha$. $c_l$ is the number of state $l$ (balls) in the oracle urn. $n_{hg}$ is the number of state $g$ (balls) in transition urn $h$.

An infinite aggregate process assumes that $d$ in (17.28) goes to infinity. Kalli & Griffin (2015) assume that the stochastic volatility follows an infinite aggregate process,

$$y_t = \beta \exp(h_t/2)u_t,$$

$$h_t = \sum_{g=1}^{\infty} h_t(\phi_g, \sigma^2 \pi_g),$$

where $u_t \sim \mathcal{N}(0,1)$ and $\pi_g$ is the proportion of the variation in $h_t$ explained by the $g$th process associated with AR parameter $\phi_g$. This parameterization flexibly models the dependence of the volatility process in financial time series. The persistence parameter of each AR(1) process is independently drawn from a distribution $F_\phi$ which is estimated nonparametrically following a DP prior. They apply the linearized model of SV (Harvey et al., 1994), and also use a finite approximation to the DP (Ishwaran & Zarepour, 2000; Neal, 2000; Kim et al., 1998). The empirical results in the daily returns of HSBC and Apple Inc. show significant difference in the distributions of the volatility persistence, suggesting different lasting effects of the information in these two sets of returns.

### 17.3.3 Infinite hidden Markov model

Recall that in a Markov switching model or hidden Markov model, the time series is modelled in terms of a finite number of discrete latent states which are governed by a transition matrix; see Chapter 13 for a review. Assuming a countable infinite number of hidden states, the infinite hidden Markov model (IHMM) extends this setting to an infinite mixture model (Beal et al., 2002; Van Gael & Ghahramani, 2011).

The hierarchical Dirichlet process (HDP) prior is introduced by Teh et al. (2006) as an extension to the DP prior. The HDP is a family of Dirichlet processes that share a common base measure. The common base measure is also distributed according to a DP prior. The HDP has a hierarchical structure which is constructed by two DPs as follows:

$$H_0 | \tau, D \sim \mathcal{DP}(\tau, D),$$

$$H_g | \alpha, H_0 \overset{i.i.d.}{\sim} \mathcal{DP}(\alpha, H_0), \quad g = 1, \ldots, \infty.$$

**TABLE 17.2**

Posterior summary statistics (results are based on 5000 MCMC iterations, after a 10,000 burn-in period)

| Parameter | Mean | Median | St. dev. | 95% HPD interval |
|---|---|---|---|---|
| $\tau$ | 1.6307 | 1.4419 | 0.9076 | (0.4096, 3.8396) |
| $\alpha$ | 0.7020 | 0.6450 | 0.3315 | (0.2390, 1.5291) |
| $G$ | 5.1108 | 5 | 1.0416 | (4, 7) |

There are group-specific probability measures $H_g$ conditional on a global probability measure $H_0$. $\alpha$ and $\tau$ are concentration parameters and $D$ is the base measure. The HDP can be used to form a prior for the rows of the infinite-dimensional transition matrix of an IHMM.

Consider the following IHMM for stock returns. If Stick($\alpha$) denotes the stick-breaking construction of the weights in (17.10) and (17.11) that defines a distribution over the natural numbers $\mathbb{N}$, then the model is

$$
\begin{aligned}
\Gamma | \tau &\sim \text{Stick}(\tau), \quad \theta_h \overset{i.i.d.}{\sim} D, \quad h = 1, 2, \ldots, \infty, \\
\xi_g | \alpha, \Gamma &\overset{i.i.d.}{\sim} \mathcal{DP}(\alpha, \Gamma), \quad g = 1, \ldots, \infty, \\
z_t | z_{t-1}, \xi_{z_{t-1}} &\sim \xi_{z_t-1}, \quad t = 1, \ldots, T, \\
y_t | z_t, \theta &\sim \mathcal{N}(\mu_{z_t}, \sigma_{z_t}^2),
\end{aligned}
\tag{17.30}
$$

where $\theta = (\mu, \sigma^2)$ and the base measure is $D \equiv \mathcal{N}(m, v^2) \times \mathcal{IG}(v_0/2, s_0/2)$. The first two equations of this model employ the HDP. Once the transition matrix $\xi$ is defined the dynamics of states $z_t$ are completely specified.

With this model there is an associated Pólya urn representation to generate states and data. There is a separate urn for each state (ball) sampled. There is a top-level *oracle* urn. Sampling states involves a two-step process. First, decide whether to sample from the existing urns (states) or to sample from the *oracle* urn. Second, sample from the respective urn. Sampling from an existing state is influenced by previous state counts and results in the recurrence of a past state. New states (balls) are only introduced from the oracle urn. This process is depicted in Figure 17.3 and shows how states and $y_t$ are generated. Note that the unknown $\Gamma$ and $\xi$ are not present and have been integrated out.

The second line of (17.30) can be represented as a stick-breaking process as well. If this is denoted by Stick2($\alpha, \Gamma$) then the two generating processes for the weights, Stick($\tau$) and Stick2($\alpha, \Gamma$), are related as

$$
\Gamma = \sum_{g=1}^{\infty} \gamma_g \delta_g, \quad \gamma_g = v_g \prod_{l=1}^{g-1} (1 - v_l), \quad v_g \overset{i.i.d.}{\sim} \mathcal{Be}(1, \tau),
$$

$$
\xi_g = \sum_{h=1}^{\infty} \xi_{gh} \delta_h, \quad \xi_{gh} = \tilde{\xi}_{gh} \prod_{l=1}^{h-1} (1 - \tilde{\xi}_{gl}), \quad \tilde{\xi}_{gh} \overset{i.i.d.}{\sim} \mathcal{Be}\left( \alpha \gamma_h, \alpha \left( 1 - \sum_{l=1}^{h} \gamma_l \right) \right).
$$

Therefore, each row of $\xi$ is a discrete distribution that governs moves over a common set of parameters $\theta_1, \theta_2, \ldots$. Each row of the transition matrix is related, $\text{E}(\xi_{ji}) = \text{E}(\gamma_i) = \tau^{i-1}/(1+\tau)^i$, but stochastic and in general will differ. However, the prior is centred around the DPM previously discussed, which is an obvious point to begin learning about state dynamics.

One approach to posterior sampling of this model is the beam sampler of Van Gael et al. (2008). Just like slice sampling, an auxiliary variable is introduced to stochastically slice the

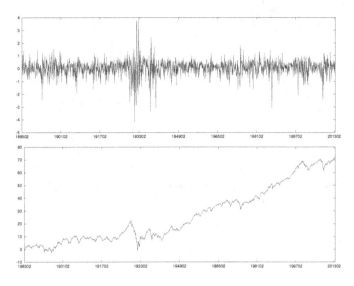

**FIGURE 17.4**
Monthly returns and log prices of the US market.

state space into a finite one for sampling purposes. Let $\mathbf{u} = (u_1, \ldots, u_T)$ denote the slice variable. A summary of the sampling steps is as follows:

1. Sample $\mathbf{z}|\mathbf{y}, \mathbf{u}, \xi$.

2. Sample $\xi_g|\mathbf{z}, \Gamma$, $g = 1, \ldots, G$.

3. Sample $\mathbf{u}|\mathbf{z}, \xi$ and update $G$.

4. Sample $\theta_g|\mathbf{z}, \mathbf{y}$, $g = 1, \ldots, G$.

5. Sample $\Gamma|\mathbf{z}$, $\tau|\mathbf{z}, \Gamma$ and $\alpha|\mathbf{z}, \Gamma$.

The latent variables $u_t$, $t = 1, \ldots, T$, are introduced such that the conditional density of $u_t$ is

$$p(u_t|z_{t-1}, z_t, \xi) = \frac{\mathbb{I}(0 < u_t < \xi_{z_{t-1},z_t})}{\xi_{z_{t-1},z_t}}.$$

The $u_t$ are sampled along with the other parameters but the sampling of the states given $u^t$ in the filter step of the FFBS becomes

$$p(z_t|y^t, u^t, \xi) \propto p(y_t|y^{t-1}, z_t) \sum_{z_{t-1}=1}^{\infty} \mathbb{I}(u_t < \xi_{z_{t-1},z_t}) p(z_{t-1}|y^{t-1}, u^{t-1}, \xi)$$

$$\propto p(y_t|y^{t-1}, z_t) \sum_{z_{t-1}:u_t < \xi_{z_{t-1},z_t}} p(z_{t-1}|y^{t-1}, u^{t-1}, \xi).$$

The $u_t$ slices out states with small $\xi_{z_{t-1},z_t}$ and results in a finite summation, since the number of states $z_{t-1}$ that satisfy $u_t < \xi_{z_{t-1},z_t}$ is finite. Once the forward pass is computed for $t = 1, \ldots, T$, the backward pass follows from

$$p(z_t|z_{t+1}, y^T, u^T) \propto p(z_t|y^t, u^t)\mathbb{I}(u_{t+1} < \xi_{z_t,z_{t+1}}), \quad t = T-1, \ldots, 1.$$

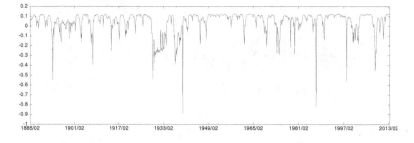

**FIGURE 17.5**
Time series plot of the posterior mean of the intercept parameter, $\mathrm{E}(\mu_{z_t}|\mathbf{y})$.

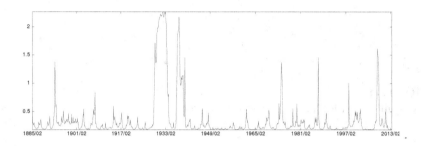

**FIGURE 17.6**
Time series plot of the posterior mean of the variance parameter, $\mathrm{E}(\sigma_{z_t}^2|\mathbf{y})$.

This is initiated with a draw of $z_T$ from the last value of the filter $p(z_T|\mathbf{y}, \mathbf{u}, \xi)$. Additional details on these sampling steps can be found in Shi & Song (2016), Maheu & Yang (2016) and Dufays (2016).

We now consider an application to monthly stock returns consisting of US capital market returns spanning from February 1885 to December 2013. From February 1885 to December 1925 the data are provided by Bill Schwert, after which they are S&P 500 index value-weighted returns excluding dividends, downloaded from CRSP. Returns and the log index are plotted in Figure 17.4. It is clear that there are periods of positive and negative growth in the index.

The following priors are used: $\tau \sim \mathcal{G}(1,1), \alpha \sim \mathcal{G}(1,1), \mu \sim \mathcal{N}(0,4)$ and $\sigma^2 \sim \mathcal{IG}(2.5, 2.5)$. Posterior summary statistics are given in Table 17.2. Here $G$ denotes the number of states in which at least one observation is allocated. On average, about five clusters of normal distributions are used, but there is considerable posterior uncertainty.

The posterior means of parameters are displayed over time in Figures 17.5 and 17.6. The posterior moments show a range of variation well beyond what a two- or three-state MS model can capture. There appear to be periods in which the mean and variance of stock returns are unique and not recurring.

Finally, we plot the predictive density from this model for two time periods along with the DPM model previously discussed. The predictive density and the log density are found in Figure 17.7. As time changes, so does the predictive density from the IHMM, while time has no effect on the DPM predictive density. It is clear from the figures that the IHMM uncovers significance changes in the density and tail structure over time compared to the DPM model.

Applications and extensions of the IHMM have become an active area of research in

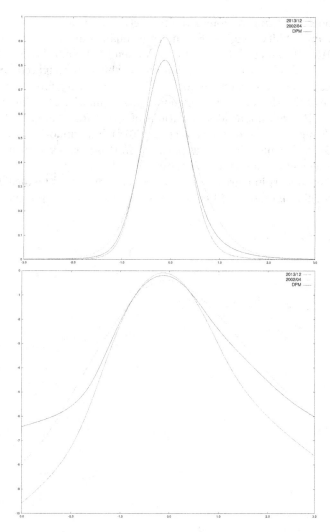

**FIGURE 17.7**
Predictive density and log density of the market returns obtained using the IHMM and DPM.

finance. Additional research in finance includes Maheu & Yang (2016), Dufays (2016), Shi & Song (2016) and Jin & Maheu (2016).

## 17.4 Concluding Remarks

This chapter has surveyed only a few examples of mixture models used in financial applications. Mixture modelling appears in many other areas of finance. Melick & Thomas (1997) use a mixture of three normals to approximate the risk-neutral density from option data, while Casarin et al. (2015) use a dynamic beta Markov random field to calibrate the risk-neutral density in option data.

Bayesian model averaging combines the predictive distribution from several models according to their probability weights. Examples in finance include Avramov (2002), Eidenberger et al. (2015), Hasan et al. (2015), Liu & Maheu (2009) and Wright (2008). Geweke & Amisano (2011) introduce optimal prediction pools, which are weighted linear combinations of predictive densities from different models to improve density forecasts.

A special case of the Markov switching model was popularized by Chib (1996) to estimate structural breaks. Related financial applications of structural change include Liu & Maheu (2008), Chib & Kang (2013) and Bauwens et al. (2014). The smooth finite mixing regression model of Geweke & Keane (2007) is applied to S&P 500 stock returns and allows state probabilities to depend on observed covariates.

Mixture models underpin a great deal of research in empirical finance. Development and use of these modelling methods will continue to be a central component of empirical work in finance.

# Bibliography

ALEXANDER, C. & LAZAR, E. (2006). Normal mixture GARCH(1, 1). Applications to exchange rate modelling. *Journal of Applied Econometrics* **21**, 307–336.

ANG, A. & CHEN, J. (2002). Asymmetric correlations of equity portfolios. *Journal of Financial Economics* **63**, 443–494.

AUGUSTYNIAK, M. (2014). Maximum likelihood estimation of the Markov switching GARCH model. *Computational Statistics & Data Analysis* **76**, 61–75.

AUSÍN, M. C., GALEANO, P. & GHOSH, P. (2014). A semiparametric Bayesian approach to the analysis of financial time series with applications to value at risk estimation. *European Journal of Operational Research* **232**, 350–358.

AVRAMOV, D. (2002). Stock return predictability and model uncertainty. *Journal of Financial Economics* **64**, 423–458.

BAI, X., RUSSELL, J. R. & TIAO, G. C. (2003). Kurtosis of GARCH and stochastic volatility models with non-normal innovations. *Journal of Econometrics* **114**, 349–360.

BATES, D. S. (2000). Post-'87 crash fears in the S&P 500 futures option market. *Journal of Econometrics* **94**, 181–238.

BAUWENS, L., BACKER, B. D. & DUFAYS, A. (2014). A Bayesian method of change-point estimation with recurrent regimes: Application to GARCH models. *Journal of Empirical Finance* **29**, 207–229.

BAUWENS, L., DUFAYS, A. & ROMBOUTS, J. V. (2011). Marginal likelihood for Markov switching and change-point GARCH models. *Journal of Econometrics* **178**, 508–522.

BAUWENS, L., HAFNER, C. M. & ROMBOUTS, J. V. K. (2007). Multivariate mixed normal conditional heteroskedasticity. *Computational Statistics & Data Analysis* **51**, 3551–3566.

BAUWENS, L., LAURENT, S. & ROMBOUTS, J. V. K. (2006). Multivariate GARCH models: A survey. *Journal of Applied Econometrics* **21**, 79–109.

BAUWENS, L., PREMINGER, A. & ROMBOUTS, J. V. K. (2010). Theory and inference for a Markov switching GARCH model. *Econometrics Journal* **13**, 218–244.

BEAL, M. J., GHAHRAMANI, Z. & RASMUSSEN, C. E. (2002). The infinite hidden Markov model. In *Advances in Neural Information Processing Systems*, T. G. Dietterich, S. Becker & Z. Ghahramani, eds. Cambridge, MA: MIT Press, pp. 577–584.

BEINE, M., LAURENT, E. & LECOURT, C. (2003). Official central bank interventions and exchange rate volatility: Evidence from a regime-switching analysis. *European Economic Review* **47**, 891–911.

BILLIO, M., CASARIN, R. & OSUNTUYI, A. (2016). Efficient Gibbs sampling for Markov switching GARCH models. *Computational Statistics & Data Analysis* **100**, 37–57.

BILLIO, M. & CAVICCHIOLI, M. (2013). Markov switching models for volatility: Filtering, approximation and duality. Working Papers 2013:24, Department of Economics, University of Venice Ca' Foscari.

BILLIO, M. & PELIZZON, L. (2000). Value-at-risk: A multivariate switching regime approach. *Journal of Empirical Finance* **7**, 531–554.

BLACKWELL, D. & MACQUEEN, J. B. (1973). Ferguson distributions via Pólya urn schemes. *Annals of Statistics* **1**, 353–355.

BOLLEN, N., GREY, S. F. & WHALEY, R. E. (2000). Regime switching in foreign exchange rates: Evidence from currency option prices. *Journal of Econometrics* **94**, 239–276.

BOLLERSLEV, T. (1986). Generalized autoregressive conditional heteroskedasticity. *Journal of Econometrics* **31**, 307–327.

CAI, J. (1994). A Markov model of switching-regime ARCH. *Journal of Business & Economic Statistics* **12**, 309–316.

CALVET, L. E. & FISHER, A. J. (2004). How to forecast long-run volatility: Regime switching and the estimation of multifractal processes. *Journal of Financial Econometrics* **2**, 49–83.

CARTER, C. K. & KOHN, R. (1994). On Gibbs sampling for state space models. *Biometrika* **81**, 541–553.

CARVALHO, C. M. & LOPES, H. F. (2007). Simulation-based sequential analysis of Markov switching stochastic volatility models. *Computational Statistics & Data Analysis* **51**, 4526–4542.

CASARIN, R., LEISEN, F., MOLINA, G. & TER HORST, E. (2015). A Bayesian beta Markov random field calibration of the term structure of implied risk neutral densities. *Bayesian Analysis* **10**, 791–819.

CHAN, W. (2008). Dynamic hedging with foreign currency futures in the presence of jumps. *Studies in Nonlinear Dynamics & Econometrics* **12**, 1558–3708.

CHAN, W. H. & MAHEU, J. M. (2002). Conditional jump dynamics in stock market returns. *Journal of Business & Economic Statistics* **20**, 377–389.

CHIB, S. (1996). Calculating posterior distributions and modal estimates in Markov mixture models. *Journal of Econometrics* **75**, 79–97.

CHIB, S. & KANG, K. H. (2013). Change-points in affine arbitrage-free term structure models. *Journal of Financial Econometrics* **11**, 302–334.

CHRISTOFFERSEN, P., JACOBS, K. & ORNTHANALAI, C. (2012). Exploring time-varying jump intensities: Evidence from S&P500 returns and options. *Journal of Financial Economics* **106**, 447–472.

CLARK, P. K. (1973). A subordinated stochastic process model with finite variance for speculative prices. *Econometrica* **41**, 135–155.

DELATOLA, E. & GRIFFIN, J. E. (2013). A Bayesian semiparametric model for volatility with a leverage effect. *Computational Statistics & Data Analysis* **60**, 97–110.

DING, Z. & ENGLE, R. F. (2001). Large scale conditional covariance matrix modeling, estimation and testing. NYU Stern School of Business.

DUAN, J.-C., RITCHKEN, P. & SUN, Z. (2006). Approximating GARCH-jump models, jump-diffusion processes, and option pricing. *Mathematical Finance* **16**, 21–52.

DUEKER, M. (1997). Switching in GARCH processes stock-market volatility and mean-reverting stock-market volatility. *Journal of Business & Economic Statistics* **15**, 26–34.

DUFAYS, A. (2016). Infinite-state Markov-switching for dynamic volatility. *Journal of Financial Econometrics* **14**, 418–460.

DURHAM, G. B. (2007). SV mixture models with application to S&P 500 index returns. *Journal of Financial Economics* **85**, 822–856.

EIDENBERGER, J., NEUDORFER, B., SIGMUND, M. & STEIN, I. (2015). What predicts financial instability? A Bayesian approach. Deutsche Bundesbank Discussion Paper 36/2014.

ENGEL, C. (1994). Can the Markov switching model forecast exchange rates? *Journal of International Economics* **36**, 151–165.

ERAKER, B. (2004). Do stock prieces and volatility jump? Reconciling evidence from spot and options prices. *Journal of Finance* **59**, 1367–1403.

ERAKER, B., JOHANNES, M. S. & POLSON, N. G. (2003). The impact of jumps in volatility and returns. *Journal of Finance* **58**, 1269–1300.

ESCOBAR, M. D. (1994). Estimating normal means with a Dirichiet process prior. *Journal of the American Statistical Association* **89**, 268–277.

ESCOBAR, M. D., MULLER, P. & WEST, M. (1994). Hierarchical priors and mixture models, with application in regression and density estimation. In *Aspects of Uncertainty: A Tribute to D.V. Lindley*, A. F. M. Smith & P. Freeman, eds. New York: Wiley.

ESCOBAR, M. D. & WEST, M. (1995). Bayesian density estimation and inference using mixtures. *Journal of the American Statistical* **90**, 577–588.

FAMA, E. F. (1965). The behavior of stock-market prices. *Journal of Business* **38**, 34–105.

FERGUSON, T. S. (1973). A Bayesian analysis of some nonparametric problems. *Annals of Statistics* **1**, 209–230.

FRANCQ, C. & ZAKOÏAN, J.-M. (2005). The L2-structures of standard and switching-regime GARCH models. *Stochastic Processes and Their Applications* **115**, 1557–1582.

GALEANO, P. & AUSÍN, M. C. (2010). The Gaussian mixture dynamic conditional correlation model: Parameter estimation, value at risk calculation, and portfolio selection. *Journal of Business & Economic Statistics* **28**, 559–571.

GEWEKE, J. & AMISANO, G. (2011). Optimal prediction pools. *Journal of Econometrics* **164**, 130 – 141.

GEWEKE, J. & KEANE, M. (2007). Smoothly mixing regressions. *Journal of Econometrics* **138**, 252–290.

GLOSTEN, L. R., JAGANNATHAN, R. & RUNKLE, D. E. (1993). On the relation between the expected value and the volatility of the nominal excess return on stocks. *Journal of Finance* **48**, 1779–1802.

GRANGER, C. W. (1980). Long memory relationships and the aggregation of dynamic models. *Journal of Econometrics* **14**, 227–238.

GRAY, S. (1996). Modeling the conditional distribution of interest rates as a regime-switching process. *Journal of Financial Economics* **42**, 27–62.

GUIDOLIN, M. & TIMMERMANN, A. (2007). International asset allocation under regime switching, skew, and kurtosis preferences. *Review of Financial Studies* **21**, 889–935.

HAAS, M. & MITTNIK, S. (2008). Multivariate regime switching GARCH with an application to international stock markets. CFS Working Paper, 2008/08.

HAAS, M., MITTNIK, S. & PAOLELLA, M. (2004). A new approach to Markov switching GARCH models. *Journal of Financial Econometrics* **2**, 493–530.

HAMILTON, J. D. (1988). Rational expectations econometric analysis of changes in regime. An investigation of the term structure of interest rates. *Journal of Economic Dynamics and Control* **12**, 385–423.

HAMILTON, J. D. (1989). A new approach to the economic analysis of nonstationary time series and the business cycle. *Econometrica* **57**, 357–384.

HAMILTON, J. D. (1990). Analysis of time series subject to changes in regime. *Journal of Econometrics* **45**, 39–70.

HAMILTON, J. D. (1994). *Time Series Analysis*. Princeton University Press, Princeton, NJ.

HAMILTON, J. D. & SUSMEL, R. (1994). Autoregressive conditional heteroskedasticity and changes in regime. *Journal of Econometrics* **64**, 307–333.

HARVEY, A., RUIZ, E. & SHEPHARD, N. (1994). Multivariate stochastic variance models. *Review of Economic Studies* **61**, 247–264.

HASAN, I., HORVATH, R. & MARES, J. (2015). What type of finance matters for growth? Bayesian model averaging evidence. Bank of Finland Research Discussion Paper.

HE, Z. & MAHEU, J. M. (2010). Real time detection of structural breaks in GARCH models. *Computational Statistics & Data Analysis* **54**, 2628–2640.

HENNEKE, J. S., RACHEV, S. T., FABOZZI, F. J. & NIKOLOV, M. (2011). MCMC-based estimation of Markov switching ARMA-GARCH models. *Applied Economics* **43**, 259–271.

ISHWARAN, H. & JAMES, L. F. (2001). Gibbs sampling methods for stick-breaking priors. *Journal of the American Statistical Association* **96**, 161–174.

ISHWARAN, H. & ZAREPOUR, M. (2000). Markov Chain Monte Carlo in approximate Dirichlet and Beta two-parameter process hierarchical models. *Journal of Computational and Graphical Statistics* **87**, 371–390.

JACQUIER, E., POLSON, N. G. & ROSSI, P. E. (2004). Bayesian analysis of stochastic volatility models with fat-tails and correlated errors. *Journal of Econometrics* **122**, 185–212.

JENSEN, M. & MAHEU, J. M. (2010). Bayesian semiparametric stochastic volatility modeling. *Journal of Econometrics* **157**, 306–316.

JENSEN, M. & MAHEU, J. M. (2013). Bayesian semiparametric multivariate GARCH modeling. *Journal of Econometrics* **176**, 3–17.

JENSEN, M. & MAHEU, J. M. (2014). Estimating a semiparametric asymmetric stochastic volatility model with a Dirichlet process mixture. *Journal of Econometrics* **178**, 523–538.

JENSEN, M. J. & FISHER, M. (2018). Nonparametric inference and prediction in a panel of multiple-structural-break models. *Journal of Econometrics* , accepted for publication.

JIN, X. & MAHEU, J. M. (2016). Bayesian semiparametric modeling of realized covariance matrices. *Journal of Econometrics* **192**, 19–39.

KALLI, M. & GRIFFIN, J. E. (2015). Flexible modeling of dependence in volatility processes. *Journal of Business & Economic Statistics* **33**, 102–113.

KALLI, M., GRIFFIN, J. E. & WALKER, S. G. (2009). Slice sampling mixture models. *Statistics and Computing* **21**, 93–105.

KALLI, M., WALKER, S. G. & DAMIEN, P. (2013). Modeling the conditional distribution of daily stock index returns: An alternative Bayesian semiparametric model. *Journal of Business & Economic Statistics* **31**, 371–383.

KAUFMANN, S. & FRÜHWIRTH-SCHNATTER, S. (2002). Bayesian analysis of switching ARCH-models. *Journal of Time Series Analysis* **23**, 425–458.

KIM, C.-J. & NELSON, C. R. (1999). *State-space models with regime switching: Classical and Gibbs sampling approaches with applications.* Cambridge, MA: MIT Press.

KIM, S., SHEPHARD, N. & CHIB, S. (1998). Stochastic volatility: Likelihood inference and comparison with ARCH models. *Review of Economic Studies* **65**, 361–393.

KLAASSEN, F. (2002). Improving GARCH volatility forecasts with regime-switching GARCH. *Empirical Economics* **27**, 363–394.

LAU, J. W. & CRIPPS, E. (2012). Bayesian non-parametric mixtures of GARCH(1,1) models. *Journal of Probability and Statistics* **2012**, Article ID 167431.

LIU, C. & MAHEU, J. M. (2008). Are there structural breaks in realized volatility? *Journal of Financial Econometrics* **6**, 326–360.

LIU, C. & MAHEU, J. M. (2009). Forecasting realized volatility: A Bayesian model-averaging approach. *Journal of Applied Econometrics* **24**, 709–733.

LO, A. Y. (1984). On a class of Bayesian nonparametric estimates: I. Density estimates. *Annals of Statistics* **12**, 351–357.

LOPES, H. F. & POLSON, N. G. (2010). Extracting SP500 and Nasdaq volatility: The credit crisis of 2007-2008. In *The Oxford Handbook of Applied Bayesian Analysis*, A. O'Hagan & M. West, eds. Oxford: Oxford University Press.

MAHEU, J. M. & MCCURDY, T. H. (2000a). Identifying bull and bear markets in stock returns. *Journal of Business & Economic Statistics* **18**, 100–112.

MAHEU, J. M. & MCCURDY, T. H. (2000b). Volatility dynamics under duration-dependent mixing. *Journal of Empirical Finance* **7**, 345–372.

MAHEU, J. M. & MCCURDY, T. H. (2004). News arrival, jump dynamics, and volatility components for individual stock returns. *Journal of Finance* **59**, 755–794.

MAHEU, J. M., MCCURDY, T. H. & ZHAO, X. (2013). Do jumps contribute to the dynamics of the equity premium? *Journal of Financial Economics* **110**, 457–477.

MAHEU, J. M. & YANG, Q. (2016). An infinite hidden Markov model for short-term interest rates. *Journal of Empirical Finance* **38**, 202–220.

MANDELBROT, B. B. (1997). *The Variation of Certain Speculative Prices*. New York: Springer.

MELICK, W. R. & THOMAS, C. P. (1997). Recovering an asset's implied PDF from option prices: An application to crude oil during the Gulf crisis. *Journal of Financial and Quantitative Analysis* **32**, 91–115.

NAKAJIMA, J. & OMORI, Y. (2009). Leverage, heavy-tails and correlated jumps in stochastic volatility models. *Computational Statistics & Data Analysis* **53**, 2335–2353.

NEAL, R. M. (2000). Dirichlet process mixture models. *Journal of Computational and Graphical Statistics* **9**, 249–265.

OMORI, Y., CHIB, S., SHEPHARD, N. & NAKAJIMA, J. (2007). Stochastic volatility with leverage: Fast and efficient likelihood inference. *Journal of Econometrics* **140**, 425–449.

PAGAN, A. R. & SCHWERT, W. G. (1990). Conditional stock volatility. *Journal of Econometrics* **45**, 267–290.

PRESS, S. J. (1967). A compound events model for security prices. *Journal of Business* **40**, 317–335.

ROBINSON, P. M. (1978). Statistical inference for a random coefficient autoregressive model. *Scandinavian Journal of Statistics* **5**, 163–168.

RYDÉN, T., TERÄSVIRTA, T. & ÅSBRINK, S. (1998). Stylized facts of daily return series and the hidden Markov model. *Journal of Applied Econometrics* **13**, 217–244.

SETHURAMAN, J. (1994). A constructive definition of Dirichlet priors. *Statistica Sinica* **4**, 639–650.

SHEPHARD, N. (1994). Partial non-Gaussians state space. *Biometrika* **81**, 115–131.

SHI, S. & SONG, Y. (2016). Identifying speculative bubbles using an infinite hidden Markov model. *Journal of Financial Econometrics* **14**, 159–184.

SO, M., LAM, K. & LI, W. (1998). A stochastic volatility model with Markov switching. *Journal of Business & Economic Statistics* **16**, 244–253.

SUSMEL, R. (2000). Switching volatility in private international equity markets. *International Journal of Finance and Economics* **283**, 265–283.

TAYLOR, S. J. (1994). Modeling stochastic volatility: A review and comparative study. *Mathematical Finance* **4**, 183–204.

TEH, Y. W., JORDAN, M. I., BEAL, M. J. & BLEI, D. M. (2006). Hierarchical Dirichlet Processes. *Journal of the American Statistical Association* **101**, 1566–1581.

TODOROV, V. & TAUCHEN, G. (2011). Volatility jumps. *Journal of Business & Economic Statistics* **29**, 356–371.

TURNER, C. M., STARTZ, R. & NELSON, C. R. (1989). A Markov model of heteroskedasticity, risk, and learning in the stock market. *Journal of Financial Economics* **25**, 3–22.

VAN GAEL, J. & GHAHRAMANI, Z. (2011). Nonparametric hidden Markov models. In *Bayesian Time Series Models*, D. Barber, A. T. Cemgil & S. Chiappa, eds. Cambridge: Cambridge University Press.

VAN GAEL, J., SAATCI, Y., TEH, Y. W. & GHAHRAMANI, Z. (2008). Beam sampling for the infinite hidden Markov model. In *Proceedings of the 25th International Conference on Machine Learning – ICML '08*. New York: ACM Press.

VIRBICKAITE, A., AUSÍN, M. C. & GALEANO, P. (2015). Bayesian inference methods for univariate and multivariate GARCH models: A survey. *Journal of Economic Surveys* **29**, 76–96.

VLAAR, P. J. G. & PALM, F. C. (1993). The message in weekly exchange rates in the European monetary system: Mean reversion, conditional heteroscedasticity, and jumps. *Journal of Business & Economic Statistics* **11**, 351–360.

WALKER, S. G. (2007). Sampling the Dirichlet mixture model with slices. *Communications in Statistics Simulation and Computation* **36**, 46–57.

WRIGHT, J. H. (2008). Bayesian model averaging and exchange rate forecasts. *Journal of Econometrics* **146**, 329–341.

# 18

---

# Applications in Genomics

**Stéphane Robin and Christophe Ambroise**

*UMR MIA-Paris, AgroParisTech, INRA, Université Paris-Saclay, France; UMR MIA-Paris, AgroParisTech, INRA, Université Paris-Saclay, Laboratoire de Mathématiques et Modélisation d'Évry (LaMME), Université d'Évry-Val-d'Essonne, UMR CNRS 8071, ENSIIE, France*

## CONTENTS

## 18.1   Introduction

Mixture models are intensively used in genetics and genomics either for identifying latent structures or for modeling densities. According to the type of mixture component and the nature of the hypothesis about latent structures, mixture models may be relevant in numerous different frameworks. In this chapter, the use of mixture models in genetic and genomics is presented by increasing complexity of the latent structure. Section 18.2 considers applications with independent latent variable structures to genome and transcriptome analysis. Section 18.3 illustrates the use of hidden Markov models (HMMs) in genomics, presenting a variety of problems with their associated translation in terms of emission distributions and hidden states. Finally, Section 18.4 introduces more complex dependency structures used in genomics such as the hidden Markov random field and stochastic block model with their associated parameter estimation difficulties.

## 18.2 Mixture Models in Transcriptome and Genome Analysis

### 18.2.1 Analyzing the genetic structure of a population

Identifying the underlying structure of populations is a recurrent task in genetics. It allows one to correct population stratification in genetic association studies (Bouaziz et al., 2011), to study the evolutionary relationships between populations as well as to learn about their demographic histories (Cavalli-Sforza et al., 1994).

In this context, mixture models emerge as a natural strategy to infer the structure of the population or the structure of the genome itself. Indeed, there is a two-level structure: each individual can be considered as belonging to a subpopulation, but regions of the genome of a given individual can themselves be considered as having different origins. The latter case is known as genetic admixture. It occurs when individuals from two or more previously separated populations begin interbreeding.

Many different parametric approaches exist, differing mainly in the estimation method, but relying on the same basic mixture of multinomials. The so-called Structure algorithm (Pritchard et al., 2000) proposes a mixture of multinomial distributions in a Bayesian framework with Markov chain Monte Carlo (MCMC) inference. FRAPPE (Tang et al., 2005) uses a maximum likelihood approach associated with an EM algorithm, and Admixture (Alexander et al., 2009) computes the same estimates using a sequential quadratic programming algorithm with a quasi-Newton scheme. We also mention the Bayesian Analysis of Population Structure(BAPS; Corander et al., 2003) which includes the number of subpopulations in the model.

Let us describe the reference Structure model in more detail. Structure uses Bayesian statistical inference to cluster individuals from genotype data or to determine admixture proportions (Pritchard et al., 2000). Different statistical models are associated with each endgame of the method. Both BAPS and Structure models assume the estimated subpopulations to be in Hardy–Weinberg equilibrium. The model without admixture assumes $G$ subpopulations from which were sampled the $n$ diploid individuals genotyped at $L$ multi-allelic (but often bi-allelic) loci $y_\ell^i = (y_\ell^{(i,1)}, y_\ell^{(i,2)})$, for $\ell = 1, \ldots, L$ and $i = 1, \ldots, n$. The parameters of the mixture are the allele frequencies $\theta = (\theta_{gj\ell})$, where $\theta_{gj\ell} = \mathrm{P}(y_\ell^{(i,a)} = j | z_i = g, \theta)$ is the frequency of allele $j$ at locus $\ell$ in population $g$. The conditional distribution for the data $y_i = (y_1^i, \ldots, y_L^i)$ for individual $i$ is

$$p(y_i | z_i = g) = \prod_{\ell=1}^{L} \prod_{a=1,2} \theta_{g, y_\ell^{(i,a)}, \ell}.$$

This resulting mixture has numerous parameters since each locus (often in the order of a million) of each class is described by a different vector of proportions. Bayesian inference is used to obtain the distribution of $p(\mathbf{z}, \theta | \mathbf{y})$ of $\mathbf{z} = (z_1, \ldots, z_n)$ and $\theta$, given data $\mathbf{y} = (y_1, \ldots, y_n)$. In the original paper (Pritchard et al., 2000), the posterior distribution

$$p(\mathbf{z}, \theta | \mathbf{y}) \propto p(\mathbf{z}) p(\theta) p(\mathbf{y} | \mathbf{z}, \theta)$$

considers a uniform prior for $\mathbf{z}$ and a Dirichlet prior for $\theta$.

To account for admixture, a new parameter is introduced in the model. Let $\pi_i = (\pi_{i1}, \ldots, \pi_{iG})$, where $\pi_{ig}$ represents the proportion of individuals $i$ that originated from population $g$,

$$p(z_i | \pi_i) = \prod_{\ell} \prod_{a=1,2} \pi_{i, z_\ell^{(i,a)}},$$

where $z_\ell^{(i,a)}$ is the population of origin of allele copy $y_\ell^{(i,a)}$ and, defining $z_\ell^i = (z_\ell^{(i,1)}, z_\ell^{(i,2)})$ for each locus $\ell$, $z_i = (z_1^i, \ldots, z_L^i)$. The conditional distribution then becomes

$$p(y_i | z_i, \pi_i) = \prod_\ell \prod_a \theta_{z_\ell^{(i,a)}, y_\ell^{(i,a)}, \ell} .$$

For each individual, the parameter $\pi_i$ has a Dirichlet prior distribution as well. The estimated posterior distribution becomes $p(\mathbf{z}, \pi, \theta | \mathbf{y})$ with $\pi = (\pi_1, \ldots, \pi_n)$.

Because of the high dimensionality of the problem, this type of inference can be really slow, and variational inference offers a faster alternative in the Bayesian framework (Raj et al., 2014).

Other approaches may also consider mixture models associated with kernel methods. In that context, the use of adapted kernel functions enables one to operate in a high-dimensional, implicit feature space via Gaussian mixture models (GMMs). SHIPS (Spectral Hierarchical Clustering for the Inference of Population Structure in Genetic Studies) is an avatar of such approaches (Bouaziz et al., 2012). It is based on a spectral clustering algorithm. The algorithm first uses a kernel based on the allele sharing distance that has previously been used to identify genetic patterns among populations (Mountain & Cavalli-Sforza, 1997). In the implicit feature space, classical GMMs are used to recursively split the individuals into subpopulations.

## 18.2.2   Finding sets of co-transcribed genes

Gene expression is modulated (up or down) depending on tissue (e.g. liver vs. brain), development stage (e.g. fetal vs. adult), disease status, genotype (e.g. mutant vs. wild) or dynamically as a response to environmental signals.

A DNA microarray consists of thousands of microscopic spots of DNA oligonucleotides from a specific DNA sequence, known as probes (or reporters), that are used to hybridize a cRNA sample (called the target). The DNA microarray technology allows the expression levels of thousands of genes to be measured across different conditions. These measurements provide a "picture" of cells functioning at a given time. Such technology is thus of great importance in many applications such as functional genomics, medical and clinical diagnosis, drug discovery, targeting, and monitoring.

The data resulting from experiments of this type is a gene expression matrix whose columns describe the genes and whose rows describe the samples. Note that each sample is also described by other variables such as the conditions of the experiment. Reasonable experiments usually involve a set of replicates for each condition. Gene expression data analysis aims to highlight differences between conditions and provide an insight into global gene patterns.

Such data has many features, few observations and is usually very noisy. Statistical analysis of microarrays raises issues and challenges for statisticians. Analyzing such data usually requires a succession of steps (McLachlan et al., 2005): a normalization process to make all samples comparable, a differential analysis to pinpoint genes which have different expression across the conditions, and an exploratory data analysis step to enhance the understanding of the results. The analysis of microarray data relies on univariate and multivariate descriptive statistics at each step in order to control the process or gain insight into the data. Clustering approaches are often used, and mixture models are a classical tool in this context (Jiang et al., 2004). When dealing with DNA microarrays, data is considered continuous and Gaussian mixtures are the dominant model.

In gene expression analysis, the aim of clustering is to find a structure within the samples and/or within the genes. Both approaches bring different and relevant information about the data. The seminal paper of Eisen et al. (1998) proposes to cluster the genes by means of classical hierarchical agglomerative clustering using average linkage and an initial metric based on Euclidian distance. The authors observe that genes of similar function cluster together. This observation justifies the use of clustering for searching for hints about gene function guided by a "guilt by association" principle.

When considering the clustering of samples, high dimension (in terms of the number of genes) constitutes a problem. Indeed, classical mixture models are not able to deal with samples where the number of variables is much greater than the number of samples, unless the variables are assumed to be uncorrelated within a cluster. This problem is caused by the estimation of the covariance matrices whose number of parameters grows quadratically with the number of variables. Yeung et al. (2001) exploit the representation of the covariance matrix in terms of its eigenvalue decomposition,

$$\Sigma_g = \lambda_g D_g A_g D_g^\top,$$

where $D_g$ is the matrix of eigenvectors, $A_g$ is a diagonal matrix whose elements are proportional to the eigenvalues, and $\lambda_g$ is a scalar. Reduction of the number of free parameters in the covariance matrix can also be achieved by mixture of factor analyzers (McLachlan et al., 2002),

$$\Sigma_g = B_g B_g^\top + D_g,$$

where $B_g$ is a matrix of loading factors and $D_g$ a diagonal matrix; see also Chapter 11 above for a detailed review of mixture analysis for high-dimensional data.

Since the mid-2000s, next generation sequencing (NGS) has become the new standard tool for measuring gene expression. Compared to data produced with microarrays, NGS data are count-based measures, discrete, positive, and highly skewed. We introduce the two main alternatives that have been proposed to adapt model-based clustering approaches to such data: it is possible to change the normalization of the data to adapt to Gaussian mixtures (Kim et al., 2007; Huang et al., 2008) or develop models specifically for discrete positively skewed data.

Although a multivariate version of the Poisson distribution does exist, Rau et al. (2015) assume variables to be independent conditionally on the component indicator $z_i$; see also Chapter 9 above for a detailed review of mixture analysis for count data. Considering discrete gene expressions $y_{ij\ell}$ of gene $i$ in condition $j$ ($j = 1, \ldots, d$) for replicate $\ell$ ($\ell = 1, \ldots, r_j$), the component distribution of the expression vector of gene $i$ is a product of Poisson distributions,

$$y_i | z_i = g \sim \prod_{j=1}^{d} \prod_{\ell=1}^{r_j} \mathcal{P}(y_{ij\ell} | \mu_{ij\ell g}).$$

The authors propose further reducing the number of parameters using parameterization in the spirit of the above-mentioned Gaussian parameterization, assuming a common mean across the replicates $\ell$,

$$\mu_{ij\ell g} = w_i \lambda_{jg},$$

or adapting the mean as a function of known library sizes $s_{j\ell}$,

$$\mu_{ij\ell g} = w_i s_{j\ell} \lambda_{jg},$$

since the number of reads mapped to a gene is highly dependent on the gene size.

Biclustering is a technique from two-way data analysis, the aim of which is to find a structure of both rows and columns of a data table. This approach is popular for exploring DNA microarrays since there is often a structure both in samples and genes. Looking for a gene/sample block structure can obviously be achieved in two steps (one step for each dimension) or simultaneously in both dimensions (Ben-Dor et al., 2003). A widespread graphical representation of this approach is the classical *heatmap* which is a false color image of the data table with reordering of the rows and columns according to some identified latent structure.

There are many different types of structure and algorithms in the field of biclustering. Block structure is a possibility and can be considered as a latent structure. The so-called latent block model (LBM), a mixture model with an associated estimation procedure, has been proposed in this context by Govaert & Nadif (2003) for identifying a simultaneous partition of rows and columns. The density of $\mathbf{y}$, knowing the partition of the rows $\mathbf{z}$ and of the columns $\mathbf{w}$, is

$$p(\mathbf{y}|\mathbf{z},\mathbf{w},\alpha) = \prod_{ij}\prod_{g\ell} f(y_{ij}|\alpha_{g\ell})^{z_{ig}w_{j\ell}},$$

where $f(\cdot|\alpha)$ is a parametric distribution with parameter vector $\alpha_{g\ell}$ for block $g\ell$. In this context, the likelihood is not tractable and variational EM (see Section 18.4.3) or Bayesian strategies have been proposed for estimating the parameters of the mixture. Although model selection is a complex problem in this context, since the likelihood is not tractable, Bayesian inference offers an efficient alternative for designing model selection criteria (Keribin et al., 2015).

### 18.2.3 Variable selection for clustering with Gaussian mixture models

As mentioned above, GMMs are not identifiable in a high-dimensional setting (where the number of observations is small compared to the number of variables) and strategies have been developed for limiting the number of parameters of the model. Biological data typically fall into this high-dimensional setting, not only because of transcriptome, but also with epigenome, proteome, metabolome, molecular pathways, molecular imaging, and others. Dimension reduction is thus a key issue in the field. It can be achieved via factor analysis, regularization, and other sorts of constrained parameterization (Bouveyron & Brunet-Saumard, 2014).

A simple alternative for reducing the dimensionality involves selecting relevant variables; see also Chapter 11, Section 11.7. Variable selection is an important, well-researched topic in supervised learning, but has a more recent history in clustering. The difficulty of the problem undoubtedly plays a role in this difference of treatment. Nevertheless, variable selection in discrimination (Kohavi & John, 1997) and clustering share common aspects. Variable selection may be performed in different ways. The so-called filter approach involves selecting "informative" variables beforehand. In the transcriptome context the most widespread method is differential analysis (see Section 18.2.4). A second possibility involves selecting (or ordering) the variables after clustering (if clustering is possible) (McLachlan et al., 2002). The final possibility involves selecting the variables while estimating the mixture model parameters.

In a Bayesian paradigm, Tadesse et al. (2005) propose a method using a reversible jump MCMC algorithm to simultaneously choose the number of mixture components and select relevant variables. Raftery & Dean (2006) define two different sets of variables: relevant and irrelevant variables. They do not assume independence between the relevant and irrelevant variables for the clustering, as considered in Tadesse et al. (2005). The model of Raftery & Dean (2006) considers a partition of the variables $y_i$ into two disjoint subsets: the variables $y_i^c$ relevant for clustering, and the irrelevant ones $y_i^o$. The model skillfully mixes clustering and regression. The density of $y_i$ is decomposed into two multiplicative parts as

$$p(y_i|z_i = g) = p(y_i^c|z_i = g)p(y_i^o|y_i^c).$$

Marginalizing over $z_i$, $p(y_i^c)$ is a classical mixture model with component density $p(y_i^c|\theta_g)$. $p(y_i^o|y_i^c)$ is a multivariate regression model where the variables $y_i^o$ are explained as linear combinations of the variables $y_i^c$. Maugis et al. (2009) propose a refined version of this model avoiding non-parsimonious models by selecting the predictor variables in the linear regression part of the model in a two-step stepwise algorithm.

In supervised learning, penalized regression is a popular approach for selecting variables while estimating the parameters. The same kind of approach has been explored in the context of mixture models. Pan & Shen (2007) propose, for example, a penalized log likelihood criterion by assuming a Gaussian mixture model with common diagonal covariance matrices. The lasso-like penalty penalizes the sum of the absolute values of the means $\mu_{gj}$ of component $j$ of cluster $g$:

$$p_\lambda(\theta) = \lambda \sum_{gj} |\mu_{gj}|.$$

There are many related works in this line of research borrowing and adapting ideas developed for supervised methods (Witten & Tibshirani, 2010).

### 18.2.4 Mixture models in the specific case of multiple testing

Because of the dimensionality of most genomic data, multiple testing issues have become commonplace in genomic analyses. The most emblematic case is that of the detection of differentially expressed genes, which can be summarized as follows. Consider all (known) genes from a given species and, for each of them, perform a sample comparison test with the null hypothesis

$$H_{0i} = \text{gene } i \text{ has the same expression level in all conditions.}$$

In a frequentist setting, each gene is then associated with a test statistic, the distribution of which is known under $H_{0i}$ and from which a $p$-value $y_i$ is derived. Such a setting obviously raises a multiple testing problem about which a huge literature exists (for reviews, see Dudoit et al., 2003; Roquain, 2011). The aim of multiple testing procedures is to control some multiple Type I error rate such as the familywise error rate or false discovery rate (FDR). Most of these procedures rely on the fact that, under $H_{0i}$, $y_i$ has a uniform distribution $\mathcal{U}[0,1]$.

*Unsupervised classification point of view*

Efron et al. (2001) rephrased this problem as a clustering problem, in which one wishes to classify genes according to the latent variable $z_i$ defined as follows:

$$z_i = \begin{cases} 0, & \text{if } H_{0i} \text{ is true} \quad (\text{``null'' gene}), \\ 1, & \text{if } H_{0i} \text{ is not true} \quad (\text{differentially expressed gene}). \end{cases}$$

The unsupervised classification task can then be achieved using the mixture model

$$y_i \sim \eta_0 f_0 + (1 - \eta_0) f_1, \tag{18.1}$$

where $\eta_0$ stands for the proportion of null genes, $f_0$ for the pdf of the uniform distribution $\mathcal{U}[0,1]$, and $f_1$ for the pdf of the $p$-values under the alternative hypothesis $H_{1i}$, which is supposed to be the same for all genes. This model provides an alternative view of the problem, in which the (estimated) conditional probability

$$\tau_{i0} = \frac{\eta_0 f_0(y_i)}{\eta_0 f_0(y_i) + (1 - \eta_0) f_1(y_i)},$$

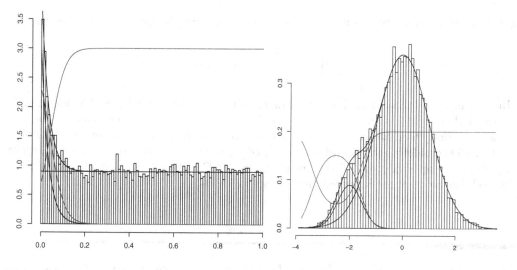

**FIGURE 18.1**

Left: histogram of $p$-values with fitted beta mixture (black solid line), $f_0$ (black dashed line), $f_1$ (black dotted line) and rescaled conditional probabilities (gray lines). Righ: histogram of probit-transformed $p$-values with fitted Gaussian mixture.

is interpreted as a *local* FDR (Efron & Tibshirani, 2002). A natural classification rule then involves classifying gene $i$ as positive (i.e. non-"null") when $\tau_{i0}$ is below a given threshold $t$. Note that, in this setting, because $f_0$ is known, the proportion $\eta_0$ of null genes can be estimated under mild conditions on $f_1$ (see, for example, Storey, 2002).

*Parametric mixture models*

A first parametric version of model (18.1) was proposed by Allison et al. (2002), where $f_1$ is supposed to be a beta distribution $\mathcal{B}e(a, b)$. In the same vein, McLachlan et al. (2006) considered a two-component mixture model similar to model (18.1), but applied to the transformed values $\widetilde{y}_i = \Phi^{-1}(y_i)$, so that the transformed null distribution $\widetilde{f}_0$ is a standard Gaussian distribution. In this approach, the alternative distribution $\widetilde{f}_1$ is supposed to be the Gaussian distribution $\mathcal{N}(\mu_1, \sigma_1^2)$. The probit transform $\Phi^{-1}$ turns out to be efficient, as it zooms into the region where $p$-values are close to 0, which improves the identification of the positive genes (see Figure 18.1). McLachlan et al. (2006) further elaborate on the estimation of the FDR and suggest the estimate

$$\widehat{FDR}(t) = \frac{\sum_i \tau_{i0} I_0(\tau_{i0} < t)}{\sum_i I_0(\tau_{i0} < t)},$$

for a given threshold $t$.

*Semi- and nonparametric mixture models*

One of the key features in model (18.1) is that the distribution $f_0$ is known, which allows for a higher flexibility for $f_1$. Allison et al. (2002) propose a "semi-parametric" extension of it, taking $f_1$ as a mixture of beta distributions itself. Equation (18.1) then takes the standard form $y_i \sim \sum_{g=0}^{G} \eta_g f_g$, but, in terms of classification, one is only interested in the distinction between group 0 and the union of all other groups, so gene $i$ is classified according to

$$\tau_{i+} := \sum_{g=1}^{G} \tau_{ig} = 1 - \tau_{i0}.$$

In a nonparametric setting, Bordes et al. (2006) prove that model (18.1) is identifiable provided that $f_1(\cdot|\theta_1) = \psi(\cdot - \theta_1)$, where $\psi$ is some even function. They consider both a symmetrization-based and a moment-based estimate for the location parameter $\theta_1$. Robin et al. (2007) consider a kernel density estimate of $f_1$, for the estimation of which a convergent algorithm is proposed, provided that $\eta_0$ is known; see Chapter 14, Section 14.3 above for a detailed discussion of the identifiability of such mixtures.

## 18.3 Hidden Markov Models in Genomics: Some Specificities

Because many genomic data sets are collected at loci (probe, nucleotide) located along the genome, hidden Markov models have become a standard tool in that field (Durbin et al., 1999; Seifert, 2013). As discussed in Chapter 13 above, an HMM deals with data collected in a sequential manner and is similar to a mixture model, except that the $\{z_t\}_{1 \leq t \leq n}$ form a Markov chain on $\{1, \ldots, G\}$ with transition matrix $\xi$. The group $z_t$ to which the observation collected at locus $t$ belongs to is called its (hidden) *state*. In most HMMs, the observations $\{y_t\}_{1 \leq t \leq n}$ are supposed to be independent conditionally on the hidden states, the conditional distribution being named the *emission* distribution. In this section, we first present a typical genomic problem where HMMs can be used and then introduce a series of special cases where the genomic context requires us to consider more sophisticated models in terms of hidden states, emission distribution, and dependency structure.

### 18.3.1 A typical case: Copy number variations

Many diseases are associated with genomic alterations which consist of either the loss or the amplification of some regions of the genome (Albertson et al., 2003). As a result, some regions of the genome are not present in two copies (as expected in a normal cell of a diploid species, such as humans), but in fewer (zero or one copy, named "loss") or more (three, four or even more copies, named "gain"). A series of technologies has been developed in recent decades to obtain a measure $y_t$ that is related to the number of genomic copies at locus $t$. These technologies range from microarrays to sequencing technologies (NGS; Albertson & Pinkel, 2003). A typical example of the signal at hand is displayed in Figure 18.2 (left).

Hidden Markov models are especially well suited to address the task of both finding the location at which the number of copies changes and to classify each of the segments according to the number of copies as "normal", "loss" or "gain". Fridlyand et al. (2004) first proposed using an HMM with Gaussian emission distribution for the detection of copy number variation (CNV) based on microarray data. The classification step, which involves retrieving the hidden path $\mathbf{z} = (z_1, \ldots, z_n)$, is referred to as 'CNV calling' and can be achieved using the Viterbi algorithm (Viterbi, 1967).

The detection of the loci where the copy number varies can be seen as a change-point detection problem, as proposed by Picard et al. (2005), but this approach does not address the calling step. As an alternative, Picard et al. (2007) introduced a mixture model where each *region* belongs to a certain group $g$, the data $y_t$ being independent with the same Gaussian distribution $\phi_g$. The inference of such a model can be made using a specific EM

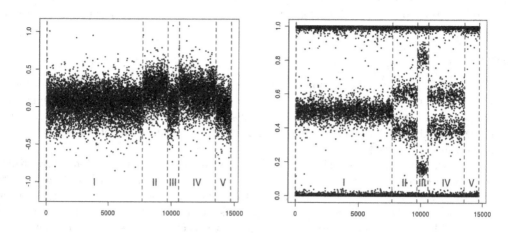

**FIGURE 18.2**

SNP data from Popova et al. (2009), chromosome 11. Left: log R ratio $\log_2[(a_t + b_t)/2]$ as a function of $t$. Right: B-allele frequency $b_t/(a_t + b_t)$ as a function of $t$.

algorithm where the M step includes a segmentation step that can be efficiently solved using dynamic programming.

The size of the data has varied a lot in the last decade, from few hundred probes per chromosome on original comparative genomic hybridization (CGH) arrays (Albertson & Pinkel, 2003) to the number of nucleotides per chromosome (i.e. about $10^8$ for the longest human one) for sequencing technologies. Indeed, the forward-backward recursion used in the E step of the EM algorithm is linear, whereas the regular dynamic programming is quadratic. Still, the number of iterations of the EM algorithm is not known and recent advances have dramatically reduced the complexity of the segmentation algorithms (see, for example, Rigaill, 2010; Killick et al., 2012). Hence, many CNV analyses are done using segmentation approaches (using post-processing for calling) for which dynamic programming-type approaches turn out to perform well (Lai et al., 2005; Hocking, 2012).

### 18.3.2   Complex emission distributions

As in any domain, HMMs need to be adapted to the nature of the signal under study. Because of the variety of both biological objects and technologies, a huge variety of emission distributions have been considered in genomics. First, the emission distribution must accommodate to the fact that some technologies (e.g. microarrays) provide a continuous signal whereas others (e.g. NGS) lead to a discrete signal (counts). As expected, Gaussian and Poisson emission distributions are respectively the most popular parametric distributions although, in the latter case, the negative binomial seems preferable. Less easy-to-handle distributions such as the negative binomial have turned out to be more relevant for a series of applications using NGS data; see, for example, Robinson et al. (2010) and Rashid et al. (2011).

Also, the efficiency of many molecular technologies depends on local properties of the genomic sequence such as the GC content (proportion of g and c nucleotides). In order to correct this bias, the emission distribution can account for some local covariate $x_t$, so

$f_g(y_t) = f(y_t|z_t = g)$ becomes $f_g(y_t|x_t) = f(y_t|z_t = g, x_t)$; see, for example, Bérard et al. (2013) and Rashid et al. (2011). Chapter 12 above provides a comprehensive review of mixture models of this type.

## DNA sequences

HMMs were used early on for genome annotation, typically to determine the boundaries of isochores (regions with different nucleotide composition) or to distinguish between gene coding regions and non-coding regions. In such analyses, the observed vector $\mathbf{y} = (y_1, y_2, \ldots, y_n)$ is made up of the DNA sequence itself, each $y_t$ being one of the elements of the nucleotide alphabet, $y_t \in \{a, c, g, t\}$.

The most naive model consists of an HMM with multinomial emission distributions with parameter $\theta_g = (\theta_{ga})_a$, where $\theta_{ga}$ represents the frequency of nucleotide $a$ in state $g$. However, such a simple model turns out to be far too poor to account for the local complexity of the DNA sequence. This model has been generalized in Muri (1998), who considered Markov chains as emission distributions to account for the local frequency of di-, tri-, or any oligo-nucleotide. The sufficient statistics for a Markov chain of order $m$ (denoted M$m$) are the frequencies of all sequences of $(m+1)$-nucleotides. As a consequence, a Markov model M$m$ with transition probabilities $\theta_g$ depending on the past $m$ states,

$$\theta_g((a_m, \ldots a_1); b) = P(y_t = b | z_t = g, (y_{t-m}, \ldots, y_{t-1}) = (a_m, \ldots, a_1)),$$

accounts for the frequency of the $(m + 1)$-nucleotides in state $g$. Such a model is denoted M1-M$m$ in Muri (1998) as the hidden states $(z_t)$ follow an M1 model and the observed sequence $(y_t)$ conditionally arises from an M$m$ model. As an example, coding regions are composed of triplets of nucleotides (*codons*) that are ultimately translated into amino acids, which constitute the building block of a protein. An M1-M2 model can typically account for this triplet structure (Nicolas et al., 2002).

## Multivariate signals

Several molecular technologies are intrinsically comparative in the sense that, at each locus $t$, they provide a pair of measures. This holds for CGH arrays, which compare the genomic material from a normal sample with a test sample to detect genomic alterations. This is also true for microarrays, which allow comparison of the level of transcription of a given locus $t$ in two different conditions. Single nucleotide polymorphism (SNP) arrays, which will be discussed below, also yield a bivariate signal $(a_t, b_t)$. In some situations, one of the signals is simply considered as a covariate (Bérard et al., 2013). In other cases, one may choose to consider a summary variable such as $y_t = b_t - a_t$, although this obviously involves a loss of information.

## Copy number variation and loss of heterozygosity

An interesting case is that of the joint detection of CNV and loss of heterozygosity (LOH) using SNP arrays. "Single nucleotide polymorphism" refers to single nucleotide loci $t$ spread along the genome where two alternative nucleotides are observed in the human population, whereas the neighborhood of the locus is very conserved.

The most frequent allele is arbitrarily named $A_t$ and the minor allele $B_t$. SNP arrays provide a signal $(a_t, b_t)$ where $a_t$ $(b_t)$ are proportional to the abundance of $A$ $(B)$ in the sample. At a normal locus, one should have $a_t + b_t \approx 2$ because two copies of each chromosome exist. This sum is often transformed into the log R ratio $LRR_t = \log_2[(a_t + b_t)/2]$, which is close to 0 in the normal case (see region I in Figure 18.2 (left)). Furthermore, in a normal situation, the B-allele frequency, defined as $BAF_t = b_t/(a_t + b_t)$, should be close to either 0, 1/2 or 1, corresponding to the three possible normal genotypes, $AA$, $AB$ and $BB$ (region I in Figure 18.2 (right)). Note that observations of $(a_t, b_t)$ and $(LRR_t, BAF_t)$ are equivalent.

As described in Section 18.3.1, CNVs are revealed by the $LRR$ profile. However, the joint analysis of the two profiles may reveal more complex patterns, such as region V in Figure 18.2 where $LRR_t$ seems normal (left) and where no heterozygosity is observed (right). Such a pattern suggests that one copy of this region has been lost and has been then rebuild by copying the remaining copy, so that this region is now made of two identical copies, making all loci homozygous. Although such events do not affect the copy number, they may be of interest as their genomic diversity has been reduced by one half and as favorable alleles may have been lost. More complex patterns may arise (see regions II, III and IV in Figure 18.2), when the total number of copies is not 2 (normal case) but more (say 3) resulting in a $BAF$ around 0, 1/3, 2/3 or 1. Note that these ratios remain theoretical as the sample is often contaminated with normal cells, which shrinks the empirical $BAF$ toward 0, 1/2 or 1 (Popova et al., 2009).

The analysis of such data aims to classify regions with respect to both the copy number and the heterozygosity status. The fully normal case corresponds to a copy number of 2 and three possible genotypes at each locus: $AA$, $AB$ and $BB$. As a consequence, in this state, the observed signal $y_t = (a_t, b_t)$ is distributed according to a bivariate mixture with three components (corresponding to each possible genotype). From a general point of view, this problem can be modeled with an HMM where the emission distributions themselves are mixtures (see Volant et al., 2014):

$$f_g(y_t) = \sum_k w_{gk} f(y_t | \gamma_k), \quad \text{with} \sum_k w_{gk} = 1,$$

where $f(y_t | \gamma)$ is some parametric distribution with parameter $\gamma$. Note that, in the present case, the parameter $\gamma_k$ does not depend on $g$ as several states $g$ may involve the same component (see, for example, regions I and V in Figure 18.2). The determination of the number of components for any given number of copies and heterozygosity status is left to the reader. Greenman et al. (2010) and Chen et al. (2011) provide a more extensive description of this problem.

### *Nonparametric emission distributions*

Not surprisingly, the classification performance of an HMM strongly relies on the choice of the emission distributions it involves. In the case of a bivariate continuous signal, bivariate Gaussian emission distributions are attractive. A careful modeling of the respective covariance matrix following Fraley & Raftery (2002) and Biernacki et al. (2006) can dramatically improve the performance, as shown in Bérard et al. (2011).

However, fully parametric emission distributions may not be flexible enough. As mentioned above, mixtures can be used as emission distributions (Volant et al., 2014; Lange et al., 2014). The identifiability of such a model was not addressed in these references, but has since been proven in Gassiat et al. (2016), who propose a nonparametric HMM, considering a kernel-based shape for each emission distribution:

$$f_g(y_t) = \sum_s w_{gs} \psi(y_t - y_s),$$

where $\psi$ is some centered pdf and $w_{gs}$ is the contribution of the data point $s$ to the definition of $f_g$. Note that, as a downside of flexibility, nonparametric HMMs may provide less interpretable results, as the estimate of $f_g$ does not always reveal to which biological regime the state $g$ corresponds.

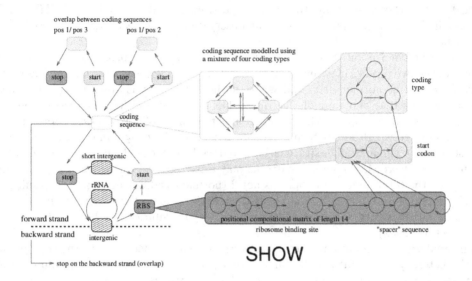

**FIGURE 18.3**
Structure of the hidden Markov chain used for gene detection (Ibrahim et al., 2007).


### 18.3.3   Complex hidden states

Most HMMs in genomics are used for classification purposes, for which the hidden space reduces to very few, easily interpretable states. However, the hidden space can be much refined in order to include prior knowledge. Genome annotation is an emblematic example: the aim is to detect coding regions in a genome, based only on the genomic sequence itself, but also taking advantage of all the knowledge we have about the structure of genes.

*Gene detection from genomic sequences*

In prokaryotic organisms, the sequence coding for a given gene is often divided into non-adjacent regions called *exons* (in between the regions called *introns*). An HMM dedicated to the detection of gene-coding regions must therefore have at least three states corresponding to non-coding (i.e. between-gene) regions, exons and introns, respectively. In addition to this, all gene coding regions start with a so-called "start" codon `atg`. So a fourth state should be added, through which the hidden Markov chain must necessarily transit when going from the non-coding state to the exon state. Figure 18.3 depicts the graph of the hidden Markov chain proposed by Ibrahim et al. (2007) to account for a series of such characteristics, including the fact that genes can be coded in both directions of the sequence.

A general property of Markov chains is that the sojourn time of the hidden chain in state $g$ has a geometric distribution with failure probability $\xi_{gg}$. Side information (e.g. the empirical distribution of the length of known exons) may suggest that this property is not desirable (Melodelima et al., 2006). However, the geometric distribution is a side product of the Markov assumption, which allows for the use of the forward-backward algorithm for inference. A typical trick to keep the Markov structure while modifying the sojourn time distribution involves building "macro-states", that is, splitting state $g$ into sub-states $g_1, g_2, \ldots, g_b$, forcing the transition from sub-state $g_{k-1}$ to sub-state $g_k$. As a result, the sojourn time has a negative binomial distribution with parameters $b$ and $\xi_{gg}$. The parameter

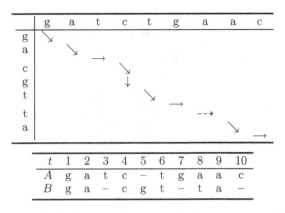

| $t$ | 1 | 2 | 3 | 4 | 5 | 6 | 7 | 8 | 9 | 10 |
|---|---|---|---|---|---|---|---|---|---|---|
| $A$ | g | a | t | c | – | t | g | a | a | c |
| $B$ | g | a | – | c | g | t | – | t | a | – |

**FIGURE 18.4**
An example of a paired HMM for sequence alignment. Top: most probable hidden path ($\searrow$ ... match; $\longrightarrow$ ... deletion; $\downarrow$ ... insertionl $--\rightarrow$ ... mismatch). Bottom: resulting alignment.

$b$ can be fitted (or manually tuned) to fit the distribution length of the exons known in species similar to the one under study.

*Gene expression profiles*

In the same vein, sub-states can be used to distinguish between a main classification step and a more refined behavior of the signal. Nicolas et al. (2009) and Mirauta et al. (2014), for example, are interested in understanding the transcriptional landscape, which means both detecting transcribed regions (main task) and the way the level of transcription varies within each of these regions (secondary task). To this end, the "transcribed" main state is divided into a series of secondary hidden states corresponding to different levels of transcription and among which Markov transitions also occur. The secondary hidden structure allows one to account for the dynamic dimension of the transcriptional process. Indeed, a given gene is typically transcribed from the "start" to the "stop" so, at a given time, a fraction of the "start" end has already started to be degraded (after translation), whereas a fraction of the "stop" end has still not been synthesized (before the end of transcription).

The distribution of the hidden states can itself be modeled to account for exogenous information such as the annotation of the genome (Bérard et al., 2011).

### 18.3.4 Non-standard hidden Markov structures

As mentioned above, HMMs are quite popular in genomics because of the one-dimensional structure that underlines many "omic" data. However, more complex hidden structures can be encountered, two examples of which are given below.

*Paired HMMs for sequence alignment*

Sequence alignment is one of the oldest problems in bioinformatics, the aim of which is to compare genomics regions (e.g. genes) observed in two different species. Suppose we observe the sequences $A = (\text{gatctgaac})$ and $B = (\text{gacgtta})$. The first step in comparing them is to align them, that is, to make them match as well as possible. Such an alignment can be viewed as an HMM with bivariate observed data $y_t = (a_t, b_t) \in \{-, \text{a}, \text{c}, \text{g}, \text{t}\}^2$, where $a_t$ ($b_t$) stands for the letter from sequence $A$ ($B$) observed at the aligned position $t$. Four hidden

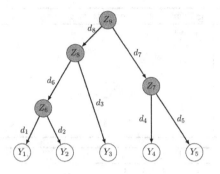

**FIGURE 18.5**
Example of an evolutionary tree. Only present nodes are observed $(y_i)$, whereas ancestor nodes $(z_j)$ are hidden.

states are then typically considered, the corresponding emission distribution having disjoint support:

*match:* $\{(a, a), (c, c), (g, g), (t, t)\}$;

*mismatch:* $\{(a, c), (a, g), (a, t), (c, a), (c, g), \ldots, (g, t)\}$;

*insertion:* $\{(-, a), (-, c), (-, g), (-, t)\}$;

*deletion:* $\{(a, -), (c, -), (g, -), (t, -)\}$.

Note that the aligned positions $t$ are not observed in advance. In practice, the whole inference process is rarely carried out. Most often, the transition matrix is given in advance and its entries are interpreted as costs of each possible transition, the emission distributions being uniform over their respective supports (except for mismatches). The alignment algorithm then simply consists of the Viterbi algorithm. Figure 18.4 gives a representation of the most probable path (top) and the resulting alignment (bottom).

*Tree-structured models*

Trees are often used to described the past evolution of a population, trait or genome. The tree structure is indeed consistent with many evolutionary scenarios. In many situations, only present observations are available, although we are interested in the past evolution. As a consequence, we are faced with situations as depicted in Figure 18.5 where the data at all past nodes (also referred to as ancestor nodes) are unobserved. In this framework, most models assume that the trait (or the genome sequence) evolves as a Markov process along the branches of a given phylogenetic tree. The aim is then to infer the parameters that governs this evolutionary process, which typically requires some insight into the value of the trait at the ancestor nodes.

EM can be used to infer the value of the trait in an ancestor node. As is often the case, the critical step is the E step where moments of the conditional distribution $p(\mathbf{z}|\mathbf{y})$ need to be computed. The case of the tree is similar to HMMs where forward-backward recursion enables us to compute these moments. Indeed, an 'upward-downward' recursion can derived in a similar way (Felsenstein, 1981) to get the conditional distribution $p(z_j|\mathbf{y})$ for each internal node $z_j$.

We only give a flavor of the upward recursion, based on the example of Figure 18.5. The upward recursion goes from the leaves $(y_i)$ to the root and involves computing the conditional distribution of each ancestor node given its offspring. First, the distributions $p(z_7|y_4, y_5)$ and $p(z_6|y_1, y_2)$ are computed directly. The remaining conditional distributions are then obtained as

$$p(z_8|y_1, y_2, y_3) = \int p(z_8|z_6, y_3)\, p(z_6|y_1, y_2)\, \mathrm{d}z_6,$$

$$p(z_9|y_1, y_2, y_3, y_4, y_5) = \iint p(z_9|z_8, z_7)\, p(z_8|y_1, y_2, y_3)\, p(z_7|y_4, y_5)\, \mathrm{d}z_7\, \mathrm{d}z_8,$$

which ends the upward recursion. We refer to Lartillot (2014) and Bastide et al. (2017) for two applications of this type of modeling.

## 18.4 Complex Dependency Structures

In the preceding section we presented models for which a genuine EM can be applied because the dependency structure is either sequential or tree structured. More complex structures may be required in some applications.

### 18.4.1 Markov random fields

For example, on the genetic structure of a population (see Section 18.2.1), georeferenced individuals may be considered. Spatial and genetic information are related in some way, and two individuals living close by are more likely to be genetically close. To take this prior information into account, it is relevant to consider a hidden Markov random field (HMRF) as prior distribution for the latent variable describing the population structure (François et al., 2006).

Markov fields extend the dependence structure to stochastic processes with indexes belonging to a multidimensional space, rather than simply to a subset of $\mathbb{R}$. There are two types of Markov random field: Markov fields with continuous indexes, commonly used in theoretical physics, and Markov fields with discrete indexes, used, *inter alia*, as models for statistics of a spatial nature; see also Chapter 16, Section 16.2 for a review. In applications to genomics, we are mainly concerned with the second category.

When the domain of the index is a subset of $\mathbb{R}^d$ rather than a subset of $\mathbb{R}$, the idea of left and right in relation to an index no longer applies, and Markov random fields must revert to a more general concept of neighborhood. The neighborhood system can be modeled via a contiguity graph where each node corresponds to an index and each vertex to a neighborhood relationship.

In the population structure example, nodes are not distributed regularly and Markov modeling requires that relations of contiguity are explicitly defined. One solution is to draw a Voronoi tessellation and to specify that two sites are contiguous if their respective Voronoi tiles have an edge in common. The Strauss model represents a natural prior distribution for the latent cluster variable. It can be considered as a generalization of the Ising model, in the case where the variables take $G \geq 2$ discrete values. In the isotropic case (i.e. where there is no particular spatial direction) the Gibbs distribution is defined by the energy function

$$H(\mathbf{z}) = -\beta \sum_{r,s \in S: r \sim s} \mathbb{I}(z_s = z_r) = -\beta \sum_{r,s \in S: r \sim s} z_s \cdot z_r, \tag{18.2}$$

where the binary vector $z_s$ denotes the class of node $s$ ($z_{sk} = 1$ if node $s$ belongs to class

$g$) and $r \sim s$ means that $r$ and $s$ are neighbors. This energy function counts the number of pairs of contiguous nodes which have the same value, and is maximized when the variables of the entire set of nodes are identical. In physics, this is referred to as the Potts model. It has been used as a prior for the spatial normalization of CGH array data (Neuvial et al., 2006). In the latter case, the hidden variable was coding a type of experimental artifact. Identifying this spatial experimental bias allowed the bias adjustment to be tuned for each area of the microarray.

This type of prior about graph neighborhoods can also be found in differential analysis. The classical approach for differential analysis relies on univariate test statistics for selecting a short list of genes. Then links from selected genes to known biological pathways through gene set enrichment analysis can be performed in order to identify involved pathways (Wei & Li, 2007). In that context, directly assuming a prior model considering that neighboring genes in the network are more likely to have a joint effect may improve the relevance of the differential analysis.

### 18.4.2 Stochastic block model

The analysis of biological networks has become commonplace in bioinformatics. Such networks typically describe the interactions between a set of entities such as genes, proteins or, at a higher level, bacterial species. Such data typically consist of a set of $n$ nodes, each corresponding to an entity, and the value $y_{ij}$ of the edge between nodes $i$ and $j$. Because of the diversity of the interactions, the form of $y_{ij}$ ranges from binary (simple presence or absence of the edge), to continuous uni- or multivariate.

Understanding the topology of such a network has been one of the primary tasks. The clustering point of view can be used to this end, assigning each node to a specific class with a typical role in the network. This results in the well-known stochastic block model (SBM) (Frank & Harary, 1982; Holland et al., 1983), in which the hidden classes $\{z_i\}$ are drawn independently for each node and the edges are drawn independently conditionally on the $z_i$. Importantly, the distribution of the edge $y_{ij}$ is conditional on the class of both nodes $i$ and $j$:

$$p(y_{ij}|z_i, z_j) = f_{z_i, z_j}(y_{ij}).$$

In the binary case, conditional on $z_i = g$ and $z_j = g'$, edge $y_{ij}$ is present with probability $\gamma_{gg'}$. Picard et al. (2009) give a series of examples of the use of SBMs for various biological networks.

### 18.4.3 Inference issues

Both HMRFs and SBMs raise inference issues. Indeed, in both cases the hidden labels $z_i$ are not independent conditionally on the observed data $\mathbf{y}$. In HMRFs the conditional dependency structure is given by the corresponding graph, and for SBMs this turns out to be the complete graph (Matias & Robin, 2014). The latent block model introduced in Section 18.2.2 raises similar issues. In such cases, no factorization exists to compute efficiently the moments of the conditional distribution $p(\mathbf{z}|\mathbf{y})$ that are required in a regular EM algorithm.

Because of the size of the data, variational approximations are often used, as they result in deterministic and reasonably fast algorithms. The general principle of these techniques is to replace the hard-to-compute distribution $p(\mathbf{z}|\mathbf{y})$ with an approximate distribution $q(\mathbf{z})$ which is easier to handle. So the regular E step is replaced by an approximation step,

$$q^{(s+1)} = \arg \min_{q \in \mathcal{Q}} D(q(\mathbf{z})||p(\mathbf{z}|\theta^{(s)}, \mathbf{y})),$$

where $\mathcal{Q}$ is a restricted class of distributions (typically factorizable) and $D$ is a divergence measure between probability measures. In the specific case where $D$ is equal to the Kullback–Leibler divergence, it can be seen that the variational EM (VEM) algorithm aims to maximize a lower bound of the likelihood of the data simply because, denoting by $\mathrm{E}_q$ the expectation with respect to $q$,

$$\log p(\mathbf{y}|\theta) \geq \log p(\mathbf{y}|\theta) - KL\left(q(\mathbf{z}), p(\mathbf{z}|\theta, \mathbf{y})\right) = \mathrm{E}_q\left(\log p(\mathbf{y}, \mathbf{z}|\theta)\right) - \mathrm{E}_q\left(\log q(\mathbf{z})\right).$$

A huge literature exists on these techniques. We refer to Jaakkola (2000) for a tutorial, to Minka (2005) for a discussion on the choice of $D$, and to Wainwright & Jordan (2008) for a comprehensive review.

This approach can be extended to a Bayesian setting, resulting in so-called variational Bayes inference. In this case, an approximation of the joint conditional distribution $p(\theta, \mathbf{z}|\mathbf{y})$ is sought; see, for example, Beal & Ghahramani (2003) for an introduction and Latouche et al. (2012) and Aicher et al. (2013) for applications to SBMs.

## 18.5   Concluding Remarks

In this chapter, we introduced a series of examples illustrating the versatility of mixture models for applications in genomics and genetics. For many of these applications, maximum likelihood inference can be carried out via the expectation-maximization algorithm. The computational efficiency of EM often depends on the E step, which relates to how intricate the dependency structure between observed and latent variables is. Efficient recursions exist, typically when this dependency is tree-structured.

For the most complex cases, only approximate versions of the E step are available at the time of writing. These approximations work in practice, especially in terms of classification. However, the price paid for this approximation is that the resulting estimates do not enjoy the general properties of maximum likelihood estimates. The analysis of modern molecular data will require more complex models, accounting for pedigree, spatial, and/or temporal dependencies. There is still room for improvement in this direction, in order to combine computational efficiency with statistical guaranties.

# Bibliography

AICHER, C., JACOBS, A. & CLAUSET, A. (2013). Adapting the stochastic block model to edge-weighted networks. *ICML Workshop on Structured Learning (SLG)*.

ALBERTSON, D. G., COLLINS, C., McCORMICK, F. & GRAY, J. W. (2003). Chromosome aberrations in solid tumors. *Nature Genetics* **34**, 369–376.

ALBERTSON, D. G. & PINKEL, D. (2003). Genomic microarrays in human genetic disease and cancer. *Human Molecular Genetics* **12**, R145–R152.

ALEXANDER, D. H., NOVEMBRE, J. & LANGE, K. (2009). Fast model-based estimation of ancestry in unrelated individuals. *Genome Research* **19**, 1655–1664.

ALLISON, D. B., GADBURY, G., HEO, M., FERNANDEZ, J., LEE, C.-K., PROLLA, T. A. & WEINDRUCH, R. A. (2002). Mixture model approach for the analysis of microarray gene expression data. *Computational Statistics & Data Analysis* **39**, 1–20.

BASTIDE, P., MARIADASSOU, M. & ROBIN, S. (2017). Detection of adaptive shifts on phylogenies using shifted stochastic processes on a tree. *Journal of the Royal Statistical Society, Series B* **79**, 1067–1093.

BEAL, M. J. & GHAHRAMANI, Z. (2003). The variational Bayesian EM algorithm for incomplete data: With application to scoring graphical model structures. In *Bayesian Statistics 7*, J. M. Bernardo, J. O. Berger, A. P. Dawid & A. F. M. Smith, eds. Oxford: Oxford University Press.

BEN-DOR, A., CHOR, B., KARP, R. & YAKHINI, Z. (2003). Discovering local structure in gene expression data: The order-preserving submatrix problem. *Journal of Computational Biology* **10**, 373–384.

BÉRARD, C., MARTIN-MAGNIETTE, M.-L., BRUNAUD, V., AUBOURG, S., ROBIN, S. et al. (2011). Unsupervised classification for tiling arrays: ChIP-chip and transcriptome. *Statistical Applications in Genetics and Molecular Biology* **10**, 1–22.

BÉRARD, C., SEIFERT, M., MARY-HUARD, T. & MARTIN-MAGNIETTE, M.-L. (2013). MultiChIPmixHMM: An R package for ChIP-chip data analysis modeling spatial dependencies and multiple replicates. *BMC Bioinformatics* **14**, 271.

BIERNACKI, C., CELEUX, G., GOVAERT, G. & LANGROGNET, F. (2006). Model-based cluster and discriminant analysis with the mixmod software. *Computational Statistics & Data Analysis* **51**, 587–600.

BORDES, L., DELMAS, C. & VANDEKERKHOVE, P. (2006). Semiparametric estimation of a two-component mixture model where one component is known. *Scandinavian Journal of Statistics* **33**, 733–752.

BOUAZIZ, M., AMBROISE, C. & GUEDJ, M. (2011). Accounting for population stratification in practice: A comparison of the main strategies dedicated to genome-wide association studies. *PLoS One* **6**, e28845.

BOUAZIZ, M., PACCARD, C., GUEDJ, M. & AMBROISE, C. (2012). SHIPS: Spectral hierarchical clustering for the inference of population structure in genetic studies. *PLoS One* **7**.

BOUVEYRON, C. & BRUNET-SAUMARD, C. (2014). Model-based clustering of high-dimensional data: A review. *Computational Statistics & Data Analysis* **71**, 52–78.

CAVALLI-SFORZA, L. L., MENOZZI, P. & PIAZZA, A. (1994). *The History and Geography of Human Genes*. Princeton, NJ: Princeton University Press.

CHEN, H., XING, H. & ZHANG, N. R. (2011). Estimation of parent specific DNA copy number in tumors using high-density genotyping arrays. *PLoS Computational Biology* **7**, e1001060.

CORANDER, J., WALDMANN, P. & SILLANPÄÄ, M. J. (2003). Bayesian analysis of genetic differentiation between populations. *Genetics* **163**, 367–374.

DUDOIT, S., SHAFFER, J. P. & BOLDRICK, J. C. (2003). Multiple hypothesis testing in microarray experiments. *Statistical Science* **18**, 71–103.

DURBIN, R., EDDY, S., KROGH, A. & MITCHISON, G. (1999). *Biological Sequence Analysis: Probabilistic Models of Proteins and Nucleic Acids*. Cambridge: Cambridge University Press.

EFRON, B. & TIBSHIRANI, R. (2002). Empirical Bayes methods and false discovery rates for microarrays. *Genetic Epidemiology* **23**, 70–86.

EFRON, B., TIBSHIRANI, R., STOREY, J. D. & TUSHER, V. (2001). Empirical Bayes analysis of a microarray experiment. *Journal of the American Statistical Association* **96**, 1151–1160.

EISEN, M. B., SPELLMAN, P. T., BROWN, P. O. & BOTSTEIN, D. (1998). Cluster analysis and display of genome-wide expression patterns. *Proceedings of the National Academy of Sciences* **95**, 14863–14868.

FELSENSTEIN, J. (1981). Evolutionary trees from DNA sequences: A maximum likelihood approach. *Journal of Molecular Evolution* **17**, 368–376.

FRALEY, C. & RAFTERY, A. E. (2002). Model-based clustering, discriminant analysis and density estimation. *Journal of the American Statistical Association* **97**, 611–631.

FRANÇOIS, O., ANCELET, S. & GUILLOT, G. (2006). Bayesian clustering using hidden Markov random fields in spatial population genetics. *Genetics* **174**, 805–816.

FRANK, O. & HARARY, F. (1982). Cluster inference by using transitivity indices in empirical graphs. *Journal of the American Statistical Association* **77**, 835–840.

FRIDLYAND, J., SNIJDERS, A. M., PINKEL, D., ALBERTSON, D. G. & JAIN, A. N. (2004). Hidden Markov models approach to the analysis of array CGH data. *Journal of Multivariate Analysis* **90**, 132–153.

GASSIAT, E., CLEYNEN, A. & ROBIN, S. (2016). Inference in finite state space non parametric hidden Markov models and applications. *Statistics and Computing* **26**, 61–71.

GOVAERT, G. & NADIF, M. (2003). Clustering with block mixture models. *Pattern Recognition* **36**, 463–473.

GREENMAN, C. D., BIGNELL, G., BUTLER, A., EDKINS, S., HINTON, J. et al. (2010). Picnic: An algorithm to predict absolute allelic copy number variation with microarray cancer data. *Biostatistics* **11**, 164–175.

HOCKING, T. D. (2012). *Learning Algorithms and Statistical Software, with Applications to Bioinformatics.* Ph.D. thesis, École Normale Supérieure de Cachan.

HOLLAND, P., LASKEY, K. & LEINHARDT, S. (1983). Stochastic blockmodels: Some first steps. *Social Networks* **5**, 109–137.

HUANG, H., CAI, L. & WONG, W. H. (2008). Clustering analysis of SAGE transcription profiles using a Poisson approach. In *Serial Analysis of Gene Expression (SAGE): Methods and Protocols*, K. L. Nielsen, ed. Totowa, NJ: Humana Press, pp. 185–198.

IBRAHIM, M., NICOLAS, P., BESSIÈRES, P., BOLOTIN, A., MONNET, V. & GARDAN, R. (2007). A genome-wide survey of short coding sequences in streptococci. *Microbiology* **153**, 3631–3644.

JAAKKOLA, T. (2000). Tutorial on variational approximation methods. In *Advanced Mean Field Methods: Theory and Practice*, M. Opper & D. Saad, eds. Cambridge, MA: MIT Press.

JIANG, D., TANG, C. & ZHANG, A. (2004). Cluster analysis for gene expression data: A survey. *IEEE Transactions on Knowledge and Data Engineering* **16**, 1370–1386.

KERIBIN, C., BRAULT, V., CELEUX, G. & GOVAERT, G. (2015). Estimation and selection for the latent block model on categorical data. *Statistics and Computing* **25**, 1201–1216.

KILLICK, R., FEARNHEAD, P. & ECKLEY, I. (2012). Optimal detection of changepoints with a linear computational cost. *Journal of the American Statistical Association* **107**, 1590–1598.

KIM, K., ZHANG, S., JIANG, K., CAI, L., LEE, I.-B., FELDMAN, L. J. & HUANG, H. (2007). Measuring similarities between gene expression profiles through new data transformations. *BMC Bioinformatics* **8**, 29.

KOHAVI, R. & JOHN, G. H. (1997). Wrappers for feature subset selection. *Artificial Intelligence* **97**, 273–324.

LAI, W. R., JOHNSON, M. D., KUCHERLAPATI, R. & PARK, P. J. (2005). Comparative analysis of algorithms for identifying amplifications and deletions in array CGH data. *Bioinformatics* **21**, 3763–3770.

LANGE, H., ZUBER, H., SEMENT, F. M., CHICHER, J., KUHN, L. et al. (2014). The RNA helicases AtMTR4 and HEN2 target specific subsets of nuclear transcripts for degradation by the nuclear exosome in *Arabidopsis thaliana*. *PLoS Genetics* **10**, e1004564.

LARTILLOT, N. (2014). A phylogenetic Kalman filter for ancestral trait reconstruction using molecular data. *Bioinformatics* **30**, 488–496.

LATOUCHE, P., BIRMELÉ, E. & AMBROISE, C. (2012). Variational Bayesian inference and complexity control for stochastic block models. *Statistical Modelling* **12**, 93–115.

MATIAS, C. & ROBIN, S. (2014). Modeling heterogeneity in random graphs through latent space models: A selective review. *ESAIM: Proceedings* **47**, 55–74.

MAUGIS, C., CELEUX, G. & MARTIN-MAGNIETTE, M.-L. (2009). Variable selection for clustering with Gaussian mixture models. *Biometrics* **65**, 701–709.

McLACHLAN, G., BEAN, R. & BEN-TOVIM JONES, L. (2006). A simple implementation of a normal mixture approach to differential gene expression in multiclass microarrays. *Bioinformatics* **22**, 1608–1615.

McLACHLAN, G. J., BEAN, R. & PEEL, D. (2002). A mixture model-based approach to the clustering of microarray expression data. *Bioinformatics* **18**, 413–422.

McLACHLAN, G. J., Do, K.-A. & AMBROISE, C. (2005). *Analyzing Microarray Gene Expression Data*. Hoboken, NJ: John Wiley.

MELODELIMA, C., GUÉGUEN, L., PIAU, D. & GAUTIER, C. (2006). A computational prediction of isochores based on hidden Markov models. *Gene* **385**, 41–49.

MINKA, T. (2005). Divergence measures and message passing. Tech. Rep. MSR-TR-2005-173, Microsoft Research Ltd.

MIRAUTA, B., NICOLAS, P. & RICHARD, H. (2014). Parseq: Reconstruction of microbial transcription landscape from RNA-Seq read counts using state-space models. *Bioinformatics* **30**, 1409–1416.

MOUNTAIN, J. L. & CAVALLI-SFORZA, L. L. (1997). Multilocus genotypes, a tree of individuals, and human evolutionary history. *American Journal of Human Genetics* **61**, 705–718.

MURI, F. (1998). Modelling bacterial genomes using hidden Markov models. In *COMPSTAT: Proceedings in Computational Statistics*, R. Payne & P. Green, eds. Heidelberg: Physica-Verlag.

NEUVIAL, P., HUPÉ, P., BRITO, I., LIVA, S., MANIÉ, É. et al. (2006). Spatial normalization of array-CGH data. *BMC Bioinformatics* **7**, 264.

NICOLAS, P., BIZE, L., MURI, F., HOEBEKE, M., RODOLPHE, F. et al. (2002). Mining bacillus subtilis chromosome heterogeneities using hidden Markov models. *Nucleic Acids Research* **30**, 1418–1426.

NICOLAS, P., LEDUC, A., ROBIN, S., RASMUSSEN, S., JARMER, H. & BESSIÈRES, P. (2009). Transcriptional landscape estimation from tiling array data using a model of signal shift and drift. *Bioinformatics* **25**, 2341–2347.

PAN, W. & SHEN, X. (2007). Penalized model-based clustering with application to variable selection. *Journal of Machine Learning Research* **8**, 1145–1164.

PICARD, F., MIELE, V., DAUDIN, J.-J., COTTRET, L. & ROBIN, S. (2009). Deciphering the connectivity structure of biological networks using mixnet. *BMC Bioinformatics* **10** **(Suppl. 6)**, S17.

PICARD, F., ROBIN, S., LAVIELLE, M., VAISSE, C. & DAUDIN, J.-J. (2005). A statistical approach for array CGH data analysis. *BMC Bioinformatics* **6**, 1.

PICARD, F., ROBIN, S., LEBARBIER, E. & DAUDIN, J.-J. (2007). A segmentation/clustering model for the analysis of array CGH data. *Biometrics* **63**, 758–766.

POPOVA, T., MANIÉ, E., STOPPA-LYONNET, D., RIGAILL, G., BARILLOT, E., STERN, M.-H. et al. (2009). Genome Alteration Print (GAP): A tool to visualize and mine complex cancer genomic profiles obtained by SNP arrays. *Genome Biology* **10**, R128.

PRITCHARD, J., STEPHENS, M. & DONNELLY, P. (2000). Inference of population structure using multilocus genotype data. *Genetics* **155**, 945–959.

RAFTERY, A. E. & DEAN, N. (2006). Variable selection for model-based clustering. *Journal of the American Statistical Association* **101**, 168–178.

RAJ, A., STEPHENS, M. & PRITCHARD, J. K. (2014). fastSTRUCTURE: Variational inference of population structure in large SNP data sets. *Genetics* **197**, 573–589.

RASHID, N. U., GIRESI, P. G., IBRAHIM, J. G., SUN, W. & LIEB, J. D. (2011). ZINBA integrates local covariates with DNA-seq data to identify broad and narrow regions of enrichment, even within amplified genomic regions. *Genome Biology* **12**, R67.

RAU, A., MAUGIS-RABUSSEAU, C., MARTIN-MAGNIETTE, M.-L. & CELEUX, G. (2015). Co-expression analysis of high-throughput transcriptome sequencing data with Poisson mixture models. *Bioinformatics* **31**, 1420–1427.

RIGAILL, G. (2010). Pruned dynamic programming for optimal multiple change-point detection. Preprint, arXiv:1004.0887.

ROBIN, S., BAR-HEN, A., DAUDIN, J.-J. & PIERRE, L. (2007). A semi-parametric approach for mixture models: Application to local false discovery rate estimation. *Computational Statistics & Data Analysis* **51**, 5483–93.

ROBINSON, M. D., McCARTHY, D. J. & SMYTH, G. K. (2010). edgeR: A Bioconductor package for differential expression analysis of digital gene expression data. *Bioinformatics* **26**, 139–140.

ROQUAIN, E. (2011). Type I error rate control in multiple testing: A survey with proofs. *Journal de la Société Française de Statistique* **152**, 3–38.

SEIFERT, M. (2013). *Hidden Markov Models with Applications in Computational Biology*. Saarbrücken: SVH-Verlag.

STOREY, J. D. (2002). A direct approach to false discovery rates. *Journal of the Royal Statistical Society, Series B* **64**, 479–498.

TADESSE, M. G., SHA, N. & VANNUCCI, M. (2005). Bayesian variable selection in clustering high-dimensional data. *Journal of the American Statistical Association* **100**, 602–617.

TANG, H., PENG, J., WANG, P. & RISCH, N. J. (2005). Estimation of individual admixture: Analytical and study design considerations. *Genetic Epidemiology* **28**, 289–301.

VITERBI, A. J. (1967). Error bounds for convolutional codes and an asymptotically optimal decoding algorithm. *IEEE Transactions in Information Theory* **13**, 260–269.

VOLANT, S., BÉRARD, C., MARTIN-MAGNIETTE, M.-L. & ROBIN, S. (2014). Hidden Markov models with mixtures as emission distributions. *Statistics and Computing* **24**, 493–504.

WAINWRIGHT, M. J. & JORDAN, M. I. (2008). Graphical models, exponential families, and variational inference. *Foundations and Trends of Machine Learning* **1**, 1–305.

WEI, Z. & LI, H. (2007). A Markov random field model for network-based analysis of genomic data. *Bioinformatics* **23**, 1537–1544.

WITTEN, D. M. & TIBSHIRANI, R. (2010). A framework for feature selection in clustering. *Journal of the American Statistical Association* **105**, 713–726.

YEUNG, K. Y., FRALEY, C., MURUA, A., RAFTERY, A. E. & RUZZO, W. L. (2001). Model-based clustering and data transformations for gene expression data. *Bioinformatics* **17**, 977–987.

# 19

## Applications in Astronomy

### Michael A. Kuhn and Eric D. Feigelson

*Millennium Institute of Astrophysics, Universidad de Valparaíso, Chile; Pennsylvania State University, USA*

## CONTENTS

## 19.1   Introduction

Astronomy is the scientific study of objects beyond Earth: planets, stars, galaxies, and the cosmos itself. Observations are made with ground-based and satellite-borne telescopes spanning the entire electromagnetic spectrum from radio through gamma rays. Data structures and scientific problems are diverse, so that many statistical techniques are needed to advance our understanding of cosmic objects and phenomena. Mixture models have played a significant role in such analyses, though not always under this name. The method is used for many purposes, ranging from the classification of objects in a multidimensional parameter space to the study of spatial clustering patterns of stars or galaxies. This second problem has attracted attention among statisticians. A galaxies data set (Postman et al., 1986), made up of recessional velocities of 83 galaxies in units of kilometers per second, has served as a challenging test case for estimating the number of components in a mixture model.

Astronomical problems involving mixture models often differ from situations familiar from social or biological sciences. For example, astronomical data sets may have unusual forms of the probability density function. Some distributions may be fitted with the more conventional log-normal distributions (e.g. masses of globular clusters), Pareto distributions (e.g. initial masses of high-mass stars), or gamma distributions (e.g. galaxy luminosities). However, other, more unusual examples may originate from physical and astrophysical processes: the distribution of photon energies from an X-ray source is dictated by thermal and quantum physics; the approximate distribution of stars in a dynamically relaxed star cluster can be derived from Newtonian gravity; and the distribution of different populations of stars in a galaxy is based on the galaxy's star-formation history.

Other problems apply normal mixture models to "big data" produced by wide-field surveys. Here, a telescope can produce exabytes of images from repeated scans of the sky, from which catalogs of billions of sources (rows) with tens of measured properties (columns) are generated. Dozens of diverse cosmic populations may be present in the survey. The most famous of such surveys has been the Sloan Digital Sky Survey (SDSS; Abazajian et al., 2003), which initially provided spectra for several million stars and galaxies and photometry (brightness measurements) for about 500 million objects in five filters. The scale of surveys continues to grow, with the planned Large Synoptic Survey Telescope (LSST; Ivezić et al., 2008) intended to monitor more than 30 billion objects over a thousand epochs during a ten-year time frame.

Our presentation of astronomical uses of mixture models here is not systematic or comprehensive, but is designed to give a sense of the scope and challenges arising in a variety of settings. We hope this review and commentary will encourage statisticians to share their expertise with astronomers, advancing the characterization and understanding of many facets of the universe around us.

## 19.2    Clusters of Stars and Galaxies

Statistical methods have been important for modeling the spatial distributions of astronomical objects that are physically associated. Examples include star clusters and galaxy clusters, both of which are held together by gravity but may exhibit anisotropic and intertwined structures inherited from a complicated formation process. A variety of statistical methods have been used to examine these systems, including hierarchical clustering methods for identifying individual clusters, and spatial autocorrelative methods for understanding stochastic patterns.

### 19.2.1    Galaxy clusters

The strongly clustered spatial distribution of galaxies was recognized from galaxy counts of wide-field photographic plates during the mid-twentieth century. Many prominent astronomers conducted studies on this problem (including Edwin Hubble, Harlow Shapley, Fritz Zwicky, Gérard de Vaucouleurs, and P. J. E. Peebles). Several statistical approaches were taken (see the review by de Vaucouleurs, 1971). One early result was that the frequency distribution of galaxy counts in quadrats followed a log-normal distribution rather than the Poisson distribution expected from spatial randomness. An "index of clumpiness", the ratio of observed to expected variance in number counts, was investigated. The spatial autocorrelation function was found to have signal out to several degrees in the sky. Shane & Wirtanen (1954) produced contour maps of equal surface density based on a uniform kernel, remarking: "So many aggregations stand out prominently that one is tempted to speculate that clustering may be a predominant characteristic of nebular [galactic] distri-

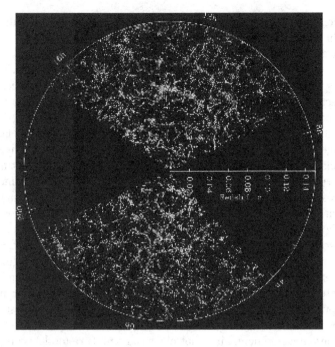

**FIGURE 19.1**
Projection into two dimensions of a portion of the three-dimensional Sloan Digital Sky Survey galaxy redshift survey showing the difficulties of mixture modeling of the "cosmic web" of galaxies in space. https://www.sdss3.org/science/gallery_sdss_pie2.php

bution." Shane teamed with Berkeley statisticians Jerzy Neyman and Elizabeth Scott to develop statistical models (such as a double Poisson model) of the distribution (Neyman et al., 1953).

Some analyses can be viewed as mixture models for the galaxy distribution in the nearby universe, although they are not usually described in this way. Abell (1958) conducted a heroic visual survey of the Palomar Observatory Sky Survey plates to identify several thousand individual galaxy clusters using a decision tree applied to visual galaxy counts. Turner & Gott (1976) constructed a catalog of galaxy groups based on a single-linkage agglomerative clustering algorithm. (Their procedure became very popular in the astronomical community under the label "friends-of-friends" or percolation algorithm without awareness of its widespread use in other fields.) Tully (1987) identified clouds, associations and groups of nearby galaxies in three dimensions from a dendrogram procedure with linkages based on the gravitational forces between galaxies.

Statistical approaches to galaxy clustering as a stationary stochastic process were initiated by P. J. E. Peebles in the 1960s based on the two-point (pair) correlation function and the Fourier power spectrum. These were particularly important as they were linked to the astrophysical theory of structure formation in an expanding universe (see the review by Fall, 1979). For example, Bardeen (1986) examined the correlation functions of peaks in three-dimensional Gaussian random fields arising from gravitational attraction in an initial spectrum of weak density fluctuations arising from the radiation-dominated era after the Big Bang. Peebles's approaches are still in wide use today; for instance, the faint signal expected from baryon acoustic oscillations, an important test for standard cosmological theory, was recently discovered using the two-point galaxy correlation function from SDSS data (Eisenstein, 2005).

All of these early studies treated galaxy clustering as an isotropic process. But this assumption was radically invalidated as larger telescopes devoted observing time to galaxy redshift surveys. Redshifts represent an approximate measure of galaxy distance in the expanding universe and, when combined with location in the sky, give a three-dimensional view of the galaxy distribution. When about 1000 redshifts were obtained, the distribution was found to resemble "a slice through the suds in the kitchen sink". The language of galaxy clustering changed: "clusters" were now viewed as the intersections of "filaments", "sheets" and "Great Walls" of galaxies that surround "voids". The volume by Martínez & Saar (2002) lays the foundation between three-dimensional galaxy statistics and cosmological theory.

Increasing resources were devoted to constructing the three-dimensional map of galaxies, most notably with the acquisition of more than 2 million galaxy redshifts by the SDSS (Alam et al., 2015). Figure 19.1 shows a two-dimensional projection of a small portion of this data set. The links to cosmological theory are strong. Not only does the SDSS Fourier power spectrum agree well with the standard $\Lambda$CDM cosmological model (Tegmark et al., 2004), but also massive simulations of structure formation in a dark matter dominated expanding universe accurately reproduce the soap bubble or "cosmic web" appearance of the galaxy distribution (Springel, 2005). Examination of the structure of galaxy clusters using mixture models has shown that most have clumpy, complex structures (Einasto et al., 2012).

In light of these developments, it is not clear that mixture models can play a significant role in the characterization or understanding of the galaxy distribution. It is not clear either that the concept of distinct galaxy "clusters" is meaningful. Simple one-dimensional treatments of small data sets, as examined by a number of statisticians (Roeder, 1990; Escobar & West, 1995; Carlin & Chib, 1995; Phillips & Smith, 1996; McLachlan & Peel, 1997; Roeder & Wasserman, 1997; Richardson & Green, 1997; Aitkin, 2001, 2011), are no longer appropriate. The multi-scale, high-amplitude, anisotropic, web-like structure of the three-dimensional galaxy clustering is difficult to treat using standard methods of mixtures, multivariate analysis, or spatial point processes. A number of heuristic algorithms for finding filaments or voids are in use, but with little foundation in statistical theory. There is thus a real need for development of statistical tools – such as two-sample tests for comparing observations with simulations of different cosmological models – that are well suited to the complexities of galaxy clustering.

## 19.2.2    Young star clusters

Star formation is another topic in astronomy where the spatial clustering of objects, in this case young stars, can have important implications. Star formation is an ongoing process in many galaxies, including within our own Milky Way Galaxy, where the current rate is approximately 1 star per year (Robitaille & Whitney, 2010). Stars form in molecular clouds – the coldest, densest phase of interstellar gas, which are mostly composed of $H_2$ – when these clouds becomes unstable to gravitational collapse and contracts to form stars. However, gravitational collapse must compete with phenomena that resist collapse, such as cloud turbulence and thermal and magnetic pressures. Thus, star formation is restricted to the densest cloud cores, within which stars form in groups that often merge into temporary rich clusters (Bate et al., 2003). On galactic scales, new-born stars are concentrated in large complexes known as star-forming regions, which often lie within the spiral arms in many galaxies. These complexes last for several million years, after which most of the stars will have dispersed into the galaxy. However, some of the stars may remain in gravitationally bound groups known as open clusters.

In star-forming regions, the locations of stars reflect the structure of the natal molecular

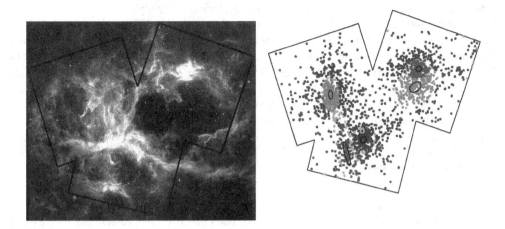

**FIGURE 19.2**
Left: the mid-infrared view of NGC 6357 seen by the *Spitzer Space Telescope*. The molecular clouds, forming several bubbles, are seen prominently in these images but the star clusters are not immediately evident. Right: the cluster members identified from the X-ray/infrared MYStIX study, which are color-coded by group from the mixture model. Light gray stars have ambiguous cluster memberships, and dark green stars are members of a distributed population in the model. The core regions of the various mixture components are shown as black ellipses (Feigelson et al., 2013; Broos et al., 2013; Kuhn et al., 2013, 2014).

cloud, with the exact relation between cloud and star properties a matter of active research (Lada et al., 2013). Gravitational collapse of the clouds causes the gas in star-forming regions to collapse and fragment into multiple clumps and filaments, so new-born stars will typically be distributed in several subclusters within a star-forming region. An example region is shown in Figure 19.2; the image from NASA's *Spitzer Space Telescope* shows both gas clouds and stars, while the spatial distribution of the stars, selected using data from NASA's *Chandra X-ray Observatory*, reveals multiple subclusters.

Individual stars can be identified and their spatial distributions analyzed in a number of star-forming regions in the Galaxy within a distance of several kiloparsecs – a section of the Galaxy that includes part of our own spiral arm and neighboring spiral arms. In this chapter, we discuss star clusters in 18 different star-forming regions included in the MYStIX study (Feigelson et al., 2013) and related studies (Townsley et al., 2014; Kuhn et al., 2017b). The mixture model analysis for star clusters was performed by Kuhn et al. (2014), and similar methods were used by Kuhn et al. (2017a) and Getman et al. (2018). For the mixture model analysis, Kuhn et al. (2014) only used the two variables, right ascension (RA) and declination (Dec), which describe the stars' angular coordinates on the sky. The data are also limited by the irregular fields of view observed by the *Chandra* telescope. Information about the third radial dimension of stellar positions in the clusters is also not available. In this example, we will refer to these two spatial coordinates as $x$ and $y$.

### 19.2.2.1  Star-cluster models

Older star clusters that have reached a quasi-equilibrium dynamical state are relatively well understood, but young star clusters, which are still affected by the initial conditions of star formation, are not. Several families of spherically symmetric model have been used to fit the density profiles of star clusters. These include the isothermal sphere, the King profile, and the Plummer sphere, which are all approximations to quasi-equilibrium distributions of stars in gravitationally bound groups (Binney & Tremaine, 2008).

For cluster analysis, Kuhn et al. (2014) used the isothermal sphere, which has been shown to provide a good empirical description of the distribution of stars in some young stellar clusters (e.g. Hillenbrand & Hartmann, 1998; Wang et al., 2008; Kuhn et al., 2014, 2017a). The model is unphysical at large distances from the cluster center because the number of stars diverges when integrated over all space. Nevertheless, this model provides an adequate fit to clusters within the observed fields of view.

The isothermal sphere has a characteristic "core" radius, $r_c$, which defines the size of the cluster. The distribution of stars projected in two dimensions on the sky (the surface density distribution) can be approximated out to several core radii by an analytic expression known as the Hubble model,

$$f_h(R) = \frac{A}{1 + (R/r_c)^2},$$

where $A$ is a constant and $R$ is the distance from the center of the cluster. However, many young stellar clusters are not spherically symmetric, but show significant ellipticity (Hillenbrand & Hartmann, 1998; Kuhn et al., 2017a). Generalization of this model to allow for elliptical contours of equal density requires the introduction of two new model parameters: ellipticity $\epsilon$ and the ellipse orientation $\varphi$ on the sky. The resulting surface density for the "isothermal ellipsoid" at the coordinates $r = (x, y)$ is described by the equation

$$f_{\text{ie}}(r; x_0, y_0, r_c, \varphi, \epsilon) =$$

$$A \left[ 1 + \left| \left[ \begin{array}{cc} (1-\epsilon)^{-1/2} \cos\phi & (\epsilon-1)^{1/2} \sin\phi \\ (1-\epsilon)^{-1/2} \sin\phi & (1-\epsilon)^{1/2} \cos\phi \end{array} \right] \left[ \begin{array}{c} \Delta x \\ \Delta y \end{array} \right] \right|^2 / r_c^2 \right]^{-1}, \quad (19.1)$$

where $r_0 = (x_0, y_0)$ is the center of the ellipsoid, and $\Delta x = x - x_0$ and $\Delta y = y - y_0$. Thus, this cluster-component model has five parameters, $x_0$, $y_0$, $r_c$, $\epsilon$, and $\varphi$, and $A$ is a normalization constant. Kuhn et al. (2014) call this model the "isothermal ellipsoid". However, this model is meant merely as an empirical description of the projected spatial distribution of stars in a cluster, since information about a cluster's dynamical states is lacking. Furthermore, this model is only applicable to the field of view observed by the telescope, which we denote the window $W$.

The "isothermal ellipsoid" model provides a closer match to observed young stellar clusters than other, better-understood, distributions like the multivariate normal distribution or Student's $t$-distribution. Figure 19.3 shows the cluster NGC 6231 fitted with two models, the isothermal ellipsoid model on the left and a multivariate normal distribution on the right. These models show a slice through the two-dimensional surface density maps, with the nonparametrically smoothed data shown in black and the models shown in gray. Surface densities (ordinate) are shown with logarithmic values. While the isothermal ellipsoid provides a good match to the data, with only minor deviations at large distances from the center, the multivariate normal model misses both the cluster center and the wings of the distribution. When using normal functions to fit stellar surface density distributions, the modeling tries to compensate for this mismatch by using several, approximately concentric, normal mixture models to fit a single star cluster.

In addition to the "isothermal ellipsoid" components, Kuhn et al. (2014) used an additional component, $f_U(r)$, to model stars distributed uniformly in the field of view. These can either be young stars that are not part of clusters (e.g. a "distributed population") or contaminants in the sample which are expected to exhibit complete spatial randomness. This approach is also used by the well-known normal mixture model procedure mclust to deal with noise and outliers (Fraley et al., 2012).

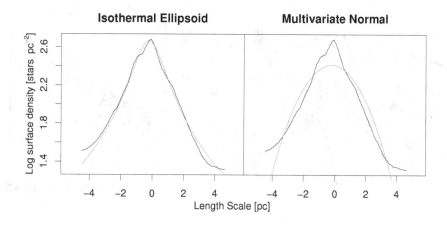

**FIGURE 19.3**

The black lines in each plot show the density of stars in the region NGC 6231 estimated through adaptive smoothing. The gray lines show the models that have been fitted. The left panel shows the "isothermal ellipsoid" model and the right panel shows the normal distribution model. (The $y$-axes of these plots are shown with a logarithmic scale.) Clearly, the isothermal ellipsoid model provides a better description of the data (Kuhn et al., 2017a).

The mixture model for the spatial distribution of the stellar population will be the sum of the isothermal-ellipsoid models for $G$ clusters plus the unclustered component, each of which is weighted by mixing coefficients, $\eta_g$. This model is given by the equation

$$p(r|\theta) = \sum_{g=1}^{G+1} \eta_g f_g(r|\theta_g) = \sum_{g=1}^{G} \eta_g f_{\text{ie}}(r|x_{0,g}, y_{0,g}, r_{c,g}, \varphi_g, \epsilon_g) + \eta_{G+1} f_{\text{U}}(r),$$

where $\theta = \{\eta_1, x_{0,1}, y_{0,1}, r_{c,1}, \varphi_1, \epsilon_1, \ldots, \eta_G, x_{0,G}, y_{0,G}, r_{c,G}, \varphi_G, \epsilon_G, \eta_{G+1}\}$ denotes the model parameters. The model thus has six parameters for each ellipsoidal component and one for the uniform component mixing parameter, but one degree of freedom fewer because the model must be normalized, yielding $6G$ dimensions for the full model.

### 19.2.2.2 Model fitting and validation

The log likelihood for a point pattern within a finite window ($W$) is given by the equation

$$\ell_o(\theta) = \log p(r_1, \ldots, r_n|\theta) = \sum_{i=1}^{n} \log p(r_i|\theta), \qquad (19.2)$$

under the assumption that the pattern of points, $\{r_1, \ldots, r_n\}$, is generated by an inhomogeneous Poisson point process containing $n$ points. The mixture model $p(r|\theta)$ must be normalized in the window $W$, which is done by numerical integration due to the irregular shape of the window. The R package spatstat also makes use of irregular windows for analysis of spatial point processes (Baddeley et al., 2015).

Kuhn et al. (2014) carried out the maximum likelihood estimation (MLE) by directly searching for the maximum of $\ell_o(\theta)$, rather than using the typical expectation–maximization (EM) approach. This method was used because there is no maximum likelihood formula for the parameters of the isothermal-ellipsoid model for points within an irregular window

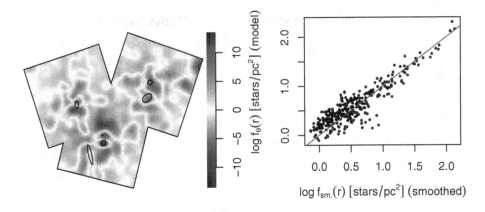

**FIGURE 19.4**
Left: residual surface density for NGC 6357. Negative residuals are shown in blue and positive residuals are shown in red. The peak residuals are roughly ~10% of the peak surface density in a smoothed map of the observed star distribution. Right: density obtained from the mixture model is plotted against density from the adaptively smoothed surface density maps (Kuhn et al., 2014).

$W$. Direct searching can be computationally challenging due to the high dimensionality of the parameter space. While standard optimizers like the EM algorithm treat all of the variables equally, in this case the scientific motivation requires that the clusters be present in two variables $(x, y)$ while the other variables $(r_c, \varphi, \epsilon)$ are secondary. Kuhn et al. (2014) started the MLE computation with a superset of possible clusters obtained from bumps in an adaptively smoothed surface density map of the point process. A Nelder–Mead optimization algorithm, implemented in the R function *optim*, was used to find the global maximum likelihood. In R, even for complicated distributions, with $n \approx 1000$ stars and $G \approx 10$ cluster components, the Nelder–Mead algorithm produces a reasonable solution in less than 15 CPU minutes.

Kuhn et al. (2014) based model selection on minimizing the Akaike information criterion,

$$\mathrm{AIC}(G) = -2\ell_o(\hat{\theta}_G) + 2(6G);$$

see Chapter 7, Section 7.2.2 above for a review of information criteria for model selection. Although there has been much debate over which penalized likelihood to use for model selection (e.g. Lahiri, 2001; Konishi & Kitagawa, 2008; Burnham & Anderson, 2002; Kass & Raftery, 1995), for this problem the AIC has several advantages. A typical star-forming region may have a large dynamic range in the numbers of stars in young stellar clusters. For example, a subcluster of ~20 stars may reside next to a rich cluster with ~500 stars. In addition, clusters may be superimposed on each other, either due to the projection of multiple discrete clusters along the same line of sight or astrophysical cases of core–halo structure. The AIC has greater sensitivity at probing these effects than, say, the Bayesian information criterion (BIC).

To validate the accuracy of the mixture model MLE, Kuhn et al. (2014) examined kernel smoothed residuals of the mixture model. The construction of these residual maps is described by Baddeley et al. (2005, 2008) and implemented using the *diagmose.ppm* tool in the R package spatstat for statistical analysis of spatial point processes. Residual maps can indicate the amplitude of residuals and give insight into physical deviations from the model assumptions. Better fits will have lower amplitudes and lack coherent structures in

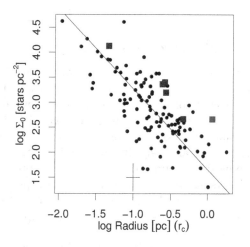

**FIGURE 19.5**

Scatter plot showing properties of clusters identified by the mixture model analysis in 18 star-forming regions. The ordinate is the density of stars at the center of a cluster, $\Sigma_0$, and the abscissa is the core radius of the cluster, $r_c$, both of which are derived from the mixture model. Blue points are clusters from NGC 6357 and the red point is from NGC 6231. The black cross indicates typical $1\sigma$ uncertainties on the model's parameters (Kuhn et al., 2017a).

residual maps. The identification of possible missing clusters using the residual maps is similar to the use of "final prediction error" to fit a model recommended by Rao et al. (2001). Figure 19.4 shows the kernel smoothed residual map for the star-forming region NGC 6357, while Figure 19.3 shows a comparison between the model prediction and the smoothed data in the young stellar cluster NGC 6321.

### 19.2.2.3 Results from the mixture model approach

Several nonparametric methods have been used in the astronomical literature for identifying clusters of stars in star-forming regions based on the minimal spanning tree, Voronoi tessellations, kernel density estimation, and nearest neighbor distributions (Schmeja, 2011). But the parametric mixture model approach offers a decisive advantage: an astrophysical model from which astrophysical inferences may be devised. Three important quantities obtained from the isothermal ellipsoidal fit to each subcluster are the core radius of the cluster (in parsecs), the central star density (in stars per cubic parsec), and the total number of stars in the cluster. None of these parameters may be obtained from nonparametric methods: the core radius is not clearly defined without a model, inferring three-dimensional properties from two-dimensional data requires a model, and overlapping clusters impede counting the number of stars in a cluster.

Figure 19.5 shows, for the full set of 18 star-forming regions, the relations between two of these variables: $\Sigma_0$, the density of stars at the center of a cluster; and $r_c$, the cluster's core radius. This plot is of astrophysical interest because it can be interpreted with astrophysical models about how star clusters form and evolve. For example, decreasing density with increasing radius can be interpreted as an effect of cluster expansion (Pfalzner, 2009). Subsequent investigation of the ages of stars in these clusters supports this interpretation (Getman et al., 2014; Kuhn et al., 2015, 2017a).

## 19.3   Classification of Astronomical Objects

Classification is very important to astronomers: a full-text search of the astronomical literature published in 2015 shows that 30% of all papers are concerned with classification. Using
modern observatories, it is common for astronomers to gather data from a large number of
objects of different types. This can arise from collections of images containing thousands
of stars or galaxies in a single field of view, or from fiber-fed spectrographs capable of obtaining dozens of spectra simultaneously. Imaging and spectroscopic observations provide
information that can be used to group objects with similar properties, including photometry (the brightness of objects at various wavelength bands), spectroscopy (showing emission
and absorption lines from atoms and molecules that are signatures of chemical, thermal,
and electromagnetic properties), time-variability data, morphology from structure in the
images, and motions on the sky.

  Different classes are sometimes clearly distinguished. For example, gamma-ray bursts
have been subdivided into two classes, long and short gamma-ray bursts, based on distinct
peaks in the distribution of prompt burst durations, with a relatively clear division at 2
seconds (Kouveliotou et al., 1993). However, in other cases, there can be significant overlap
in the classes' properties and it may not be possible, with limited data, to reliably classify
an object.

  One of the first problems in astronomy that used the concept of mixtures was groups
of co-moving stars in the Galaxy. These groups, now called stellar associations or moving
groups, are important because they can represent groups of stars that all formed in one star-
formation episode. In an investigation of star streams in the Galaxy with different kinematic
properties (now recognized to be based on spurious data), Eddington (1906) wrote: "We
cannot, as a rule, pick out an individual star and decide (from its motion) to which drift
[moving group] it belongs. However, we may roughly separate out stars typical of the two
drifts and examine their characteristics." Mixture models are commonly used in astronomy
for distinguishing a single class of objects from a mixture of multiple classes of objects. The
most-cited papers on mixture models in astronomy, by Ashman et al. (1994) and Muratov
& Gnedin (2010), describe methods for hypothesis tests to distinguish a unimodal normal
distribution from mixtures of multiple normal distributions. Astronomers may interpret the
empirical appearance of multimodality as evidence for physically distinct classes, and this
is often confirmed through follow-up studies.

### 19.3.1   Tests for multiple components

A variety of tests for multimodality are used by astronomers, typically based on assumptions
of multivariate normal distributions. This often involves the classical likelihood ratio test,
AIC, BIC, or full calculation of the Bayes factor. The null hypothesis of a single class can
be rejected in favor of an alternate hypothesis of a mixture of multiple classes (e.g. Jeffreys,
1939; Kass & Raftery, 1995). However, there is no consistency on the choice of AIC, BIC,
or other model selection approach within the astronomical literature.

  The use of normal mixture models to perform a hypothesis test of bimodality by Ashman
et al. (1994) has been influential in the astronomical literature. They test whether univariate
data, $\{y_0, \ldots, y_n\}$, are consistent with a normal distribution (the null hypothesis) or are
better described by $G$ equal-width normal distributions (the alternate hypothesis). For the

$G$-component model, the pdf and the complete-data log likelihood are given by the equations

$$p(y|\mu, \sigma^2) = \sum_{g=1}^{G} \eta_g \phi(y|\mu_g, \sigma_g^2),$$

$$\ell_c(\mu, \sigma^2) = \sum_{g=1}^{G} \sum_{i=1}^{n} z_{gi}(\log \eta_g + \log \phi(y_i|\mu_g, \sigma_g^2)),$$

where the $\mu_g$ are the component means, the $\sigma_g$ are the standard deviations, the $\eta_g$ are the component mixing parameters, and the $z_{gi}$ are indicator variables with value 1 when the $j$th point is assigned to group $g$ and 0 otherwise. The assignments of points to groups are not known *a priori*, so the EM algorithm is used, as presented in Chapter 2 above. We recall here that, during the E step, the $\hat{z}_{gi}$ are calculated using

$$\hat{z}_{gi} = \frac{\eta_g \phi(y_i|\mu_g, \sigma_g^2)}{p(y_i|\mu, \sigma^2)}.$$

Then, during the M step, maximum likelihood values of mixing parameters, $\eta_g$, the group means, $\mu_g$, and the common variance of the groups, $\sigma_g^2$, are obtained with the following equations:

$$\hat{\eta}_g = \frac{1}{n} \sum_{i=1}^{n} \hat{z}_{gi},$$

$$\hat{\mu}_g = \frac{1}{\hat{\eta}_g n} \sum_{i=1}^{n} y_i \hat{z}_{gi},$$

$$\hat{\sigma}_g^2 = \frac{1}{\hat{\eta}_g(n-1)} \sum_{i=1}^{n} (y_i - \mu_g)^2 \hat{z}_{gi}.$$

For homoscedastic cases where all components have a common variance, the standard deviation in each M step can be obtained from a weighted mean of the $\hat{\sigma}_g^2$ values,

$$\hat{\sigma}^2 = \sum_{g=1}^{G} \eta_g \hat{\sigma}_g^2.$$

Ashman et al. (1994) used the log likelihood ratio test statistic, $-2 \log \lambda = -2(\ell^1 - \ell^G)$, to test for statistical significance. For mixture models in general, $-2 \log \lambda$ does not have its usual asymptotic $\chi^2$ distribution due to a breakdown in the regularity condition in the case of finite mixture models (Ghosh & Sen, 1985); see also Chapter 7, Section 7.2.1 above. However, for mixture models where the variances of all components are equal, Wolfe (1971) found empirically that the statistic $-2C \log \lambda$ approximately follows a $\chi^2$ distribution. With $d$-dimensional data modeled with $G_0$ (null hypothesis) and $G_1$ (alternate hypothesis) components, the correction factor is $C = (n - 1 - d - G_1/2)/n$ and the number of degrees of freedom of the $\chi^2$ distribution is equal to $2d(G_1 - G_0)$. Ashman et al. (1994) were unaware of this subtlety and report $p$-values based on the $\chi^2$ distribution with the approximation $C = 1$.

A statistically significant identification of multiple components is not sufficient to identify multimodal distributions, because mixtures of closely spaced components may have a single mode, as discussed in Chapter 1 above. Ashman et al. (1994) used the statistic

$$\Delta = \frac{|\mu_1 - \mu_2|}{\sigma}, \tag{19.3}$$

to measure the separation between components. In the case of equal mixing parameters, a two-component distribution will be multimodal when $\Delta > 2$. Hartigan's dip test may be used to test for the existence of bimodality of a data set (Hartigan & Hartigan, 1985). However, Muratov & Gnedin (2010) argue that, for bimodal distributions with equal mixing parameters and the same value of $\Delta$, the log likelihood ratio test is more sensitive to multiple normal distribution components than the dip test is to bimodality. Nevertheless, in astronomy, the identification of multiple components is often more scientifically interesting than the identification of bimodality in a particular set of variables (Taylor et al., 2015).

Muratov & Gnedin (2010) presented a new, more general, code for establishing the presence of heteroscedastic mixture components. Three statistics are investigated: the log likelihood ratio test statistic, the kurtosis of the distribution (a negative kurtosis is a necessary condition of a bimodal distribution produced by a two-component normal mixture model), and the separation between the two components, now defined as

$$\Delta = \frac{|\mu_1 - \mu_2|}{\sqrt{(\sigma_1^2 + \sigma_2^2)/2}}.$$

Once the model is fitted with the EM algorithm, nonparametric bootstrap resampling is performed to estimate uncertainties on the model parameters and $\Delta$. Finally, the parametric bootstrap is run to estimate the $p$-value for a unimodal distribution.

The statistical tests provided by these codes have been used in many astronomical studies. Most commonly, this test is used to study the distribution in the color of astronomical sources. In astronomy, color refers to the ratio of the amount of light observed in one band (e.g. the $V$ band) to the amount of light in another band (e.g. the $B$ band). Brightness is usually measured in magnitudes, a logarithmic unit where larger values indicate dimmer sources, so a color would be written as $B - V$, with a larger value indicating relatively more light in the $V$ band than in the $B$ band. Hundreds of papers have referenced these codes in investigations of colors of stars, globular clusters, and galaxies (see, for example, Weldrake et al., 2007; Larsen et al., 2001; Kundu & Whitmore, 2001).

We now present three individual studies that illustrate a variety of common characteristics of astronomical classification problems.

### 19.3.2   Two or three classes of gamma-ray bursts?

The two major classes of gamma-ray bursts (GRBs), short and long GRBs, were identified by a distinct bimodality in the distribution of burst durations (Kouveliotou et al., 1993). Today, there is ancillary evidence that the classes are physically distinct: long-duration GRBs are produced by collapsars, the implosion of a massive star at the end of its life, while short-duration GRBs originate from the merger of binary neutron stars (Nakar, 2007; Abbott et al., 2017). The early analyses revealing these two classes were based on univariate or bivariate distributions, but observations of GRBs provide a larger variety of properties that can be used for classification.

Mukherjee et al. (1998) examined the clustering of GRB properties using a larger set of variables than were typically included in previous studies to provide a fuller picture of the classes of GRBs. Their sample consisted of 797 GRB events from the BATSE instrument on NASA's *Compton Gamma-Ray Observatory*, each of which is described by 15 variables. This analysis suggested the presence of a third group of GRBs with intermediate properties.

The variables in the study included several measures of burst duration, fluence (the total amount of light observed from the burst), and spectral hardness (the average energy of the observed photons). Analysis of these variables showed that several of them are redundant: some mainly add noise to the clustering process, while others were not astrophysically meaningful. A reduced set of five variables was obtained.

Cluster analysis was performed using two methods: hierarchical average-linkage clustering and normal mixture models. The cluster analysis used unit-free variables obtained through logarithmic transformation of the variables. This practice is more natural for use in astronomy than standardization by the sample standard deviation, since measurements often vary by several powers of 10, and in many cases measurements can be approximated by log-normal distributions. The hierarchical clustering was performed using a Euclidean distance metric (which is unit-dependent), and the number of clusters was selected based on the squared correlation coefficient (the fraction of the total variance accounted for by a partition into $G$ clusters) and the squared semi-partial correlation coefficient (the difference in the variance between the resulting cluster and the immediate parent clusters normalized by the total sample variance). Each analysis approach suggested that three classes of bursts are present: Class I (long, bright, soft bursts), Class II (short, faint, hard bursts), and Class III (intermediate, intermediate, soft bursts). The first two categories reproduce the long GRB and short GRB classes, but the third class was new.

The mixture model analysis was performed using the mclust software (Fraley & Raftery, 1998) for normal mixture models where the number of clusters is evaluated using the BIC. The set of variables was further reduced to three for this analysis. The best value of BIC was found for three clusters, with the difference between two and three clusters being $\Delta \text{BIC} = 68$, strongly supporting the results from hierarchical clustering that more than two classes of GRB exist.

It is still not certain whether this third class of GRB is astrophysically distinct. NASA's more recent *Swift* Gamma Ray Burst and *Fermi* missions have discovered most known GRBs. Evidence for this third class of GRB has been weaker or absent in these later samples. For example, a multivariate analysis of GRBs detected by *Fermi* finds that a two-component mixture model is highly favored (Narayana Bhat et al., 2016). It is possible that the presence of a third component in the original data set may have been an effect of sample selection caused by uninteresting properties of the BATSE instrument, rather than a distinct astrophysical class.

### 19.3.3 Removal of contaminants

It is often desirable to obtain large samples of astronomical objects of a particular type. However, source lists obtained from observations may include contaminants, which are objects of a different type that masquerade as objects of the desired class. It is often difficult to completely eliminate contaminants from a large study without extensive follow-up observations, but in many cases some level of contamination is acceptable if the contaminant rate is kept sufficiently low.

Jordán et al. (2009) performed a study of globular star clusters within nearby galaxies observed by NASA's *Hubble Space Telescope* (HST). Sources of light were identified within the images taken by HST's Advanced Camera for Surveys, which include globular clusters, as well as contaminant foreground stars and background galaxies. (The host galaxy, which contains the globular clusters, was ignored in the analysis.) For each of the sources, photometric $g$ and $z$ magnitudes and a characteristic radius, $r_h$, were measured. Foreground stars may be easily distinguished and removed from the catalogs because their radii in the image are nearly zero. Globular clusters typically have smaller radii and brighter $z$ magnitudes than background galaxies, but these two populations overlap. These observations were made for 100 host galaxies, each observation having its own population of globular clusters and background galaxy contaminants.

To distinguish between globular clusters and background galaxies, Jordán et al. (2009) used a mixture model strategy. The distribution of globular cluster properties $(r_h, z)$ was assumed to be universal for all cases, and was taken from prior knowledge of the well-studied globular cluster properties. The only free parameter is the mean radius, $\mu_{r_h}$, of the clusters. The distribution of background galaxy properties $(r_h, z)$ was estimated for each observation, using a separate "control field" near to the original observation on the sky. All the sources in these control fields (once stars are removed) were assumed to be background galaxies, and the distribution of these sources was estimated using kernel density estimation. This density distribution has no free parameter to be fitted in the mixture model analysis.

The mixture models were fitted using the EM algorithm to find the mixing parameters and $\mu_{r_h}$ for each of the 100 observations. This method has two main advantages over a more typical method used by astronomers of using a fixed boundary between globular clusters and background galaxies in $(r_h, z)$-space. The mixture method accounts for variation in size of globular clusters from one host galaxy to another. In addition, the soft classifications provided by the mixture models allow samples of probable globular clusters to be obtained with different levels of contamination, depending on the needs of different science questions.

### 19.3.4  Red and blue galaxies

Galaxies generally fall into two groups: one class known as early-type galaxies (which are smaller, older, redder, and less likely to have star formation) and the other class known as late-type galaxies (which are larger, younger, bluer, and more likely to have star formation). The presence of two classes can be seen in distributions in galaxy colors, for example the Sloan $g - i$ color index, with a group of "blue" galaxies and a group of "red" galaxies. The two populations have color distributions that overlap, which means that some "red" galaxies have bluer colors than some "blue" galaxies, and vice versa. A number of studies have dealt with this distribution using a line on the color-magnitude diagram to separate both classes, with objects falling on one side of the line being assigned to one class and objects on the other side assigned to the other class (e.g. Bell et al., 2003; Baldry et al., 2004; Peng et al., 2010). However, when different proposed dividing lines are applied, the properties of the resulting samples, for example their galaxy mass distributions, will differ (Taylor et al., 2015, Figures 3 and 4).

Taylor et al. (2015) instead used a mixture model approach for this problem. Their galaxy data originated from the GAMA project (Driver et al., 2011), from which they derived a subset containing more than 23,000 objects pruned for reliability and to avoid biases from selection effects. Two variables were included in the analysis, galaxy mass $M_\star$ and $g - i$ color (corrected for redshift). A complex model with 40 parameters was used to describe the "red" and "blue" galaxies in the $(\log M_\star, g - i)$-space. This model used gamma functions – known in astronomy as the Schechter (1976) function – to describe the distribution of galaxy masses, and models describing the color–mass relations, the scatter around these relations, and outlying data points. This model has more parameters than are necessarily demanded by the data. However, the purpose of the model is to provide a sufficiently flexible model that will describe the data, not an in-depth study of the model parameters; and the authors stated that, based on their analysis, they were not "grossly overfitting the data". Fitting was performed with a Markov chain Monte Carlo method with uniform or uninformative priors using the Python software package EMCEE (Foreman-Mackey et al., 2013).

The result of the analysis was a soft classification of galaxies into two populations, with no evidence for the existence of an intermediate "green" population of galaxies. For both the "red" galaxies and the "blue" galaxies, the mass functions were similar to single Schechter functions. Small deviations included an excess of "red" galaxies with low mass and a deficit of "blue" galaxies with high mass. The mixture model mass functions avoid some of the unexpected artifacts present in the mass functions produced by hard classification methods. The models of the two populations show that the colors of "blue" galaxies do not depend strongly on mass, but that the colors of "red" galaxies do vary strongly with mass. The most massive "red" galaxies have red $g - i$ colors indicative of very little star formation. However, the mixture model fitted suggests that these galaxies are only one end of a broader distribution, which includes lower-mass "red" galaxies that are not as different in color from the "blue" population (Taylor et al., 2015, Figures 10 and 11).

## 19.4 Advanced Mixture Model Applications

For more advanced statistical modeling of data, it is often convenient if a distribution can be described by a flexible parametric model, and mixtures of normal distributions are one such possibility. Many probability density functions can be mimicked by normal mixture models (see, for example, McLachlan & Peel, 2000), so these models can often be used to estimate distributions of data even in cases in which there is no theoretical reason to suspect that the data should originate from multiple components. Here, we describe two applications in astronomy where advanced methods for dealing with problems such as heteroscedastic measurement errors or missing values become feasible when it is assumed that underlying distributions are described by normal mixture models. These examples demonstrate how a mixture model can be incorporated into a hierarchical statistical model, facilitating the computation of a likelihood for complicated scenarios that may arise in astronomy.

### 19.4.1 Regression with heteroscedastic uncertainties

Astronomers are often interested in the relationship between two properties of a cosmic population, but must infer results from samples that are limited by telescope sensitivity or subject to significant heteroscedastic measurement uncertainties. Fortunately, measurement uncertainties can usually be directly measured from calibration tests conducted under identical conditions to the true observation, and can thus enter the data set rather than be parameterized in the model.

A widely cited treatment of such problems in astronomy is a bivariate regression procedure involving semi-parametric density estimation using mixture models by Kelly (2007). This approach uses normal mixture models as part of a hierarchical model of the problem, a strategy developed by generalizing a model presented by Carroll et al. (1999) to allow for heteroscedastic measurement error. In this case, the mixture model is an internal part of an algorithm, rather than a fundamental property of the input or the output of the statistical analysis. Thus, the properties of the mixture model, such as number of components and component parameters, are not important so long as the model can provide an adequate approximation of the underlying distribution. By using a mixture model framework, Kelly (2007) is able to construct a likelihood for a hierarchical model, which can then be used to perform MLE or Bayesian inference.

For a case of linear regression with measurement errors on both the independent and dependent variables, Kelly (2007) construct a hierarchical model. Note that in this chapter we have altered the notation used by Kelly (2007) to be consistent with usage in the rest of this book. We denote the intrinsic value of the independent variable $\xi$ and the dependent variable $\iota$ with the relation

$$\iota_i = \alpha + \beta \xi_i + \epsilon_i,$$

where $(\alpha, \beta)$ are the regression coefficients and the error term $\epsilon_i$ is a random variable drawn from a normal distribution with variance $\sigma^2$. However, observational effects will yield measurement errors (which may be correlated) on both $\xi$ and $\iota$. We model the relation between observed values $(x, y)$ and intrinsic values $(\xi, \iota)$ with the standard errors-in-variables formulation,

$$
\begin{aligned}
x_i &= \xi_i + \epsilon_{x,i}, \\
y_i &= \iota_i + \epsilon_{y,i},
\end{aligned}
$$

where the errors for each measurement $(\epsilon_{y,i}, \epsilon_{x,i})$ are drawn from a multivariate normal distribution with known covariance matrix

$$
\Sigma_i = \begin{pmatrix} \sigma_{y,i}^2 & \sigma_{xy,i} \\ \sigma_{xy,i} & \sigma_{x,i}^2 \end{pmatrix}.
$$

Finally, we assume that the distribution of the intrinsic variable $\xi$ has the probability distribution of a $G$-component univariate normal mixture model, with mixing components $\eta_g$, means $\mu_g$, and standard deviations $\tau_g$. This scenario can be described by the following hierarchical model:

$$
\xi_i | \eta, \mu, \tau^2 \sim \sum_{g=1}^{G} \eta_g \mathcal{N}(\mu_g, \tau_g^2), \tag{19.4}
$$

$$
\iota_i | \xi_i, \alpha, \beta, \sigma^2 \sim \mathcal{N}(\alpha + \beta \xi_i, \sigma^2), \tag{19.5}
$$

$$
y_i, x_i | \iota_i, \xi_i \sim \mathcal{N}((\iota_i \; \xi_i)^\top, \Sigma_i). \tag{19.6}
$$

The observed-data likelihood can be found by integrating the complete-data likelihood over the missing data $(\xi, \iota)$. Thus, we are able to obtain an equation for the observed-data likelihood in terms of observed quantities, $\mathbf{x} = (x_1, \ldots, x_n)$, $\mathbf{y} = (y_1, \ldots, y_n)$, the known covariance matrices $\Sigma = (\Sigma_1, \ldots, \Sigma_n)$, and the model parameters, which we write as $\psi = (\eta, \mu, \tau^2)$ and $\theta = (\alpha, \beta, \sigma^2)$:

$$
\begin{aligned}
p(\mathbf{x}, \mathbf{y} | \theta, \psi) &= \prod_{i=1}^{n} \iint p(x_i, y_i, \xi_i, \iota_i | \theta, \psi) \mathrm{d}\xi_i \, \mathrm{d}\iota_i \\
&= \prod_{i=1}^{n} \iint p(x_i, y_i | \xi_i, \iota_i) p(\iota_i | \xi_i, \theta) p(\xi_i | \psi) \mathrm{d}\xi_i \, \mathrm{d}\iota_i.
\end{aligned}
$$

This separation into components is possible due to the hierarchical nature of the model. Substituting the probability distributions from equations (19.4)–(19.6) and integrating over $\xi$ and $\iota$ yields

$$
p(\mathbf{x}, \mathbf{y} | \theta, \psi) = \prod_{i=1}^{n} \sum_{g=1}^{G} \frac{\eta_g}{2\pi |V_{g,i}|^{1/2}} \exp\left\{ -\frac{1}{2} (w_i - \zeta_g)^T V_{g,i}^{-1} (w_i - \zeta_g) \right\}, \tag{19.7}
$$

where

$$
w_i = \begin{pmatrix} y_i \\ x_i \end{pmatrix}, \quad \zeta_g = \begin{pmatrix} \alpha + \beta \mu_g \\ \mu_g \end{pmatrix},
$$

$$
V_{g,i} = \begin{pmatrix} \beta^2 \tau_g^2 + \sigma^2 + \sigma_{y,i}^2 & \beta \tau_g^2 + \sigma_{xy,i} \\ \beta \tau_g^2 + \sigma_{xy,i} & \tau_g^2 + \sigma_{x,i}^2 \end{pmatrix}.
$$

Equation (19.7) expresses the likelihood of the hierarchal model in terms of the observational data, the mixture model parameters, and the regression model parameters.

Kelly (2007) provides an example from Kelly & Bechtold (2007) where this method is

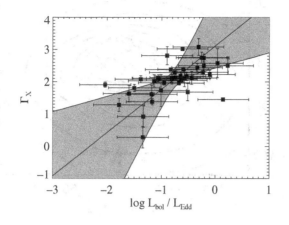

**FIGURE 19.6**

Regression of $\log L_{\rm bol}/L_{\rm Edd}$ as a function of X-ray photon flux $\Gamma_x$ fitted using the hierarchical model using Bayesian inference. The individual uncertainties on the dependent and independent variables are indicated by the error bars (marking one standard deviation). Spread in the observed $\Gamma_x$ is dominated by measurement error rather than intrinsic spread (Kelly, 2007).

useful for regression analysis. The data consist of 39 quasars that had been observed with the *Chandra* X-ray Observatory and SDSS. Figure 19.6 is a scatter plot showing two quantities obtained from these data: $\Gamma_x$, the "X-ray photon index" (a measure of the distribution of energies of X-ray photons from the quasars); and $\log L_{\rm bol}/L_{\rm Edd}$, an estimate of the rate of inflow of matter onto the central black hole in these quasars in terms of the theoretical maximum rate. In this example, the uncertainties on both estimates are independent. Using Bayesian methods with Markov chain Monte Carlo sampling, a reasonable set of priors, and $G = 2$ mixture components, Kelly (2007) estimates $\hat{\alpha} = 3.12 \pm 0.41$, $\hat{\beta} = 1.35 \pm 0.54$, and $\hat{\sigma} = 0.26 \pm 0.11$. From the mixture model parameters, it turns out that the scatter in the independent variable, $\Gamma_x$, is dominated by measurement error.

Implementation of the *linmix_err* algorithm in IDL is available from the IDL Astronomy User's Library[1] and in Python from GitHub.[2]

## 19.4.2 Deconvolution of distributions from data with heteroscedastic errors and missing information

The remarkable ability of normal mixture models to closely model many different probability density functions is also used by Bovy et al. (2011) in their Extreme Deconvolution algorithm. This algorithm is used to recover underlying distribution functions in cases where individual measurements have known heteroscedastic uncertainties and/or some variables are missing due to projection effects. For example, for astronomers studying the motions of stars by measuring their position in images of the sky (proper motion), the velocity of a star in three-dimensional space would be projected onto the two-dimensional plane of the sky. In this case, the component of velocity parallel to the line of sight is the missing data.

---

[1] https://idlastro.gsfc.nasa.gov
[2] https://github.com/jmeyers314/linmix

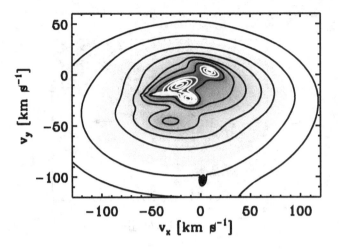

**FIGURE 19.7**
The recovered three-dimensional distribution of stellar velocities for *Hipparcos* stars in the solar neighborhood is projected onto the $(x, y)$-plane. Density is indicated by the grayscale and contour lines. Clumps in the image are moving groups (Bovy et al., 2011).

The model created by Bovy et al. (2011) describes the intrinsic distribution of quantities subject to these observational limitations, with the assumption that the intrinsic distribution is a mixture of multivariate normal distributions. The method addresses the common challenge of "deconvolution" in astronomy, but, unlike the conventional deconvolution methods used by astronomers, it allows each measurement to have individual uncertainties.

In this model, the observed data $\mathbf{w}$ are related to the intrinsic values $\mathbf{v}$ through the addition of measurement error $\epsilon_i$. Here, the use of boldface notation indicates that variables $\mathbf{w}$ and $\mathbf{v}$ may be elements of multidimensional vector spaces. We assume that the measurement error for each observation $\mathbf{w}_i$ is drawn from a multivariate normal distribution with a known covariate matrix $\mathbf{S}_i$. In addition to uncertainty, Bovy et al. (2011) also include transformation and projection of the data with a projection matrix $\mathbf{R}_i$ to account for missing data.

When including all these effects, the observed data are related to the intrinsic values by the equation

$$\mathbf{w}_i = \mathbf{R}_i \mathbf{v}_i + \epsilon_i.$$

The next assumption is that the intrinsic variable $\mathbf{v}$ has the probability distribution of a $G$-component normal mixture model,

$$\mathbf{v} | \eta, \mathbf{m}, \mathbf{V} \sim \sum_{g=1}^{G} \eta_g \mathcal{N}(\mathbf{m}_g, \mathbf{V}_g),$$

where the $\eta_g$ are the mixing coefficients, the $\mathbf{m}_g$ are the means of the components, and the $\mathbf{V}_g$ are their covariance matrices. The estimation of these model parameters, denoted by $\theta$, is the desired output of the algorithm.

The observed likelihood equation for this model is then

$$p(\mathbf{w}_i|\mathbf{R}_i, \mathbf{S}_i, \theta) = \sum_{g=1}^{G} \int_{\mathbf{v}} p(\mathbf{w}_i, \mathbf{v}, g|\theta)\, \mathrm{d}\mathbf{v}$$

$$= \sum_{g=1}^{G} \int_{\mathbf{v}} p(\mathbf{w}_i|\mathbf{v})p(\mathbf{v}|g, \theta)p(g|\theta)\, \mathrm{d}\mathbf{v},$$

for which

$$p(\mathbf{w}_i|\mathbf{v}) = \mathcal{N}(\mathbf{w}_i|\mathbf{R}_i\mathbf{v}, \mathbf{S}_i),$$
$$p(\mathbf{v}|g, \theta) = \mathcal{N}(\mathbf{v}|\mathbf{m}_g, \mathbf{V}_g),$$
$$p(g|\theta) = \eta_g.$$

Integrating over $\mathbf{v}$ gives a likelihood in the form of a normal mixture model, which is written

$$p(\mathbf{w}_i|\theta) = \sum_{g=1}^{G} \eta_g \mathcal{N}(\mathbf{w}_i|\mathbf{R}_i\mathbf{m}_g, \mathbf{T}_{ig}),$$

where the covariance matrices are

$$\mathbf{T}_{ig} = \mathbf{R}_i\mathbf{V}_g\mathbf{R}_i^\top + \mathbf{S}_i.$$

The log likelihood for $\theta$, given all $n$ observations $\mathbf{w}_i$, is then

$$\ell_o(\theta) = p(\mathbf{w}_1, ..., \mathbf{w}_n|\theta) = \sum_{i=1}^{n} \log p(\mathbf{w}_i|\theta) = \sum_{i=1}^{n} \log \sum_{g=1}^{G} \eta_g \mathcal{N}(\mathbf{w}_i|\mathbf{R}_i\mathbf{m}_g, \mathbf{T}_{ig}).$$

This log likelihood equation can be used for MLE or Bayesian inference to estimate the parameters $\eta_g$, $\mathbf{m}_g$, and $\mathbf{V}_g$ of the intrinsic distribution. Bovy et al. (2011) use an EM approach to address this problem. Implementation of the Extreme Deconvolution algorithm in Python is available from GitHub.[3]

The Extreme Deconvolution method has been useful in assessing the structure and kinematics (star motions) of our Milky Way Galaxy. Bovy et al. (2009) used the algorithm to model the three-dimensional kinematics of $10^4$ stars in the solar neighborhood based on their two-dimensional tangential velocities as measured in the plane of the sky by the *Hipparcos* satellite. In this example, the radial velocities of individual stars are unknown and the uncertainties on the tangential velocities depend on the stars' brightnesses and distances. Figure 19.7 shows the three-dimensional velocity distributions recovered for the *Hipparcos* stars, projected onto the $(x, y)$-plane. Contours and grayscale show density of sources, with the clumps in the distribution being likely kinematic moving groups. In this example, the Extreme Deconvolution algorithm was successful at identifying major known moving groups based on three-dimensional kinematics.

## 19.5  Concluding Remarks

We have reviewed here both traditional normal mixture models and some unusual types of mixture modeling in astronomical research. Several lines of reasons suggest there is considerable potential for the expansion of such statistical applications in the future. First, the

---

[3]https://github.com/jobovy/extreme-deconvolution

appearance of the phrase "mixture model" in astronomical research papers has increased considerably in recent years from a negligible level prior to 2000, to a dozen studies annually around 2010 and several dozen studies around 2015. This is mostly due to promulgation of a few astrostatistical papers in certain subfields (Ashman et al. 1994; Muratov & Gnedin 2010; Kelly 2007; Bovy et al. 2011). Knowledge of the broader methodology among astronomers is weak; out of 20,000 papers published annually barely two refer to the authoritative monograph by McLachlan & Peel (2000). But the actual use is undoubtedly much greater, as most astronomers are not familiar with the label "mixture models" and simply view them as a class of multi-component regression models. The occurrence of mixtures (overlapping populations in a survey, groups of stars or galaxies from different classes, clouds of gas with different velocities in a single structure, etc.) is very common in astronomy.

Second, astronomers have often grouped points in a $p$-dimensional space (where the dimensions can represent either spatial location or values of some space of observed properties) using heuristic methods. A common procedure is to construct a decision tree classifier by visual examination without any algorithm. Long-standing classifications of galaxy morphology, active galactic nuclei, supernovae and other classes of cosmic objects are based on these subjective procedures. In cases where the property distributions can be represented by multivariate normals or similar distributions, mixture modeling can improve definitions of classes and allow new objects to be associated with these classes in an objective fashion. For example, de Souza et al. (2017) have refined a traditional heuristic three-cluster division of galaxy by emission line properties (the Baldwin–Phillips–Terlevich diagram) into a four-cluster structure using Gaussian mixture models.

Third, in cases where astronomers have used objective clustering procedures such as the single-linkage "friends-of-friends" algorithm, the resulting classifications are often unstable in the face of arbitrary choices of nonparametric procedures. Shifting to maximum likelihood estimation of parametric mixture models can lead to more stable and reproducible classifications. In particular, the number of components emerges from quantitative model selection measures such as the BIC rather than from subjective decisions.

Finally, in the early twenty-first century, enormous resources are being devoted to wide-field sky surveys in many spectral bands. Like commercial fishing trawlers that draw a wide variety of sea creatures in a single haul, these surveys collect vast numbers of stars, galaxies, active galactic nuclei, and transient phenomena. Mixture models should be promoted as important classification tools to treat these problems arising from SDSS, LSST and similar large surveys. These models may be used in advanced statistical modeling of big data, such as the methods by Roberts et al. (2017) for obtaining cosmological parameters from LSST.

Statisticians can not only help to improve interpretation of astronomical data, but also find a rich world of data sets in astronomy for testing methodological developments in mixture modeling. Nearly the entire research literature is publicly available online in full text through the NASA/Smithsonian Astrophysical Observatory Astrophysics Data System.[4] A considerable proportion of the data underlying these studies are publicly available. Petabytes of calibrated images, spectra, and time series are provided by archive centers for both space-based and ground-based observatories. But derived tabular materials may be more useful for the statistician. These include large billion-object catalogs from surveys such the SDSS, and smaller catalogs and tables from specialized studies. The Vizier Web service[5] gives access to thousands of such tables, including a dozen with more than $10^8$ objects. Catalogs of cosmic populations with specified properties can be obtained from the NASA/IPAC Extragalactic Database, SIMBAD database, and the International Virtual Observatory Alliance. However, due to the complexities of the scientific questions and the

---

[4]http://adswww.harvard.edu/abstract_service.html
[5]http://vizier.u-strasbg.fr

individual peculiarities of each survey, we recommend that statisticians exercise their expertise in collaboration with astronomers to address important astronomical and astrophysical questions in a reliable fashion.

---

## Acknowledgements

Michael A. Kuhn is grateful to the Chilean Millennium Institute for Astrophysics (MAS) and CONICET agency for a post-doctoral fellowship, and to the Department of Astronomy at the University of Valparaíso for hospitality. Eric D. Feigelson's research in astrostatistics is supported by NSF grant AST-1614690 and travel for this work was supported by the MAS.

# Bibliography

ABAZAJIAN, K., ADELMAN-MCCARTHY, J. K., AGÜEROS, M. A., ALLAM, S. S., AN-DERSON, S. F. et al. (2003). The first data release of the Sloan Digital Sky Survey. *Astronomical Journal* **126**, 2081–2086.

ABBOTT, B. P., ABBOTT, R., ABBOTT, T. D., ACERNESE, F., ACKLEY, K. et al. (2017). Multi-messenger observations of a binary neutron star merger. *Astrophysical Journal, Letters* **848**, L12.

ABELL, G. O. (1958). The distribution of rich clusters of galaxies. *Astrophysical Journal, Supplement* **3**, 211.

AITKIN, M. (2001). Likelihood and Bayesian analysis of mixtures. *Statistical Modelling* **1**, 287–304.

AITKIN, M. (2011). How many components in a finite mixture. In *Mixtures: Estimation and Applications*, K. Mengersen, C. P. Robert & D. M. Titterington, eds. Hoboken, NJ: John Wiley.

ALAM, S., ALBARETI, F. D., ALLENDE PRIETO, C., ANDERS, F., ANDERSON, S. F. et al. (2015). The eleventh and twelfth data releases of the Sloan Digital Sky Survey: Final data from SDSS-III. *Astrophysical Journal, Supplement* **219**, 12.

ASHMAN, K. M., BIRD, C. M. & ZEPF, S. E. (1994). Detecting bimodality in astronomical datasets. *Astronomical Journal* **108**, 2348–2361.

BADDELEY, A., MØLLER, J. & PAKES, A. G. (2008). Properties of residuals for spatial point processes. *Annals of the Institute of Statistical Mathematics* **60**, 627–649.

BADDELEY, A., RUBAK, E. & TURNER, R. (2015). *Spatial Point Patterns: Methodology and Applications with R.* Boca Raton, FL: CRC Press.

BADDELEY, A., TURNER, R., MØLLER, J. & HAZELTON, M. (2005). Residual analysis for spatial point processes (with discussion). *Journal of the Royal Statistical Society, Series B* **67**, 617–666.

BALDRY, I. K., GLAZEBROOK, K., BRINKMANN, J., IVEZIĆ, Ž., LUPTON, R. H., NICHOL, R. C. & SZALAY, A. S. (2004). Quantifying the bimodal color-magnitude distribution of galaxies. *Astrophysical Journal* **600**, 681–694.

BARDEEN, J. M. (1986). Galaxy formation in an Omega = 1 cold dark matter universe. In *Inner Space/Outer Space: The Interface between Cosmology and Particle Physics*, E. W. Kolb, M. S. Turner, D. Lindley, K. Olive & D. Seckel, eds. Chicago: University of Chicago Press, pp. 212–217.

BATE, M. R., BONNELL, I. A. & BROMM, V. (2003). The formation of a star cluster: Predicting the properties of stars and brown dwarfs. *Monthly Notices of the Royal Astronomical Society* **339**, 577–599.

BELL, E. F., McINTOSH, D. H., KATZ, N. & WEINBERG, M. D. (2003). The optical and near-infrared properties of galaxies. I. Luminosity and stellar mass functions. *Astrophysical Journal, Supplement* **149**, 289–312.

BINNEY, J. & TREMAINE, S., eds. (2008). *Galactic Dynamics*. Princeton, NJ: Princeton University Press, 2nd ed.

BOVY, J., HOGG, D. W. & ROWEIS, S. T. (2009). The velocity distribution of nearby stars from *Hipparcos* data. I. The significance of the moving groups. *Astrophysical Journal* **700**, 1794–1819.

BOVY, J., HOGG, D. W. & ROWEIS, S. T. (2011). Extreme deconvolution: Inferring complete distribution functions from noisy, heterogeneous and incomplete observations. *Annals of Applied Statistics* **5**, 1657–1677.

BROOS, P. S., GETMAN, K. V., POVICH, M. S., FEIGELSON, E. D., TOWNSLEY, L. K. et al. (2013). Identifying young stars in massive star-forming regions for the MYStIX project. *Astrophysical Journal, Supplement* **209**, 32.

BURNHAM, K. P. & ANDERSON, D. A. (2002). *Model Selection and Multimodel Inference: A Practical Information-Theoretical Approach*. New York: Springer.

CARLIN, B. P. & CHIB, S. (1995). Bayesian model choice via Markov chain Monte Carlo methods. *Journal of the Royal Statistical Society, Series B* **57**, 473–484.

CARROLL, R. J., MACA, J. D. & RUPPERT, D. (1999). Nonparametric regression in the presence of measurement error. *Biometrika* **86**, 541–554.

DE SOUZA, R. S., DANTAS, M. L. L., COSTA-DUARTE, M. V., FEIGELSON, E. D., KILLEDAR, M. et al. (2017). A probabilistic approach to emission-line galaxy classification. *Monthly Notices of the Royal Astronomical Society* **472**, 2808–2822.

DE VAUCOULEURS, G. (1971). The large-scale distribution of galaxies and clusters of galaxies. *Publications of the Astronomical Society of the Pacific* **83**, 113.

DRIVER, S. P., HILL, D. T., KELVIN, L. S., ROBOTHAM, A. S. G., LISKE, J. et al. (2011). Galaxy and mass assembly (GAMA): Survey diagnostics and core data release. *Monthly Notices of the Royal Astronomical Society* **413**, 971–995.

EDDINGTON, A. S. (1906). Systematic motions of the stars. *Monthly Notices of the Royal Astronomical Society* **67**, 34–63.

EINASTO, M., VENNIK, J., NURMI, P., TEMPEL, E., AHVENSALMI, A. et al. (2012). Multimodality in galaxy clusters from SDSS DR8: Substructure and velocity distribution. *Astronomy & Astrophysics* **540**, A123.

EISENSTEIN, D. J. (2005). Dark energy and cosmic sound [review article]. *New Astronomy Review* **49**, 360–365.

ESCOBAR, M. D. & WEST, M. (1995). Bayesian density estimation and inference using mixtures. *Journal of the American Statistical Association* **90**, 577–588.

FALL, S. M. (1979). Galaxy correlations and cosmology. *Reviews of Modern Physics* **51**, 21–43.

FEIGELSON, E. D., TOWNSLEY, L. K., BROOS, P. S., BUSK, H. A., GETMAN, K. V. et al. (2013). Overview of the Massive Young Star-Forming Complex Study in Infrared and X-Ray (MYStIX) project. *Astrophysical Journal, Supplement* **209**, 26.

FOREMAN-MACKEY, D., HOGG, D. W., LANG, D. & GOODMAN, J. (2013). emcee: The MCMC hammer. *Publications of the Astronomical Society of the Pacific* **125**, 306–312.

FRALEY, C. & RAFTERY, A. E. (1998). How many clusters? Which clustering method? Answers via model-based cluster analysis. *Computer Journal* **41**, 578–588.

FRALEY, C., RAFTERY, A. E., MURPHY, T. B. & SCRUCCA, L. (2012). mclust version 4 for R: Normal mixture modeling for model-based clustering, classification, and density estimation. Tech. Rep. 597, Department of Statistics, University of Washington.

GETMAN, K. V., FEIGELSON, E. D., KUHN, M. A., BROOS, P. S., TOWNSLEY, L. K. et al. (2014). Age gradients in the stellar populations of massive star forming regions based on a new stellar chronometer. *Astrophysical Journal* **787**, 108.

GETMAN, K. V., KUHN, M. A., FEIGELSON, E. D., BROOS, P. S., BATE, M. R. & GARMIRE, G. P. (2018). Young star clusters in nearby molecular clouds. *Monthly Notices of the Royal Astronomical Society* **477**, 298–324.

GHOSH, M. A. & SEN, A. J. (1985). On the asymptotic performance of the log-likelihood ratio statistic for the mixture model and related results. In *Proceedings of the Berkeley Conference in Honor of Jerzy Neyman and Jack Kiefer*, vol. 2 of *Astrophysics and Space Science Proceedings*. Monterey, CA: Wadsworth.

HARTIGAN, J. A. & HARTIGAN, P. (1985). The dip test of unimodality. *Annals of Statistics* **13**, 70–84.

HILLENBRAND, L. A. & HARTMANN, L. W. (1998). A preliminary study of the Orion Nebula Cluster structure and dynamics. *Astrophysical Journal* **492**, 540–553.

IVEZIĆ, Z., TYSON, J. A., ABEL, B., ACOSTA, E., ALLSMAN, R. et al. (2008). LSST: From science drivers to reference design and anticipated data products. Preprint, arXiv:0805.2366.

JEFFREYS, H. (1939). *Theory of Probability.* Oxford: Clarendon Press.

JORDÁN, A., PENG, E. W., BLAKESLEE, J. P., CÔTÉ, P., EYHERAMENDY, S. et al. (2009). The ACS Virgo Cluster Survey XVI. Selection procedure and catalogs of globular cluster candidates. *Astrophysical Journal, Supplement* **180**, 54–66.

KASS, R. E. & RAFTERY, A. E. (1995). Bayes factors. *Journal of the American Statistical Association* **90**, 773–795.

KELLY, B. C. (2007). Some aspects of measurement error in linear regression of astronomical data. *Astrophysical Journal* **665**, 1489–1506.

KELLY, B. C. & BECHTOLD, J. (2007). Virial masses of black holes from single epoch spectra of active galactic nuclei. *Astrophysical Journal, Supplement* **168**, 1–18.

KONISHI, S. & KITAGAWA, G. (2008). *Information Criteria and Statistical Modeling.* New York: Springer.

KOUVELIOTOU, C., MEEGAN, C. A., FISHMAN, G. J., BHAT, N. P., BRIGGS, M. S. et al. (1993). Identification of two classes of gamma-ray bursts. *Astrophysical Journal, Letters* **413**, L101–L104.

KUHN, M. A., FEIGELSON, E. D., GETMAN, K. V., BADDELEY, A. J., BROOS, P. S. et al. (2014). The Spatial Structure of Young Stellar Clusters. I. Subclusters. *Astrophysical Journal* **787**, 107.

KUHN, M. A., FEIGELSON, E. D., GETMAN, K. V., SILLS, A., BATE, M. R. & BORISSOVA, J. (2015). The spatial structure of young stellar clusters. III. Physical properties and evolutionary states. *Astrophysical Journal* **812**, 131.

KUHN, M. A., GETMAN, K. V., FEIGELSON, E. D., SILLS, A., GROMADZKI, M. et al. (2017a). The structure of the young star cluster NGC 6231. II. Structure, formation, and fate. *Astronomical Journal* **154**, 214.

KUHN, M. A., MEDINA, N., GETMAN, K. V., FEIGELSON, E. D., GROMADZKI, M., BORISSOVA, J. & KURTEV, R. (2017b). The structure of the young star cluster NGC 6231. I. Stellar population. *Astronomical Journal* **154**, 87.

KUHN, M. A., POVICH, M. S., LUHMAN, K. L., GETMAN, K. V., BUSK, H. A. & FEIGELSON, E. D. (2013). The massive young star-forming complex study in infrared and X-ray: Mid-infrared observations and catalogs. *Astrophysical Journal, Supplement* **209**, 29.

KUNDU, A. & WHITMORE, B. C. (2001). New insights from HST studies of globular cluster systems. I. Colors, distances, and specific frequencies of 28 elliptical galaxies. *Astronomical Journal* **121**, 2950–2973.

LADA, C. J., LOMBARDI, M., ROMÁN-ZÚÑIGA, C., FORBRICH, J. & ALVES, J. F. (2013). Schmidt's conjecture and star formation in molecular clouds. *Astrophysical Journal* **778**, 133.

LAHIRI, P., ed. (2001). *Model Selection.* IMS Lecture Notes – Monograph Series, Vol. 38. Beachwood, OH: Institute of Mathematical Statistics.

LARSEN, S. S., BRODIE, J. P., HUCHRA, J. P., FORBES, D. A. & GRILLMAIR, C. J. (2001). Properties of globular cluster systems in nearby early-type galaxies. *Astronomical Journal* **121**, 2974–2998.

MARTÍNEZ, V. J. & SAAR, E. (2002). *Statistics of the Galaxy Distribution.* Boca Raton, FL: Chapman & Hall.

MCLACHLAN, G. & PEEL, D. (1997). On a resampling approach to choosing the number of components in normal mixture models. In *Computing Science and Statistics*, L. Billard & N. I. Fisher, eds. Fairfax Station, VA: Interface Foundation of North America, pp. 260–268.

MCLACHLAN, G. & PEEL, D. (2000). *Finite mixture models.* New York: John Wiley.

MUKHERJEE, S., FEIGELSON, E. D., JOGESH BABU, G., MURTAGH, F., FRALEY, C. & RAFTERY, A. (1998). Three types of gamma-ray bursts. *Astrophysical Journal* **508**, 314–327.

MURATOV, A. L. & GNEDIN, O. Y. (2010). Modeling the metallicity distribution of globular clusters. *Astrophysical Journal* **718**, 1266–1288.

NAKAR, E. (2007). Short-hard gamma-ray bursts. *Physics Reports* **442**, 166–236.

NARAYANA BHAT, P., MEEGAN, C. A., VON KIENLIN, A., PACIESAS, W. S., BRIGGS, M. S. et al. (2016). The third Fermi GBM gamma-ray burst catalog: The first six years. *Astrophysical Journal, Supplement* **223**, 28.

NEYMAN, J., SCOTT, E. L. & SHANE, C. D. (1953). On the spatial distribution of galaxies: A specific model. *Astrophysical Journal* **117**, 92.

PENG, Y.-J., LILLY, S. J., KOVAČ, K., BOLZONELLA, M., POZZETTI, L. et al. (2010). Mass and environment as drivers of galaxy evolution in SDSS and zCOSMOS and the origin of the Schechter function. *Astrophysical Journal* **721**, 193–221.

PFALZNER, S. (2009). Universality of young cluster sequences. *Astronomy and Astrophysics* **498**, L37–L40.

PHILLIPS, D. B. & SMITH, A. F. M. (1996). Bayesian model comparison via jump diffusions. In *Markov Chain Monte Carlo in Practice*, W. Gilks, S. Richardson & D. Spiegelhalter, eds. London: Chapman & Hall, pp. 215–239.

POSTMAN, M., HUCHRA, J. P. & GELLER, M. J. (1986). Probes of large-scale structure in the Corona Borealis region. *Astronomical Journal* **92**, 1238–1247.

RAO, C. R., WU, Y., KONISHI, S. & MUKERJEE, R. (2001). On model selection. In *Model Selection*, P. Lahiri, ed., Lecture Notes – Monograph Series Vol. 38. Beachwood, OH: Institute of Mathematical Statistics, pp. 1–64.

RICHARDSON, S. & GREEN, P. J. (1997). On Bayesian analysis of mixtures with an unknown number of components (with discussion). *Journal of the Royal Statistical Society, Series B* **59**, 731–792.

ROBERTS, E., LOCHNER, M., FONSECA, J., BASSETT, B. A., LABLANCHE, P.-Y. & AGARWAL, S. (2017). zBEAMS: A unified solution for supernova cosmology with redshift uncertainties. *Journal of Cosmology and Astroparticle Physics* **10**, 036.

ROBITAILLE, T. P. & WHITNEY, B. A. (2010). The present-day star formation rate of the Milky Way determined from Spitzer-detected young stellar objects. *Astrophysical Journal, Letters* **710**, L11–L15.

ROEDER, K. (1990). Density estimation with confidence sets exemplified by superclusters and voids in the galaxies. *Journal of the American Statistical Association* **85**, 617–624.

ROEDER, K. & WASSERMAN, L. (1997). Practical Bayesian density estimation using mixtures of normals. *Journal of the American Statistical Association* **92**, 894–902.

SCHECHTER, P. (1976). An analytic expression for the luminosity function for galaxies. *Astrophysical Journal* **203**, 297–306.

SCHMEJA, S. (2011). Identifying star clusters in a field: A comparison of different algorithms. *Astronomische Nachrichten* **332**, 172.

SHANE, C. D. & WIRTANEN, C. A. (1954). The distribution of extragalactic nebulae. *Astronomical Journal* **59**, 285–304.

SPRINGEL, V. (2005). The cosmological simulation code GADGET-2. *Monthly Notices of the Royal Astronomical Society* **364**, 1105–1134.

TAYLOR, E. N., HOPKINS, A. M., BALDRY, I. K., BLAND-HAWTHORN, J., BROWN, M. J. I. et al. (2015). Galaxy and mass assembly (GAMA): Deconstructing bimodality – I. Red ones and blue ones. *Monthly Notices of the Royal Astronomical Society* **446**, 2144–2185.

TEGMARK, M., BLANTON, M. R., STRAUSS, M. A., HOYLE, F., SCHLEGEL, D. et al. (2004). The three-dimensional power spectrum of galaxies from the Sloan Digital Sky Survey. *Astrophysical Journal* **606**, 702–740.

TOWNSLEY, L. K., BROOS, P. S., GARMIRE, G. P., BOUWMAN, J., POVICH, M. S. et al. (2014). The massive star-forming regions omnibus X-ray catalog. *Astrophysical Journal, Supplement* **213**, 1.

TULLY, R. B. (1987). Nearby groups of galaxies. II – An all-sky survey within 3000 kilometers per second. *Astrophysical Journal* **321**, 280–304.

TURNER, E. L. & GOTT, III, J. R. (1976). Groups of galaxies. I. A catalog. *Astrophysical Journal, Supplement* **32**, 409–427.

WANG, J., TOWNSLEY, L. K., FEIGELSON, E. D., BROOS, P. S., GETMAN, K. V., ROMÁN-ZÚÑIGA, C. G. & LADA, E. (2008). A Chandra study of the Rosette star-forming complex. I. The stellar population and structure of the young open cluster NGC 2244. *Astrophysical Journal* **675**, 464–490.

WELDRAKE, D. T. F., SACKETT, P. D. & BRIDGES, T. J. (2007). A deep wide-field variable star catalog of $\omega$ Centauri. *Astronomical Journal* **133**, 1447–1469.

WOLFE, J. H. (1971). A Monte Carlo study of the sampling distribution of the likelihood ratio for mixtures of multinormal distributions. Technical Bulletin STB 72-2, Navel Personnel and Training Research Laboratory, San Diego, CA.

# Index